Conformal Invariance and String Theory

PERSPECTIVES IN PHYSICS

Huzihiro Araki, A. Libchaber, and Giorgio Parisi, editors

Conformal Invariance and String Theory

Edited by

Petre Dita *and* Vladimir Georgescu
Theoretical Physics Department
Central Institute of Physics
Bucharest, Romania

ACADEMIC PRESS, INC.
Harcourt Brace Jovanovich, Publishers

Boston San Diego New York
Berkeley London Sydney
Tokyo Toronto

ACADEMIC PRESS, INC.
1250 Sixth Avenue, San Diego, CA 92101

United Kingdom Edition published by
ACADEMIC PRESS INC. (LONDON) LTD.
24–28 Oval Road, London NW1 7DX

Library of Congress Cataloging-in-Publication Data

Conformal invariance and string theory/edited by Petre Dita and Vladimir Georgescu.
p. cm. — (Perspectives in physics)
Lectures delivered at the Summer School on Conformal Invariance and String Theory, held at Poiana Brasov, Romania, Sept. 1–12, 1987, sponsored by the Central Institute of Physics–Bucharest.
Bibliography: p.
ISBN 0-12-218100-X (alk. paper)
1. Statistical mechanics—Congresses. 2. Conformal invariants—Congresses. 3. String models—Congresses. I. Dita, P. (Petre), Date- . II. Georgescu, V. (Vladimir), Date- . III. Summer School on Conformal Invariance and String Theory (1987 : Poiana Brasov, Romania) IV. Institutul Central de Fizică (Romania)
V. Series
QC174.7.C66 1989
530.1'3—dc20 89-6941
CIP

Printed in the United States of America
89 90 91 92 9 8 7 6 5 4 3 2 1

Contents

Contributors

Numbers in parentheses refer to the pages on which the authors' contributions begin.

I. Ya. Aref'eva (165), *Steklov Mathematical Institute, Ul. Vavilova 42, Moscow 117 333, USSR*

Peter Bouwknegt (115), *Institute for Theoretical Physics, University of Amsterdam, Valckenierstraat 65, 1018 XE Amsterdam, The Netherlands*

P. Cartier (443), *Ecole Polytechnique, Centre de Mathématiques F-91128, Palaiseau, France*

P. di Francesco (63), *Service de Physique Théorique, Institut de Recherche Fondamentale, CEA-CEN Saclay, 91191 Gif-sur Yvette Cédex, France*

R. Flume (89), *Physikalisches Institut der Universität Bonn, Nussallee 12, 5300 Bonn 1, F.R. Germany*

Jean-Loup Gervais (205), *Physique Théorique, Ecole Normale Superieure, 24 rue Lhomond, 75231 Paris Cédex 05, France*

Malte Henkel (127), *Fachbereich Physik, Universitat Essen, Postfach 103764, D-4300 Essen 1, F.R. Germany*

M. Karowski (147), *Theoretical Physics, ETH-Honggerberg, CH-8093 Zurich, Switzerland*

Richard Kerner (243), *LPTPE, Université Pierre et Marie Curie, 4 place Jussieu, 75005 Paris, France*

W. Lerche (367), *CERN, 1211 Geneva 23, Switzerland*

Yu. I. Manin (285, 293), *Steklov Mathematical Institute, Ul. Vavilova 42, Moscow 117 333, USSR*

Hans-Peter Nilles (305), *CERN, 1211 Geneva 23, Switzerland*

Sylvie Paycha (335), *Mathematics Department, Ruhr-University, Bochum, F.R. Germany*

V. Rittenberg (37), *Physikalisches Institut der Universität Bonn, Nussallee 12, 5300 Bonn 1, F.R. Germany*

H. Saleur (63), *Service de Physique Théorique, Institut de Recherche Fondamentale, CEA-CEN Saclay, 91191 Gif-sur-Yvette Cédex, France*

A.N. Schellekens (367), *CERN, 1211 Geneva 23, Switzerland*

I.T. Todorov (3), *Institute for Nuclear Research and Nuclear Energy, Bulgarian Academy of Sciences, P.O. Box 373, 1090 Sofia, Bulgaria*

LIST OF PARTICIPANTS

ANDRICA, D. - University of Cluj-Napoca
ANGHEL, V. - INPR - Pitești
ANTOHI, O. - INPR - Pitești
APOSTOL, M. - Central Institute of Physics, Bucharest
AREF'EVA, I.Ya. - Steklov Institute of Mathematics, Moscow
AVRAM CRISTEA, A. - Central Institute of Physics, Bucharest

BADEA, M. - Institute of Medicine, Bucharest
BANICA, C. - INCREST - Bucharest
BATTLE ARNAU, C. - University of Barcelona
BECIU, M. - Institute of Constructions, Bucharest
BOCIORT, F. - Central Institute of Physics, Bucharest
BOGATU, N. - Central Institute of Physics, Bucharest
BOLDEA, Venera - Central Institute of Physics, Bucharest
BOUTET de MONVEL, Anne - Universite Paris VI
BOUWKNEGT, P. - University of Amsterdam
BRANZANESCU, V. - Polytechnical Institute Bucharest
BUNDARU, M. - Central Institute of Physics, Bucharest
BUZATU, Daniela - ICSIT - Bucharest
BUZATU, F. - Central Institute of Physics, Bucharest

CAO, C.T.Y. - Trinity College, University of Cambridge
CAPRINI, Irinel - Central Institute of Physics, Bucharest
CARTIER, P. - Ecole Polytechnique, Palaiseau
CATA, G. - Central Institute of Physics, Bucharest
CEAUSESCU, M. - Central Institute of Physics, Bucharest
CEAUSESCU, V. - Central Institute of Physics, Bucharest
CIONGA, Aurelia - Central Institute of Physics, Bucharest
CIONGA, V. - Central Institute of Physics, Bucharest
CIUBOTARU, Luminita - Central Institute of Physics, Bucharest
COSTACHE, G. - Central Institute of Physics, Bucharest

COSTIN, O.	-	Central Institute of Physics, Bucharest
COSTIN, Rodica	-	INCREST - Bucharest
CRISTEA, O.	-	University of Cluj-Napoca
CULETU, H.	-	ICECHIM - Bucharest
DEMUTH, M.	-	Institute of Mathematics, Berlin
DI FRANCESCO, P.	-	CEN - Saclay
DITA, P.	-	Central Institute of Physics,Bucharest
DITA, Sanda	-	Central Institute of Physics, Bucharest
DJAMO, V.	-	INPR - Piteşti
DOMITIAN, V.	-	High School Ploieşti
DUDAS, E.	-	University of Bucharest
DULEA, M.	-	Central Institute of Physics, Bucharest
FAZAKAS, A.	-	Central Institute of Physics, Bucharest
FLUME, R.	-	University of Bonn
FÖRSTER, D.	-	Freie Universität Berlin
FUCHS, J.	-	Inst. for Theoretical Physics, Heidelberg
GADIDOV, R.	-	INCREST - Bucharest
GEORGESCU, V.	-	Central Institute of Physics, Bucharest
GERVAIS, J.-L.	-	Ecole Normale Superieure, Paris
GHEORGHE, A.C.	-	Central Institute of Physics, Bucharest
GHEORGHIU, H.	-	INPR - Piteşti
GRIGORE, R.D.	-	Central Institute of Physics, Bucharest
GUITTER, E.	-	CEN - Saclay
GUSSI, G.	-	INCREST - Bucharest
HARABOR, Ana	-	University of Craiova
HARABOR, V.	-	University of Craiova
HENKEL, M.	-	University of Bonn
HRISTEA, M.R.	-	ICECHIM - Bucharest
IANCU, E.	-	University of Bucharest
IMBROANE, A.	-	Polytechnical Institute, Cluj-Napoca
INCULET, V.	-	University of Bucharest
IONISEI, G.	-	University of Iaşi
IORDANESCU, R.	-	Central Institute of Physics, Bucharest
ISAR, A.	-	Central Institute of Physics, Bucharest

KALITZIN, S.	-	INRNE - Sofia
KAROWSKI, S.	-	Freie Universität Berlin
KERNER, P.	-	Universite Paris VI
KISS-TOTH, T.	-	Osijek University
LUDU, A.	-	University of Bucharest
LUKIERSKY, J.	-	University of Wroclaw
LUPU, C.	-	ICPE - Bucharest
MAGNON, Anne	-	Universite Clermont II
MANDA, H.	-	Central Institute of Physics, Bucharest
MANIN, Yu.I.	-	Steklov Institute of Mathematics, Moscow
MANOLIU, Mihaela	-	Central Institute of Physics, Bucharest
MARCHETTI, P.	-	University of Padova
MARTIN, M.	-	INCREST - Bucharest
MAZILU, D.	-	Central Institute of Physics, Bucharest
MANDESCU, R.	-	IPMP - Buzau
MANTOIU, M.	-	High School Brasov
MINNAERT, P.	-	Universite de Bordeaux
MINTI, H.	-	ICPE - Bucharest
MOCANU, Raluca	-	Central Institute of Physics, Bucharest
MUNOZ, C.	-	University of Madrid
MUNTEANU, V.	-	University of Bucharest
NAIDIN, Andreea	-	IPGG - Bucharest
NICULA, M.	-	University of Bucharest
NILLES, H.P.	-	CERN - Geneva
NYKANEN, E.	-	NORDITA - Copenhagen
PANTEA, D.	-	Central Institute of Physics, Bucharest
PASCU, E.	-	INCREST - Bucharest
PATEOPOL, L.	-	IPA - Bucharest
PAYCHA, Silvie	-	Ruhr Universität Bochum
PISO, M.	-	ICPE - Bucharest
POENARU, D.	-	Central Institute of Physics, Bucharest
POPESCU, S.	-	University of Bucharest
POPP, O.T.	-	Central Institute of Physics, Bucharest
PRIPOAE, G.	-	ICSIT - Bucharest

PURICE, R.	-	Central Institute of Physics, Bucharest
PUTINAR, M.	-	INCREST - Bucharest
RADULESCU, F.	-	INCREST - Bucharest
RICHERT, J.	-	CRN - Strasbourg
RITTENBERG, V.	-	University of Bonn
ROSU, H.	-	Central Institute of Physics, Bucharest
SALEUR, H.	-	CEN - Saclay
SALIU, L.	-	University of Craiova
SARU, D.	-	ICPE - Bucharest
SASU, G.	-	Central Institute of Physics, Bucharest
SANDULESCU, A.	-	Central Institute of Physics, Bucharest
SANDULESCU, N.	-	Central Institute of Physics, Bucharest
SCHELLEKENS, A.N.	-	CERN - Geneva
SOFONEA, M.	-	INCREST - Bucharest
SPINEANU, F.	-	Central Institute of Physics, Bucharest
STAN-SION, C.	-	Central Institute of Physics, Bucharest
STEINBRECHER, G.	-	University of Craiova
STRATAN, G.	-	Central Institute of Physics, Bucharest
TARINA, M.	-	University of Cluj-Napoca
TATARU, L.	-	University of Craiova
TODOROV, I.T.	-	INRNE - Sofia
TOPOR-POP, V.	-	Central Institute of Physics, Bucharest
TOPOR-POP, Rodica	-	High School Bucharest
VISINESCU, Anca	-	Central Institute of Physics, Bucharest
VLAD, Madalina	-	Central Institute of Physics, Bucharest
VONSOVICI, A.	-	University of Bucharest
WEIGT, G.	-	IHEP Berlin-Zeuthen
WOLLENBERG, M.	-	Institute of Mathematics, Berlin

Preface

This book contains the lectures delivered at the Summer School on Conformal Invariance and String Theory, held at Poiana Brasov, Romania, 1–12 September 1987. The aim of the School was to review some of the most important problems and results in these two strongly interrelated areas of theoretical physics, which are currently extremely active and in which much progress is being made. Relations between the two domains were emphasized for the benefit of students in each field, who were thereby enabled to interact and get accustomed to the concepts and methods of the other field. The first part of the volume presents the conferences dealing with the implications of conformal invariance in the study of two-dimensional systems. The second part contains several lectures on recent advances in string theory and related topics.

The School was sponsored by the Central Institute of Physics-Bucharest, and acknowledges a grant from UNESCO.

We owe special thanks to all the invited lecturers for their contributions to the success of the School; Professor Vladimir Rittenberg's assistance was decisive in the early stage of the School's organization.

Many of our colleagues from the Central Institute of Physics helped us in various ways. We are especially grateful to Drs. Valentin Ceausescu and Marin Ivascu for their inestimable help. Dr. Gabriel Costache's experience was a great benefit and we thank him for his permanent support.

Ms. Geta Uglai, the technical secretary of the School, handled most of the tedious work related to the School with high competence and efficiency, for which we owe our warm gratitude.

THE EDITORS
Bucharest

I. CONFORMAL INVARIANCE IN STATISTICAL MECHANICS

FINITE TEMPERATURE 2-DIMENSIONAL QFT MODELS OF CONFORMAL CURRENT ALGEBRA*)

I.T. Todorov

1. Introduction

   Synopsis on Gibbs states and KMS boundary condition

2. Two Dimensional Conformal QFT and CCA on the Circle

2A. *Conserved chiral currents. Compact picture. CCA commutation relations*

2B. *Lowest weight (LW) unitary representations (URs) of CCA. The Sugawara formula*

3. KMS States on the CCA

3A. *KMS correlation functions in terms of charge distribution*

3B. *Local QFT model, associated with a free complex Weyl field*

4. Classification of 2-Dimensional Conformal Models of Chiral U(1) Current Algebra

4A. *Construction of primary charged fields from localized morphisms of the CCA*

4B. *Field algebras with single and double valued representations on the circle*

5. Concluding Remarks

   References

*) The present notes are a preview of a work in progress by D.Buchholz, G.Mack and the author and are based on the lectures presented at the International School on Group Theoretical Methods, Varna, Bulgaria June 1987 and Poiana Braşov Summer School, September 1987.

ISBN 0-12-218100-X

1. Introduction

Conformal quantum field theory (QFT) models in 1+1 dimensions give rise to infinite dimensional Lie algebras of observables whose lowest weight (positive energy) representations have been classified. They yield explicit Gibbs states and provide a new method of computing Virasoro characters and partition functions. A discrete series of models associated with the chiral U(1) conformal current algebra (CCA) $\mathcal{V}U(1) + \mathcal{V}U(1)$, is singled out by a "univalence" requirement (related to the Bose-Fermi alternative) for the basic charged fields. The first terms of this series correspond to previously studied special points (characterized by a higher symmetry) of the critical line of the Ashkin-Teller model (see, e.g., [Ya 1, 3] and references to earlier work cited there).

The rigorous approach to finite temperature equilibrium QFT (see e.g. [Ha 1, Ka 5] as well as the book [Br 1] and further references cited there) and the work on algebraic methods in the study of 2-dimensional conformal models (see [De 1, Fe 2, Fu 1, Lü 1, Ca 5, Be 1, Wi 1, Kn 1, Go 2,3, To 1,2,3 , Ge 3, Fu 2] among others) have been pursued apart from each other. One objective of these lectures is to show how the Kubo-Martin-Schwinger (KMS) boundary condition can be used to evaluate finite temperature correlation functions of fields satisfying Lie (super)algebraic (anti) commutation relations. Combined with small distance operator product expansions (OPEs) they allow to compute the internal energy and hence the partition function of the system.

These computations appear as a part of a wider program: the construction and classification of 2-dimensional local QFT models with a given algebra of observables, the CCA, each theory having a discrete additive set of superselection sectors.

Thus one obtains a (comparatively) simple example in which the "algebraic QFT" program of Haag et al. ([Ha 2, Do 1, ..., Do 2]) is realized. In particular, we present what appears to us a neat construction of the charged fields as localized morphisms of the CCA - a construction which has appeared under many disguises (as "bosonization", "vertex operator", "Coulomb gas" representation ...) and at various levels of rigour (for a small sample of references, see [Sk 1, St 1,2, Ma 3, Fr 1,2, Ca 3,4, Di 2]).

Regarded from the Euclidean point of view these models appear as (critical) statistical mechanical systems defined on a torus. Imposing modular (or global reparametrization) SL(2, Z)-invariance we end up with a family of 2-dimensional conformal QFT models involving integer helicities only.

The framework of QFT models on a (compact) 2-dimensional torus may appear rather trivial to the protagonists of the algebraic QFT. We feel, however, that there are enough newly discovered beautiful structures in this simple case to justify its careful study even without invoking the hope that it will serve the now fashionable (super)string program. One should have the patience to first master the theory of a single complex variable even if one is chiefly interested in the many variable case. We therefore endeavour to make this notes accessible not only to insiders of 2-dimensional conformal theory and devote most of Sec.2 to a review of the current status of the subject before going to the above mentioned new developments. Conversely, for the benefit of "conformal theorists", not acquainted with finite temperature QFT we shall devote the rest of this introduction to a

Synopsis on Gibbs states and KMS boundary condition.

A characteristic feature of quantum relativistic dynamics being the creation and annihilation of pairs, it is natural to work with the *grand canonical ensemble* in which neither the energy nor the particle number are fixed.

We shall be dealing with a field algebra F (which contains the observable algebra *) A) with a time evolution automorphism α_t. A Gibbs equilibrium state on the algebra F is labeled by the inverse temperature β and is characterized by the following *KMS condition*. For each $A,B \in F$ the function $\langle A\alpha_t B\rangle_\beta$ can be regarded as the boundary value of an analytic function $\langle A\alpha_\zeta B\rangle_\beta$ of the complex time ζ holomorphic in the strip $0 < \text{Im}\zeta < \beta$. It also has a boundary value at $\text{Im}\zeta = \beta$ given by

$$\langle A\alpha_{t+i\beta}B\rangle_\beta = \langle(\alpha_t B)A\rangle_\beta \quad . \tag{1.1}$$

*) Departing from the tradition of algebraic QFT we won't demand that either of these algebras is normed. In effect, we shall be dealing typically with unbounded operators in our examples.

This property implies, in particular, that the temperature mean is stationary: $<\alpha_t A>_\beta = <A>_\beta$.

The KMS condition reflects stability (or energy positivity) and the cyclic property of the trace. This is easy to see in the case of interest, in which the configuration space is compact. In this case the time evolution is given by a bounded below Hamiltonian H,

$$\alpha_t A = e^{itH} A e^{-itH} \tag{1.2}$$

which can be assumed to have a discrete spectrum such that $e^{-\beta H}$ is a trace class operator for (sufficiently large) positive β. Defining then the temperature state by

$$<A>_\beta = \frac{1}{Z(\beta)} \operatorname{tr}(e^{-\beta H} A) \quad , \tag{1.3}$$

where $Z(\beta)$ is the *partition function*

$$Z(\beta) = \operatorname{tr} e^{-\beta H} \quad , \tag{1.4}$$

and the trace is carried over the state space $\mathcal{H}$ of the theory, one can deduce the KMS condition. (In the above cited work of Haag et al. the main interest is in the infinite volume "thermodynamic" limit in which neither the Hamiltonian H nor the partition function Z can, in general, be assumed to exist, but the automorphism α_t (1.2) and the expectation value (1.3) still make sense and satisfy the KMS condition. We shall see, however, that even in the comparatively trivial case of regular temperature states for which one can use the density matrix representation (1.3), the KMS condition still appears as a handy technical tool).

2. Two Dimensional Conformal QFT and CCA on the Circle

2A. *Conserved chiral currents. Compact picture. CCA commutation relations*

A QFT, invariant under infinitesimal conformal(spin (D,2)) or having Schwinger functions invariant under global Euclidean Spin (D + 1, 1) -

transformations, can be extended to a theory on the universal cover $\tilde{M}$ of compactified Minkowski space $\bar{M}$, invariant under global transformations of the universal covering of the Minkowski space conformal group - see [Lü 2]. In D > 2 dimensions $\tilde{M}$ is the cylinder space $S^{D-1} \times R^1$ - the "Einstein universe", singled out by Segal [Se 1] for its conformally invariant global causal structure. We shall consider here a QFT on the 2-dimensional cylinder $\tilde{M}_2 = S^1 \times R^1$, although it is not simply connected. (Doing this means that we restrict our attention to half integer "helicity" fields; we recall that there is no proper notion of spin or helicity, in 1 + 1 dimension - since there is no rotation group left - and the formal counterpart of helicity can, in principle, take arbitrary real values). If $\xi_\pm = X^0 \pm X^1$ are the light cone coordinates in M_2 then the corresponding compact picture coordinates are

$$Z_\pm = \frac{1 + i/2\xi_\pm}{1 - i/2\xi_\pm} = e^{i(t\pm\sigma)} \qquad \left(i\xi_\pm = 2\,\frac{Z_\pm - 1}{Z_\pm + 1}\right) . \tag{2.1}$$

(For this realization of compact picture QFT and for the underlying understanding of local fields on the torus as images of fields in real Minkowski space - rather than Euclidean fields - see [To 1,2,3] ; the last two papers also contain a more leisurely discussion of what is just briefly mentioned in this section, and provide further references).

Free fields and conserved chiral currents can be split into mutually commuting right and left moving components. For the traceless stress-energy tensor $T^\mu{}_\nu$ and the $u(1)_R \oplus u(1)_L$ - conserved current J^μ (satisfying $\partial_\mu J^\mu = 0 = \partial_\mu \varepsilon^{\mu\nu} J_\nu$) these chiral components are

$$T_\pm(Z_\mp) = \frac{1}{2}\,(T^0{}_0 \pm T^1{}_0) \quad , \quad J_\pm(Z_\mp) = \frac{1}{2}\,(J^0 \pm J^1) \quad . \tag{2.2}$$

Periodicity in σ for such right and left movers also implies periodicity in t.

It is a common feature of soluble 2-dimensional models [Fu 1, Be 1, Wi 1, Kn 1, Za 1, Ge 3] that even interacting fields of anomalous dimension split into finite sums of right and left movers [So 1]. We shall restrict our attention in most of what follows to right movers and shall omit the subscript - and + on arguments and field components. In the few places in which $T_-(Z_+)$ will reappear it will be denoted by $\bar{T}(\bar{Z})$.

Lüscher and Mack [Lü 3] established the following *theorem*.

Let $U(\rho)$ be a unitary representation of the dilation group acting in the Hilbert space of physical states and let

$$U(\rho)|0\rangle = |0\rangle \quad , \quad U(\rho)\,\Theta(\xi)U^{-1}(\rho) = \rho^2\,\Theta(\rho\xi) \qquad (\rho > 0) \quad (2.3)$$

where $|0\rangle$ is the unique translation invariant vacuum state. Let, further, $\Theta(\xi)$ be the density of the (positive definite) ξ-translation generator:

$$\int[\Theta(\xi),\, \Theta(\eta)]\, d\eta = i\,\Theta'(\xi) \quad . \qquad (2.4)$$

Then the most general local commutator of Θ is given by

$$-i[\Theta(\xi_1),\, \Theta(\xi_2)] = -\frac{c}{24\pi}\,\delta'''(\xi_{12}) + \delta'(\xi_{12})(\Theta(\xi_1) + \Theta(\xi_2)), \qquad \xi_{12} = \xi_1 - \xi_2 \ , \qquad (2.5)$$

where c is a *central charge* ($[c, \Theta(\xi)] = 0$).

The vacuum expectation values of products of Θ's are computed from (2.5) and are found to be invariant under (fractional linear) $SL(2, R)$ transformations. This allows to view $\Theta(\xi)$ as an expression in a local chart of a field T defined on compactified Minkowski space $\overline{M}$.

We shall use two alternative descriptions of compact picture fields: the analytic or Z - *picture* in which the field argument is the complex variable Z (2.1) on the unit circle and

$$T(Z) = 2\pi\left(i\frac{d\xi}{dZ}\right)^2 \Theta(\xi(Z)) \qquad \left(i\frac{d\xi}{dZ} = \left(\frac{2}{1+Z}\right)^2\right) \qquad (2.6)$$

and the *real angle*, or ϑ - *picture* in which

$$T(\vartheta) = 2\pi\left(\frac{d\xi}{d\vartheta}\right)^2 \cdot \Theta(\xi(\vartheta)) - \frac{c}{12}\{\xi,\vartheta\} \qquad (2.7a)$$

$$\xi(\vartheta) = 2\,\mathrm{tg}\,\vartheta/2 \quad (\vartheta = t - \sigma,\ Z = e^{i\vartheta}) \qquad \frac{d\xi}{d\vartheta} = \frac{1}{\cos^2\vartheta/2} \qquad (2.7b)$$

and $\{\xi,\vartheta\}$ is the Schwarz derivative:

$$\{\xi,\vartheta\} = \frac{\xi'''}{\xi'} - \frac{3}{2}\left(\frac{\xi''}{\xi'}\right)^2 \quad \left(= \frac{1}{2} \text{ for } \xi = 2\,\mathrm{tg}\,\frac{\vartheta}{2}\right) \tag{2.8}$$

which vanishes for fractional linear transformations (so that $\{\xi,Z\} = 0$). The Fourier(-Laurent) expansion in either of the two compact pictures allows to introduce the Virasoro generators L_n:

$$T(Z) = \sum_n L_n Z^{-n-2} \quad , \quad \mathcal{T}(\vartheta) + \frac{c}{24} = \sum_n L_n\, e^{-in\vartheta} \quad . \tag{2.9a}$$

Hermicity of $\mathcal{T}$ (or T) implies

$$L_n^* = L_{-n} \quad . \tag{2.9b}$$

The L_n satisfy, in view of (2.5), the commutation relations of the Virasoro algebra Vir:

$$[L_n, L_m] = (n-m)L_{n+m} + \frac{c}{12}\, n(n^2-1)\delta_{n+m} \quad . \tag{2.10}$$

A (Z-picture) field $\Psi(Z)$ is said to be *primary* (with respect to Vir) of conformal weight Δ if

$$[L_n,\Psi(Z)] = Z^n\left(Z\frac{d}{dZ} + (n+1)\Delta\right)\Psi(Z) \tag{2.11}$$

for all $n \in \mathbb{Z}$; it is said to be *quasiprimary* if this condition is only satisfied for $n = 0,\pm 1$ - that is for the SU(1,1) subalgebra of Vir. (The translation of these notions to the ϑ-picture is obtained by setting $\varphi(\vartheta) = e^{i\Delta\vartheta}\Psi_Z(e^{i\vartheta})$, where we have added the subscript Z for the Z-picture field). J(Z) provides an example of a primary field of weight 1; T(Z) is an example of a quasiprimary field of weight 2, which is not primary. (The derivative of a (quasi)primary field is, in general, not a quasiprimary field).

Following the argument of Lüscher and Mack one can also prove that a dilation invariant current J(Z) with normalized 2-point function

$$\begin{aligned} \langle J(Z_1)\, J(Z_2)\rangle &= Z_{12}^{-2} \\ \langle \mathcal{J}(\vartheta_1)\mathcal{J}(\vartheta_2)\rangle &= -\left(2\sin\frac{\vartheta_{12} - i0}{2}\right)^{-2} \end{aligned} \tag{2.12}$$

satisfies the Heisenberg commutation relations for a free Bose field:

$$[J(Z_1)J(Z_2)] = -\delta'(Z_{12})\,;\left(\oint_{S^1} \delta(Z_{12})f(Z_2)\,\frac{dZ_2}{2\pi i} = f(Z_1)\right) . \tag{2.13}$$

If we use the mode expansion

$$\begin{aligned} J(Z) &= \sum_n J_n Z^{-n-1} \\ J(\vartheta) &= \sum_n J_n e^{-in\vartheta} \qquad (n \in \mathbb{Z}) \end{aligned} \tag{2.14}$$

then we obtain the following discrete form of (2.13) and of the condition that J is primary:

$$[J_n, J_m] = n\delta_{n+m} \tag{2.15a}$$

$$[J_n, L_m] - nJ_{n+m} . \tag{2.15b}$$

Eqs. (2.10) and (2.15) give the canonical form of the CCA $\mathcal{V}U(1)$ (which appears as a *graded* infinite dimensional Lie algebra). The time evolution automorphism of the CCA is given by

$$\begin{aligned} \alpha_t T(\vartheta) &= T(\vartheta+t) \quad , \quad \alpha_t J(\vartheta) = J(\vartheta+t) \\ \alpha_t L_n &= L_n e^{-int} \quad , \quad \alpha_t J_n = J_n e^{-int} . \end{aligned} \tag{2.16}$$

Clearly, if A is any element of the right movers'CCA then α_t can be viewed as a Hamiltonian evolution operator (of type (1.2)) with Hamiltonian identified with L_o:

$$\alpha_t A = e^{iL_o t} A e^{-iL_o t} \quad \text{for} \quad A \in \mathcal{V}U(1) . \tag{2.17}$$

(The total Hamiltonian H of the 2-dimensional QFT will be given by $H = L_o + \bar{L}_o + C$, where C is a constant and we denote by $\bar{L}_n$ the generators of the left mover's Vir). We shall define the *affine* U(1) - *algebra* as the subalgebra $\widehat{u(1)}$ of $\mathcal{V}U(1)$ generated by J_n and L_o.

2B. *Lowest weight (LW) unitary representations (URs) of CCA. The Sugawara formula.*

The discovery of Kac's determinant formula [Ka 1] opened the way to the classification of all irreducible*) positive energy representations of Vir, and in singling out the unitary ones([Ka 1,3, Fr 3, Go 1]). If the algebra of right moving observables is generated by a current J satisfying (2.12-15), then the classification of all its (irreducible) LWURs is much simpler. Every LWUR of $\mathcal{V}U(1)$ that remains irreducible when restricted to $\hat{u}(1)$ is generated by a representation of the Heisenberg algebra (2.15a) (extended by L_0) characterized by a LW vector $|e\rangle$ satisfying

$$(J_0 - e)|e\rangle = 0 = (L_0 - \Delta)|e\rangle \ , \quad J_n|e\rangle = 0, \qquad n=1,2,\ldots \tag{2.18}$$

The representation space V_e is spanned by vectors of the form $J_{-n_1}^{\nu_1} \ldots$ $\ldots J_{-n_k}^{\nu_k} |e\rangle$ with positive integer powers $\nu_1, \ldots, \nu_k$ (which form an orthogonal basis in V_e with respect to the algebraically defined inner product).

The extension of such a representation to a LWUR of $\mathcal{V}U(1)$ is achieved via the *Sugawara construction* which we proceed to describe.

The normal product

$$:J(Z_1)J(Z_2): \;=\; J(Z_1)J(Z_2) - Z_{12}^{-2} \tag{2.19a}$$

where Z_{12}^{-2} is defined as the boundary value of a function analytic in the domain where $|Z_1| > |Z_2|$ so that

$$Z_{12}^{-2} - Z_{21}^{-2} = -\delta'(Z_{12}) \tag{2.19b}$$

has a finite limit for $Z_1 \to Z_2$. The field $T(Z) - \frac{1}{2}:J^2(Z):$ (which is, thus, well defined) commutes with $J(Z_1)$ and is therefore a multiple of the identity (in view of the irreducibility assumption). A simple argument (given in [To 2]) shows that it should actually vanish. This gives the Sugawara relation whose mode expression is

$$L_n = \frac{1}{2}\Big(\sum_{l\geq 1} + \sum_{l\geq -n}\Big) J_{-l}J_{l+n} \tag{2.20a}$$

*) A representation π of an algebra A in a (Banach) space H is said to be irreducible if every (bounded) operator B in H which commutes with $\pi(A)$ is a multiple of the identity.

or

$$L_{2n} = \frac{1}{2} J_n^2 + \sum_{l\geq 1} J_{n-l} J_{n+l}$$
$$L_{2n+1} = \sum_{l\geq 0} J_{n-l} J_{n+l+1} \quad . \tag{2.20b}$$

From (2.20) and (2.15a) one easily derives (2.15b) and then (2.10) with

$$c = c_J = 1 \quad . \tag{2.21}$$

On the other hand, comparison of (2.20b) with (2.18) tells us that the conformal LW is determined by the charge:

$$\Delta = \Delta_e = \tfrac{1}{2}e^2 \quad . \tag{2.22}$$

In a QFT realization of the CCA one is usually interested in representations of an extension of $\mathcal{V}U(1)$ which includes the charge conjugation automorphism γ_C defined by

$$\gamma_C\, J(Z) = - J(Z) \quad , \qquad \gamma_C T(Z) = T(Z) \quad . \tag{2.23}$$

LW charges come then in pairs, $\pm$ e.

3. KMS States on the CCA

3A. *KMS correlation functions in terms of charge distribution*

We begin by showing that the correlation functions of the Fourier coefficients $\tilde{L}_n$ of $T(\vartheta)$ (2.9a)

$$T(\vartheta) = \sum_n \tilde{L}_n e^{-in\vartheta} \quad , \quad \tilde{L}_n = L_n - \frac{c}{24}\delta_n \tag{3.1}$$

which satisfy, due to (2.10), the commutation relations

$$[\tilde{L}_n, \tilde{L}_m] = (n - m)\tilde{L}_{n+m} + \frac{c}{12} n^3 \delta_{n+m} \tag{3.2}$$

(the 2-cocycle $n^3\delta_{n+m}$ differing from $n(n^2-1)\delta_{n+m}$ by a coboundary), are expressed in terms of (c and) the expectation values $\langle\tilde{L}_o^k\rangle_\beta$. Indeed, stationarity implies

$$\langle\tilde{L}_n\rangle_\beta = \langle\tilde{L}_o\rangle_\beta\, \delta_n \tag{3.3a}$$

while the KMS condition yields the recurrence relation

$$(e^{n_k\beta} - 1) \langle \tilde{L}_{n_1} \ldots \tilde{L}_{n_k} \rangle_\beta =$$

$$\langle \{(n_k - n_1)\tilde{L}_{n_k+n_1} + \frac{c}{12} n_k^3 \delta_{n_k+n_1}\}\tilde{L}_{n_2} \ldots \tilde{L}_{n_{k-1}} + \ldots \tag{3.3b}$$

$$\tilde{L}_{n_1} \ldots \tilde{L}_{n_{k-2}} \{(n_k - n_{k-1}) \tilde{L}_{n_k+n_{k-1}} + \frac{c}{12} n_k^3 \delta_{n_k+n_{k-1}}\}\rangle_\beta \quad .$$

Since Eqs.(3.3) imply the conservation law

$$\langle \tilde{L}_{n_1} \ldots \tilde{L}_{n_k} \rangle_\beta (1 - \delta_{n_1 + \ldots + n_k}) = 0 \tag{3.4}$$

we deduce that for $n_k = 0$ both sides of (3.3b) vanish. If, on the other hand, $n_k \neq 0$ we can express $\langle \tilde{L}_{n_1} \ldots \tilde{L}_{n_k} \rangle_\beta$ in terms of (k-1) and (k-2)-point functions. If at least one $n_j \neq 0$ $(1 \leq j \leq k)$ then we can use the commutation relations (3.2) to transfer L_n to the right and then apply (3.3b). We find, in particular, for $n \neq 0$

$$\langle \tilde{L}_n \tilde{L}_{-n} \rangle_\beta = (1 - e^{-\beta n})^{-1} \{2n \langle \tilde{L}_0 \rangle_\beta + \frac{c}{12} n^3\} \quad . \tag{3.5}$$

The expectation values $\langle \tilde{L}_0^k \rangle_\beta$ can, in principle, be determined given a partition function of the type

$$Z(\beta,\bar{\beta}) = \text{tr } e^{-\beta\tilde{L}_0 - \bar{\beta}\bar{\tilde{L}}_0} \quad . \tag{3.6}$$

We would have in this case

$$\langle \tilde{L}_0 \rangle_\beta = - \frac{\partial}{\partial\beta} \ln Z(\beta,\bar{\beta}) \tag{3.7}$$

and hence

$$\langle \tilde{L}_0^2 \rangle_\beta = (\langle \tilde{L}_0 \rangle_\beta)^2 - \frac{\partial}{\partial\beta} \langle \tilde{L}_0 \rangle_\beta \qquad \text{etc.} \tag{3.8}$$

Simple examples of partition functions are given by the irreducible Virasoro characters [Ka 1,2, Fe 1]. They are, however, insufficient to fit a realistic QFT which is assumed to involve a (unique) vacuum representation (of LW $\Delta = 0$) and one or more primary fields of positive weight.

A slightly richer and more realistic structure is provided by the representations of the CCA.

We shall now demonstrate that the finite temperature many point correlation functions of the current modes are expressed in terms of the momenta of the charge distribution, $\langle J_o^l \rangle_\beta$.

The KMS condition implies the recurrence relation

$$(1 - e^{-n_o\beta}) \langle J_{n_o} J_{n_1} \dots J_{n_l} \rangle_\beta = n_o \langle \{ \delta_{n_o+n_1} J_{n_2} \dots J_{n_l} + \delta_{n_o+n_2} J_{n_1} J_{n_3} \dots J_{n_l} + \dots + \delta_{n_o+n_l} J_{n_1} \dots J_{n_{l-1}} \} \rangle_\beta \quad . \tag{3.9}$$

For $n_o \neq 0$ it allows to reduce the $(\ell + 1)$-point function to a linear combination of $(\ell - 1)$-point functions; for $n_o = 0$ both sides vanish. If we consider equilibrium states that are charge conjugation invariant,

$$\langle \gamma_C A \rangle_\beta = \langle A \rangle_\beta \quad , \tag{3.10}$$

then Eq. (2.23) will imply that odd point current correlation functions vanish

$$\langle J_o \rangle_\beta = 0 = \langle J_{n_1} \dots J_{n_{2k+1}} \rangle_\beta \quad . \tag{3.11}$$

The 2-point function, on the other hand, is then given by (3.9) as

$$\langle J_n J_{-n} \rangle_\beta = \frac{n}{1 - e^{-\beta n}} \qquad (\text{for } n \neq 0) \quad , \tag{3.12}$$

so that

$$\langle J(\vartheta_1) J(\vartheta_2) \rangle_\beta = \sum_{n \neq 0} \frac{n}{1-e^{-\beta n}} e^{-in\vartheta_{12}} + \langle J_o^2 \rangle_\beta = - \left(2 \sin \frac{\vartheta_{12} - i0}{2}\right)^{-2} + \langle J_o^2 \rangle_\beta + 2 \sum_{n=1}^{\infty} \frac{n}{e^{\beta n} - 1} \cos n\vartheta_{12} \quad . \tag{3.13}$$

Here we have used the Fourier expansion for the current 2-point function (2.12)

$$\sum_{n=1}^{\infty} n e^{-in\vartheta_{12}} = \frac{d}{d\vartheta} \left(\frac{e^{-i\vartheta_{12}/2}}{2 \sin \frac{\vartheta_{12} - i0}{2}} \right) = - \left(2 \sin \frac{\vartheta_{12} - i0}{2}\right)^{-2} \tag{3.14}$$

Combining (3.13) with the small distance OPE

$$J(\vartheta_1) J(\vartheta_2) = - \left(2 \sin \frac{\vartheta_{12} - i0}{2}\right)^{-2} + 2T(\vartheta) + \frac{1}{12} + O(\sin^2 \frac{\vartheta_{12}}{2}) \tag{3.15}$$

(which follows from (2.9) (2.20) or using (3.12) and the Sugawara formula (2.20))we find

$$\langle L_0 \rangle_\beta = \langle \tilde{L}_0 \rangle_\beta + \frac{1}{24} = \frac{1}{2} \langle J_0^2 \rangle_\beta + \sum_{n=1}^{\infty} \frac{n}{e^{\beta n}-1} \quad . \tag{3.16}$$

Thus, both the correlation functions of J and the internal energy are expressed in terms of the (moments of the) charge distribution.

There is just one example of an irreducible LWUR of $\mathcal{V}U(1) \oplus \mathcal{V}U(1)$ which corresponds to a QFT. This is the vacuum representation. Naively, we should set $\langle J_0^2 \rangle_\beta = 0$ in that case. A more interesting theory appears, however, if we take for $\langle J_0^2 \rangle_\beta$ the limit $1/\beta$ of the ratio $n(1-e^{-\beta n})^{-1}$ for $n \to 0$, which only vanishes at zero temperature (i.e. for $\beta \to \infty$). In fact, the partition function

$$Z_0(\tau) = \frac{1}{\sqrt{\operatorname{Im}\tau}} \quad |\eta(\tau)|^{-2} \tag{3.17}$$

where η is the Dedekind η-function,

$$\eta(\tau) = q^{1/24} \prod_{n=1}^{\infty} (1 - q^n) \quad , \qquad q = e^{2\pi i \tau} \, , \qquad \operatorname{Im}\tau > 0 \tag{3.18}$$

and the variable τ is related to the (complexified) inverse temperature β by

$$2\pi\tau = i\beta \tag{3.19}$$

is shown in [It 1] to correspond to the theory of a free logarithmic scalar field on the (2-dimensional) torus.

3B. *Local QFT model associated with a free complex Weyl field*

More interesting realizations of the CCA are obtained in a theory of charged fields. Already the simplest example of a free complex Fermi field of helicity 1/2 on the circle displays some intriguing properties. We shall consider this example in detail as a preparation for a systematic study of charged QFT models of CCA in Sec. 4.

The theory of a charged Fermi field Ψ of weight 1/2 is summarized by the small distance OPE

$$\Psi^*(Z_1)\Psi(Z_2) = \frac{1}{Z_{12}} + J(Z) + Z_{12}T(Z) + O(Z_{12}^2) \qquad (Z=\sqrt{Z_1 Z_2}) \tag{3.20a}$$

or, equivalently (for $\psi(\vartheta) = \sqrt{Z}\,\psi(Z)$, $J(\vartheta) = ZJ(Z)$, $T(\vartheta) = Z^2T(Z) - \frac{1}{24}$, $Z = e^{i\vartheta}$)

$$\psi^*(\vartheta_1)\psi(\vartheta_2) = \left(2i\sin\frac{\vartheta_{12}-i0}{2}\right)^{-1} + J(\vartheta) + 2i\sin\tfrac{1}{2}\vartheta_{12}\left(T(\vartheta)+\frac{1}{24}\right)+\dots \tag{3.20b}$$

$$\vartheta = \frac{1}{2}(\vartheta_1 + \vartheta_2) \quad .$$

Indeed, Eq. (3.20) allows us to read off the 2-point function and the canonical anticommutation relations of the conjugate charged fields ψ and ψ^* and to express the current and the stress energy tensor in terms of them. Clearly, all these relations remain invariant under phase transformations $e^{-i\alpha J_0}\,\psi e^{i\alpha J_0} = e^{i\alpha}\psi$, $e^{-i\alpha J_0}\,\psi^* e^{i\alpha J_0} = e^{-i\alpha}\psi^*$.

There are two *spin structures* on the circle (cf.[Se 2]): the state space H of the theory can be split into a direct sum of two subspaces corresponding to two superselection sectors, the Neveu-Schwarz (NS) and the Ramond (R) sector defined by the superselection operator

$$\alpha_{2\pi}\psi(\vartheta) = \psi(\vartheta+2\pi) \qquad (\alpha_{2\pi}\psi(Z) = -\psi(e^{2\pi i}Z)) \tag{3.21}$$

as follows: $H = H_{NS} \oplus H_R$, where

$$(\alpha_{2\pi}\psi)H_{NS} = -\psi H_{NS} , \qquad (\alpha_{2\pi}\psi)H_R = \psi H_R \quad . \tag{3.22}$$

We shall write the mode expansion of ψ in the two sectors in a unified form:

$$\psi(Z) = \sum_n \psi_{n-\kappa}\, Z^{\kappa-n-\frac{1}{2}} , \qquad \psi(\vartheta) = \sum_n \psi_{n-\kappa}\, e^{i(\kappa-n)\vartheta} , \tag{3.23}$$

where κ is 1/2 in the NS and 0 in the R sector:

$$\alpha_{2\pi}\psi = (-1)^{2\kappa}\psi , \qquad (\kappa - \tfrac{1}{2})H_{NS} = 0 = \kappa H_R \quad . \tag{3.24}$$

The canonical anticommutation relations then read

$$[\psi_\rho,\psi_{\rho'}]_+ = 0 \quad , [\psi_{n-\kappa},\overset{*}{\psi}_{\kappa-m}]_+ = \delta_{n+m} \qquad (\overset{*}{\psi}_{-\rho} = (\psi_\rho)^*) \quad . \tag{3.25}$$

The KMS states on the $\psi^{(*)}$ are *quasifree*: the 2n-point correlation function is expressed as product of 2-point functions. Noting the relation

$\alpha_t \overset{(*)}{\Psi}_\rho = \overset{(*)}{\Psi}_\rho e^{-i\rho t}$ we find that the only non-zero 2-point function is

$$<\Psi_{n-\kappa} \overset{*}{\Psi}_{\kappa-n}>_{\beta,\kappa} = (1 + e^{-\beta(n-\kappa)})^{-1} = <\overset{*}{\Psi}_{n-\kappa} \Psi_{\kappa-n}>_{\beta,\kappa} \tag{3.26a}$$

or, using $\sum_{n=0}^{\infty} e^{i(n+\frac{1}{2})\vartheta} = (2i \sin\frac{1}{2}(\vartheta - i0))^{-1}$ - cf. (3.14) -

$$<\Psi^*(\vartheta_1)\Psi(\vartheta_2)>_{\beta,\kappa} = \frac{\cos(\frac{1}{2}-\kappa)\vartheta_{12}}{2i\sin\frac{1}{2}(\vartheta_{12}-i0)} + 2i \sum_{n=1}^{\infty} \frac{\sin(n-\kappa)\vartheta_{12}}{e^{(n-\kappa)\beta} + 1} . \tag{3.26b}$$

Unlike the finite temperature means of the stress energy and the current, there is no freedom left in these expectation values. Moreover, combined with the small distance OPE they determine the charge distribution and the internal energy. Indeed, comparing (3.26) with (3.20) we find $<J(\vartheta)>_{\beta,\kappa} = 0$ in accord with charge conjugation invariance (3.10,11) -and

$$<T(\vartheta) + \frac{1}{24}>_{\beta,\kappa} = <L_0>_{\beta,\kappa} = \frac{1}{2}(\frac{1}{2} - \kappa)^2 - 2\frac{d}{d\beta} \ln\{ \prod_{n=1}^{\infty} (1+e^{-(n-\kappa)\beta})\}. \tag{3.27}$$

It follows from here (and from (3.24)) that the lowest eigenvalue of L_0 in the R sector is 1/8. It is instructive to rederive this result by studying the algebra of the charged fields' zero modes that is characteristic for the R sector.

The operators Ψ_0 and $\overset{*}{\Psi}_0$, defined in H_R, are isotropic generators of the 2-dimensional Clifford algebra:

$$\Psi_0^2 = 0 \; (= \overset{*}{\Psi}{}_0^2) \qquad \Psi_0\overset{*}{\Psi}_0 + \overset{*}{\Psi}_0\Psi_0 = 1 \tag{3.28}$$

Completed with the commutator

$$\Gamma = [\overset{*}{\Psi}_0, \Psi_0] \quad (= \Gamma^*) \tag{3.29}$$

satisfying

$$\Gamma\Psi_0 = -\Psi_0\Gamma = -\Psi_0, \qquad \Gamma\overset{*}{\Psi}_0 = -\overset{*}{\Psi}_0\Gamma = \overset{*}{\Psi}_0 \;, \qquad \Gamma^2 = 1 \tag{3.30}$$

this algebra has a unique (up to equivalence) irreducible representation, which is 2-dimensional. The corresponding representation space $H_R^{(0)}$ can be identified with the subspace of LW vectors (with respect to Vir) of H_R. Choosing a basis $|\underline{+}\rangle$ of eigenvectors of Γ in $H_R^{(0)}$

$$\Gamma|\varepsilon\rangle = \varepsilon|\varepsilon\rangle \qquad (\langle\varepsilon|\varepsilon'\rangle = \delta_{\varepsilon\varepsilon'}\,, \quad \varepsilon,\varepsilon' = \underline{+}) \tag{3.31}$$

we find (by fixing appropriately a relative phase factor)

$$\Psi_0|+\rangle = |-\rangle\,, \qquad \overset{*}{\Psi}_0|+\rangle = 0, \quad \Psi_0|-\rangle = 0, \quad \overset{*}{\Psi}_0|-\rangle = |+\rangle\,. \tag{3.32}$$

The OPE (3.20) implies, on the other hand, the following expression for the charge operator:

$$J_0 = (\tfrac{1}{2} - \kappa)(\overset{*}{\Psi}_{-\kappa}\Psi_\kappa - \Psi_{-\kappa}\overset{*}{\Psi}_\kappa) + \sum_{l=1}^{\infty}(\overset{*}{\Psi}_{\kappa-l}\Psi_{l-\kappa} - \Psi_{\kappa-l}\overset{*}{\Psi}_{l-\kappa})\,. \tag{3.33}$$

Since $|\underline{+}\rangle$ are LW vectors of Vir they are annihilated by all negative frequency modes:

$$\overset{(*)}{\Psi}_l|\underline{+}\rangle = 0 \; (= J_l\;|\underline{+}\rangle) \qquad l = 1,2, \ldots \tag{3.34}$$

Eq. (3.33)(for $\kappa = 0$) and the Sugawara formula (2.20) then give

$$(J_0 \mp \tfrac{1}{2})\;|\underline{+}\rangle = 0 = (L_0 - \tfrac{1}{8})\;|\underline{+}\rangle \quad . \tag{3.35}$$

Remark. The operator Γ, defined on the subspace $H_R^{(0)}$ by (3.29) can be extended to the entire state space $H = H_R + H_{NS}$ to coincide with the (right movers') *fermion parity operator*

$$\Gamma = (-1)^F = e^{i\pi(J_0 - 2J_0^2)} \quad . \tag{3.36}$$

Noting that the charge spectrum in H covers $1/2\,\mathbb{Z}$ and the $e-2e^2$ is integer whenever e is half integer, we derive the following characteristic properties for Γ:

$$\Gamma\Psi^{(*)}(\vartheta) = -\Psi^{(*)}(\vartheta)\Gamma \quad , \qquad \Gamma^2 = 1 \tag{3.37a}$$

$$\Gamma|0\rangle = |0\rangle\,, \quad \Gamma|\underline{+}\rangle = \underline{+}|\underline{+}\rangle \quad , \quad \Gamma|e\rangle = (-1)^{e-2e^2}\;|e\rangle. \tag{3.37b}$$

Let $\varphi(Z, \pm\frac{1}{2})$ be the primary fields carrying charge $\pm\frac{1}{2}$ and satisfying the local current-field commutation relations

$$[J(Z_1), \varphi(Z_2,e)] = e\delta(Z_{12})\, \varphi(Z_2,e) \quad . \tag{3.38}$$

(We shall prove in the next section that one can construct a $\varphi(Z,e)$ with these properties for any real value e of the charge. It is the appearance of these "spin fields" which interpolate between the vacuum and the R sector which makes the theory of a canonical Weyl field on S^1 non-trivial).

We should like to display the restrictions coming from locality on the 2-dimensional fields, built as bilinears of the primary fields 1, $\psi^{(*)}$ and φ and their left moving counterparts. Strict locality (in the sense of the Bose-Fermi alternative - see [De 1]) is essentially equivalent to the requirement that the Schwinger functions $<\phi_1(Z_1,\bar{Z}_1)\ldots\phi_n(Z_n,\bar{Z}_n)>_E$ are single valued in the Euclidean region (where $\bar{Z}_k$ is complex conjugate to Z_k - cf. [Be 1]).

Only fields of (half)integer helicity $\Delta - \bar{\Delta}$ can satisfy the above condition. Therefore, φ can only enter a scalar field combination (in which it is multiplied by a $\bar{\varphi}$). We consider a hermitian 2-dimensional field of fixed combined parity $\Gamma\bar{\Gamma}$ (assuming, as usual, $\varphi^*(Z,e) = \varphi(Z,-e)$):

$$\sigma(Z,\bar{Z};e\bar{e}) = \frac{1}{\sqrt{2}}\{\varphi(Z,e)\bar{\varphi}(\bar{Z},\bar{e}) + \varphi(Z,-e)\bar{\varphi}(\bar{Z},-\bar{e})\} \qquad (e,\bar{e} = \pm\tfrac{1}{2}) \tag{3.39}$$

where φ and $\bar{\varphi}$ are assumed to commute. (The notation should suggest that σ is the counterpart of the magnetization field of the Ising model). The helicity 1/2 fields $\psi^{(*)}(Z)$ are not local relative to σ because of the small distance OPEs

$$\psi(Z_1)\varphi(Z_2, \tfrac{1}{2})|0> = \frac{1}{\sqrt{Z_{12}}}\, \varphi(Z_2, -\tfrac{1}{2})|0> + \ldots \tag{3.40a}$$

$$\psi^*(Z_1)\varphi(Z_2, -\tfrac{1}{2})|0> = \frac{1}{\sqrt{Z_{12}}}\, \varphi(Z_2, \tfrac{1}{2})|0> + \ldots \tag{3.40b}$$

which follow from (3.32) (the dots stand for terms vanishing as $\sqrt{Z_{12}}$). It is,however, local with respect to the real field

$$\varepsilon(Z,\bar{Z};\, 4e\bar{e}) = \frac{1}{\sqrt{2}}\{\psi(Z,2e)\bar{\psi}(\bar{Z},2\bar{e}) + \psi(Z,-2e)\bar{\psi}(\bar{Z},-2\bar{e})\} \quad . \tag{3.41a}$$

where

$$\Psi(Z,-1) = \Psi(Z) \quad , \quad \Psi(Z, 1) = \Psi^*(Z) \quad . \tag{3.41b}$$

A 2-dimensional field theory whose state space is generated from the vacuum by polynomials of σ and ε has a partition function [Pa 1]

$$Z(\beta, \bar{\beta}) = \frac{1}{2} \{|K_0(\tau,1)|^2 + |\tilde{K}_0(\tau,1)|^2 + |K_1(\tau,1)|^2\} \quad , \tag{3.42}$$

where τ is the complex time variable related to the inverse temperature β by (3.19) and $K_\nu(\tau,1)$ can be determined from (3.27)

$$K_0(\tau,1) = q^{-1/24} \prod_{n=1}^{\infty} (1 + q^{n-\frac{1}{2}})^2 = (\chi_{\frac{1}{2},0}(\tau) + \chi_{\frac{1}{2},\frac{1}{2}}(\tau))^2 \tag{3.43a}$$

$$K_1(\tau,1) = 2q^{1/12} \prod_{n=1}^{\infty} (1 + q^n)^2 = 2(\chi_{\frac{1}{2},1/16}(\tau))^2 \quad , \tag{3.43b}$$

$$\tilde{K}_0(\tau,1) = e^{i\pi/12} K_0(\tau + 1,1) = (\chi_{\frac{1}{2},0}(\tau) - \chi_{\frac{1}{2},\frac{1}{2}}(\tau))^2 \tag{3.43c}$$

for q given in (3.18). The functions $\chi_{c,\Delta}$ in the right hand side of (3.43) are the irreducible Virasoro characters, corresponding to the Ising (c=1/2)-model. (The complex field Ψ is replaced by a real Majorana-Weyl field in the Ising model which is described by the small distance OPE

$$Z_{12}\Psi(Z_1)\Psi(Z_2) = 1 + 2Z_{12}^2 T(Z) + O(Z_{12}^4) \qquad (Z=\sqrt{Z_1 Z_2}) \tag{3.44}$$

instead of (3.20). The characters $\chi_{\frac{1}{2},1/16}$ and $\chi_{\frac{1}{2},0} + \chi_{\frac{1}{2},\frac{1}{2}}$ are again evaluated from the finite temperature 2-point correlation function of Ψ which is related to the expectation of T by the above OPE).

There is an alternative way to evaluate the reducible Virasoro characters $K_\nu(\tau,1)$ associated with the OPE algebra of the free charged field $\psi^{(*)}$. It is based on the representation (3.16) of $<\tilde{L}_0>_\beta$ in terms of the charge square distribution

$$\begin{aligned} <J_0^2>_{\beta,e} &= (\mathrm{tr}_{H_e}\, e^{-\beta L_0})^{-1}\, \mathrm{tr}_{H_e}(J_0^2 e^{-\beta L_0}) \\ &= \{\sum_n e^{-\frac{1}{2}(e+n)^2\beta}\}^{-1} \sum_n (e+n)^2\, e^{-\frac{1}{2}(e+n)^2\beta} \end{aligned} \tag{3.45a}$$

where the sum is over all integers $(n \in \mathbb{Z})$,

$$e = 0, \pm\frac{1}{2}, \quad H_0 = H_{NS}, \ H_{\frac{1}{2}} = H_{-\frac{1}{2}} = H_R \quad . \tag{3.45b}$$

The result (for $e_0 = 0$, $e_1 = \frac{1}{2}$) is

$$\langle \tilde{L}_0 \rangle_{\beta, e_\nu} = -\frac{\partial}{\partial\beta} K_\nu(\tau, 1) \qquad (\tau = \frac{i\beta}{2\pi}, \ \nu = 0,1) \quad , \tag{3.46}$$

$$K_\nu(\tau,1) = \frac{1}{\eta(\tau)} \sum_n \exp\{-\frac{1}{2}(n + \frac{\nu}{2})^2 \beta\} \quad , \tag{3.47a}$$

$$\tilde{K}_0(\tau,1) = \frac{1}{\eta(\tau)} \sum_n (-1)^n q^{\frac{1}{2}n^2} \quad , \tag{3.47b}$$

where $\eta(\tau)$ is defined by (3.18). Putting together (3.43) and (3.47) we reproduce two known (albeit, non-trivial) identities.

The partition function (3.42) regarded as a function of τ is invariant under the group PSL $(2, \mathbb{Z})$ of modular transformations generated by the translation

$$T: \quad \tau \rightarrow \tau + 1 \tag{3.48a}$$

and the inversion

$$S: \quad \tau \rightarrow -\frac{1}{\tau} \tag{3.48b}$$

(which is only an involution in $PSL(2, \mathbb{Z}) = SL(2, \mathbb{Z})/\mathbb{Z}_2$, but has an image $S \longleftrightarrow S_1 = \begin{pmatrix} 0 & 1 \\ -1 & 0 \end{pmatrix}$ in $SL(2, \mathbb{C})$ satisfying $S_1^2 = -1$). The form (3.47) of the reducible characters is better suited for verifying this property. Indeed, we have (in accord with (3.43c))

$$K_0(\tau + 1,1) = e^{-i\pi/12} \tilde{K}_0(\tau,1), \qquad \tilde{K}_0(\tau + 1,1) = e^{-i\pi/12} K_0(\tau,1) \ ,$$
$$K_1(\tau + 1,1) = e^{i\pi/6} K_1(\tau,1) \quad ; \tag{3.49a}$$

using the celebrated Poisson formula we obtain further ([Pa 1])

$$K_0(-\frac{1}{\tau}, 1) = K_0(\tau,1) \ , \qquad K_1(-\frac{1}{\tau}, 1) = \tilde{K}_0(\tau,1) \ ,$$
$$\tilde{K}_0(-\frac{1}{\tau}, 1) = K_1(\tau,1) \quad . \tag{3.49b}$$

The verification that the partition function (3.42) is invariant under transformations (3.49) is immediate.

We thus see that modular invariance of the partition function is related to the locality property of the underlying QFT model. The concept of modular invariance, on the other hand, appears most naturally if one starts with an Euclidean theory defined on a torus (see [Ca 2], [It 1]). It is also interpreted in the context of string theory as a condition for the absence of global reparametrization anomalies (see, e.g., [Se 2] and [Ge 3]). At any rate it serves as an effective tool in classifying consistent 2-dimensional conformal QFT models. We shall also refer to it in the next section in our classification of $c = 1$ theories.

4. Classification of 2-Dimensional Conformal Models of a Chiral U(1) Current Algebra

4A. Construction of primary charged fields from localized morphisms of the CCA

We start with the following generalization of the example of the preceding section.

Let $\Psi(Z, \pm g)$ be a pair of hermitian conjugate charged primary fields (of dimension Δ, satisfying (2.11)). Let them also have local current-field commutation relations of type (3.38) with $e = \pm g$, or in terms of the current's Laurent coefficients J_n, introduced in (2.14),

$$[J_n, \Psi(Z,g)] = gZ^n \Psi(Z,g) \quad . \tag{4.1}$$

Such a field will be called VU(1)-primary. (The non-abelian counterpart of $\widehat{u(1)}$, corresponding to what is now called the Wess-Zumino-Witten model, was studied in similar terms in [Kn 1]; the general case, including the U(1) x U(1) Thirring model was considered in [To 1,2] where the notion of a Vir primary field of [Be 1] was extended to the case of a field transforming homogeneously with respect to the entire CCA. The local Minkowski space version of the current-field commutation relations, similar to (3.38), has been a starting point for a current algebra approach to the (generalized) Thirring model, developed long ago in [De 1]. In fact, the Thirring field appears as a product, $\Psi(Z,g)\overline{\Psi}(\overline{Z},\overline{g})$ of a right moving and a left moving field of the above type). The following statement has been proven in [To 1,2].

Proposition 4.1. The commutation relations (2.11) and (4.1) are only consistent with each other and with the Sugawara formula (2.20) if

$$\Delta = \frac{1}{2} g^2 \tag{4.2}$$

and the field Ψ satisfies the equation

$$\frac{d\Psi}{dZ}(Z,g) = g\colon J(Z)\Psi(Z,g)\colon \tag{4.3}$$

where the normal product is defined by

$$\begin{aligned}\colon J(Z_1)\Psi(Z_2,g)\colon &= \left(J(Z_1) - \frac{g}{Z_{12}}\right)\Psi(Z_2,g) = J_{(+)}(Z_1)\Psi(Z_2,g) + \\ &+ \Psi(Z_2,g)\, J^{(-)}(Z_1)\end{aligned} \tag{4.4}$$

and the frequency parts of the current are

$$J_{(\pm)}(Z) = \sum_{n=1}^{\infty} J_{\mp n} Z^{\pm n-1} \quad , \qquad J^{(-)}(Z) = J_{(-)}(Z) + \frac{J_0}{Z} \tag{4.5}$$

We shall now construct an operator solution of Eq. (4.3), acting in a space H in which the charge spectrum is symmetric with respect to translation by $\pm$ g. Ultimately, we shall deal with a direct sum of eigenspaces of the charge operator J_0 of the following structure:

$$H = \bigoplus_{\substack{n \in \mathbb{Z} \\ \nu}} H_{e_\nu + ng} \tag{4.6}$$

(the sum over ν will turn out to be finite and the set of e_ν will include $e_0 = 0$, but we shall not need these properties at present). Each eigenspace of J_0 carries an irreducible LWUR of $\mathcal{V}U(1)$.

To begin with one can define in H a (Z-independent) unitary charge shift operator U_g satisfying

$$J(Z)U_g = U_g\left(J(Z) + \frac{g}{Z}\right) \qquad \text{or} \qquad J_n U_g = U_g(J_n + g\delta_n) \tag{4.7a}$$

$$U_g^* = U_{-g} (= U_g^{-1}) \tag{4.7b}$$

Indeed, if $|e_\nu + ng\rangle$ is a complete set of (normalized) LW vectors with respect to $\mathcal{V}U(1)$ (each such vector being determined uniquely up to an inessential phase factor by this property then we define $U_{\pm g}$ by

$$U_{\pm g}|e_\nu + ng\rangle = |e_\nu + (n\pm 1)g\rangle \quad . \tag{4.8}$$

Since each such vector is cyclic in $\mathcal{H}_{e_\nu+ng}$ with respect to J_{-n} (n=1,2,...) and U_g commutes with J_n for $n\neq 0$, it is thus extended to the entire space.

Consider more general automorphisms of the U(1)-current algebra of the type

$$J(Z) \to J(Z) + \alpha(Z) \quad , \qquad \alpha(Z) = \sum_n \alpha_n Z^{-n-1} \quad , \quad \alpha_n^* = \alpha_{-n} \tag{4.9}$$

(α is a c-number function). Such an automorphism is unitarily implementable in $\mathcal{H}$ if the total charge shift is a multiple of g:

$$(\alpha_0 =) \oint_{S^1} \alpha(Z) \frac{dZ}{2\pi i} = ng \qquad\qquad (n \in \mathbb{Z}) \quad . \tag{4.10}$$

In the special case of n = 0, i.e. for

$$\alpha(Z) = if'(Z) \; , \qquad f^*(\tfrac{1}{Z}) = f(Z) \quad , \tag{4.11}$$

it is inner. Indeed, if we set

$$J(f) = \oint_{S^1} J(Z)f(Z) \frac{dZ}{2\pi i} \; (= \sum_n J_n f_{-n} \text{ for } f(Z) = \sum f_n Z^{-n}) \quad , \tag{4.12}$$

then

$$e^{iJ(f)} J(Z) e^{-iJ(f)} = J(Z) + if'(Z) \quad . \tag{4.13}$$

Eq. (4.7) provides a special case of an implementable outer automorphism, satisfying (4.10) with n = 1. The field $\Psi(Z,g)$, we wish to construct, differs from U_g by an operator, $U_g^*\Psi(Z,g)$ which does not change the charge and can, hence, be presented as a function of the current.

Proposition 4.2. The solution of Eq.(2.11),(4.2),(4.3) normalized by the condition

$$Z_{12}^{g^2} \langle\Psi(Z_1,g)\, \Psi^*(Z_2,g)\rangle = 1 \qquad (\Psi^*(Z,g) = \Psi(Z,-g)) \tag{4.14}$$

differs at most by a constant phase factor from

$$\Psi(Z,g) = e^{ig\phi_{(+)}(Z)} U_g Z^{gJ_o} e^{ig\phi_{(-)}(Z)} \quad , \quad i\frac{d}{dZ}\,\phi_{(\pm)}(Z) = J_{(\pm)}(Z) \ , \tag{4.15}$$

where U_g is defined by (4.7, 8) and

$$i\phi_{(+)}(Z) = \int_0^Z J_{(+)}(\zeta)d\zeta = \sum_{n=1}^{\infty} J_{-n}\frac{Z^n}{n} \quad ,$$
$$i\phi_{(-)}(Z) = -\int_Z^{\infty} J_{(-)}(\zeta)d\zeta = -\sum_{n=1}^{\infty} J_n\frac{Z^{-n}}{n} \quad . \tag{4.16}$$

The fields $\Psi(Z,\pm g)$ satisfy the OPE

$$Z_{12}^{g^2}\,\Psi(Z_1,g)\Psi^*(Z_2,g) = :\exp\{-g\int_{Z_1}^{Z_2} J(Z)dZ\}: \tag{4.17a}$$

$$= 1 + gZ_{12}J(Z) + g^2Z_{12}^2\,T(Z) + O(Z_{12}^3) \qquad (Z=\sqrt{Z_1Z_2}) \quad . \tag{4.17b}$$

(All $J^{(-)}$ should be to the right of the $J_{(+)}$ in each term of the Taylor expansion of the normal order exponential).

The *proof* is immediate, Eq. (4.3) is verified by a simple differentiation of (4.15). Eq. (4.17a) is obtained by applying the known formula.

$$e^Ae^B = e^{A+B}e^{\frac{1}{2}[A,B]} \qquad \text{for} \qquad [A,[A,B]] = 0 = [B,[A,B]] \ . \tag{4.18}$$

Remark. For non-integer g^2 (and $e_\nu g$) the factor Z^{gJ_o} in (4.15) is not a single valued function of Z. In general, the field Ψ is function on the infinite cover of the circle and only the ϑ-picture (in which $Z = e^{i\vartheta}$ and the expression for Ψ is multiplied by $e^{ig^2\vartheta/2}$) has a precise meaning. In what follows we shall deal with theories which live on a finite covering of S^1 and will use the above compact notation.

Eq.(4.17) is a special case of a simple fusion rule for the product of two charged fields of arbitrary charges g_1 and g_2 which reflects the charge conservation law. The fields $\Psi(Z,g)$ also satisfy group theoretic type generalized free commutation relations. The two properties are combined in the relation

$$Z_{12}^{-g_1g_2}\,\Psi(Z_1,g_1)\Psi(Z_2,g_2) = Z_{21}^{-g_1g_2}\,\Psi(Z_2,g_2)\Psi(Z_1,g_1) \underset{\text{for } Z_1,Z_2\to Z}{\to} \Psi(Z,g_1+g_2). \tag{4.19}$$

In deriving (4.19) we use the additivity of the charge spectrum and the possibility to extend $U_{\pm g}$ to an abelian group of unitary operators whose elements are labeled by the eigenvalues of J_o and satisfy $U_{g_1} U_{g_2} = U_{g_1+g_2}$.

As another application of the construction (4.15) we derive the expression for the Wightman function of n fields $\Psi(Z_i, g_i)$. Using repeatedly the Weyl commutation relations

$$e^{ig_1\phi_{(-)}(Z_1)} e^{ig_2\phi_{(+)}(Z_2)} = \left(\frac{Z_{12}}{Z_1}\right)^{g_1g_2} e^{ig_2\phi_{(+)}(Z_2)} e^{ig_1\phi_{(-)}(Z_1)}, \quad (4.20)$$

which follow from (4.16), (4.18) and from

$$[J_{(-)}(\zeta_1), J_{(+)}(\zeta_2)] = \langle J(\zeta_1)\, J(\zeta_2)\rangle = \zeta_{12}^{-2} \quad , \quad (4.21)$$

and the implication of charge conservation

$$\langle U_{g_1} \dots U_{g_n}\rangle = \delta_{g_1 + \dots + g_2} \quad , \quad (4.22)$$

we find

$$\langle\Psi(Z_1,g_1) \dots \Psi(Z_n,g_n)\rangle = \prod_{1\le i<j\le n} Z_{ij}^{g_ig_j}\, \delta_{g_1+ \dots +g_n} \quad . \quad (4.23)$$

4A. *Field algebras with single and double valued representations on the circle*

The field algebra $F(g^2)$ generated by the pair of primary charged fields $\Psi(Z,\pm g)$ given by (4.15) appears as a deformation of the free Weyl field algebra of Sec.3B (reproduced for g = 1). (In particular, the small distance OPE (4.17b) is a deformation, parametrized by g, of the OPE (3.20), which, as we have stressed, can serve as a definition of the free field model). Exploiting the parallel with the free field case we shall study cyclic positive energy representation of $F(g^2)$ characterized by a LW vector $|e\rangle$ of charge e and conformal weight (minimal eigenvalue of L_o) $1/2\ e^2$. We first note that e cannot be arbitrary. Since the states $U_{\pm g}|e\rangle=|e\pm g\rangle$ have weights $\frac{1}{2}(e\pm g)^2$ which should not be smaller that $1/2\ e^2$, it follows that $e^2 \le \frac{1}{4} g^2$. The representation space is then a direct sum of LW spaces of $\mathcal{V}U(1)$:

$$H(e;g^2) = \bigoplus_n H_{e+ng} \qquad (-\tfrac{1}{2}g \leq e \leq \tfrac{1}{2}g). \tag{4.24}$$

The *generalized canonical fields* - the fields of (half)integer dimension - are singled out by the property to acquire a constant (rather than an operator valued) phase shift under the superselection automorphism $\alpha_{2\pi}$ (where α_t is given by (2.17)) in each $H(e,g^2)$. According to (4.15) we have

$$\alpha_{2\pi}\Psi(Z,g) = e^{i\pi g^2}\Psi(Z,g)\, e^{2\pi i g J_0} , \tag{4.25a}$$

hence

$$\alpha_{2\pi}\Psi(Z,g)|e+ng\rangle = e^{\pi i(g^2+2eg)}\, e^{2\pi i n g^2}\, \Psi(Z,g)|e+ng\rangle . \tag{4.25b}$$

The result is independent of n iff g^2 is an integer. For e = 0 the value of the phase shift is $(-1)^{g^2}$. We shall restrict the allowed values of e by demanding that the range of the phase shift remains ± 1. Combined with the inequality $e^2 \leq \frac{1}{4}g^2$ (4.24) this leads to the spectrum

$$e_\nu = \frac{\nu}{2g} , \qquad -g^2 \leq \nu < g^2 \qquad (g \geq 1) . \tag{4.26}$$

(We exclude the charge $e_{g^2} = \frac{1}{2}g$, since it is obtained from e_{-g^2} by the action of $\Psi(Z,g)$. Writing the mode expansion of $\Psi(Z,g)$ in the space $H(-\frac{1}{2}g, g^2) = H(\frac{1}{2}g, g^2) \equiv H(\frac{1}{2}g)$ as

$$\Psi(Z,\pm g)H(\tfrac{1}{2}g) = \sum_n \Psi_n(\pm g)Z^{-n-\frac{1}{2}g^2}\, H(\tfrac{1}{2}g) \tag{4.27}$$

this means that we include in the definition of the LW state $|-\frac{1}{2}g\rangle$ the requirement

$$\Psi_0(-g)|-\tfrac{1}{2}g\rangle = 0 . \tag{4.28}$$

The LW subspace of Vir of weight $1/8g^2$ is 2-dimensional and it is a matter of convention which of the two charge eigenstates will be taken as a LW state in $H(\frac{1}{2}g)$ (cf. our discussion of the R-sector of the $g^2 = 1$ model in Sec. 3B, where we treated the two charges $e=\pm\frac{1}{2}$ in a symmetric fashion).

Generalizing the discussion of the complex (helicity 1/2) Weyl field in Sec.3B we shall define the *R sector* of an odd g^2 model as the direct sum of H_{e_ν} for odd ν. It corresponds to eigenvalue 1 for the "univalence phase shift", defined by (4.25). Similarly, we shall say that the odd ν $F(g^2)$ LW modules define the *twisted sector* for even g^2. The set of even ν's (including the vacuum representation for $\nu=0$) determines in both cases the *NS sector*.

The experience with the Weyl field suggests that the construction (and classification) of local 2-dimensional models which provide a QFT realization of the field algebra $F(g^2)$ should be based on the construction of modular invariant partition functions out of the reducible Virasoro characters

$$K_\nu(\tau,g^2) = \mathrm{tr}_{H(e_\nu,g^2)}\, q^{\tilde{L}_0} \tag{4.29a}$$

$$= \frac{1}{\eta(\tau)} \sum_n q^{\frac{1}{2}(gn+\frac{\nu}{2g})^2} \qquad (q=e^{2\pi i\tau}=e^{-\beta}) \tag{4.29b}$$

and the associated functions

$$\tilde{K}_\nu(\tau,g^2) = \frac{1}{\eta(\tau)} \sum_n (-1)^n\, q^{\frac{1}{2}(gn+\frac{\nu}{2g})^2} \quad . \tag{4.30}$$

We note that fuller information is provided by the affine characters of [Ka 2,4] which involve what is called a chemical potential (in the context of a grand canonical ensemble; in our case ζ is conjugate to the charge operator). We have

$$\mathrm{tr}_{H(e_\nu,g^2)}\{q^{\tilde{L}_0}\, e^{2\pi i(\zeta J_0+ug^2)}\} = \frac{1}{\eta(\tau)}\, \Theta_{\nu,g^2}(\tau,\zeta,u) \tag{4.31}$$

where (q being the same as in (4.29b))

$$\Theta_{\nu,g^2}(\tau,\zeta,u) = e^{2\pi i g^2 u} \sum_n q^{\frac{1}{2}(gn+\frac{\nu}{2g})^2}\, e^{2\pi i(gn+\frac{\nu}{2g})\zeta} \tag{4.32}$$

is a classical Θ-function. (Obviously, the K_ν- character (4.29) is recovered as a special case of (4.31) for $\zeta=u=0$). Under a modular transformation

$$\begin{pmatrix} a & b \\ c & d \end{pmatrix} \in SL(2,\mathbb{Z})$$

the arguments of Θ change as follows:

$$\tau \to \frac{a\tau + b}{c\tau + d} , \qquad \zeta \to (c\tau + d)^{-1}\zeta , \quad u \to \frac{c}{2}(c\tau + d)^{-1}\zeta^2 + u . \tag{4.33}$$

The classification of modular invariant partition functions corresponding to the $F(g^2)$-models [Pa 1] (based on the analysis of $c = 1$ theories, given in [Di 1]) can be extended to partition functions depending on all three variables τ, ζ, u.

We shall content ourselves with one remark and a couple of examples.

The remark concerns the relation between theories of multiple charges. If α is a positive integer, then

$$\Theta_{\nu,g^2}(\tau,\xi,u) = \sum_{r=0}^{\alpha-1} \Theta_{\alpha\nu+2\alpha g^2 r ,\alpha^2 g^2} (\tau,\zeta, \frac{u}{\alpha^2}) \tag{4.34}$$

(since $\alpha g n + \frac{1}{2\alpha g}(\alpha\nu + 2\alpha g^2 r) = g(\alpha n + r) + \frac{\nu}{2g}$ and $\alpha n + r$ takes every integer value just once for n running through $\mathbb{Z}$ and $r = 0,1,\dots \alpha - 1$). An identity of this type can be also written for the *indefinite* Θ-*function*.

$$\tilde{\Theta}_{\nu,g^2}(\tau,\xi,u) = e^{2\pi i g^2 u} \sum_n q^{\frac{1}{2}(gn + \frac{\nu}{2g})^2} e^{2\pi i(gn + \frac{\nu}{2g})\zeta} (-1)^n . \tag{4.35}$$

In particular, for $\alpha = 2$, there is a one-to-one correspondence between the set of all Θ-functions of the g^2-model and the NS functions of the $4g^2$-model. Eq. (4.34) is complemented in that case by

$$\tilde{\Theta}_{\nu,g^2}(\tau,\zeta,u) = \Theta_{2\nu,4g^2}(\tau,\zeta,u/4) - \Theta_{2\nu\pm 4g^2,4g^2}(\tau,\zeta, u/4) \tag{4.36}$$

(The sign of $4g^2$ in the index of the last Θ- function is irrelevant because of the periodicity condition

$$\Theta_{\nu,g^2}(\tau,\zeta, u) = \Theta_{\nu+2g^2,g^2}(\tau,\zeta,u) \tag{4.37}$$

which follows from the definition (4.32) of Θ, extended beyond the range (4.26) of ν). There is, thus, a correspondence between the modular invariant partition functions of the two theories. This is not surprising, since the charge spectrum (4.26) is the same in the two cases. For example, the NS sector of the $g^2 = 4$

model is equivalent to the $g^2 = 1$ theory of the charged Weyl field considered in Sec. 3B.

We proceed to the first two examples associated with a prime g^2: $g^2 = 2$ and $g^2 = 3$.

For $g^2 = 2$ the charged fields $\Psi(Z,\pm g)$ have helicity 1 and hence can be identified with (conserved) charged currents. Moreover, in view of the Sugawara formula (2.20) and the small distance OPE (4.17b) (for $g = \sqrt{2}$), they generate (together with $J(Z)$) a level 1 representation of the affine Kac-Moody algebra $\hat{A}_1 = \widehat{su}(2)$:

$$[\Psi(Z_1,\sqrt{2}),\ \Psi(Z_2,\ -\sqrt{2})] = \sqrt{2}\delta(Z_{12})J(Z_2) - \delta'(Z_{12}) \tag{4.38}$$

$$T(Z) = \frac{1}{2}:J^2(Z): = \frac{1}{6}:\{J^2(Z) + \Psi(Z\ ,\sqrt{2})\Psi(Z,-\sqrt{2}) + \Psi(Z,\ -\sqrt{2})\Psi(Z,\sqrt{2})\}: \tag{4.39}$$

The NS sector of this model contains two LW states: the vacuum $|0>$and the state $|-1/\sqrt{2}>$ of dimension 1/4. This can be regarded as the simplest example of a Wess-Zumino-Witten model (see [Wi 1, Kn 1]). It has been realized as an SO(4) "quark model" with an SU(2)- gauge group and the corresponding local 2-dimensional field of weight $(\Delta,\bar{\Delta}) = (\frac{1}{4}, \frac{1}{4})$ has been constructed in [To 4]. What we call (at variance with the mathematical terminology - see the discussion in Sec. 3.5 of [Go 3]) the twisted sector of the $g^2 = 2$-model has two LW vectors of charge $\pm 1/2\sqrt{2}$ and dimension 1/16. The product of two (say, primary) fields of this sector is expanded into fields of the NS sector.

Our second example, $g^2 = 3$, is still richer in content. The small distance OPE (4.17b) here takes the form

$$\Psi(Z_1,\sqrt{3})\Psi(Z_2,-\sqrt{3}) = \frac{1}{Z_{12}^3} + \frac{\sqrt{3}}{Z_{12}^2}J(Z) + \frac{3}{Z_{12}}\ T(Z) + O(1)\ . \tag{4.40}$$

It shows that the local anticommutator of the two helicity 3/2 fields involves the current J and the stress tensor T (apart from the most singular c-number term, proportional to $-\delta'''(Z_{12})$). It follows that $\Psi(Z,\pm\sqrt{3})$, $J(Z)$ and $T(Z)$ span the $c = 1$ level of the $N = 2$ extended super

Virasoro algebra - see [Ad 1, Bo 1, Sc 1, Wa 1, Ya 1,2]. There are two independent modular invariant partition functions which mix the NS- and R-characters of this model. They can be written for a general odd $g^2=2M+1$, but only exhaust the possibilities if g^2 is prime (for detail see [Pa 1]). The $c = 1$ twisted sector of the $N = 2$ supersymmetric model (see, e.g., [Bo 1]) is not incorporated by the present approach, since the current J is no longer a single valued (observable) field in that sector.

The examples $g^2 = 1,2,3$, singled out here, are the only ones in the g^2 - series in which the local (anti)commutator

$$[\Psi(Z_1,g),\Psi(Z_2,-g)]_{(-)^{g^2+1}} = \Psi(Z_1,g)\Psi(Z_2,-g)-(-1)^{g^2}\Psi(Z_2,-g)\Psi(Z_1,g) \tag{4.41}$$

gives rise to a Lie (super)algebra involving a finite number of conformal fields.

5. Concluding Remarks

The aim of the present lectures has been twofold: elaboration on a method and working out a step in the construction and classification of two-dimensional conformal QFTs.

The method involves a systematic use of the KMS boundary condition combined with Lie (super)algebraic (anti)commutation relations among conformal quantum fields to compute finite temperature correlation functions. Local QFT models are described by modular invariant partition functions which summarize the information about the energy and charge distribution, and, as a corollary, of the operator content of the theory. (The possibility to code the charge distribution has not been fully exploited). We have also attempted - with no claim of originality - to provide a clear understanding of the much discussed in the literature "vertex operator" (in our case - charged field) construction.

Our discussion (in Secs. 3 and 4 of these lectures and in [Pa 1]) covers Bose/Fermi charged field models with a chiral U(1)-symmetry, and (as a consequence of the Sugawara formula) a Virasoro central charge $c = 1$. Here is a comment about the place of this step in the general program of constructing and classifying 2-dimensional conformal models.

The classification of minimal conformal theories with $c < 1$, undertaken in [Be 1], has been essentially completed in [Ca 1]. What is still lacking is a good understanding of the (finite)symmetry group of each of these models. (The work in this direction - see in particular [Zu 1] and [So 1] - is yet inconclusive; there is in particular an erroneous statement in the latter reference concerning the symmetry of the $c = 4/5$ Potts model).

The classification of the minimal $c < 1$ theories appeared to be closely related and was done in parallel with the study of affine $A_1^{(1)}$-models [Ge 1] with $1 \leq c_k = \frac{3k}{k+2} < 3$ ($k = 1,2,\ldots$ being the Kac-Moody level of $A_1^{(1)} = \widehat{su}(2)$. (The last reference continued the work of a number of authors: [Kn 1, Go 1, Ge 3]).

The $c = 1$ modular invariant partition functions were studied (and, essentially, classified) in [Di 1]. Our work provides them with a field theoretic interpretation. There is, in fact, a wider class of chiral U(1) theories, obtained by relaxing the univalence condition which led to the restriction to integer g^2. Although the class of theories thus obtained is both physically interesting and natural from a mathematical point of view (being related to the classification of modular forms of weight 1/2 - see, e.g. [Ka 4] and references therein) there may be also meaningful theories corresponding to more general rational values of g^2. A first step in this direction would be the inclusion of $g^2 \in \frac{1}{2}\mathbb{Z}_+$. The superselection operator $\alpha_{2\pi}$ would then have just two eigenvalues, ± 1, in the representation space $\mathcal{H}(e,g^2)$, splitting it into an even and odd part. Such an extension would include, for $g^2 = \frac{3}{2}$, the $\mathbb{Z}_4$ - model of the parafermionic series of [Za 1] (see also [Ge 2] and [Ya 2]) characterized by Virasoro central charge

$$c_p = 2\,\frac{p-1}{p+2}\,, \qquad p = 1,2,\ldots \quad (c_1 = 0,\ c_2 = \frac{1}{2},\ c_3 = \frac{4}{5}\,,\ c_4 = 1,\ldots) \tag{5.1}$$

We stop short in our discussion before reaching the class of superconformal models which, too, is under intensive study (for a recent work and references - see [Ra 1]).

One of the authors (I.T.) would like to thank J.-B.Zuber for discussions.

References

[Ad 1] M.Ademollo et al. Supersymmetric string and colour confinement. Phys.Lett. 62B (1976) 105-110; Dual string with U(1) colour symmetry, Nucl.Phys. B111 (1976) 77-110.

[Be 1] A.A.Belavin, A.M.Polyakov, A.B.Zamolodchikov, Infinite conformal symmetry in two-dimensional quantum field theory. Nucl.Phys. B241 (1984) 333-380.

[Bo 1] W.Boucher, D.Friedan, A.Kent, Determinant formulae and unitarity for the N=2 superconformal algebras in two dimensions or exact results on string compactification. Phys.Lett. 172B (1986) 316.

[Br 1] O.Bratteli, D.W.Robinson, *Operator Algebras and Quantum Statistical Mechanics*. Springer, N.Y., Heidelberg,1981, (see, in particular, p.77 ff.).

[Ca 1] A.Capelli, C.Itzykson, J.-B.Zuber, Modular invariant partition functions in two dimensions. Nucl.Phys. B280 [FS 18] (1987)445-465; The A-D-E classification of minimal and $A_1^{(1)}$ conformal invariant theories. Saclay preprint S.Ph-T/87-59, Submitted to Commun. Math.Phys.

[Ca 2] J.L.Cardy, Effect of boundary conditions on the operator content of two-dimensional conformally invariant theories. Nucl.Phys. B275 [FS 17] (1986) 200-218.

[Ca 3] A.L.Carey, C.A.Hurst, A note on the boson-fermion correspondence and infinite dimensional groups, Commun.Math.Phys. 98(1985)435-448.

[Ca 4] A.L.Carey, S.N.M.Ruijsenaars, On fermion gauge groups, current algebras and Kac-Moody algebras.Dept.of Math., Australian Nat. Univ. (1985).

[Ca 5] A.L.Carey, S.N.M.Ruijsenaars, J.D.Write, The massless Thirring model: positivity of Klaiber's n-point functions. Commun.Math.Phys. 99 (1985) 347-364.

[De 1] G.F.Dell'Antonio, Y.Frishman, D.Zwanziger, Thirring model in terms of currents: solution and light-cone expansion.Phys.Rev. D6 (1972) 988-1007.

[Di 1] P.Di Francesco, H.Saleur, J.-B.Zuber, Modular invariance in non minimal two-dimensional conformal theories. Saclay preprint S.Ph-T/86-184,Nucl.Phys.B285 [FS 19](1987) 454-480 .

[Di 2] P.Di Francesco, H.Saleur, J.-B.Zuber, Relations between the Coulomb gas picture and conformal invariance of 2-dimensional critical models. Saclay preprint S.Ph.-T/87-31, J.Stat.Phys.

[Do 1] S.Doplicher, R.Haag, J.E.Roberts, Fields, observables and gauge transformations. I and II. Commun.Math.Phys. 13 (1969) 1-23; 15 (1969) 173-200.

[Do 2] S.Doplicher, J.E.Roberts, C^*-algebra and duality of compact groups: why there is a compact group of internal symmetries in particle physics, Proc.Int.Conf.Math.Phys. Marseille, 1986.

[Fe 1] B.L.Feigin, D.B.Fuchs, Verma modules over the Virasoro algebra. in Lecture Notes in Math. 1060, Springer, Berlin 1984, pp.230-245.

[Fe 2] S.Ferrara, R.Gatto, A.Grillo, Conformal algebra in two space-time dimensions and the Thirring model. Nuovo Cim. 12A (1972) 959-968.

[Fr 1] I.B.Frenkel, Two constructions of affine Lie algebra representations and Boson-Fermion correspondence in quantum field theory, J.Funct.Anal. 44 (1981) 259-327.

[Fr 2] I.B.Frenkel, V.G.Kac, Basic representations of affine Lie algebras and dual resonance models, Invent.Math. 62 (1980) 23-66.

[Fr 3] D.Friedan, Z.Qiu, S.Shenker, Conformal invariance, unitarity and critical exponents in two dimensions. Phys.Rev.Lett. 151B (1985) 37-43; Details of the non-unitarity proof for highest weight representations of the Virasoro algebra.Commun.Math.Phys. 107(1986) 535-542.

[Fu 1] S.Fubini, A.J.Hanson, R.Jackiw, New approach to field theory, Phys.Rev. D7 (1973) 1732-1760.

[Fu 2] P.Furlan, G.Sotkov, Ya.Stanev, Minimal models of U(1) conformal current algebra. ISAS Trieste preprint 16/87/E.P.

[Ge 1] D.Gepner, On the spectrum of 2D conformal field theories. Nucl. Phys. B287 (1987) 111-130.

[Ge 2] D.Gepner, Z.Qiu, Modular invariant partition functions for parafermionic field theories, Nucl.Phys.B285[FS19](1987) 423-453.

[Ge 3] D.Gepner, E.Witten, String theory on group manifolds. Nucl.Phys. B278 (1986) 493-549.

[Ge 4] G.von Gehlen, V.Rittenberg, Superconformal invariance in the Ashkin-Teller quantum chain with free boundary conditions, J.Phys. A: Math.Gen. 19 (1986) L1039-L1945.

[Go 1] P.Goddard, A.Kent, D.Olive, Unitary representation of the Virasoro and super-Virasoro algebras. Commun.Math.Phys. 103(1986) 105-119.

[Go 2] P.Goddard, D.Olive, Kac-Moody algebras, conformal symmetry and critical exponents. Nucl.Phys. B257 (1985) 226-240.

[Go 3] P.Goddard, D.Olive, Kac-Moody and Virasoro algebras in relation to qunatum physics. Int.J.Mod.Phys. 1 (1986) 303-414.

[Ha 1] R.Haag, N.M.Hugenholz, M.Winnink, On the equilibrium states in quantum statistical mechanics. Commun.Math.Phys. 5(1967) 215-236.

[Ha 2] R.Haag, D.Kastler, An algebraic approach to quantum field theory. J.Math.Phys. 5 (1964) 848-861.

[It 1] C.Itzykson, J.-B.Zuber, Two-dimensional conformal invariant theories on a torus, Nucl.Phys. B275[FS 17],(1986) 580-616.

[Ka 1] V.G.Kac, Contravariant form for infinite dimensional Lie algebras and superalgebras. in Lecture Notes in Physics, Springer, Berlin(1979),pp. 441-445.

[Ka 2] V.G.Kac, *Infinite Dimensional Lie Algebras*, Birkhäuser, Boston 1983, second edition, Cambridge Univ.Press (1985).

[Ka 3] V.G.Kac, Highest weight representations of conformal current algebras, in: *Topological and Geometrical Methods in Field Theory*. Symposium in Espoo, Finland, June 1986. Eds.J.Hietarinta, J. Westerholm, World Scientific, Singapore (1986) pp.3-15.

[Ka 4] V.G.Kac, M.Wakimoto, Modular and conformal invariance constraints in representation theory of affine algebras. Cambridge MIT preprint (1987).

[Ka 5] D.Kastler, Fondements de la mécanique statistique de l'équilibre. (Exposé à la 23ème Rencontre entre Physiciens et Mathematicièns, Strasbourg, November, 1976).

[Kn 1] V.G.Knizhnik, A.B.Zamolodchikov, Current algebra and Wess-Zumino model in two-dimensions. Nucl.Phys. B247 (1984) 83-103.

[Lü 1] M.Lüscher, Operator expansions on the vacuum in conformal QFT in two space-time dimensions. Commun.Math.Phys. 50 (1976) 23-52.

[Lü 2] M.Lüscher, G.Mack, Global conformal invariance in quantum field theory. Commun.Math.Phys. 41 (1975) 203-234.

[Lü 3] M.Lüscher, G.Mack, The energy momentum tensor of critical quantum field theory in 1 + 1 dimensions (unpublished manuscript, Hamburg 1976) (for a review see To 2).

[Ma 1] G.Mack, Duality in quantum field theory. Nucl.Phys. B118 (1977) 445-457.

[Ma 2] G.Mack, K.Symanzik, Currents, stress tensor and generalized unitarity in conformal invariant quantum field theory, Commun.Math. Phys. 27 (1972) 247-281.

[Ma 3] S.Mandelstam, Soliton operators for the quantized sine-Gordon equation. Phys.Rev. D11 (1975) 3026-3030.

[Pa 1] R.R.Paunov, I.T.Todorov, Modular invariant QFT models of U(1) conformal current algebra, Phys.Lett. B196 (1987) 519-526.

[Ra 1] F.Ravanini, S.-K.Yang, Modular invariance in N=2 superconformal field theories, Nordita preprint 87/25. Phys.Lett. B195 (1987)202.

[Sc 1] A.Schwimmer,N.Seiberg,Comments on the N=2,3,4 superconformal algebra in two dimensions, Phys.Lett. B184 (1987) 191-

[Se 1] I.E.Segal, Causally oriented manifolds and groups. Bull.Amer.Math. Soc. 77 (1971) 958-959.

[Se 2] N.Seiberg, E.Witten, Spin structures in string theories. Nucl. Phys. B276 (1986) 272-290.

[Sk 1] T.H.R.Skyrme, Particle state of a quantized meson field, Proc.Roy. Soc. A262 (1961) 237-245.

[So 1] G.M.Sotkov, I.T.Todorov, M.S.Stanishkov, V.Yu.Trifonov, Higher symmetries in conformal QFT models, in:*Topological and Geometrical Methods in Field Theory*. Symposium in Espoo, Finland, June 1986. Eds. J.Hietarinta, J.Westerholm, World Scientific, Singapore (1986), pp. 195-217.

[St 1] R.F.Streater, Charges and currents in the Thirring model, in: *Physical Reality and Mathematical Description*. Eds. P.Enz, J.Mehra, Reidel, Dordrecht(1974), pp. 375-386.

[St 2] R.F.Streater, I.F.Wilde, Fermion states of a boson field, Nucl. Phys. B24 (1970) 561-575.

[To 1] I.T.Todorov, Algebraic approach to conformal invariant 2-dimensional models, Bulg.J.Phys.12(1985) 3-19; Current algebra approach to conformal invariant two-dimensional models. Phys.Lett.153B (1985)77-81.

[To 2] I.T.Todorov, Infinite Lie algebras in 2-dimensional conformal field Theory, in: *Differential Geometric Methods in Theoretical Physics*. Eds.H.D.Doebner, T.D.Palev, World Scientific, Singapore (1986), pp. 297-347.

[To 3] I.T.Todorov, Infinite dimensional Lie algebras in conformal QFT models, in: *Conformal Groups and Related Symmetries, Physical Results and Mathematical Background*. Eds. A.O.Barut, H.D.Doebner, Springer, Berlin(1986), pp.387-443.

[To 4] I.T.Todorov, Two dimensional conformal models as constraint dynamical systems, in: *Constraint's Theory and Relativistic Dynamics*. Eds.G.Longhi, L.Lusanna, World Scientific, Singapore (1987) pp.321-334.

[Wa 1] G.Waterson, Bosonic construction of an N=2 extended superconformal theory in two dimensions. Phys.Lett.B171 (1986),77-80.

[Wi 1] E.Witten, Non-abelian bosonization in two dimensions. Commun.Math. Phys. 92 (1984) 455-472.

[Ya 1] S.-K.Yang, Modular invariant partition function of the Ashkin-Teller model on the critical line and N=2 superconformal invariance, Nucl.Phys. B285 [FS 19](1987) 183-203.

[Ya 2] S.-K.Yang, $Z_4 x Z_4$ symmetry and parafermion operators in the self-dual critical Ashkin-Teller model, Nucl.Phys.B285[FS 19](1987) 639-650.

[Ya 3] S.-K.Yang, H.B.Zheng, Superconformal invariance in two dimensional Ashkin-Teller model. Nucl.Phys. B285 [FS 19](1987) 410-422.

[Za 1] A.B.Zamolodchikov, V.A.Fateev, Representations of "parafermi currents" of spin 2/3 in 2-dimensional conformal field theory. Minimal models and the 3-critical Z_3-Potts model. Zh.Eksp.Teor. Fiz. 89 (1985) 380-399 (transl.Sov.Phys. JETP 62 (1985) 215).

[Za 2] A.B.Zamolodchikov, V.A.Fateev, Disorder fields in two-dimensional conformal quantum field theory and N=2 extended supersymmetry. Zh.Eksp.Teor.Fiz. 90 (1986) 1553-1566.

[Zu 1] J.-B.Zuber, Discrete symmetries of conformal theories.Phys.Lett. 176B (1986) 127- ; Conformal invariant theories on a torus and their modular invariance, in: *Topological and Geometrical Methods in Field Theory*. Symposium in Espoo, Finland, June 1986. Eds. J. Hietarinta, J.Westerholm, World Scientific, Singapore (1986), pp. 219-228.

Institute for Nuclear Research and Nuclear Energy
Bulgarian Academy of Sciences, Sofia

Operator Content of Some c = 1 Models in Statistical Mechanics

V. Rittenberg
Physikalisches Institut der Universität Bonn
Nußallee 12, 5300 Bonn 1, West-Germany

I. Introduction

In the last five years an enormous progress was achieved in the understanding of the phase transitions of two-dimensional systems. The concept of universality which appears in all text-books has changed and got more precise and one is not far away from a classification of the possible structures at the critical point. This implies the knowledge of all the anomalous dimensions of energy-density like, order, disorder and parafermionic operators as well as the surface exponents. More than that, the n-point functions for various operators can be computed and from this knowledge one can write a perturbation theory which describes the behaviour of the system in the vecinity of the critical point. In this way the effect of changes of the temperature, magnetic fields or finite-size effects can be studied. All this progress was achieved due to the special proporties of conformal invariance in two-dimensions. At the same time, the experimental verifications of even the most simple theoretical predictions have not yet been done. For example, to my knowledge, the behaviour of the three-point function which is one of the most transparent predictions of conformal invariance was not yet checked experimentally.

The aim of this lectures is to describe the critical behaviour of a few "classical" systems in statistical mechanics. We first give the necessary mathematical background

ISBN 0-12-218100-X

presenting the irreducible representations of the Virasoro, U(1) and SU(2) Kac-Moody algebras for c = 1. We then define the D_n symmetric n-states self-dual one-dimensional quantum chains. The n = 2 system corresponds to the Ising model which has a phase transition with a central charge of the Virasoro algebra c = 1/2, n = 3 corresponds to the three-states Potts model (c = 4/5), for n = 4 we have the Ashkin-Teller model which has c=1. Starting with n=5 not only the physical picture of the phase-diagram changes but for a given n there are different types of phase transitions with different values of c starting with c=1 and ending with a maximum value c_{max} which increases with n. In these lectures we concentrate on the c=1 phase transitions only. We next consider the well known one-dimensional Heisenberg model which again has c=1. For all the systems mentioned above we will give all the critical exponents. There are other systems in statistical mechanics which are known to have c=1 phase transitions but we will not consider them here. Examples of this kind are the Ising model with 3-spins interactions [1], the magnetic hard squares model [2], the restricted SOS models [3] and their generalisations [4].

A different and complementary approach to the problem is to derive directly the possible partition functions with c=1 at the critical point looking for modular invariant solutions. In this way one derives part of the critical exponents (one does not get the anomalous dimensions of the disorder and para-fermionic operators). For the conformal field theory approach to the c=1 case see Refs. [5] - [9], I will not touch upon this subject in these lecture.

II. Representations of the Virasoro and the U(1) and SU(2) Kac-Moody algebras for c=1

The unitary irreps. of the Virasoro algebra with c=1

$$[L_m, L_n] = L_{m+n} + 1/12\, m(m^2-1)\,\delta_{m+n,0} \quad (m,n \in Z) \qquad (2.1)$$

are defined by the conditions:

$$L_o|\Delta> = \Delta|\Delta> \quad , \; L_m|\Delta> = 0 \;(m>0) \qquad (2.2)$$

The character functions

$$\chi_\Delta(z) = \mathrm{Tr}(z^{L_o}) \tag{2.3}$$

are [10]

$$\chi_\Delta(z) = z^\Delta \Pi_V(z) \qquad (\Delta \neq \frac{t^2}{4}) \tag{2.4a}$$

$$\chi_{\frac{t^2}{4}}(z) = z^{t^2/4}(1-z^{t+1})\Pi_V(z) \qquad (\Delta = \frac{t^2}{4}) \tag{2.4b}$$

where t is an integer and

$$\Pi_V(z) = \prod_{m=1}^{\infty} \frac{1}{1-z^m} \tag{2.5}$$

We denote by (Δ) an irrep. of the Virasoro algebra. We now consider the U(1) Kac-Moody algebra which is obtained adding to Eq. (2.1) the commutations relations:

$$\begin{aligned} [L_m, T_n^3] &= -\, n/2\; T^3_{m+n} \\ & \qquad\qquad\qquad\qquad (m,n \in Z) \\ [T_m^3, T_n^3] &= m\, \delta_{m+n,0} \end{aligned} \tag{2.6}$$

The irrep $[\Delta, q]$ of the U(1) Kac-Moody algebra is defined by

$$T_o^3|\Delta, q> = q|\Delta, q> \; , \; L_o|\Delta, q> = \Delta|\Delta, q> \tag{2.7}$$

and has a character function:

$$\chi_{\Delta,q}(z,y) = \mathrm{Tr}(z^{L_o} y^{T_o}) = z^{q^2}\, \Pi_V(z) y^q \tag{2.8}$$

with $\Delta = q^2$. Since the charge q is given up to a sign by Δ, sometimes we will denote the representation $[\Delta, q]$ simply by $[\Delta]$. The U(1) twisted Kac-Moody algebra [11] is obtained if we replace T_m^3 $(m \in Z)$ in Eq. (2.6) by $T^3_\mu (\mu \in Z + 1/2)$. This new algebra has only one irrep. that we denote by $(1/16)_T$ with a character expression:

$$\chi_{(1/16)_T}(z) = \mathrm{Tr}(z^{L_o}) = \sum_{m \in Z} z^{(4m+1)^2/16}\; \Pi_V(z) \tag{2.9}$$

It is often useful to separate the irrep. $(1/16)_T$ into two parts:

$$(1/16)_T = \{1/16\} \oplus \{9/16\} \qquad (2.10)$$

where

$$\begin{aligned} \chi_{\{1/16\}}(z) &= \sum_{m \in Z} z^{(8m+1)^2/16} \Pi_v(z) \\ \chi_{\{9/16\}}(z) &= \sum_{m \in Z} z^{(8m+3)^2/16} \Pi_v(z) \end{aligned} \qquad (2.11)$$

Finally let us consider the level one (c=1) SU(2) Kac-Moody algebra. We add to Eqs. (2.1) and (2.6) the commutation relations:

$$\begin{aligned} &[L_m, T_n^{\pm}] = -nT^{\pm}_{m+n}, \quad [T_m^3, T_n^{\pm}] = \pm T^{\pm}_{m+n} \\ &[T_m^+, T_n^-] = 2T^3_{m+n} + m\,\delta_{m+n,0}, \quad [T_m^{\pm}, T_n^{\pm}] = 0 \end{aligned} \qquad (2.12)$$

This algebra has two irreducible representations corresponding to an SU(2) isoscalar ($\varepsilon = 0$) and an SU(2) isospinor ($\varepsilon = 1/2$) with the corresponding character expressions [12]

$$\chi_\varepsilon(z,y) = \mathrm{Tr}(z^{L_o} y^{T_o^3}) = \sum_{m \in Z} z^{(m+\varepsilon)^2} y^{m+\varepsilon} \Pi_v(z) \quad (\varepsilon = 0, 1/2) \qquad (2.13)$$

We have also to consider the <u>ξ-shifted</u> SU(2) Kac-Moody which is defined through the transformation:

$$\begin{aligned} &\tilde{L}_m = L_m + \xi\, T_m^3 + \xi^2/4\, \delta_{m,0}, \quad \tilde{T}_m^3 = T_m^3 + \xi/2\, \delta_{m,0}, \\ &\tilde{T}^{\pm}_{m_{\mp}} = T_m^{\pm} \quad (m_{\mp} = m \mp \xi) \end{aligned} \qquad (2.14)$$

Part of the commutation relations which define the ξ-shifted algebra are changed as compared to the unshifted one:

$$\begin{aligned} &[\tilde{L}_m, \tilde{T}^{\pm}_{n_{\mp}}] = -n_{\mp} \tilde{T}^{\pm}_{m+n_{\mp}} \quad ; \quad [\tilde{T}_m^3, \tilde{T}^{\pm}_{n_{\mp}}] = \pm \tilde{T}^{\pm}_{m+n_{\mp}} \\ &[\tilde{T}_m^+, \tilde{T}_n^-] = 2\tilde{T}^3_{m_-+n_+} + m_-\, \delta_{m_-+n_+,0} \end{aligned} \qquad (2.15)$$

For each ξ-shifted we have again two irreps. with the character functions:

$$\chi_{\varepsilon}^{(\xi)}(z,y) = \mathrm{Tr}(z^{\tilde{L}_o} y^{\tilde{T}_o^3}) = \sum_{m\in Z} z^{(m+\varepsilon+\xi/2)^2} y^{m+\varepsilon+\xi/2} \Pi_v(z) \quad (2.16)$$

Comparing the expressions (2.13) and (2.16) we observe that the character expressions of the shifted and unshifted algebras can all be written in the form:

$$\chi_{\rho}(z,y) = \sum_{m\in Z} z^{(m+\rho)^2} y^{m+\rho} \Pi_v(z) \quad (2.17)$$

i.e., all the representations can be seen as a shift of the vacuum representation ($\varepsilon = 0$) of the unshifted algebra ($\xi = 0$). We will denote by $<\rho>$ the irrep. with the character function (2.17). That much for the necessary background.

III. n-states self-dual quantum chain with D_n symmetry

The n-states model have a long history in statistical mechanics [13]. Their transfer-matrices are related to a one-dimensional quantum chains. Here we consider their self-dual version. The Hamiltonians are:

$$H = \frac{1}{\xi} \sum_{j=1}^{N} \sum_{k=1}^{n-1} a_k (\sigma_j^k + \lambda \Gamma_j^k \Gamma_{j+1}^{n-k}) \quad (3.1)$$

where $a_k = a_{n-k}$ are real coupling constants, λ plays the role of the inverse of the temperature, N represents the number of sites and the n x n matrices σ and Γ are:

$$\sigma = \begin{pmatrix} \omega^0 & 0 & & 0 \\ 0 & \omega^1 & & 0 \\ & & \ddots & \\ 0 & & & \omega^{n-1} \end{pmatrix} , \quad \Gamma = \begin{pmatrix} 0 & 0 & 0 & 1 \\ 1 & 0 & 0 & 0 \\ 0 & 1 & 0 & 0 \\ 0 & & \ddots 1 & 0 \end{pmatrix} \quad (3.2)$$

Here $\omega = e^{\frac{2\pi i}{n}}$ and ξ is a normalisation factor (which fixes the euclidean time scale) to be discussed later. The Hamiltonian H is self-dual. i.e.

$$H(\lambda) = \lambda H(1/\lambda) \quad (3.3)$$

Eq. (3.3) is valid for the infinite system and connects various sectors of the models if one considers finite systems with various boundary conditions. We now specify the boundary conditions (b.c.) for the Hamiltonian H taking:

$$\Gamma^{k}_{N+1} = B^{km}\,\Gamma^{m}_{1} \tag{3.4}$$

where B is a certain (n-1) x (n-1) matrix. First we consider <u>free</u> b.c., this corresponds to taking

$$\Gamma_{N+1} = 0 \tag{3.5}$$

and denote the corresponding Hamiltonian by H^F, The global symmetry of H^F (and of the infinite system) is given by the transformations

$$(\Gamma'_j)^m = A^{mn}\,\Gamma^n_j \tag{3.6}$$

where then (n-1) x (n-1) matrices A form a representation (in general reducible) of the symmetry group G. This symmetry group depends on the choice of the coupling constants a_k in the Hamiltonians (3.1), it is at least D_n (due to the choice $a_k = a_{n-k}$ to make the Hamiltonian hermitian) but it is in general larger [14].

<u>Toroidal</u> b.c. are obtained when one takes for the matrix B in Eq. (3.4) one of the matrices A of Eq. (3.6) corresponding to the symmetry group G. We denote by H^B the Hamiltonian with the boundary condition B. One can easily see that two Hamiltonians H^{B_1} and H^{B_2} have the same spectra if B_1 and B_2 belong to the same conjugacy class of the group G. The global symmetry of the Hamiltonian H^B is in general smaller than G. It is given by all matrices A of G which commute with B.

For the purpose of these lectures we are interested in the symmetries D_n (for any n) and in S_4 (the permutation group of four objects) which occurs in the four-states Potts model to be discussed later. For other cases see Ref. [15]. The group D_n is given by the 2n matrices Σ^{ℓ} and $\Sigma^k C$ ($\ell, k = 0, 1, \ldots, n-1$) where

$$\Sigma = \begin{pmatrix} \omega & 0 & & 0 \\ 0 & \omega^2 & & 0 \\ & & \ddots & \\ 0 & & & \omega^{n-1} \end{pmatrix} \qquad C = \begin{pmatrix} 0 & 0 & & 0 & 1 \\ 0 & 0 & & 1 & 0 \\ & & \ddots & & \\ 0 & 1 & & 0 & 0 \\ 1 & 0 & & 0 & 0 \end{pmatrix} \tag{3.7}$$

Let us write n = 2p+1 for n odd and n = 2p+2 for n even. If the symmetry group of the finite Hamiltonian is D_n itself (this happens for periodic or antiperiodic b.c.) we have to organize the spectra in terms of irreps. of D_n. We have p two-dimensional representations:

$$D_Q(\Sigma^\ell) = \begin{pmatrix} \omega^{Q\ell} & 0 \\ 0 & \omega^{-Q\ell} \end{pmatrix} \qquad D_Q(C) = \begin{pmatrix} 0 & 1 \\ 1 & 0 \end{pmatrix}$$
$$(Q = 1, \ldots, p) \tag{3.8}$$

and for n odd there are 2 one-dimensional representations

$$\begin{aligned} D_{0,+}(\Sigma^\ell) &= 1, \qquad & D_{0,+}(C) &= 1 \\ D_{0,-}(\Sigma^\ell) &= 1, \qquad & D_{0,-}(C) &= -1 \end{aligned} \tag{3.9}$$

For n even we have two more one-dimensional representations

$$\begin{aligned} D_{n/2,+}(\Sigma^\ell) &= (-1)^\ell \qquad & D_{n/2,+}(C) &= 1 \\ D_{n/2,-}(\Sigma^\ell) &= (-1)^\ell \qquad & D_{n/2,-}(C) &= -1 \end{aligned} \tag{3.10}$$

We now give the separation of the 2n group elements of D_n into conjugacy classes. For n=2p+1 we have p+2 conjugacy classes

$$\begin{aligned} &\{\Sigma^o\}, \ \{\Sigma^k, \Sigma^{n-k}\} \quad (k = 1,2,\ldots, p) \\ &\{C, \Sigma C, \ldots, \Sigma^{n-1}C\} \end{aligned} \tag{3.11}$$

For n = 2p+2 we have p+4 conjugacy classes:

$$\{\Sigma^{o}\}\ ,\ \{\Sigma^{n/2}\}\ ,\ \{\Sigma^{k},\ \Sigma^{n-k}\}\quad (k = 1,2,\ldots,\ p)\ ,$$
$$\{C,\ \Sigma^{2}C,\ \ldots\ \Sigma^{n-2}C\}\ ,\ \{\ C,\ \Sigma^{3}C,\ \ldots,\ \Sigma^{n-1}C\} \tag{3.12}$$

In the case of toroidal b.c. if we take the matrix B in Eq. (3.4) as one of the matrices defining D_n, the global symmetries can be found in Table I.

Later on we will be interested into the four-states Potts model. It is obtained taking n = 4 and $a_1=a_2=a_3$ in Eq. (3.1). In this case the symmetry of the infinite system is S_4 given by the transformations $\Sigma^{\ell} C^{m} \Omega^{n}$ ($\ell=0,1,2,3$ $m=0,1,$ $n=0,1,2$) where

Table 1   Symmetry of the n-states Hamiltonian with toroidal boundary conditions corresponding to D_n

	Boundary condition	Group	Number of elements	Elements
	Σ^{0}	D_n	2n	$\Sigma^{\ell},\ \Sigma^{m}C$ ($\ell,m = 0,1,\ldots n-1$)
	$\Sigma^{\tilde{Q}}$	Z_n	n	Σ^{ℓ} ($\ell = 0,1,\ldots n-1$)
n=2p+2	Σ^{p+1}	D_n	2n	$\Sigma^{\ell},\ \Sigma^{m}C$ ($\ell,m = 0,1,\ldots n-1$)
n=2p+2	$\Sigma^{R}C$ (R=0,...n-1)	$Z_2 \otimes Z_2$	4	$\Sigma^{0}, \Sigma^{p+1}, \Sigma^{R}C, \Sigma^{R+p+1}C$
n=2p+1	$\Sigma^{R}C$ (R=0,...n-1)	Z_2	2	$\Sigma^{0},\ \Sigma^{R}C$

$$\Omega = \begin{pmatrix} \frac{-i}{2} & \frac{1+i}{2} & \frac{1}{2} \\ \frac{1+i}{2} & 0 & \frac{1-i}{2} \\ \frac{1}{2} & \frac{1-i}{2} & \frac{i}{2} \end{pmatrix} \tag{3.13}$$

The symmetry of the Hamiltonian H^{Ω} is Z_3 and is given by the transformations Ω^o, Ω^1 and Ω^2. That much on the symmetry properties of the n-states model.

IV. Phase transitions in the n-states self-dual quantum chains

At a second order phase transition all the energy gaps of the infinite system vanish. Conformal invariance makes however predictions on the finite-size corrections. Let us first consider free boundary conditions (Hamiltonian H^F) and let E^F by the ground-state energy and $E(r)$ some excited state. For large numbers of sites N one has [16]:

$$-E^F = A_o N + A_1 + \frac{\pi c}{24N} + \ldots \tag{4.1}$$

where A_o and A_1 are non-universal constants and c is the central charge of the Virasoro algebra corresponding to the phase transition. For the energy gaps one has [17]

$$\lim_{N\to\infty} \frac{N}{\pi} (E(r) - E^F) = \Delta + r \tag{4.2}$$

where Δ corresponds to the lowest weight of an irrep. of the Virasoro algebra and r is a higher level with a degeneracy given by the corresponding character function $\chi_{\Delta}(z)$ given by Eq. (2.4). Thus the finite-size scaling spectrum defined by Eq. (4.2) can be organized in terms of irreps. of the Virasoro algebra. We now consider toroidal b.c. (Hamiltonian H^B). Let E^P be the ground-state energy for periodic b.c. For large numbers of sites N one has [16], [18]:

$$-E^P = A_o N + \frac{\pi c}{6N} + \ldots \tag{4.3}$$

where A_o is the same as in Eq. (4.1) and c is again the central charge of the Virasoro algebra. Let's now take a certain b.c. B and let $E_i^B(P)$ be an energy level corresponding to the momentum P (toroidal b.c. keep translational invariance). The finite-size scaling spectra are [19]:

$$\lim_{N\to\infty} \frac{N}{2\pi} (E_i^B(P) - E^P) = (\Delta + r) + (\bar{\Delta} + r)$$
$$P = (\Delta + r) - (\bar{\Delta} + \bar{r}) \tag{4.4}$$

where Δ and $\bar{\Delta}$ are the lowest weights of two irreps. of two commuting Virasoro algebras with a central charge c corresponding to the phase transition.

The factors ξ in the Hamiltonians (3.1) have to be chosen such that the Eqs. (4.2) and (4.4) are verified. Let us now turn to the description of the phase structure of the Hamiltonians (3.1). For n=2 we have a second-order Ising phase transition with c=1/2 at λ =1. For n=3 again we have a phase-transition at λ =1 with c=4/5, this corresponds to the 3-states Potts model [19] - [20]. In both cases we have only one coupling constant which coincides with the normalisation factor . Starting with n=4 (there are two coupling constants in this case) the phase structure is more complex. Let us first describe this case. The properly normalised Hamiltonian is:

$$H = \frac{1-4h}{4\sqrt{\lambda}\sin \pi/4h} \sum_{j=1}^{N} (\sigma_j + \sigma_j^3 + \varepsilon\sigma_j^2) + \lambda(\Gamma_j\Gamma_{j+1}^3 + \Gamma_j^3\Gamma_{j+1} + \varepsilon\Gamma_j^2\Gamma_{j+1}^2) \tag{4.5}$$

where $-1< \varepsilon \lesssim 1$ and

$$h = \frac{\pi}{4\cos^{-1}(-\varepsilon)} \tag{4.6}$$

This Hamiltonian corresponds to the Ashkin-Teller model. The phase-diagram was established by both analytic and numerical methods. The system is massless for λ =1 in the whole domain $-1 < \varepsilon \lesssim 1$. For $\varepsilon < -1/\sqrt{2}$ it is also massless in a domain $\frac{1}{\lambda_{max}(\varepsilon)} \lesssim \lambda \lesssim \lambda_{max}(\varepsilon)$. This is the <u>critical fan</u>. The central charge of the Virasoro algebra is c=1 everywhere. The critical exponents to be discussed in the next section depend on ε but are independent on λ (inside the critical fan).

Starting with n=5 the massless regions in the space of the coupling constants and λ is only partially known. It was known for a long time [13] that there are regions which are described by the Coulomb gas (this is c=1 in the modern language). The news are that there are regions with higher values of the central charge [23]. Let us now concentrate on the regions with c=1. The picture here is "orthogonal" to the one the Ashkin-Teller model. In the massless region for λ =1 the critical exponents are independent of the coupling constants. The critical exponents are however dependent on the position inside the critical fan. More precisely [24], for a certain choice of the coupling constants, the system is massless for $\frac{1}{\lambda_{max}} \leq \lambda \leq \lambda_{max}$ where λ_{max} depends on the coupling constants. There is a function $g(\lambda)$ such that

$$g(\tfrac{1}{\lambda}) = \frac{1}{g(\lambda)} \quad , \quad g(\lambda_{max}) = \frac{n}{4} \tag{4.7}$$

and the critical exponents depend only of g for <u>all</u> the coupling constants in the domain with c=1.

V. <u>Critical exponents of the Ashkin-Teller model</u>

a) <u>Free boundary conditions [25] (surface exponents)</u>

The finite-size scaling spectra which can be obtained from numerical estimates as well as analytic results for the lower levels, are:

$$\begin{aligned}
D_{0,+} &: \bigoplus_{k\geq 0} (4k^2)^+ \oplus \bigoplus_{k\geq 1} \left(\frac{4k^2}{h}\right)^+ \\
D_{0,-} &: \bigoplus_{k\geq 0} \left(\frac{(2k+1)^2}{h}\right)^+ \\
D_{2,+} &: \bigoplus_{k\geq 0} ((2k+1)^2)^+ \oplus \bigoplus_{k\geq 1} \left(\frac{4k^2}{h}\right)^- \\
D_{2,-} &: \bigoplus_{k\geq 0} \left(\frac{(2k+1)^2}{h}\right)^+ \\
D_1 &= \bigoplus_{k\geq 0} \left(\frac{(2k+1)^2}{4h}\right)^+
\end{aligned} \tag{5.1}$$

Here we have given the irreps. $(\Delta)^{\pm}$ of the Virasoro algebra occuring in the sectors defined by the irreps. of D_4 (see Eqs. (3.8) - (3.10)). The signs $(\ )^{\pm}$ corresponds to even (+) or odd (-) parity states (the Hamiltonian with free b.c. is invariant under parity). The parameter h in Eq. (5.1) is defined by Eq. (4.6).

b) Toroidal boundary conditions [26]

The finite-size spectra are organized according to the eight b.c. and for each b.c. one obtains different sectors corresponding to the irreps. of the symmetry of the Hamiltonian with the given b.c. As shown in Table 2 there are althogether 13 independent spectra. Two types of irreps. $(\Delta,\overline{\Delta})$ occur. Some are h independent and some are not. The latter have the structure:

$$(\Delta,\overline{\Delta}) = \left(\left(\frac{(M+Nh)^2}{16h}\right), \left(\frac{(M-Nh)^2}{16h}\right)\right), \quad M,\ N \in Z \tag{5.2}$$

We now enumerate the operator content of the 13 independent sectors.

(a) Sectors with $P \in Z$ (integer momentum)

$$A = (\{0\},\{0\}) \oplus (\{1\},\{1\}) \oplus A_1$$

$$B = (\{0\},\{1\}) \oplus (\{1\},\{0\}) \oplus A_1$$

$$A_1 = \bigoplus_{n\geq 0}\left(((n+1)^2h,\ (n+1)^2h) \oplus \left(\frac{(n+1)^2}{h},\frac{(n+1)^2}{h}\right)\right) \oplus R(4,4;4,4|h) \oplus R(4,4;\ -4,\ -4|h)$$

$$C = R(4,2:4,2|h) \oplus R(4,2;\ -4,\ -2|h)$$

$$D = \bigoplus_{n\geq 0}\left(\frac{(2n+1)^2}{16}h,\frac{(2n+1)^2}{16}h\right) \oplus R(4,4;2,1|h) \oplus R(4,4;\ -2,\ -1|h)$$

$$E(h) = D\left(\frac{1}{h}\right)$$

$$F = \bigoplus_{n\geq 0}\left(\frac{(2n+1)^2}{16},\frac{(2n+1)^2}{16}\right) \oplus R(4,2;4,4|h) \oplus R(4,2;\ -4,\ -4|h) \tag{5.3a}$$

$$G(h) = F\left(\frac{1}{h}\right)$$

$$H = \left(\left\{\frac{1}{16}\right\},\left\{\frac{1}{16}\right\}\right) \oplus \left(\left\{\frac{9}{16}\right\},\left\{\frac{9}{16}\right\}\right).$$

(b) Sectors with $P \in Z + 1/2$ (half-integer momentum)

$$J = R(4,2;2,1|h) \oplus R(4,2;-2,-1|h) \;, J(h) = I(\tfrac{1}{h})$$
$$K = (\{\tfrac{1}{16}\}, \{\tfrac{9}{16}\}) \oplus (\{\tfrac{9}{16}\}, \{\tfrac{1}{16}\}). \quad (5.3b)$$

(c) Sector with $P \in Z + 1/4$

$$L = R(4,1;4,1|h) \oplus (4,3;4,3|h) \oplus R(4,1;-4.-3|h) \oplus R(4,3;-4,-1|h). \quad (5.3c)$$

(d) Sectors with $P \in Z + 3/4$

$$\tilde{L} = R(4,1;-4,-1|h) \oplus R(4,3;-4,-3|h) \oplus R(4,1;4,3|h) \oplus R(4,3;4,1|h) \quad (5.3d)$$

where

$$R(p,q;r,s|h) = \bigoplus_{\substack{m\geq 0\\ n\geq 0}} \left(\frac{(pm+q+(rn+s)h)^2}{16h}, \frac{(pm+q-(rn+s)h)^2}{16h}\right). \quad (5.4)$$

Here

$$\{0\} = \bigoplus_{k\geq 0}(4k^2) \;, \{1\} = \bigoplus_{k\geq 0}((2k+1)^2) \quad (5.5)$$

and $\{\frac{1}{16}\}$, $\{\frac{9}{16}\}$ are given by Eq. (2.11). We also notice the relations

$$A(h) = A(\tfrac{1}{h}), \quad B(h) = B(\tfrac{1}{h}), \quad C(h) = C(\tfrac{1}{h}),$$
$$L(h) = L(\tfrac{1}{h}), \quad \tilde{L}(h) = \tilde{L}(\tfrac{1}{h}) \quad (5.6)$$

Table 2 The sectors of the Ashkin-Teller model

Sector	Boundary condition	
	$\mathbb{1}$	Σ^2
$D_{0,+}$	A	G
$D_{0,-}$	C	F
$D_{2,+}$	G	B
$D_{2,-}$	F	C
D_1	H	K

Sector	Boundary Condition	
Σ	Σ	Σ^3
1	H	H
i	L	$\tilde{L}$
-1	K	K
-i	$\tilde{L}$	L

Sector		Boundary Condition	
Σ^2	ΣC	ΣC	$\Sigma^3 C$
+	+	H	H
+	-	K	K
+	+	E	J
-	-	J	E

Sector		Boundary Condition	
Σ^2	$\Sigma^2 C$	C	$\Sigma^2 C$
+	+	D	D
+	-	I	I
-	+	H	K
-	-	K	H

VI. The finite-size spectra of the four-states Potts model

We leave it as an exercise for the student to show that for h=1 ($\varepsilon = -\frac{1}{\sqrt{2}}$) and $h=\frac{1}{4}$ ($\varepsilon=1$) the finite-size spectra of the Ashkin-Teller model can be organized in terms of irreps. of the SU(2) Kac-Moody algebra given in Sec. 2 (for more details see Ref. [27]). Here we would like to complete the description of the four-states Potts model ($\varepsilon=1$) [28]. For this special values of the coupling constant the symmetry of the system is S_4 instead of D_4 (see Sec. 3) and one has a conjugacy class with 8 elements one of them being the transformation Ω (see Eq. 3.13). As noticed already the Hamiltonian H^{Ω} has Z_3 symmetry. We denote by $\mathcal{E}_0$, $\mathcal{E}_1$ and $\mathcal{E}_2$ the sectors with charge 0, 1 and 2. We will do this exercise in more detail because in this way we can understand better how higher symmetries play a role and how numerical estimates can confirm theoretical guesses.

First we notice that a Z_3 symmetry implies the appearance of momenta

$$P = \pm \frac{1}{3} + N \qquad (N \in Z) \tag{6.1}$$

in Eq. (4.4). They occur if we perform a shift $\xi = \frac{1}{3}$ in the Kac-Moody algebra (see Eqs. (2.14) and (2.16)). With the definition (2.17) we thus expect that we have:

$$\mathcal{E}_0 + \mathcal{E}_1 + \mathcal{E}_2 = (\langle\tfrac{1}{6}\rangle, \langle\tfrac{1}{6}\rangle) + (\langle\tfrac{1}{3}\rangle, \langle\tfrac{1}{3}\rangle) \tag{6.2}$$

Now in the sector $\mathcal{E}_0$ the momenta are integer, in $\mathcal{E}_1$ they are $\frac{1}{3} + N$ $(N \in Z)$ and in $\mathcal{E}_2$ they are $-\frac{1}{3} + N (N \in Z)$ and all we have to do is to find the right combinations inside Eq. (6.2). Taking y=1 in Eq. (2.17) we have the following decomposition of $\langle\frac{1}{6}\rangle$ and $\langle\frac{1}{3}\rangle$ in terms of irreps. of the Virasoro algebra:

$$\begin{aligned} \langle\tfrac{1}{6}\rangle &= \{\tfrac{1}{36}\} \oplus \{\tfrac{25}{36}\} \oplus \{\tfrac{49}{36}\} \\ \langle\tfrac{1}{3}\rangle &= \{\tfrac{1}{9}\} \oplus \{\tfrac{4}{9}\} \oplus \{\tfrac{16}{9}\} \end{aligned} \tag{6.3}$$

where

$$\begin{aligned} \{\tfrac{1}{36}\} &= \bigoplus_{k \in Z} \left(\frac{(1+18k)^2}{36}\right) & \{\tfrac{1}{9}\} &= \bigoplus_{k \in Z} \left(\frac{(1+9k)^2}{9}\right) \\ \{\tfrac{25}{36}\} &= \bigoplus_{k \in Z} \left(\frac{(5+18k)^2}{36}\right) & \{\tfrac{4}{9}\} &= \bigoplus_{k \in Z} \left(\frac{(2+9k)^2}{9}\right) \\ \{\tfrac{49}{36}\} &= \bigoplus_{k \in Z} \left(\frac{(7+18k)^2}{36}\right) & \{\tfrac{16}{9}\} &= \bigoplus_{k \in Z} \left(\frac{(4+9k)^2}{9}\right) \end{aligned} \tag{6.4}$$

The operator content for the three charge sectors is then given by

$$
\begin{aligned}
\mathcal{E}_o &= (\{\tfrac{1}{36}\}, \{\tfrac{1}{36}\}) \oplus (\{\tfrac{25}{36}\}, \{\tfrac{25}{36}\}) \oplus (\{\tfrac{49}{36}, \{\tfrac{49}{36}\}) \oplus \\
&\quad (\{\tfrac{1}{9}\}, \{\tfrac{1}{9}\}) \oplus (\{\tfrac{4}{9}\}, \{\tfrac{4}{9}\}) \oplus (\{\tfrac{16}{9}\}, \{\tfrac{16}{9}\}) \\
\mathcal{E}_1 &= (\{\tfrac{1}{36}\}, \{\tfrac{25}{36}\}) \oplus (\{\tfrac{49}{36}\}, \{\tfrac{1}{36}\}) \oplus (\{\tfrac{25}{36}\}, \{\tfrac{49}{36}\}) \oplus \\
&\quad (\{\tfrac{4}{9}\}, \{\tfrac{1}{9}\}) \oplus (\{\tfrac{1}{9}\}, \{\tfrac{16}{9}\}) \oplus (\{\tfrac{16}{9}\}, \{\tfrac{4}{9}\}) \\
\mathcal{E}_2 &= (\{\tfrac{25}{36}\}, \{\tfrac{1}{36}\}) \oplus (\{\tfrac{1}{36}\}, \{\tfrac{49}{36}\}) \oplus (\{\tfrac{49}{36}\}, \{\tfrac{25}{36}\}) \oplus \\
&\quad (\{\tfrac{1}{9}\}, \{\tfrac{4}{9}\}) \oplus (\{\tfrac{16}{9}\}, \{\tfrac{1}{9}\}) \oplus (\{\tfrac{4}{9}\}, \{\tfrac{16}{9}\})
\end{aligned}
\tag{6.5}
$$

In order to check this hypothesis Grimm has taken chains with up to 8 sites and has determined estimates for the finite-size scaling spectra (see Eq. (4.4)) using the Burlish-Stoer approximants [29] which have been proven very useful for these type of applications [30]. In Table 3 we show the numerical estimates together with the predictions of Eq. (6.5) for $\mathcal{E}_2$. The lowest level was actually predicted earlier [31] using the Coulomb gas approach.

Table 3 Finite-size scaling estimates for momenta 2/3, 5/3 and 8/3 in the charge 2 sector. The numerical date (with estimated errors) are compared with the spectra obtained from the conjectured operator content.

P	$\Delta + r + \bar{\Delta} + \bar{r}$	$(\frac{25}{36},\frac{1}{36})$	$(\frac{1}{9},\frac{4}{9})$	$(\frac{49}{36},\frac{25}{36})$	$(\frac{16}{9},\frac{1}{9})$	$(\frac{1}{36},\frac{49}{36})$	$(\frac{4}{9},\frac{16}{9})$	$(\frac{25}{9},\frac{1}{9})$	$(\frac{121}{36},\frac{25}{36})$	$\mathcal{E}_2(P)$("exp")
$\frac{2}{3}$	$\frac{26}{36} = 0.722$	1	-	-	-	-	-	-	-	0.70(2)
	$\frac{5}{9}+1 = 1.556$	-	1	-	-	-	-	-	-	1.49(5)
	$\frac{74}{36} = 2.056$	-	-	1	-	-	-	-	-	1.87(5)
	$\frac{26}{36}+2 = 2.722$	1	-	-	-	-	-	-	-	2.67(8)
	$\frac{17}{9}+1 = 2.889$	-	-	-	1	-	-	-	-	2.96(8)
	$\frac{50}{36}+2 = 3.389$	-	-	-	-	2	-	-	-	3.37(5)3.47(8)
	$\frac{5}{9}+3 = 3.556$	-	2	-	-	-	-	-	-	3.51(9)3.60(9)
	$\frac{74}{36}+2 = 4.056$	-	-	1	-	-	-	-	-	3.82(10)
	$\frac{20}{9}+2 = 4.222$	-	-	-	-	-	2	-	-	3.9(2)4.0(2)
	$\frac{26}{36}+4 = 4.722$	4	-	-	-	-	-	-	-	4.5(3)
$\frac{5}{3}$	$\frac{26}{36}+1 = 1.722$	1	-	-	-	-	-	-	-	1.70(2)
	$\frac{17}{9} = 1.889$	-	-	-	1	-	-	-	-	1.95(4)
	$\frac{5}{9}+2 = 2.556$	-	2	-	-	-	-	-	-	2.49(5)2.58(5)
	$\frac{74}{36}+1 = 3.056$	-	-	1	-	-	-	-	-	2.83(7)
	$\frac{26}{36}+3 = 3.722$	2	-	-	-	-	-	-	-	3.61(8)3.66(8)
	$\frac{17}{9}+2 = 3.889$	-	-	-	1	-	-	1	-	3.76(8)3.95(9)
	$\frac{50}{36}+3 = 4.389$	-	-	-	-	3	-	-	-	4.34(5)4.3(1)4.3(1)
	$\frac{5}{9}+4 = 4.556$	-	3	-	-	-	-	-	-	4.5(2)4.6(2)4.7(2)
	$\frac{74}{36}+3 = 5.056$	-	-	2	-	-	-	-	-	4.9(2)
$\frac{8}{3}$	$\frac{26}{36}+2 = 2.722$	2	-	-	-	-	-	-	-	2.70(2)2.70(2)
	$\frac{17}{9}+1 = 2.889$	-	-	-	1	-	-	1	-	2.79(4)2.96(3)
	$\frac{5}{9}+3 = 3.556$	-	3	-	-	-	-	-	-	3.44(8)3.50(8)3.58(8
	$\frac{74}{36}+2 = 4.056$			2	-	-	-	-	1	3.79(8)3.84(8)4.22(9
	$\frac{26}{36}+4 = 4.722$	3	-	-	-	-	-	-	-	4.6(1)4.7(1)

VII. Operator content for the n-states (n>5) quantum chains in the c=1 region

a) Free boundary conditions [24]

Painful numerical studies (up to n=12) are compatible with the following content of Virasoro irreps. in the various sectors given by irreps. of D_n (see Eqs. (3.8) - (3.10)):

(a) n = 2p+2 (n even > 4):

$$D_{0,+} \quad : \bigoplus_{k\geq 0}(4k^2)^+ \oplus \bigoplus_{k\geq 1}\left(\frac{nk^2}{g}\right)^+ \tag{7.1a}$$

$$D_{0,+} \quad : \bigoplus_{k\geq 0}((2k+1)^2)^+ \oplus \bigoplus_{k\geq 1}\left(\frac{nk^2}{g}\right)^- \tag{7.1b}$$

$$D_{n/2,+} : \bigoplus_{k\geq 0}\left(\frac{n(2k+1)^2}{4g}\right)^+ \tag{7.1c}$$

$$D_{n/2,-} : \bigoplus_{k\geq 0}\left(\frac{n(2k+1)^2}{4g}\right)^+ \tag{7.1d}$$

$$D_Q \quad : \bigoplus_{k\in Z}\left(\frac{(nk+Q)^2}{ng}\right)^+ \qquad (Q=1,\ldots,p) \tag{7.1e}$$

(b) n = 2p+1 (n odd > 3):

$$\begin{aligned} D_{0,+} \quad &: \bigoplus_{k\geq 0}(4k)^2)^+ \oplus \bigoplus_{k\geq 1}\left(\frac{nk^2}{g}\right)^+ \\ D_{0,+} \quad &: \bigoplus_{k\geq 0}((2k+1)^2)^+ \oplus \bigoplus_{k\geq 1}\left(\frac{nk^2}{g}\right)^- \\ D_Q \quad &: \bigoplus_{k\in Z}\left(\frac{(nk+Q)^2}{ng}\right)^+ \qquad (Q=1,\ldots,p) \end{aligned} \tag{7.2}$$

These expressions take a compact form if we combine the sectors of opposite charge conjugation (like $D_{n/2,+}$ with $D_{n/2,-}$) and split the spectra only according to the Z_n charge Q, we have

$$\mathcal{E}_Q : \bigoplus_{k\in Z}\left[\frac{(nk+Q)^2}{ng}\right] \qquad (Q=0,\ldots,n-1) \tag{7.3}$$

where we have used irreps. of the U(1) Kac-Moody algebra (Eq. (2.8)). Notice the relation:

$$\mathcal{E}_Q = \mathcal{E}_{n-Q} \tag{7.4}$$

b) Toroidal boundary condition [24]

The lowest excitations are known since a long time [13] here we give the complete picture. We denote by $\mathcal{E}$ the sector with Z_n charge Q corresponding to the boundary condition $\Sigma^{\tilde{Q}}$ (see Eq. (3.7)). In this case we can give the operator content in terms of irreps. of the U(1) Kac-Moody algebra. We have:

$$\mathcal{E}_Q^{\tilde{Q}} : \bigoplus_{\alpha,\beta \in Z} \left(\left[\frac{(Q+n\alpha+q(\tilde{Q}+n\beta))^2}{4nq}\right] , \left[\frac{(Q+n\alpha-q(\tilde{Q}+n\beta))^2}{4nq}\right] \right) \qquad (7.5)$$

The expression (7.5) includes the older results [13], is compatible with numerical estimates [24] and with extended modular invariance [32].

The operator content of the chains with periodic b.c. ($\tilde{Q}=0$) and without splitting into sectors according to the Z_n charge is

$$\mathcal{E} = \bigoplus_{Q=0}^{n-1} \mathcal{E}_Q^0 = \bigoplus_{\alpha,\beta \in Z} \left(\left[\frac{(\alpha+4h\beta)^2}{16h}\right] , \left[\frac{(\alpha-4h\beta)^2}{16h}\right] \right) \qquad (7.6)$$

with

$$h = \frac{nq}{4} \qquad (7.7)$$

This implies that the operator content has a unique form independent on n. We now describe the remaining sectors of the models. They correspond to take as boundary condition $\Sigma^R C$. The symmetry of the Hamiltonians is $Z_2 \otimes Z_2$ for n even and Z_2 for n odd. We have with obvious notations:

n <u>even</u> (R=0,1,...,n-1)

$$\begin{aligned}
\mathcal{E}^{\Sigma^R C}_{\Sigma^{n/2}=+,\Sigma^R C=+} = \mathcal{E}^{\Sigma^R C}_{\Sigma^{n/2}=-,\Sigma^R C=+} &= (\{\tfrac{1}{16}\},\{\tfrac{1}{16}\}) \oplus (\{\tfrac{9}{16}\},\{\tfrac{9}{16}\}) \\
\mathcal{E}^{\Sigma^R C}_{\Sigma^{n/2}=+,\Sigma^R C=-} = \mathcal{E}^{\Sigma^R C}_{\Sigma^{n/2}=-,\Sigma^R C=-} &= (\{\tfrac{1}{16}\},\{\tfrac{9}{16}\}) \oplus (\{\tfrac{9}{16}\},\{\tfrac{1}{16}\})
\end{aligned} \qquad (7.8)$$

n odd (R=0,1,...,n-1)

$$\begin{aligned}
\mathcal{E}^{\Sigma^R C}_{\Sigma^R C=+} &= 2\,(\{\tfrac{1}{16}\},\{\tfrac{1}{16}\}) \oplus 2(\{\tfrac{9}{16}\},\{\tfrac{9}{16}\}) \\
\mathcal{E}^{\Sigma^R C}_{\Sigma^R C=-} &= 2\,(\{\tfrac{1}{16}\},\{\tfrac{9}{16}\}) \oplus 2(\{\tfrac{9}{16}\},\{\tfrac{1}{16}\})
\end{aligned} \qquad (7.9)$$

Notice that if we combine the sectors together one obtains

$$2n\,((\tfrac{1}{16})_T\,,\,(\tfrac{1}{16})_T) \qquad (7.10)$$

for n both even and odd (we have used here the notations (2.9) - (2.11)). We observe that the sectors with $Z_2 \otimes Z_2$ (respectively Z_2) symmetry have an operator content independent on g and are given by the twisted U(1) Kac-Moody algebra. This completes the description of the finite-size scaling spectra of the n-states ($n \geq 5$) Hamiltonians in the c=1 region.

VIII. Operator content of the XXZ chain

The XXZ model is defined by the Hamiltonian

$$H = -\frac{\gamma}{2\pi\sin\gamma} \sum_{j=1}^{N} (\sigma_j^x \sigma_{j+1}^x + \sigma_j^y \sigma_{j+1}^y - \cos\gamma\, \sigma_j^z \sigma_{j+1}^z) \tag{8.1}$$

$$(0 \leq \gamma < \pi)$$

where σ^x, σ^y and σ^z are Pauli matrices. This model has again a long history in statistical mechanics [33]. The properties of its finite-size scaling spectrum have been investigated by several authors using both analytic and numerical methods [34]. For γ real the system is critical with c=1. We give here a conjecture for the whole operator content of the model for various boundary conditions. This conjecture is based on all known results (Bethe ansatz calculations with numerical solutions of the transcendental equations as well as finite-size calculations up to 18 sites and extended modular invariance).

The global symmetry of the infinite system is O(2). In the basis

$$\sigma_j^{\pm 1} = \sigma_j^{+} \pm i\sigma_j^{y} \quad , \quad \sigma_j^{0} = \sigma_j^{z} \tag{8.2}$$

the Hamiltonian (8.1) is invariant under the transformations

$$\sigma'^{m}_j = \sum_{n=-1}^{1} A^{mn} \sigma_j^n \tag{8.3}$$

where the matrices A form the O(2) group

$$\mathcal{G} = \{G(\theta)C^{\alpha} \mid \theta \in [0, 2\pi), \alpha = 0, 1\} = O(2) \tag{8.4}$$

where

$$G(\theta) = \begin{pmatrix} e^{-i\theta} & 0 & 0 \\ 0 & 1 & 0 \\ 0 & 0 & e^{i\theta} \end{pmatrix} , \quad C = \begin{pmatrix} 0 & 0 & 1 \\ 0 & -1 & 0 \\ 1 & 0 & 0 \end{pmatrix} \tag{8.5}$$

We now specify the boundary conditions for the Hamiltonian.

(a) Free boundary conditions (H^F)

$$\sigma^x_{N+1} = \sigma^y_{N+1} = \sigma^z_{N+1} = 0 \tag{8.6}$$

(b) Toroidal boundary conditions (H^B)

$$\sigma^m_{N+1} = \sum_{n=-1}^{1} B^{mn} \sigma^n_1 \tag{8.7}$$

where B is one of the matrices of (8.4). The conjugacy classes of O(2) are as follows:

$$\text{(I)} \ \{\mathbb{1}\} \quad \text{(II)} \ \{G(\pi)\} \quad \text{(III)} \ \{G(\theta), G(-\theta)\} \ \forall \theta \neq 0, \pi$$
$$\text{(IV)} \ \{G(\theta)C \mid \theta \in [0, 2\pi)\} \tag{8.8}$$

The global symmetry of the Hamiltonian with free b.c. (H^F) is O(2). The same is true for periodic ($B=\mathbb{1}$) and antiperiodic ($B=G(\pi)$) b.c. The symmetry for the elements of the conjugacy classes (III) is $SO(2) = \{G(\theta) \mid \theta \in [0, 2\pi]\}$. For the conjugacy class (IV), e.g. $G(\theta)C$, the symmetry is only $Z_2 \otimes Z_2 = \{\mathbb{1}, G(\theta)C, G(\pi), G(\pi+\theta)C\}$. (These are the symmetries for an even number of sites).

The finite-size scaling limits of the spectra have to be computed separately for even number of sites (N) and odd number of sites. The operator content for an odd number of sites will be given taking as reference energy the ground-state energy for an even number sites. For example, if one takes free boundary conditions and the ground-state energies for N sites are $E^F(N)$ we take as reference energy for 2N+1 sites

$$E_R(2N+1) = \frac{1}{2} (E^F(2N) + E^F(2N+2)) \tag{8.9}$$

The charge operator which commutes with the Hamiltonian (except in the case when the b.c. belongs to the conjugacy class (IV)) is:

$$\hat{Q} = \frac{1}{2} \sum_{j=1}^{N} \sigma_j^z \tag{8.10}$$

Notice that the eigenvalues (q) are integer (half-integer) when N is even (odd).

It turns out that the dependence of the operator content of the coupling constant γ (see Eq. (8.1)) is simpler when expressed in terms of

$$h = \frac{\pi}{4(\pi-\gamma)} \qquad (\infty > h \geqslant \frac{1}{4}) \tag{8.11}$$

We now list the operator content for various b.c.

a) <u>Free boundary conditions</u>

If we denote by F_q^F the operator content for the charge sector q we have

$$F_q^F = [\frac{q^2}{4h}] \tag{8.12}$$

where q is integer or half-integer.

b) <u>Toroidal boundary conditions</u>

The operator content for the conjugacy classes (I), (II) and (III) can be given in a compact form. With the notation:

$$\sigma_{N+1}^{\pm 1} = e^{\pm 2\pi i s}\sigma_1^{\pm} \qquad 0 \leqslant s < 1 \qquad \sigma_{N+1}^{0} = \sigma_1^{0} \tag{8.13}$$

the operator content in the charge q sector is

$$F_q^s = \bigoplus_{m \in Z} ([\frac{(q+4h(s+m))^2}{16h}] , [\frac{(q-4h(s+m))^2}{16h}]) \tag{8.14}$$

where q is an integer or half integer. Here $[\Delta]$ is an irrep. of the U(1) Kac-Moody algebra.

For the conjugacy class (IV) (we take the b.c. given by the charge conjugation C transformation (8.5) as a typical element), we have for N <u>even</u>

$$F^{C}_{C=+,\ G(\pi)=+} = F^{C}_{C=+,\ G(\pi)=-} = (\{\tfrac{1}{16}\},\{\tfrac{1}{16}\}) \oplus (\{\tfrac{9}{16}\},\{\tfrac{9}{16}\})$$

$$F^{C}_{C=-,\ G(\pi)=+} \quad F^{C}_{C=-,\ G(\pi)=-} = (\{\tfrac{1}{16}\},\{\tfrac{9}{16}\}) \oplus (\{\tfrac{9}{16}\},\{\tfrac{1}{16}\}) \tag{8.15}$$

Here $G(\pi) = \pm$, $C = \pm$ are the eigenvalues of the $G(\pi)$ and C operators which generate the $Z_2 \otimes Z_2$ symmetry of the boundary condition C. In Eq. (8.15) we have used Eqs. (2.9) - (2.11). For N odd we have

$$F^{C}_{C=+} = F^{C}_{C=-} = ((\tfrac{1}{16})_T\ ,\ (\tfrac{1}{16})_T) \tag{8.16}$$

Here the eigenvalues of C and $G(\pi)$ cannot be measured simultaneously.

It is interesting to observe that the finite-size scaling spectrum of the XXZ chain contains all the previously known spectra. If we take $h = \frac{nq}{4}$ in Eq. (8.12) and take only q integer

$$\mathcal{E}_Q = \bigoplus_{k\in Z} F^{F}_{nk+Q} = \bigoplus_{k\in Z} [\frac{(nk+Q)^2}{ng}] \quad (Q=0,1,\ldots,n-1) \tag{8.17}$$

which is just the operator content of the n-states model with free boundary conditions (Eq. (7.3)). Similarly, if we sum over all the charges in (8.12) (only integers!)

$$\mathcal{E} = \bigoplus_{q\in Z} [\frac{q^2}{4h}] \tag{8.18}$$

we find the total operator content of the Ashkin-Teller model (Eq. (5.1)). A similar situation occurs for toroidal b.c. From Eq. (8.14) we have

$$\mathcal{E}^{\tilde{Q}}_Q = \bigoplus_{\alpha\in Z} F^{\tilde{Q}/n}_{n\alpha+Q} \qquad (Q,\tilde{Q}=0,1,\ldots,n-1) \tag{8.19}$$

which coincides with Eq. (7.5). (We have taken again $h=\frac{nq}{4}$). If we want to get the total operator content of the Ashkin-Teller model, we have to take n=4 in (8.19) which implies Z_4 boundary conditions together with the spectra corresponding to the boundary conditions $G(0)C$, $G(\frac{\pi}{4})C$, $G(\pi)C$, $G(\frac{3\pi}{4}C)$ which are given by Eq. (8.15). The distribution of the operators into the various sectors of the Ashkin-Teller model implies a non-trivial reshuffling.

REFERENCES

[1] F.C. Alcaraz and M.N. Barber Journ. Stat. Mech. 46 (1987) 435

[2] P.A. Pearce and D. Kim, J. Phys. A20 (1987) 6471

[3] G.E. Andrews, R.J. Baxter and P.J. Forrester, J. Stat. Phys. 35 (1984) 193

[4] E. Date, M. Jimbo, A. Kuniba, T. Miwa, and M. Okado, Nucl. Phys. B290 (1987) 231, V.V. Bazhanov and N. Yu. Reshetikhin, preprint IFVE 87-102

[5] R. Dijkgraaf, E. Verlinde and H. Verlinde, Talk presented by E. Verlinde at the 1987 Cargese Summer School on Non-perturbative Quantum Field Theory, preprint THU-87/27

[6] P. Ginspark, preprint HUTP-87/A068

[7] G. Harris, preprint EFI 87-105

[8] J.-B. Zuber, to appear in the Proceedings of the Conference "Perspectives in String Theory" Copenhagen, October 1987

[9] J. Bagger, to appear in the Proceddings of the 11th Johns Hopkins Workshop on Current Problems in Particle Theory, Lanzhou 1987

[10] V. Kac, Lectures Notes in Mathematics 94 (1979) 444

[11] E. Corrigan, Phys. Letters 169B (1986) 259

[12] P. Goddard, A. Kent, and D. Olive, Comm.Math. Phys. 103 (1986) 105

[13] J.V. José, L.P. Kadanoff, S. Kirkpatrick, and D.R. Nelson, Phys.Rev. 16 (1977) 1217, L.P. Kadanoff, Annals of Physics 120 (1979) 39, E. Fradkin and L.O. Kadanoff, Nucl. Phys. B170 (1980) 1, B. Nienhuis, J. Stat. Mech. 34 (1984) 731

[14] M. Marcu, A. Regev, and V. Rittenberg, J. Math. Phys. 22 (1981) 2740

[15] P. Chaselon, to be published

[16] H.W. Blöte , J.L. Cardy, and M.P. Nightingale, Phys. Rev. Letters 56 (1986) 742

[17] J. Cardy, Nucl. Phys. B240 (1984) 514

[18] I. Affleck, Phys. Rev. Letters 56 (1986) 746

[19] J. Cardy, Nucl. Phys. B275 (1986) 200

[20] G. v. Gehlen and V. Rittenberg, J. Phys. A19 (1986) L625

[21] M. Kohmoto, M. den Nijs and L.P. Kadanoff, Phys. Rev. B24 (1981) 5229

[22] G. v. Gehlen and V. Rittenberg, J. Phys. A 20 (1987) 227

[23] G. v. Gehlen and V. Rittenberg, J. Phys. A19 (1986) 2439, J. Phys. A20 (1987) 1309; F. C. Alcaraz, J. Phys. A19 (1986) L1085, G. Schütz, preprint Bonn HE-87-05

[24] G. v. Gehlen. V. Rittenberg, and G. Schütz, preprint Rio de Janeiro CBPF-NF-58/87

[25] V. Rittenberg, Proceedings of the Paris-Meudon Colloquium, Editors H. J. de Vega and N. Sánchez, World Scientific, p. 214, M. Baake, G. v. Gehlen and V. Rittenberg, J. Phys. A20 (1987) L479

[26] S-K Yang, Nucl. Phys. B285 (1987) 183, B285 (1987) 410, M. Baake, G. v. Gehlen and V. Rittenberg, J. Phys. A20 (1987) L487, H. Saleur, J. Phys. A20 (1987) L1127

[27] M. Baake, Ph. Christe, and V. Rittenberg, preprint Nordita-87/67P

[28] U. Grimm, preprint Bonn-HE-87-32

[29] R. Burlish and J. Stoer, Num. Mathematik 6 (1964) 413

[30] M. Henkel and G. Schütz, preprint Bonn-HE-87-18

[31] B. Nienhuis and J.F. Knops, Phys. Rev. B32 (1985) 1872

[32] P. Suranyi, preprint Bonn-HE-87-23

[33] R.J. Baxter "Exactly Solved Models in Statistical Mechanics" Academic Press 1982 Chapt. X

[34] F.C. Alcaraz, M.N. Barber and G.R.W. Quispel, J. Phys. A20 (1987) 6397, C.J. Hamer and M.T. Batchelor, J. Phys. A21 (1988) L173, C.J. Hamer, G.R.W. Quispel and M.T. Batchelor, J. Phys. A20 (1987) 5677, F. Woynarowich, Phys. Rev. Letters 59 (1987) 259

[35] F.C. Alcaraz, M. Baake, U. Grimm, and V. Rittenberg, J. Phys. A21 (1988) L117

TWO DIMENSIONAL CRITICAL MODELS ON A TORUS

H. Saleur and P. di Francesco

Service de Physique Théorique
Institut de Recherche Fondamentale, CEA-CEN Saclay,
91191 Gif-sur-Yvette cedex, France

After the general developments of conformal invariance in two dimensions, it was realized that the study of critical models in finite geometries, in addition to the practical information it could provide through finite size scaling (see Pr. Rittenberg lecture), was also of great conceptual interest. The simplest example is the case of the torus, a genus 1 surface which is thus not conformally equivalent to the plane. This geometry appears quite frequently in lattice calculations for systems with periodic boundary conditions, and is also very natural from the point of view of string theory. We will discuss briefly in these notes the main results obtained so far in this simple case.

The torus can be parametrized by a complex number τ which is defined up to a modular transformation only, and the associated partition function must be invariant under the modular group. This proves to be a very strong constraint, which determines the possible operator content of the theories ie the degeneracies of the primary fields. These were not fixed by considerations in the plane only.We discuss in part 1 the case of the minimal theories with $c < 1$, and reexpose briefly the resulting ADE classification.

From another point of view, the toroïdal geometry imposes as well, through boundary conditions effects, the operator content of theories which are mapped onto a free field (Coulomb gas). Using know information about these mappings, we rederive in part 2 results of part 1 which were obtained so far by more formal considerations, providing in this way a connection between the microscopic models and their continuum limit partition functions.

To determine completely the possible conformal theories, one would need also to obtain the structure constants of the operator algebra. This can be in principle answered in various ways, like calculating four point functions on the plane, or two point functions on the torus. Correlators in toroïdal geometry have also interest on their own, in relation in particular to various lattice calculations. We expose in part 3 the simplest attempts in this direction by calculating multi-energies or spins correlations for the Ising model on a torus.

ISBN 0-12-218100-X

1. Partition Functions on a Torus and Modular Invariance

In statistical mechanics it is quite common to study models on a lattice $L \times L'$ with periodic boundary conditions, ie on a geometry which has the topology of a torus. For a temperature larger than the critical temperature T_c, the correlation lenght ξ is finite in the bulk and the properties of the finite system become independent of the precise boundary conditions as $L, L' \to \infty$, up to $exp(-L'/\xi)$ corrections. For $T = T_c$ however, the bulk correlation length is infinite and finite size scaling theory [1] shows that for the finite system it increases linearly with the dimensions, so one expects now corrections of the kind $exp(-cst\ L'/L)$. If $L, L' \to \infty$ keeping the ratio L'/L fixed, one gets for the partition function $\mathfrak{Z}$ an asymptotic behaviour

$$\mathfrak{Z} \simeq exp\ f\ .\ LL' \times Z\left(\frac{L'}{L}\right) \tag{1.1}$$

where f is the bulk free energy per unit surface, and Z contains the finite size corrections. f is not universal but Z is expected to be [1]. Using the additional property of conformal invariance, one can get a formal expression for Z. Indeed, it is the trace of the L' power of the transfer matrix, whose spectrum is determined (see Pr. Rittenberg Lecture) using the logarithmic mapping $w = L/2\pi\ log\ z$. It is known that the ground state scales as [2]

$$f_L = f + \frac{\pi c}{6L^2} \tag{1.2}$$

where c is the central charge, and the gaps as [2]

$$m_L = 2\pi\ x/L \tag{1.3}$$

associated to eigenstates of moment

$$k_L = 2\pi\ S/L \tag{1.4}$$

x,S being the dimension and spin of the operators of the theory ($x = h + \bar{h}$, $s = h - \bar{h}$). In the continuum limit, it will be useful to generalize the calculation of Z to an arbitrary torus described by two complex numbers ω_1 , ω_2 and $\tau = \frac{\omega_1}{\omega_2} = \omega_R + i\omega_I$, $\omega_I > 0$ (fig. 1). Introducing $q = exp\ 2\pi i\ \tau$ one gets [3]

$$Z=(q\bar{q})^{-c/24} \sum_{all\ operators} q^h \bar{q}^{\bar{h}} = (q\bar{q})^{-c/24} \sum_{h\bar{h}} N_{h\bar{h}}\ \chi_h(q)\chi_{\bar{h}}(\bar{q}) \tag{1.5}$$

In the last expression the sum is restricted to the primary fields, the contribution of their descendants being resummed in the characters (see Pr. Rittenberg Lecture)

$$\chi_h(q) = q^h \sum D_{h,N}\ q^N \tag{1.6}$$

where D_N is the number of independent descendents at level N.

On the other hand it has been shown that consistent conformal theories ("minimal" ones) can be built which have a finite number of primary fields, for the central charges [4]

$$c = 1 - \frac{6(p-p')^2}{pp'} \tag{1.7}$$

the dimensions of these fields being given by Kac formula [5]

$$h_{rs} = \frac{(rp - sp')^2 - (p - p')^2}{4pp'} \tag{1.8}$$

where $1 \leqslant r \leqslant p - 1$, $1 \leqslant s \leqslant p' - 1$. For such theories, the correlation functions and the whole operator algebra can be calculated [6].However the formalism in the plane does not tell the exact operator content, ie the number $N_{h\bar{h}}$ of fields of dimensions h, $\bar{h}$ being actually present in the theory [4]. This can be answered by turning to the genus 1 surface (ie the torus) and using the expression (1.5) for partition functions. Indeed it's clear that for a given torus, ω_1 and ω_2 are not uniquely defined and one could consider as well any

$$\begin{pmatrix} \omega'_1 \\ \omega'_2 \end{pmatrix} = \begin{pmatrix} a & b \\ c & d \end{pmatrix} \begin{pmatrix} \omega_1 \\ \omega_2 \end{pmatrix}, \quad abcd \in \mathbb{Z}, \quad ad - bc = 1 \tag{1.9}$$

We must thus demand that $Z(\tau)$ is modular invariant, ie invariant under the transformation $\tau \longrightarrow (a\tau + b)/c\tau + d$, and this will turn out to be a very strong constraint [3].

It is of course sufficient to ensure invariance under the two generators

$$T : \quad \tau \longrightarrow \tau + 1 \qquad\qquad S : \quad \tau \longrightarrow -1/\tau \tag{1.10}$$

Under T, $q \longrightarrow e^{2i\pi\tau}q$ and by (1.6), $\chi_h(q) \longrightarrow e^{2i\pi h}\ \chi_h(q)$. Hence invariance under T is satisfied if

$$N_{h\bar{h}} = 0 \qquad \mathit{for} \qquad h - \bar{h} \notin \mathbb{Z} \tag{1.11}$$

ie only integer spin operators contribute to Z. This was expected since in the spectrum of a model with periodic boundary conditions on a strip, only integer momenta (in units of $\frac{2\pi}{L}$) appear. For studying the effect of S one needs the precise expression of the characters χ which has been obtained in [7]. Restricting ourselves for simplicity to the unitary series [8] $p = m + 1$, $p' = m$, $m \geqslant 2$, one has

$$\chi_h = \chi_{rs} = \frac{1}{\prod_{n=1}^{\infty} (1-q^n)} \sum_{n=-\infty}^{\infty} \left\{ q^{[2nm(m+1)+(m+1)r-ms]^2/4m(m+1)} - (s \longrightarrow -s) \right\} \tag{1.12}$$

Then, denoting $\tau' = -1/\tau$, Poisson formula gives

$$q'^{-c/24}\ \chi_h(q') = \sum_{h'} A_{hh'}\ q^{-c/24}\ \chi_{h'}(q) \tag{1.13}$$

A being the $\left(\frac{m(m-1)}{2}\right)^2$ matrix

$$A_{hh'} = A_{rs,r's'} = \left[\frac{8}{m(m+1)}\right]^{\frac{1}{2}} (-)^{(r+s)(r'+s')} \frac{\sin \pi rr'}{m} \frac{\sin \pi ss'}{m+1} \tag{1.14}$$

and the sum in (1.13) runs over $1 \leqslant s' \leqslant r' \leqslant m - 1$. A is symmetric and its square is equal to one (since $S^2 = 1$). Modular invariance translates thus into

$$A \,.\, N = N \,.\, A \tag{1.15}$$

which is a system of diophantine equations [3,9]. One shows easily that the choice $N_{h\bar{h}} = \delta_{h,\bar{h}}$ is always a solution. For $c = \frac{1}{2}$ $(m = 3)$ it's the only one and

$$Z = (q\bar{q})^{-1/48} \left(|\chi_0|^2 + |\chi_{1/16}|^2 + |\chi_{\frac{1}{2}}|^2\right) \tag{1.16}$$

which is identified as the Ising model partition function, matching the known exponents $x_\epsilon = 1$, $x_H = 1/8$ or using the exact solution [10].For $c = 4/5$ $(m = 5)$, two solutions are possible, either

$$Z = (q\bar{q})^{-1/30} \left(|\chi_0|^2 + |\chi_{1/15}|^2 + |\chi_{2/5}|^2 + |\chi_{2/3}|^2 + |\chi_{7/5}|^2 + |\chi_3|^2\right) \tag{1.17}$$

$$Z = (q\bar{q})^{-1/30} \left(|\chi_0 + \chi_3|^2 + |\chi_{2/5} + \chi_{7/5}|^2 + 2|\chi_{1/15}|^2 + 2|\chi_{2/3}|^2\right) \tag{1.18}$$

(1.17) is identified with the 3 state Potts model matching dimensions [8] and using the fact that the spin with $x_H = \dfrac{2}{15}$ msut be twice degenerate due to the $\mathbb{Z}_3$ symmetry [3]. (1.18) corresponds to the tetracritical Ising model [3].

Finding all the solutions of (1.12) is a difficult task which has been achieved only recently. Surprisingly it turns out that the modular invariants can be labelled by a pair of simply laced algebra, one of which is always of A type [11]. We refer the reader to the literature and quote only the results in table 1. In addition to two infinite series (A,A) and (D,A) generalyzing (1.17) and (1.18) one notes the appearance of three exceptional cases $(E_6\text{-}E_7\text{-}E_8, A)$. Various physical realizations can been proposed for these partition functions. Generic models are obtained by building Restricted Solid on Solid Models, the heights of which are located on the Dynkin diagram of the underlying algebra [12] (Fig. 2).

It has to be noticed that the invariants of table 1 generalize readily to non unitary theories. A simple example is obtained with $p = 5$, $p' = 2$, $c = -22/5$ corresponding to the Lee Yang singularity

$$Z = (q\bar{q})^{11/60}\left(|\chi_0|^2 + |\chi_{-1/5}|^2\right) \tag{1.19}$$

A *negative* dimension appears then in the Kac table (1.8), $h_{12} = -1/5$ for (1.19), and the formula (1.2)-(1.4) must be corrected. Indeed the small q behaviour of Z is given now by $Z \sim (q\bar{q})^{-c/24}(q\bar{q})^{h_{12}}$ so (1.2) reads instead $f_L = f + Tc'/6L^2$ where

$$c' = c - 24h_{12} = \frac{4}{5} \tag{1.20}$$

Similarly the first gap is $m_L = 2\pi x'/L$ where $x' = |x| = 2/5$. The presence of negative dimensions is general in the non unitary case, all invariants containing [11] the dimension $h = \dfrac{1 - (p - p')^2}{4pp'}$. Precise rules for the modification of the transfer matrix spectrum have been given in [13].

Attempts have been made in [14] to construct also invariants associated to central charges (1.7), but with dimensions outside the minimal grid.

2. Coulomb Gas Construction of the Partition Functions on a Torus

The results of the previous section have been obtained using very general methods, but they have the drawback of being rather formal. For more practical applications in statistical mechanics, one would like to have a method for establishing connections between the abstract partition function (1.5) associated to a conformal field theory and models formulated on the lattice. In the simplest cases it is enough to proceed by inspection, like for the Ising model [8] or the Lee & Yang singularity [15] where there is only one modular invariant reproducing results known by other methods. For the three state Potts model, we have seen in section 1 that there are already two candidates, and it is an argument about the symmetry of the model which allows the correct identification [3]. For more complicated models the procedure obviously becomes hopeless, and numerical calculations can then prove very useful.

To derive critical properties of 2D models, a somehow different field theoretic technique had been developed in the past, the so called Coulomb Gas technique [16]. In this approach, one first reformulates the model to be studied (usually as a solid on solid interface model) in such a way that it can be argued to renormalize at criticality onto a free Gaussian bosonic theory with action

$$\mathcal{A} = \frac{g}{4\pi} \int |\vec{\nabla}\varphi|^2 \, d^2x \tag{2.1}$$

This crucial property cannot in general be rigorously established, but it can nevertheless be checked in various ways, performing for instance approximate renormalization group calculations. The coupling constant of the associated Gaussian model g has also to be determined, usually using some extra information from the eight-vertex model solution [17]. Physical observables can then be translated in Gaussian terms [18], as combination of electric or spin wave (ie exponentials of the field *exp* ie φ) and magnetic or vortex operators (ie operators creating a branch cut with discontinuity of $2\pi m$ for the field φ). Their dimensions are readily calculated [18]

$$x_{em} = \frac{e^2}{2g} + \frac{gm^2}{2} \quad , \quad S_{em} = em \tag{2.2}$$

This approach which was pioneered by José et al. [18] for the XY model has been successfully used in a number of cases, such as the Q-state Potts or O(n) [19] models providing, under the above mentioned renormalization flow assumptions, an exact determination of the main exponents. It turns out, as demonstrated in [20], that one can in fact use this Coulomb gas mapping to derive also the partition functions on a torus, and then the whole operator content of these models.

The partition function for the free field (2.1) with periodic boundary conditions has been calculated in many different contexts already. A dzeta regularization [9] of the determinant of the Laplacian gives

$$Z_0 = \sqrt{\frac{g}{\tau_I}} \frac{1}{|\eta(q)|^2} \tag{2.3}$$

where τ, η and q are the same as in section 1. The dependance on g in (2.3) comes from the zero mode, the subtraction of which prevents a rescaling of φ. Z_0 has a small q behaviour corresponding to c = 1 which is known to be the central charge of a free bosonic field. The complete expansion in powers of q, $\bar{q}$ shows also the presence of a marginal operator (with x = 2) identified as $\partial_z \partial_{\bar{z}} \varphi$, but none of the dimensions (2.2) is observed.

In mapping a lattice model onto a free field, some other terms are usually generated due to the boundary conditions. We can discuss for instance the case of the XY model whose lattice action is

$$\mathcal{A} = -\frac{1}{T}\sum_{\langle ij\rangle} [\cos(\varphi_i - \varphi_j) - 1] \tag{2.4}$$

At very low temperature all spins are almost parallel, and it is justified to expand (2.4) into

$$\mathcal{A} = \frac{1}{2T}\sum_{\langle ij\rangle} (\varphi_i - \varphi_j)^2 \qquad (T \longrightarrow 0) \tag{2.5}$$

which becomes (2.1) in the continuum limit with

$$g \simeq \frac{2\pi}{T} \qquad (T \longrightarrow 0) \tag{2.6}$$

At higher temperatures, the approximation (2.5) fails, in particular due to the presence of vortices, ie configurations where local angle variations are still small, but globally φ can vary by multiples of 2π around one point [18]. It is well known however that for T smaller than a critical value T_c, these vortices are irrelevant. Under renormalization transformation, vortices of opposite charges come closer and ultimately annihilate each other [18], giving still rise to a fixed point action (2.5), but with a renormalized temperature [21]

$$g \simeq \frac{2\pi}{T}\left(1 - \frac{T}{4} - \frac{T^2}{96} + \ldots\right) \tag{2.7}$$

The end of this critical line is characterized by the marginality of the lowest dimension vortex operator : $g(T_c)\ (m=1)^2/2 = 2$, thus

$$g(T_c) = 4 \tag{2.8}$$

For $T > T_c$, the vertices do not disappear under renormalization, and the model is no more critical.

Now on a torus special care must be taken of boundary conditions. Indeed the periodic geometry allows configurations where φ varies along the generators of the torus and this topological property remains of course unchanged in a renormalization. Thus for $T < T_c$, the XY model in the continuum limit is described by the free field (2.1) where φ is not periodic but can have shifts

$$\begin{cases} \varphi(z+1,\bar{z}+1) = \varphi(z,\bar{z}) + 2\pi\, m \\ \varphi(z+\tau,\bar{z}+\bar{\tau}) = \varphi(z,\bar{z}) + 2\pi\, m' \end{cases} \tag{2.9}$$

for any $m, m' \in \mathbb{Z}$. For boundary conditions (2.9), the partition function is readily calculated introducing the classical field such that $\Delta\, \varphi_{cl} = 0$

$$\varphi_{cl} = \frac{m\bar{\tau} - m'}{\tau_I}\, z + cc \tag{2.10}$$

and

$$Z_{mm'}(g) = Z_0(g)\, exp - \pi g \frac{|m - m'\tau|^2}{\tau_I} \tag{2.11}$$

It has the modular transformation properties

$$Z_{mm'}\left(\frac{a\tau + b}{c\tau + d}\right) = Z_{am+bm',cm+dm'}(\tau) \tag{2.12}$$

expected from its definition. The complete XY model partition function is obtained by summing over m and m' [20]

$$\mathfrak{Z}_{xy} \longrightarrow Z_c(g) = \sum_{m,m' \in \mathbb{Z}} Z_{mm'}(g) = \frac{1}{|\eta|^2} \sum_{em \in \mathbb{Z}} q^{\frac{1}{4}\left(\frac{c}{\sqrt{g}} + m\sqrt{g}\right)^2} \bar{q}^{\frac{1}{4}\left(\frac{e}{\sqrt{g}} - m\sqrt{g}\right)^2} \tag{2.13}$$

this last equality being obtained after a Poisson transformation. It has still $c = 1$, and presents now the operator content expected form (2.2). Expression (2.13) will be called in the following a Coulombic partition function; it is instructive to notice that this object appeared also in the past in the context of string theory [22] as the partition function of a free field compactified on a circle of radius $R \sim \sqrt{g}$. Similar formulas can be obtained for describing the F-model, or the Ashkin Teller model critical lines. In the latter case however one must add the contribution of sectors where the field φ has antiperiodic boundary conditions, corresponding to the identification of the spin as the twist operator (creating a branch cut with $\varphi \longrightarrow -\varphi$) with dimension $x_H = \frac{1}{8}$ independent of g, we refer the redear to references [22].

To reproduce the partition functions of models with $c < 1$, an additional ingredient must be added, generalizing somehow the charge at infinity of *ref.*[6] to the torus. We shall discuss that in some detail for the O(n) model on the honeycomb lattice defined by

$$\mathcal{Z}_n = \int \prod_i d\vec{S}_i \prod_{\langle jk \rangle} \left(1 + \frac{1}{T} \vec{S}_j . \vec{S}_k\right) \tag{2.14}$$

$\vec{S}$ being a n-component spin with $|\vec{S}|^2 = n$. (2.14) can be analytically continued to $n \in \mathbb{R}$ using a high temperature expansion

$$\mathcal{Z}_n = \sum_{\text{graphs}} \left(\frac{1}{T}\right)^{\mathcal{N}_B} n^{\mathcal{N}_P} \tag{2.15}$$

In (2.15) the graphs are formed by $\mathcal{N}_P$ non intersecting self avoiding loops (or polygons) of total length $\mathcal{N}_B$. The model is known to be critical for $n \in [-2,2]$. It can be transformed into a SOS model [29] by introducing height variables φ on the centers of the hexagons. An arbitrarily oriented polygon corresponds then to a wall between two regions of constant height, with a step $\pm \varphi_0$, the highest φ being on the left of each arrow. The Boltzmann weight consists of a factor $\frac{1}{T}$ for each bond, times $e^{iv}(e^{-iv})$ for each left (right) turn. Then, since the difference between the numbers of left and right turns for a polygon on the honeycomb lattice in the plane is $n_\ell - n_r = \pm 6$, one has $\mathcal{Z}_n = \mathcal{Z}_{sos}$ if $n = 2 \cos 6v$. At criticality, this SOS model is argued to renormalize onto the free field (2.1), and for the choice $\varphi_0 = \pi$ one gets [19]

$$n = -2 \cos 4\pi g \qquad\qquad g \in \left[\frac{1}{4}, \frac{1}{2}\right] \tag{2.16}$$

On a torus however, $\mathcal{Z}_n \neq \mathcal{Z}_{sos}$ since polygons which wrap around it have $n_\ell = n_r$. The necessary correction is easy to obtain in the strip limit where one has only loops wraping around w_1. To give to these loops the weight n instead of 2, one introduces charges $+(-)e_0$ at $+(-)\infty$. Then the Boltzmann weight of a configuration has an additional *exp* $i\, e_0 [\varphi_\infty - \varphi_{-\infty}]$, each non contractbile loop contributing to it by *exp* $\pm 2i\, \pi\, e_0$ depending on its orientation. Summing over all configurations gives then the new weight $2 \cos 2\pi e_0$ which is the desired result provided

$$n = 2 \cos 2\pi e_0 \tag{2.17}$$

The ground-state is modified by these charges, and accordingly the central charge becomes [13]

$$c = 1 - 24h_{e_0} = 1 - \frac{6(2g - 1/2)^2}{g} \tag{2.18}$$

In the Ising case for instance, $n = 1$ gives $g = \frac{1}{3}$, $e_0 = \frac{1}{6}$ and $c = \frac{1}{2}$ as expected. Result (2.18) was first obtained in [2] using a somehow different language.

The situation in the case of the torus is more difficult since there is no "infinity" anymore where to put the charges. First, it is clear that since the loops-heights correspondence is only local, the presence of non contractible loops leads to shifts of φ similar to (2.9). For a given configuration these read (fig. 3)

$$\begin{cases} \delta\varphi = 2\pi n \sum \epsilon_i \\ \delta'\varphi = 2\pi n' \sum \epsilon_i \end{cases} \tag{2.19}$$

To write (2.19) we have used the property that if two unoriented polygons $\mathcal{P}$ and $\mathcal{P}'$ coexist on the torus, then they are are homotopic. Thus there is one specie of polygons $\mathcal{P}$ *only* at a given time defining two basic shifts $(2\pi n, 2\pi n')$ and $(\delta\varphi, \delta'\varphi)$ is obtained by summing over all the polygons with their orientation $\epsilon_i = \pm 1$. Now the crucial observation is that n and n' are indeed prime together [20]. Thus $\sum \epsilon_i = \pm \frac{|\delta\varphi|}{2\pi} \wedge \frac{|\delta'\varphi|}{2\pi}$ where $\wedge$ denotes the greatest common divisor, and

$$\cos\left[2\pi e_0 \sum \epsilon_i\right] = \cos\left[2\pi e_0 \frac{|\delta\varphi|}{2\pi} \wedge \frac{|\delta'\varphi|}{2\pi}\right] \tag{2.20}$$

Then, exactly as in the case of the strip, summing over all ϵ_i gives to each of these non contractible loop the weight $n = 2\cos 2\pi e_0$. One gets finally [20]

$$\mathfrak{Z}_n \longrightarrow \sum_{m,m' \in \mathbb{Z}} Z_{mm'}(g) \cos[2\pi e_0 \; m \wedge m'] \tag{2.21}$$

an expression which is clearly modular invariant since $m \wedge m'$ is. (2.21) is interesting for any n, but it presents some pathologies for $n \notin \mathbb{N}$. In the case of the Ising model, the sum (2.21) can be performed decomposing it on classes of congruence of mm' *mod* 6, and one gets in the end

$$\mathfrak{Z}_{n=1} \longrightarrow \frac{1}{2}\left[Z_c(12) - Z_c\left(\frac{4}{3}\right)\right] \tag{2.22}$$

which can be shown to be equal to (1.16). A completely similar calculation reproduces as well the Q-state Potts model partition functions, both in the critical and tricritical regime.

This shows clearly how the Coulomb gas technique can provide a link between lattice models and their critical properties such as the partition functions. Of course, it has the drawback of being much less general than conformal invariance, since a new mapping has to be established for every new model under consideration before its properties can be calculated. The two approaches must however be more deeply related. Indeed Dotsenko and Fateev [6] have shown for instance that the construction of four point functions in a free field supplemented by a charge at infinity and screening operators led naturally to the kac formula for the dimensions of the fields, and to the Feigen and Fuchs integral representation for the solutions of the BPZ equations. It is also quite puzzling to notice that *all* minimal partition functions can be written [20] as

$$Z(ADE, A_{p-1}) = \sum_n \sum_{mm' \in \mathbb{Z}} Z_{mm'}\left(g = \frac{p}{p'}\right) \cos\left[\frac{2\pi}{p'} n \; m \wedge m'\right] \tag{2.23}$$

where n belongs to the exponent of the associated ADE algebra (2.23). (2.23) is a generalization of (2.21) involving several charges.

3. Correlation Functions on a Torus

To complete the understanding of critical theories on a torus one would like to calculate also the correlation functions. These have first an interest on their own, since their knowledge provides a systematic (perturbative) route to the study of deviation from criticality [23]. Also, correlation functions on a torus can be shown to be related to partition functions on higher genus surfaces after a proper "pinching" [14].

In principle, the correlators on the torus can be calculated using the structure constants of the theory and the logarithmic mapping which give the necessary transfer matrix elements. The general formula however seem quite untractable. Even the two point function which has the simplest form in the plane

$$\left\langle A(z,\bar{z})\ A(0,0)\right\rangle = \frac{1}{z^{2h}\bar{z}^{2\bar{h}}} \tag{3.1}$$

must have a rather complicated expression on the torus. If the short distance expansion reads

$$A(z,\bar{z})\ A(0,0) = \sum_{\Delta,\bar{\Delta}} z^{\Delta-2h}\ \bar{z}^{\bar{\Delta}-2\bar{h}}\ C_{AAB}\ B_{\Delta\bar{\Delta}}(0,0) \tag{3.2}$$

one must have now

$$\left\langle A(z,\bar{z})\ A(0,0)\right\rangle = \sum_{\Delta,\bar{\Delta}} z^{\Delta-2h}\ \bar{z}^{\bar{\Delta}-2\bar{h}}\ C_{AAB}\ \left\langle B_{\Delta\bar{\Delta}}\right\rangle \tag{3.3}$$

all the mean values $\langle B\rangle$ being a priori non zero.

One would like to find a more compact method of calculation, generalizing for instance the approaches of [6], but it seems difficult to implement the notion of charge at infinity and screening operators here. One can write nevertheless the Ward identity [15]

$$\begin{aligned}
&\langle T(z)\ A_1(1)\ldots A_n(n)\rangle - \langle T\rangle\langle A_1(1) - A_n(n)\rangle \\
&= \sum_{i=1}^{n}\left[-\ h_i\left(\frac{\theta_1'}{\theta_1}\right)'(z-z_i) + \frac{\theta_1'}{\theta_1}(z-z_i)\ \partial_{z_i}\right] \\
&\qquad \langle A_1(1)\ldots A_n(n)\rangle \\
&\quad + 2i\pi\ \partial_\tau\ \langle A_1(1)\ldots A_n(n)\rangle
\end{aligned} \tag{3.4}$$

where

$$\langle T \rangle = 2i\pi\, \partial_\tau\, Log\ Z \tag{3.5}$$

and θ_1 is the Jacobi theta function (see Appendix A), generalizing the expression in the plane [4] to the doubly periodic case, with the appearance of a new term associated to the variation of modular ratio in the change of coordinates. For degenerate fields [4], differential equations follow. In the simplest case of degenerescence at level 2 one has

$$\left\{ \frac{3}{2(2h+1)}\, \partial_z^2 - 2h\, \eta_1 - \sum_{i=1}^{n} \left[-h_i \left(\frac{\theta_1'}{\theta_1} \right)' (z-z_i) + \frac{\theta_1'}{\theta_1} (z-z_i)\, \partial_{z_i} \right] - 2i\pi\, \partial\tau \right\} Z \left\langle A(z,\bar{z})\ A_1(1) \ldots A_n(n) \right\rangle = 0 \tag{3.6}$$

where

$$\eta_1 = -\frac{1}{6}\, \frac{\theta_1'''(0)}{\theta_1'(0)}$$

These equations seem unfortunately quite difficult to solve in general.

Up to now, progress has been made for the free bosonic theories like the Ashkin Teller model [26], and for the Ising model which is equivalent to a free Majarana fermion theory [27] with action

$$\mathcal{A} = \frac{1}{2\pi} \int (\Psi\, \bar{\partial}\, \Psi + \tilde{\Psi}\, \partial\, \tilde{\Psi})\ d^2x \tag{3.7}$$

We shall now discuss this case in some detail.

As a consequence of the Jordan-Wigner transformation [10] the fermion $(\Psi,\tilde{\Psi})$ must be assigned periodic (P) or antiperiodic (AP) boundary conditions along the periods ω_1 and ω_2 of the torus. This gives rise to four sectors labelled by $\nu = 1,2,3,4$ for PP, PA, AA, AP. The partition function has been evaluated by dzeta-regularization [9] and reads:

$$Z = \sum_{\nu=2}^{4} \frac{|\theta_\nu(0|\tau)|}{2|\eta(\tau)|} = \sum_\nu{}' \; Z_\nu \tag{3.8}$$

where θ_ν are Jacobi theta functions (see Appendix A), $\eta = \left(\frac{\theta_1'(0)}{2\pi}\right)^{1/3}$ is Dedekind's function. The $\nu = 1$ contribution vanishes in (3.8), due to zero mode of the Laplacian with doubly periodic boundary conditions. The expression (3.8) reproduces (1.16).

One can easily write the fermion propagator $\langle \Psi(z)\ \Psi(w) \rangle_\nu$ in each sector ν. Using the short distance expansion:

$$\Psi(z)\ \Psi(w) = \frac{1}{z-w} + 0(1) \tag{3.9}$$

and the quasi-periodicity properties of the theta functions (see Appendix A) we conclude that for $\nu \neq 1$: $\langle \Psi(z)\ \Psi(w) \rangle_\nu \times \frac{\theta_1(z-w)}{\theta_\nu(z-w)}$ is doubly periodic and has *at most* one pole, the unique zero of θ_ν on the torus (see Appendix A). By a well-known theorem on elliptic functions, it must be a constant, fixed by the short distance limit.

$$\langle \Psi(z)\ \Psi(w) \rangle_\nu = \frac{\theta_\nu(z-w)}{\theta_\nu(0)} \cdot \frac{\theta_1'(0)}{\theta_1(z-w)}, \qquad \nu \neq 1 \tag{3.10}$$

so the fermion propagator reads:

$$\langle \Psi(z)\ \Psi(w) \rangle = \frac{\sum_\nu{}' \; Z_\nu \langle \Psi(z)\ \Psi(w) \rangle_\nu}{\sum_\nu{}' \; Z_\nu} \tag{3.11}$$

Using (3.10) we deduce the two point energy correlation

$$\left\langle \varepsilon(z,\bar{z})\ \varepsilon(0)\right\rangle = \frac{\sum_{\nu}' |\theta_{\nu}(z)|^2/|\theta_{\nu}(0)|}{\sum_{\nu}' |\theta_{\nu}(0)|} \times \left|\frac{\theta_1'(0)}{\theta_1(z)}\right|^2 \tag{3.12}$$

It should be noted that in (3.11), (3.12) there is no contribution of the $\nu = 1$ sector. This can be justified [27] but is not obvious a priori, even if Z_1 itself vanishes. Indeed one can show for instance that $\langle \varepsilon \rangle \neq 0$, and that *only* the $\nu = 1$ sector contributes in this case [13]

$$\langle \varepsilon \rangle = \frac{Z_1 \langle \varepsilon \rangle_1}{Z} = \frac{2\pi|\eta|^3}{\sum_{\nu}' |\theta_{\nu}(0)|} \tag{3.13}$$

2n points correlation functions can be evaluated easily using (3.11) and Wick's theorem: they are simply modulus squares of Pfaffians of the propagator in each sector [27]. The case of 2n+1 points is more difficult because of the role of the $\nu = 1$ sector.

The calculation of spin correlators is different since due to the Jordan Wigner transformation, the spin σ is not local in terms of Ψ and $\tilde{\Psi}$ and one has

$$\begin{aligned} \Psi(z)\ \sigma(w,\bar{w}) &\sim \frac{1}{(z-w)^{1/2}}\ \mu(w,\bar{w}) \\ \tilde{\Psi}(\bar{z})\ \sigma(w,\bar{w}) &\sim \frac{1}{(\bar{z}-\bar{w})^{1/2}}\ \mu(w,\bar{w}) \end{aligned} \tag{3.14}$$

where μ is the disorder operator dual to the original spin σ. The non-integer power $\frac{1}{2}$ expresses the monodromy property [4] of insertions of spins in Ψ or $\tilde{\Psi}$ correlators, that makes them change sign when the σ -argument describes a loop around Ψ's. Expressions (3.9) and (3.14) and periodicity conditions can be used to determine completely the auxiliairy function

$$G_\nu = \frac{\left\langle \Psi(z)\ \Psi(w)\ \sigma\left(z_1, \bar{z}_1\right)\ \sigma\left(z_2, \bar{z}_2\right)\right\rangle_\nu}{\left\langle \sigma\left(z_1, \bar{z}_1\right)\ \sigma\left(z_2, \bar{z}_2\right)\right\rangle_\nu} \tag{3.15}$$

Using again properties of θ -functions, we can exhibit the following candidate for G_ν:

$$H_\nu = \frac{1}{2}\left[\frac{\theta_1'(0)}{\theta_1(z-w)}\right]^2 \times \left[\left(\frac{\theta_1(z-z_1)\theta(w-z_2)}{\theta_1(z-z_2)\theta(w-z_1)}\right)^{1/2} \frac{\theta_\nu\left(z-w+\frac{z_1-z_2}{2}\right)}{\theta_\nu\left(\frac{z_1-z_2}{2}\right)} + (z\leftrightarrow w)\right] \tag{3.16}$$

Then, for $\alpha_\nu(z) = \dfrac{\theta_\nu\left(z-\frac{z_1+z_2}{2}\right)}{[\theta_1(z-z_1)\ \theta_1(z-z_2)]^{1/2}}$ the function $(G_\nu - H_\nu)/\alpha_\nu(z)\ \alpha_\nu(w)$ is again elliptic in z and w and has at most one pole in z and w, so it is a constant, vanishing by antisymmetry in the exchange of z and w; thus $G_\nu = H_\nu$.

From the expression relating the stress tensor T(z) and the fermion:

$$T(z) = \lim_{z\to w} -\frac{1}{2}\left[\Psi(z)\ \partial_w\ \Psi(w) - \partial_z\ \Psi(z)\ \Psi(\omega)\right] + \frac{1}{(z-w)^2} \tag{3.17}$$

we get now the insertion of T in the spin correlator, generating at short distances first the conformal dimension $h = 1/16$ of operator σ and second a derivative of $\langle\sigma\sigma\rangle$, for which we have an explicit value:

$$\frac{\left\langle T(z)\sigma\left(z_1, \bar{z}_1\right)\sigma\left(z_2, \bar{z}_2\right)\right\rangle_\nu}{\left\langle \sigma\left(z_1, \bar{z}_1\right)\sigma\left(z_2, \bar{z}_2\right)\right\rangle_\nu} = \frac{1}{(z-z_1)^2}\, h_\sigma + \frac{1}{(z-z_1)}\, \partial_{z_1} \mathit{Log}\left\langle \sigma\left(z_1, \bar{z}_1\right)\sigma\left(z_2, \bar{z}_2\right)\right\rangle_\nu + \mathit{reg\ terms} \tag{3.18}$$

We get a similar differential equation for the antianalytic part of $\langle\sigma\sigma\rangle_\nu$ by replacing Ψ with $\tilde{\Psi}$ in (3.15). (3.18) can be integrated to give

$$Z_\nu\left\langle \sigma\left(z_1,\bar z_1\right)\ \sigma\left(z_2,\bar z_2\right)\right\rangle_\nu = \frac{(2\pi)^{1/3}}{2}\ \left|\theta_1'(0)\right|^{-\frac{1}{2}}\ \frac{\left|\theta_\nu\left(\frac{z_1-z_2}{2}\right)\right|}{|\theta_1(z_1-z_2)|^{1/4}} \tag{3.19}$$

Periodicity conditions impose to sum (3.9) over the four sectors, and finally

$$\left\langle \sigma(z,\bar z)\sigma(0,0)\right\rangle = \frac{\sum_{\nu=1}^{4} Z_\nu\left\langle\sigma(z,\bar z)\sigma(0,0)\right\rangle_\nu}{\sum'_\nu Z_\nu} = \frac{\sum_{\nu=1}^{4} |\theta_\nu(z/2)|}{\sum'_\nu |\theta_\nu(0)|} \times \left|\frac{\theta_1'(0)}{\theta_1(z)}\right|^{1/4} \tag{3.20}$$

Physically the insertion of σ's may be viewed as creating a branch cut joining z and 0 along which Ψ has to be antiperiodic. A translation $z \rightarrow z+1$ or $z \rightarrow z+\tau$ lets the cut wrap around the torus and changes the boundary conditions for the fermion, exchanging the contributions of different ν sectors, including $\nu = 1$ (fig.4). (3.12) as well as (3.20) satisfy modular covariance in the following sense:

$$\begin{aligned}\langle A(z)\ A(w)\rangle\ (\tau+1) &= \langle A(z)\ A(w)\rangle\ (\tau)\\ \langle A(z/\tau)\ A(w/\tau)\rangle\ (-1/\tau) &= z^{2h}\ \bar z^{2\bar h}\ \langle A(z)\ A(w)\rangle(\tau)\end{aligned} \tag{3.21}$$

One checks the short distance expansion of (3.20) against (3.3) and (3.13)

$$\left\langle \sigma(z,\bar z)\ \sigma(0,0)\right\rangle = \frac{1}{|z|^{1/4}} + C_{\sigma\sigma\varepsilon}\ |z|^{3/4}\ \langle\varepsilon\rangle + \ldots \tag{3.22}$$

where [4] $C_{\sigma\sigma\varepsilon} = 1/2$. Finally (3.12), (3.20) satisfy the expected differential equation (3.6).

The preceding approach was used also in the context of string theory on orbifolds [28]. For higher spin correlators it turns out to be more convenient to use a bosonization technique where one computes squares $\langle\sigma\ \ldots\sigma\rangle_\nu^2$ of correlation functions in each fermionic sector via a free bosonic model [27]. The general result is

$$Z_\nu^2\langle \sigma(1)\ldots\sigma(2n)\rangle = \frac{1}{|\eta|^2} \sum_{\varepsilon_i=\pm1,\sum\varepsilon_i=0} \left|\theta_\nu\left(\frac{\sum\varepsilon_i z_i}{2}\right)\right|^2 \prod_{i<j} \left|\frac{\theta_1(z_i-z_j)}{\theta_1'(0)}\right|^{\frac{\varepsilon_i\varepsilon_j}{2}} \tag{3.23}$$

(3.23) can be checked numerically for n = 2,4. Indeed in [29] Burkhardt and Derrida have considered a lattice Ising model on squares N×N and calculated using a transfer matrix technique the first moments of the magnetization $M_{2n} = \left\langle \left(\sum \sigma_i\right)^{2n} \right\rangle$. They have obtained in particular the renormalized coupling constant $V(N) = M_4/M_2^2$ for $N \leqslant 14$. In the limit $N \longrightarrow \infty$ the values converge to a universal constant evaluated as $V = 1.1670 \pm 0.0015$. Now one can give an expression of M_2, M_4 using (3.23) since $M_{2n} = \int_{square} d^2x_1 \ldots d^2x_{2n} \langle \sigma(1)\ldots\sigma(2n)\rangle$. The integrals cannot be performed analytically but estimated numerically, giving $V = 1.168 \pm 0.005$ in agreement with the above value.

Naturally these calculations generalize to mixed correlators of the most general type involving σ, μ, ε [27].

Acknowledgements

We thank C. Itzykson and J.B. Zuber who have teached us most of part 1, and collaborated with us for works described in parts 2 and 3.

APPENDIX A

Jacobi theta functions

ν =	1	2	3	4
(α,β) =	(1/2,1/2)	(1/2,0)	(0,0)	(0,1/2)

$$q = e^{2i\pi\tau}$$

$$\theta_\nu(z|\tau) = \sum_{n\in z} q^{\frac{1}{2}(n+\alpha)^2} e^{2i\pi(n+\alpha)(z+\beta)}$$

Quasi-periodicity relations

	θ_1	θ_2	θ_3	θ_4
$z \longrightarrow z+1$	$-\theta_1$	$-\theta_2$	θ_3	θ_4
$z \longrightarrow z+\tau$	$-A\theta_1$	$A\theta_2$	$A\theta_3$	$-A\theta_4$

$$A(z|\tau) = e^{-2i\pi\left(z+\frac{\tau}{2}\right)}$$

Zeroes

	θ_1	θ_2	θ_3	θ_4
z	$m+n\tau$	$m+\frac{1}{2}+n\tau$	$m+\frac{1}{2}+\left(n+\frac{1}{2}\right)\tau$	$m+\left(n+\frac{1}{2}\right)\tau$

Modular transformations

	θ_1	θ_2	θ_3	θ_4
$\tau \longrightarrow \tau+1$	$e^{i\frac{\pi}{4}}\theta_1$	$e^{i\frac{\pi}{4}}\theta_2$	θ_4	θ_3
$\tau \longrightarrow -\frac{1}{\tau};\ z \longrightarrow \frac{z}{\tau}$	$-iB\theta_1$	$B\theta_4$	$B\theta_3$	$B\theta_2$

$$B(z|\tau) = \sqrt{-i\tau}\ e^{\frac{i\pi z^2}{\tau}}$$

REFERENCES

[1] M.N. Barber in Domb & Lebowitz, Vol.8, Ac. Press London (1983).

[2] J.L. Cardy, J. phys. A16 (1984) L36.

H.W. Blöte, J.L. Cardy, M.P. Nightingale, Phys. Rev. Lett. 56 (1986) 742.

I. Affleck, Phys. Rev. Lett. 56 (1986) 746.

[3] J.L. Cardy, Nucl. Phys. B270 (1986) 186.

[4] A. Belavin, A. Polyakov, A. Zamolodchikov, Nucl. Phys. B241 (1984) 333.

[5] V.G. Kac, Lecture Notes in Physics 94 (1979) 441.

[6] V.S. Dotsenko, V.A. Fateev, Nucl. Phys. B240 (1984) 312, Nucl. Phys. B251 (1985) 691.

[7] B.L. Feigen, D.B. Fuchs, Functs. Anal. Prilozhen 17 (1983) 91.

[8] D. Friedan, Z. Qiu, S. Shenker, Phys. Rev. Lett. 52 (1984) 1575.

[9] C. Itzykson, J.B. Zuber, Nucl. Phys. B275 (1986) 580.

[10] B.M. Mc Coy, T.T. Wu, The two dimensional Ising Model, Oxford Univ. Press, 1973.

[11] A. Cappelli, C. Itzykson, J.B. Zuber, Nucl. Phys. B280 (1986) 445.

[12] V. Pasquier, Nucl. Phys. B285 (1987) 162.

[13] C. Itzykson, H. Saleur, J.B. Zuber, Europhys. Lett. 2 (1986) 91.

[14] P. di Francesco, H. Saleur, J.B. Zuber, Nucl. Phys. B285 (1987) 454.

[15] J.L. Cardy, Phys. Rev. Lett. 54 (1986) 1354.

[16] L.P. Kadanoff, J. Phys. A11 (1978) 1399; A. Luther, I. Peschel, Phys. Rev. B12 (1975) 3908.

H.J.F. Knops, Ann. Phys. 128 (1981) 448.

[17] R.J. Baxter, in Exactly Solved Models in Statistical Physics (Academic Press, 1982).

[18] J.V. José, L.P. Kadanoff, S. Kirkpatrick, D.R. Nelson, Phys. Rev. B16 (1977) 12.

[29] B. Nienhuis, J. Stat. Phys. 34 (1984) 731; MPM Den Nijs, Phys. Rev. B27 (1983) 1674; J. Phys. A17 (1984) L295.

[20] P. Di Francesco, H. Saleur and J.B. Zuber, J. Stat. Phys. 49 (1987) 57.

[21] J.M. Luck, J. Phys. A15 (1982) L169.

[22] S.K. Yang, Nucl. Phys. B285 (1987) 183.

H. Saleur, J. Phys. A20 (1987) L1127.

[23] H. Saleur, C. Itzykson, J. Stat. Phys. 48 (1987) 449.

[24] K. Miki, Phys. Lett. 191B (1987) 127

D. Bernard, Meudon Preprint 1987

R. Dijkgraaf, E. Verlinde, H. Verlinde, Preprint Utrecht THU 87/17.

[25] T. Eguchi and H. Ooguri, Nucl. Phys. B282 (1987) 308.

[26] H. Saleur, Preprint Saclay SPhT 87/134.

[27] P. Di Francesco, H. Saleur, J.B. Zuber, Nucl. Phys. B290 [FS20] (1987) 527.

[28] J.T. Attick and A. Sen, Nucl. Phys. B286 (1987) 189
L. Dixon, D. Friedan, E. Martinec and S. Shenker, Nucl. Phys. B282 (1987) 13.

[29] T.W. Burkhardt, B. Derrida, Phys. Rev. B32 (1985) 727.

TABLE I

List of known partition functions in terms of conformal characters. The unitary series corresponds to $p' = m+1$, $p = m$ or $p = m+1$, $p' = m$, $m = 3,4,\ldots$.

$$\frac{1}{2}\sum_{r=1}^{p'-1}\sum_{s=1}^{p-1} |\chi_{rs}|^2 \qquad (A_{p'-1}, A_{p-1})$$

$$p' = 4\rho+2,\ \rho\geq 1 \qquad \frac{1}{2}\sum_{s=1}^{p-1}\left\{\sum_{\substack{r\ \mathrm{odd}=1\\ r\neq 2\rho+1}}^{4\rho+1} |\chi_{rs}|^2 + 2|\chi_{2\rho+1\,s}|^2 + \sum_{r\ \mathrm{odd}=1}^{2\rho-1} (\chi_{rs}\chi^*_{p'-r\,s} + \mathrm{c.c.})\right\} \qquad (D_{2\rho+2}, A_{p-1})$$

$$p' = 4\rho,\ \rho\geq 2 \qquad \frac{1}{2}\sum_{s=1}^{p-1}\left\{\sum_{r\ \mathrm{odd}=1}^{4\rho-1} |\chi_{rs}|^2 + |\chi_{2\rho\,s}|^2 + \sum_{r\ \mathrm{even}=1}^{2\rho-2} (\chi_{rs}\chi^*_{p'-r\,s} + \mathrm{c.c.})\right\} \qquad (D_{2\rho+1}, A_{p-1})$$

$$p' = 12 \qquad \frac{1}{2}\sum_{s=1}^{p-1}\left\{|\chi_{1s}+\chi_{7s}|^2 + |\chi_{4s}+\chi_{8s}|^2 + |\chi_{5s}+\chi_{11s}|^2\right\} \qquad (E_6, A_{p-1})$$

$$p' = 18 \qquad \frac{1}{2}\sum_{s=1}^{p-1}\left\{|\chi_{1s}+\chi_{17s}|^2 + |\chi_{5s}+\chi_{13s}|^2 + |\chi_{7s}+\chi_{11s}|^2 + |\chi_{9s}|^2 + [(\chi_{3s}+\chi_{15s})\chi^*_{9s} + \mathrm{c.c.}]\right\} \qquad (E_7, A_{p-1})$$

$$p' = 30 \qquad \frac{1}{2}\sum_{s=1}^{p-1}\left\{|\chi_{1s}+\chi_{11s}+\chi_{19s}+\chi_{29s}|^2 + |\chi_{7s}+\chi_{13s}+\chi_{17s}+\chi_{23s}|^2\right\} \qquad (E_8, A_{p-1})$$

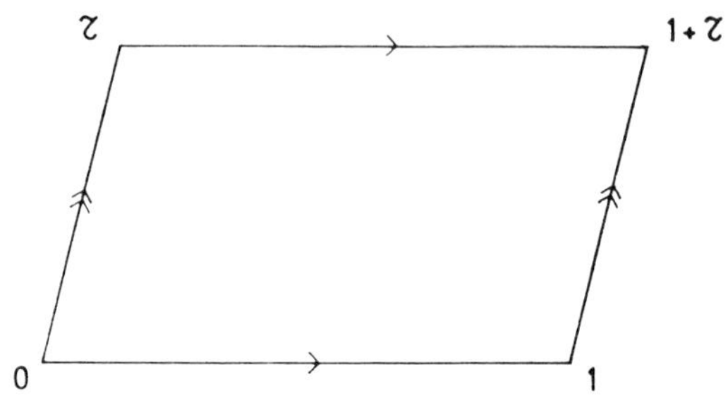

Figure 1: In the continuum limit the torus will be described by its modular ratio $\tau = \frac{\omega_2}{\omega_1}$. Without loss of generality ω_1 can be taken equal to 1.

Name of the algebra	Diagram	Coexeter number	Exponent
A_n		$n+1$	1, 2, n
D_n	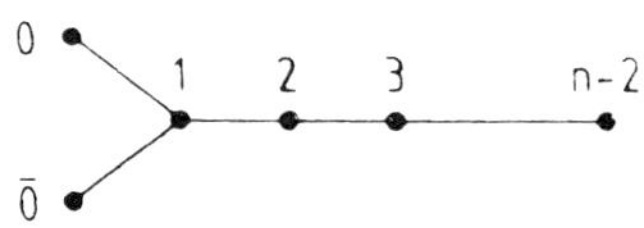	$2(n-1)$	1, 3, , 2n-3 n-1
E_6	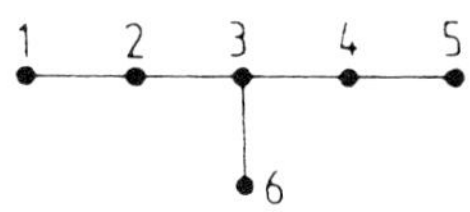	12	1, 4, 5, 7, 8, 11
E_7	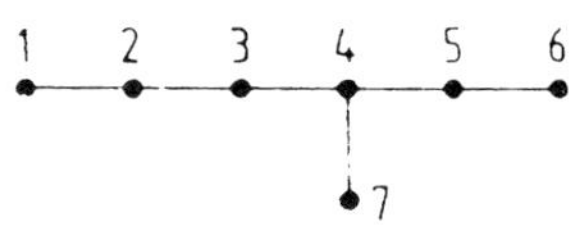	18	1, 5, 7, 9, 11, 13, 17
E_8	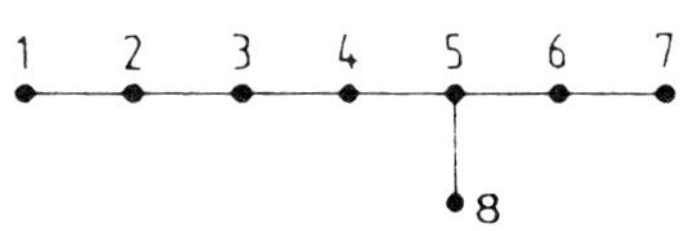	30	1, 7, 11, 13, 17, 19, 23, 29

Figure 2: The unitary models of the ADE classification can be interpreted as RSOS models the heights of which take values on the associated Dynkin diagram.

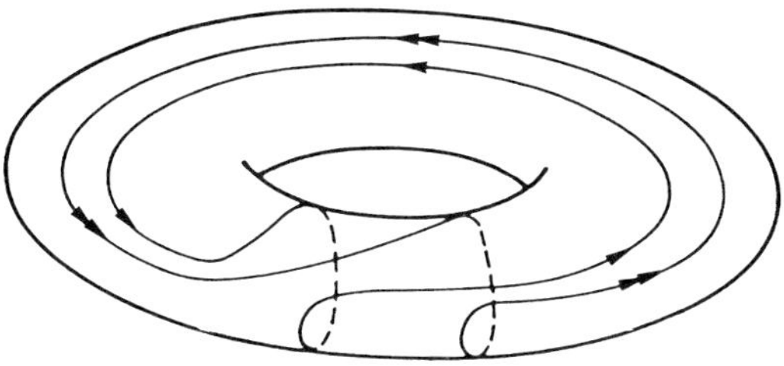

Figure 3: A typical configuration with two polygons ont he torus $\varepsilon_1 = \varepsilon_2 = 1$; $n_1 = n_2 = 1$.

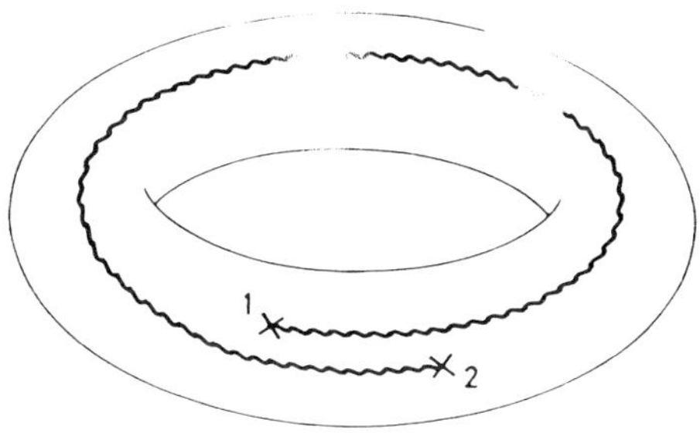

Figure 4: Translating the argument in the $\sigma\sigma$ correlation maps a frustration line around the torus and changes thus the fermionic sector.

WESS-ZUMINO-WITTEN MODELS AND OTHER TWO-DIMENSIONAL CRITICAL SYSTEMS

R. Flume

1. Introduction

σ - models on group manifolds with a "Wess-Zumino term" have been introduced by Witten [1]. We call them Wess-Zumino-Witten (WZW) models. These theories have isolated conformally invariant fixed points in their two-dimensional parameter space of coupling constants. The two simplest fixed point theories are in the cases of the group manifolds O(n) and U(n) equivalent to the free field theories of n massless Majorana and Dirac fermions, respectively. The equivalence was termed by Witten "non-abelian bosonization". The conformal fixed points are labeled by a positive integer k being the value of the central charge of the underlying Kac-Moody algebras. The theories with k=1 on the groups O(n) and U(n) render the above mentioned bosonic representation of free fermion systems. Choosing other groups than O(n) or U(n) or going in case of the latter groups to fixed points with $k > 1$ one will in general encounter systems which are not equivalent to free field theories. They may nevertheless be explicitly solvable. That is, one has a realistic chance to find concrete representations of all correlation functions in these models (at least on all compact classical groups). The key to the solution of WZW models was supplied through the work of Knizhnik and Zamolodchikov [2]. Those authors (generalizing and extending earlier work of Dashen and Frishman [3]) observed that the Sugawara form - current times current - of the energy momentum tensor together with two-dimensional conformal invariance implies a system of

ISBN 0-12-218100-X

linear matrix differential equations for the correlation functions of primary operators. The derivation of the Knizhnik-Zamolodchikov differential equations (KZDE) will be given in two different ways in the following section II. I first reproduce the argument due to Knizhnik and Zamolodchikov which is based on current algebra Ward identities and operator product expansions. I will then show, following ref. [5], that the KZDE can be obtained through partial formal summation of perturbation theory on the basis of the WZW action.

Section III is devoted to a discussion of solutions of the KZDE for the four point correlation functions of the SU(2) WZW model and implications for the operator product algebra in that model. The solutions are taken from references [4] and [5]. The material related to the operator product algebra is, what concerns the implementation of the conformal bootstrap program, taken from the articles [6] by Dotsenko and Fateev and some unpublished work of my own and, what concerns the closure of the operator algebra, taken from ref. [7].

It can by now be forseen that the conformally invariant WZW models will appear at a central position in a general classification scheme of all critical theories with a finite number of primary operators (more precisely operators being primary with respect to some graded infinite dimensional symmetry algebra which may be Virasoro, super Virasoro, Kac-Moody... or some other algebra yet unidentified).

I will collect in section IV some of the known relations of WZW models (in particular the SU(2) model) to other critical theories. These relations bolster, as I believe, the conjecture about the central role of the WZW models.

II. Wess-Zumino-Witten Models

IIa. Current algebra and the Knizhnik-Zamolodchikov differential equations

I postpone the discussion of the action of the WZW model and choose instead - with Knizhnik and Zamolodchikov - as constitutive starting point the assumption of an irreducible set of symmetry currents giving rise to two Kac-Moody algebras. The approach we are following may be considered as the non-abelian generalization of the classical work of dell'Antonio, Frishman and Zwanziger [8] on the U(1) Thirring model (see also the lectures of Todorov at this school).

We will stay throughout these lectures in the framework of Euclidean quantum field theory. A complex variable z (with complex conjugate $\bar{z}$) will be used as coordinate of the Euclidean plane. Let $\{j^a\}$ denote a set of currents related to the internal symmetry group G of some conformally invariant field theory. I suppose in the following that G is one of the compact simple groups. The currents j^a, a = 1, ... d_g, (d_g = dimension of G) have to be local operators of scaling dimension one (the canonical dimension of a bosonic symmetry current in two dimensions) and of Lorentz spin ± 1. The conservation equations read in complex notation either

$$\frac{\partial}{\partial \bar{z}} \; j^a(z) \;=\; 0 \tag{2.1a}$$

or

$$\frac{\partial}{\partial z} \, j^a(\bar{z}) \;=\; 0 \tag{2.1b}$$

where Eq. (2.1a), ((2.1b)) holds for Lorentz spin +1 ((-1)).

Exercise 1: Show that the (anti)holomorphic form of the conservation laws are a consequence of the canonical dimension and Lorentz spin of the currents.

We assume that parity is conserved in the theory under consideration. There are then as many holomorphic (Eq. (2.1a)) as antiholomorphic (Eq. (2.1b)) current operators. The charges of both sets of currents, which have to commute with each other by locality, generate isomorphic groups. This way, we have in fact to deal with a tensor product G x G of symmetries where one factor is related to the holomorphic and the other to the antiholomorphic dependence of the field theory. Let j_L^a and j_R^a denote the holomorphic and antiholomorphic current components.

We shall concentrate on the j_L^a. All deductions will equally hold for the j_R^a.

Let us assume that the ground state of our theory is invariant under conformal (that is, more precisely, projective) transformations and also under the internal symmetry transformations related to the currents j_L^a. (We will see later on that a dynamical breaking of G is in general incompatible with conformal invariance). The form of the current two point function

$$<j_L^a(z_1)j_L^b(z_2)> = -k\delta^{ab}(z_1 - z_2)^{-2} \tag{2.2}$$

is a consequence of the invariances. That the second inverse power appears is dictated by projective invariance and the canonical dimension one of the currents. The tensorial structure δ^{ab} is enforced by G-invariance. (The upper labels a, b of the currents are supposed to refer to a basis of the adjoint representation of G in which the group structure constants are completely antisymmetric). Field theoretic positivity enforces the constant k to be positive. (The currents j^a are taken to be antihermitian). The form of the n-point current correlations is found through the recursive relations

$$<j_L^a(z)j_L^{a_1}(z_1)\ldots j_L^{a_n}(z_n)> = \sum_{j,c} f^{aa_jc}(z-z_j)^{-1}<j_L^{a_1}\ldots j_L^c(z_j).>$$
$$- \sum_j k\delta^{aa_j}(z-z_j)^{-2} <j_L^{a_1} \ldots \hat{j}_L^{a_j} \ldots j_L^{a_n}> \tag{2.3}$$

where f^{abc} denote completely antisymmetric structure constants of G. k is the same constant as appearing on the r.h.s. of Eq. (2.2). The hat over $j_L^{a_j}$ shall mean deletion of that operator.

Eq. (2.3) has to be considered as solution of the current Ward identity of $j^a(z)$ in the field theoretic sector generated by the currents. The form of the r.h.s. of (2.3) is easily seen to be fixed through the following conditions:

i) $<j_L^a(z)\ j_L^{a_1}(z_1)\ \ldots\ j_L^{a_n}(z_n)> \equiv F^{aa_1\ldots a_n}(z;z_1, \ldots, z_n)$ is holomorphic in z for $z \neq z_1, \ldots, z_n$ and meromorphic at points of coincidence $z = z_i$, $i = 1, \ldots, n$.

ii) $F^{aa_1\ldots a_n}$ decreases for $|z| \to \infty$ and fixed $z_1,\ldots,z_n$ like z^{-2} since j^a is of scaling dimension one.

iii) The singular contributions to the short distance expansion of two currents is given by

$$j^a(z)j^b(0) = -k\delta^{ab}z^{-2} + f^{abc}j^c(0)z^{-1} + O(1) \qquad (2.4)$$

The most singular term $\sim z^{-2}$ represents the coupling of the unit operator. It is consistent with the current two point function (2.2). The next to leading singular term $\sim z^{-1}$ reflects the fact that the currents rotate each other in the adjoint representation of G. (The group generators are obtained from the currents through integration of the latter along closed contours). The r,h.s. of Eq. (2.3) has just to reproduce the short distance singularities prescribed by the operator product expansion (2.4) and no others.

The current commutation relations, which are implied through Eq. (2.3), are conveniently quoted in terms of a Laurent decomposition with components

$$j_n^a = (2\pi i)^{-1} \oint d\xi j_L^a(\xi)\ \xi^n$$

where $\oint$ denotes an arbitrary simply closed finite contour. With the stipulation that in products $j_n^a\ j_m^b$ the closed contour of the operator to the right is comprised by the contour of the operator to the left one finds with the help of Eq. (2.3) the Kac-Moody algebra relations

$$[j_n^a, j_m^b] = f^{abc}j_{n+m}^c - \frac{1}{2}\ k\delta^{ab}\ n\delta_{n+m,o} \qquad (2.5)$$

Let us now turn to the construction of the energy momentum tensor in terms of currents. The symmetric and traceless energy momentum tensor of a two-dimensional conformally invariant field theory has two independent components which we choose as the z-z component T_{zz} and $\bar{z}$-$\bar{z}$ component $T_{\bar{z}\bar{z}}$. We shall concentrate on T_{zz} in the following. It should be an operator of scaling dimension two carrying no charge of the internal symmetry group G and satisfying the conservation equation

$\partial_z T_{zz} = 0$. Trying for T_{zz} a local expression in currents one is uniquely lead to

$$T_{zz} \sim \sum_a :j_L^a j_L^a: \tag{2.6}$$

fulfilling therewith all the conditions quoted above. The dots : : denote here normal ordering with respect to current frequencies. We impose the solution of the projective Ward identities

$$\langle T_{zz}(z) j_L^{a_1}(z_1) \ldots j_L^{a_n}(z_n)\rangle = \sum_i ((z-z_i)^{-1} \partial_{z_i} + (z-z_i)^{-2}) \langle j_L^{a_1} \ldots j_L^{a_n}\rangle \tag{2.7}$$

to find the correct normalization in (2.6). (Integrating Eq. (2.7) in z along a contour surrounding the points $z_1, \ldots, z_n$ with weight 1, z and z^2 resp., (producing therewith a vanishing result) one arrives at the Ward identities of translation, dilation and special conformal symmetries for the correlations of the dimension one current operators).

Inserting in Eq. (2.7) the expression (2.6) and evaluating with help of Eq. (2.3) one finds the correct normalization:

$$T_{zz} = (k + c_V)^{-1} \sum_a :j_L^a j_L^a:$$

where c_V denotes the eigenvalue of the quadratic Casimir operator in the adjoint representation.

Exercise 2: Confirm the normalization by evaluating a correlation function of T_{zz} and three current operators.

Decomposing T_{zz} into a Laurent series with components

$$L_n = (2\pi i)^{-1} \oint d\xi \xi^{n+1} T_{zz}(\xi)$$

one derives from the Kac-Moody relations (2.5) the Virasoro algebra relations

$$[L_n, L_m] = (n+m)L_{n+m} + \frac{c(n^3-n)}{12} \delta_{n+m,o} \tag{2.8}$$

where the central Virasoro charge c is given by $c = \dfrac{k\, d_g}{k+c_V}$.

One also obtains from Eq. (2.5) straightforwardly

$$[L_n, j^a_m] = - m\, j^a_{m+n} \tag{2.9}$$

that is, the Kac-Moody and Virasoro algebras form a semidirect product.

A completely analogous structure is obtained by starting from j^a_R and

$$T_{\overline{zz}} = - \frac{1}{k+c_V} \sum_a :j^a_R\, j^a_R: \quad .$$

The Laurent components of j^a_R and $T_{\overline{zz}}$ will be denoted by $\bar{j}^a_n$ and $\bar{L}_n$ respectively. They commute with the above introduced j^a_n and L_n and give rise to a second semidirect product of Kac-Moody and Virasoro algebras isomorphic to those quoted in Eqs. (2.5) and (2.8).

Having constructed the energy-momentum tensor in the current sector we postulate that this is in fact the energy momentum tensor of the entire theory. To complete the preparatory material of this section I mention the concept of a primary field introduced by Belavin, Polyakov and Zamolodchikov [9] in the context of the Virasoro algebra. A local operator φ is called primary relative to the Virasoro and Kac-Moody algebras, respectively, if the relations

$$L_n\varphi = \bar{L}_n\varphi = j^a_n\varphi = \bar{j}^a_n\varphi = 0 \quad , \quad \forall n > 0 \tag{2.10a}$$

$$L_0\varphi = \Delta_\varphi\, \varphi \qquad \bar{L}_0\varphi = \bar{\Delta}_\varphi\, \varphi \tag{2.10b}$$

$$j^a_0\varphi = t^a_\varphi\varphi \quad , \qquad \bar{j}^a_0\varphi = \bar{t}^a_\varphi\, \varphi \tag{2.10c}$$

are fulfilled. We use here standard shorthand notations e.g.

$$L_n\varphi(z) = (2\pi i)^{-1} \oint d\xi(\xi-z)^{n+1} T_{zz}(\xi)\varphi(z)$$

with the contour of the integral surrounding z. Δ_φ and $\bar{\Delta}_\varphi$ in (2.10b) are the scaling dimensions of φ with respect to dilations in z and $\bar{z}$. t^a_φ and $\bar{t}^a_\varphi$ in (2.10c) denote generators of the left- and right-handed symmetry groups G in the representation carried by φ. (φ is supposed to be an operator multiplet). Let $\varphi_1,\ldots,\varphi_n$ denote primary operators. The solution of the current algebra and projective Ward identities of correlation functions of the φ_i are given by

$$<j_L^a(z)\varphi_1(z_1),\ldots,\varphi_n(z_n)> = \sum_i t_i^a(z-z_i)^{-1}<\varphi_1 \ldots\varphi_n> \quad (2.11)$$

$$<T_{zz}(z)\varphi_1(z_1) \ldots \varphi_n(z_n)> = \sum_i ((z-z_i)^{-1} \partial_{z_i} + \Delta_i(z-z_i)^{-2})<\varphi_1 \ldots \varphi_n> \quad (2.12)$$

The derivation of these equations proceeds along the same lines as the demonstration of Eq. (2.5) given above. Analogous relations hold for correlations involving j_R^a or $T_{\bar{z}\bar{z}}$. One should note that the meaning of Eqs. (2.11) and (2.12) can be expressed by saying that primary fields are completely characterized with respect to the Kac-Moody algebras through the representations of the global symmetry groups they carry and with respect to the Virasoro algebras through the scaling dimensions. The full information about a field theory of the type we are considering is contained in the specification of all primary operators in that theory and the expansion coefficients of their operator product algebra. All other conceivable information can then be deduced from (in principle known) representation theory of Kac-Moody and Virasoro algebras.

We are finally prepared to derive the KZDE. Remembering the Sugawara form of the energy momentum tensor one verifies with a sidelook to Eq. (2.11) easily the short distance expansion

$$T_{zz}(z)\varphi_i(z_i) = C_{L,i}(k + c_V)^{-1}(z-z_i)^{-2}\varphi_i(z_i) + 2(k+c_V)^{-1}(z-z_i)^{-1}:j^a t_i^a\varphi_i: + \text{terms regular in } (z-z_i) \quad . \quad (2.13)$$

$C_{L,i}$ denotes here the eigenvalue of the quadratic Casimir operator in the representation carried by φ_i(which we assume to be irreducible). The dots at the second term on the r.h.s.

stand as above for normal ordering of current frequencies (relative here to the position of φ_i).

Identification of singularities in Eqs. (2.12) and (2.13) yields first the determination

$$\Delta_i = C_{L,i}\ (k + c_v)^{-1} \tag{2.14}$$

of the z-scaling dimension of φ_i and second a system of linear differential equations, the KZDE,

$$\begin{aligned} \partial z_i \langle \varphi_1(z_1)\ldots\varphi_n(z_n)\rangle &= \sum_a (-2)(k+c_v)^{-1} \langle \varphi_1 \ldots :j^a t_i^a \varphi_i: \ldots \varphi_n\rangle \\ &= \sum_{\substack{a,j \\ j\neq i}} (-2)(k+c_v)^{-1}\ t_i^a \otimes t_j^a (z_i - z_j)^{-1} \langle \varphi_1 \ \ldots \ \varphi_n\rangle \end{aligned} \tag{2.15}$$

Use has been made once more of the Ward identity Eq. (2.11) for the second equality in (2.15). One has, replacing j_L^a by j_R^a in the preceding argument, the determination

$$\overline{\Delta}_i = C_{R,i}\ (k + c_v)^{-1} \tag{2.14'}$$

for the $\overline{z}$-scaling dimension of φ_i and one obtains a system of differential equations,analogous to Eq. (2.15), governing the antiholomorphic parts of the correlation functions.

Exercise 3: Show that the KZDE's imply the Ward identities of dilations and special conformal transformations only in the singlet sector of the internal symmetry group.

IIb. The Wess-Zumino-Witten action

The action of a σ-model on a group manifold with Wess-Zumino term,tuned to a conformal fixed point, is

$$S(g) = \frac{k}{8\pi}\left(\frac{1}{2}\int d^2x\ \mathrm{tr}\,\partial_\mu g \partial_\mu g^{-1} + \frac{1}{3}\int_{||\mathbf{x}||<1} d^3x\ \mathrm{tr}(g^{-1}dg)^3\right) \tag{2.16}$$

where g takes values in a group G. We assume as before that G is compact and simple. The differential three-form $\mathrm{tr}(g^{-1}dg)^3$ appearing in the second term on the r.h.s. of Eq. (2.16), the Wess-Zumino term,is closed. The integral over the three-dimensional volume $||x|| \leqslant 1$ can therefore be transformed by use of Stoke's theorem into an integral over the two-dimensional sphere, which in turn may be identified with the two-dimensional plane. Witten argued that the coefficient in front of the Wess-Zumino term has to be quantized in order that the path integral based on the action (2.16) makes sense. I will not go through this argument here. (We will see later on that k also becomes quantized through the requirement to have unitary representations of the Kac-Moody algebras).

I want to give some plausible evidence that the action (2.16) leads to a conformally invariant quantum theory. It is instructive in this context to look on the structure of the equation of motion. The Euler variation of the conventional σ - model action ($\sim \int d^2x \; tr \partial_\mu g \partial_\mu g^{-1}$) gives a contribution proportional to

$$\partial_{\bar{z}}(g^{-1}\partial_z g) + \partial_z(g^{-1}\partial_{\bar{z}} g) \tag{2.17}$$

or equivalently

$$\partial_z(g\partial_{\bar{z}} g^{-1}) + \partial_{\bar{z}}(g\partial_z g^{-1}) \tag{2.18}$$

with non-singular factors of proportionality. The equations of motion of the conventional group manifold σ-model are seen to be equivalent to the conservation equation of the currents $g^{-1} \partial g$ related to shifts from the right in G and also equivalent to the conservation equations of the currents $g \partial g^{-1}$ related to shifts from the left. But we have seen (cf.exercise 1) that continuous internal symmetries in two dimensional quantum field theory are compatible with conformal invariance only if the corresponding symmetry currents are not only conserved but also curl free. (This is equivalent to the requirement of a more restricted form of the conservation law, $\partial_z j_{\bar{z}}=0$ or $\partial_{\bar{z}} j_z=0$). We come to the conclusion that the conventional σ - model even though conformally invariant on the classical level cannot be conformally invariant as quantum theory since the equations of motion do not have the right structure. (One should note that in order to reach this conclusion one has not even to mention the renormalization group). The situation is changed if one introduces the Wess-Zumino term. The Euler variation of this term yields a contribution proportional to

$$\partial_{\bar{z}}(g^{-1}\,\partial_z\, g) - \partial_z(g^{-1}\,\partial_{\bar{z}}\, g) \qquad (2.19)$$

or

$$\partial_z(g\,\partial_{\bar{z}}\, g^{-1}) - \partial_{\bar{z}}(g\,\partial_z\, g^{-1}) \qquad (2.20)$$

with the same factor of proportionality as in (2.17) and (2.18) respectively. Plugging together (2.17) and (2.19) (or equivalently (2.18) and (2.20)) we arrive at the - **from** the point of conformal invariance - desirable form of the conservation laws

$$\partial_{\bar{z}}(g^{-1}\,\partial_z\, g) = 0 \quad , \qquad \partial_z(g\,\partial_{\bar{z}}\, g^{-1}) = 0 \quad .$$

The obstruction for conformal invariance in the quantum theory is therewith eliminated. For a direct proof that conformal invariance is preserved to all orders in perturbation theory I refer to the paper by Mukhi [10].

I want to show that the KZDE can be obtained through partial summation of perturbation theory. The argument, taken from the appendix of ref. [11], is heuristic, since it does not properly take wave function renormalizations into account.

Let $X_1 \ldots X_{d_g}$ be coordinates of the group manifold in a patch around the unit element of the group, with

$$g \in G\ , \quad g = g(X_1, \ldots, X_{d_g}),\quad g\big|_{X_1=\ldots X_{d_g}=0} = \mathbb{1}$$

$\mathbb{1}$ denoting here the unit element of G. Let $\{\tau^a\}$ denote a basis of the fundamental representation of the Lie algebra of G with $[\tau^a,\tau^b] = f^{abc}\tau^c$, $\mathrm{tr}\,\tau^a\tau^b \sim \delta^{ab}$. We define $j^\mu_a = \mathrm{tr}\, g^{-1}\partial_\mu g\tau^a$, $j^\mu_a = \sum_i V^a_i \partial_\mu X_i$. The V^a_i are "pure gauge fields",

$$\partial_{X_i} V^a_j - \partial_{X_j} V^a_i + f^{abc}\, V^b_i V^c_j = 0 \quad .$$

The equations of motion of the action (2.16) expressed through coordinates X_i take the form

$$\frac{\delta}{\delta X_k} S \equiv S_{,k} = \sum_{a,l} V_k^a \partial_{\bar{z}}(V_l^a \partial_z X_l) \tag{2.21}$$

Let $Y^i(X)$, $i = 1,\ldots,n$, be some operators which may, as classical objects, be arbitrary functions of the group coordinates. The aim is to show how to derive through a formal use of the equation of motion (2.21) in an iterative procedure the KZDE

$$\partial_{z_i} \langle Y^1(X(z_1))\ldots Y^n(C(z_n))\rangle \simeq \sum_{\substack{a,j\\ j\neq i}} \frac{\delta_i^a \ \delta_j^a}{z_i - z_j} \langle Y^1 \ldots Y^n\rangle \tag{2.22}$$

δ_i^a stands here for the variation of Y^i under an infinitesimal shift in G from the right along the direction of the generator τ^a: $\delta^a Y = \sum_k \partial_{X_k} Y \, V^{ak}$, where V^{ak} is defined through

$$\sum_k V^{ak} V_k^b = \delta^{ab} \tag{2.23}$$

Inspection of graphical alternatives in the Feyman graph expansion immediately gives

$$\begin{aligned}\partial_{z_1} \langle Y^1(X(z_1))\ldots Y^n(X(z_n))\rangle &= \langle Y^1_{,k} \partial_z X_k Y^2 \ldots Y^n\rangle = \\ \sum_{s=2}^{n} \left(-\frac{\partial_z}{\Box}\right)_{1,s} \langle Y^1_{,k} \ldots Y^s_{,k} \ldots Y^n\rangle &+ \langle Y^1_{,k} \frac{\partial_z}{\Box} * S^{int}_{,k} \, Y^2 \ldots Y^n\rangle\end{aligned} \tag{2.24}$$

We make here use of the notation $\partial_z = \frac{\partial}{\partial z}$, $\Box = \partial_z \partial_{\bar{z}}$,
$Y_{,k} = \frac{\partial Y}{\partial X_k}$, $\left(\frac{\partial_z}{\Box}\right)_{1,s} = \frac{\partial_z}{\Box}((z_1 - z_s))$, $Y^1_{,k} \frac{\partial_z}{\Box} * S^{int}_{,k} =$
$\int dz_1 d\bar{z} \, Y^1_{,k} \frac{\partial_{z_1}}{\Box} ((z_1 - z)) \, S^{int}_{,k}$, $S^{int}_{,k} = S_{,k} + \Box X_k$. Repeated indices are to be summed over. S^{int} is decomposed as follows:

$$S^{int}_{,k} = -(V_k^a - \delta_k^a)\partial_{\bar{z}}(V_l^a \partial_z X_l) - \delta_k^a \partial_{\bar{z}} [(V_l^a - \delta_l^a)\partial_z X_l] \tag{2.25}$$

The equations of motion can be factored out in the first term on the r.h.s. of (2.25)

$$-(V_k^a - \delta_k^a)\partial_{\bar{z}}(V_l^a \partial_z X_l) = -(\delta^{kl} - V^{kl}) \, S_{,l} \tag{2.26}$$

Eqs. (2.21) and (2.23) have been used here. Formal application of the equations of motion then gives

$$
\begin{aligned}
&<Y^1_{,k} \frac{\partial_z}{\square} * (\delta^{kl} - V^{kl}) S_{,l} \prod_{i=2}^{n} Y^i> = \\
&\sum_{j>1} - (\frac{\partial_z}{\square})_{1,j} <Y^1_{,k} Y^j_{,k} (V^{kl} - \delta^{kl}) \prod_{r \neq j} Y^r>
\end{aligned}
\tag{2.27}
$$

The contribution of the second term on the r.h.s. of (2.25) to the second term on the r.h.s. of (2.24) is brought by partial integration into the form

$$
<Y^1_k (\delta^k_l - V^k_l) \partial_z X \prod_{i>1} Y^i> \tag{2.28}
$$

The steps leading from (2.24) to (2.7) and (2.28) can now be repeated starting with the expression on the r.h.s. of (2.28). A recursive procedure is therewith set up which gives rise to a geometric series reproducing indeed the KZDE (2.23).

III. Four Point Functions

IIIa. Solutions of the Knizhnik-Zamolodchikov differential equations

We select the instance where the KZDE's appear in their simplest possible form (still containing more information than the conformal Ward identities) which is constituted by the four point functions of the SU(2) WZW model. The analysis of correlation functions with more than four points and/or the consideration of groups of higherrank imply a considerable increase of technical difficulties. Some progress into both directions has been made recently. This will be reported elsewhere [11].

We denote by $\varphi_{j,\bar{j}}$ a primary operator multiplet carrying the isospin j and $\bar{j}$ representations of the left and right shifts respectively in the SU(2) group manifold. $\varphi_{j,\bar{j}}$ has according to Eqs. (2.14) and (2.14') the scaling dimensions

$$
\Delta_j = \frac{c_j}{k+2} \quad , \qquad \bar{\Delta}_{\bar{j}} = \frac{c_{\bar{j}}}{k+2} \tag{3.1}
$$

c_j and $c_{\bar{j}}$ being SU(2) Casimir eigenvalues, $c_x = x(x+1)$.

We will ignore for the moment the $\bar{z}$-dependence of the correlation functions and suppress consequently also reference to the isospins $\bar{j}$ going along with it.

Let us consider four point correlations $\langle\varphi_{j_1}(z_1)\ \ldots\ \varphi_{j_4}(z_4)\rangle$ with the isospins $j_1, \ldots, j_4$ satisfying the relations

$$j_1 \leq j_2 \leq j_3 \leq j_4 \quad ; \quad j_4 = \sum_{i=1}^{3} j_i - l \quad ;$$

$$l \leq 2j_1 \ , \quad l = \text{non-negative integer.}$$

The isospins $j_1 \ldots j_4$ are to be combined inside the correlation into SU(2) singlets. (One breaks otherwise, besides the internal symmetry, also projective invariance (cf. exercise 3)). The given set up of isospins allows the formation of $(l+1)$ different singlet combinations. One possible basis of singlet combinations is obtained by projecting the tensor product of representations with isospin j_1 and j_2 into the representation with isospin $j_1 + j_2 - (k-1)$, $k = 1, \ldots, l+1$ and contracting with the appropriate representation fusing the isospins j_3 and j_4. We collect the so constructed set of SU(2) singlets into an $(l+1)$-component vector $\underset{\sim}{X} = (X_1, \ldots, X_{l+1})$. The KZDE becomes an $(l+1)$- dimensional matrix differential equation for $\underset{\sim}{X}$. Projective invariance restricts the functional form of the components X_k to

$$X_k = \prod_{l<m} (z_l - z_m)^{-\gamma_{lm}} \ Y_k(\eta) \qquad .$$

The constants γ_{lm} here are given as linear combinations of the scaling dimensions of the operators $\varphi_i \cdot Y_k(\eta)$ denote so far undetermined functions of the projective invariant

$$\eta = \frac{(z_1 - z_2)\ (z_3 - z_4)}{(z_1 - z_4)\ (z_3 - z_2)} \qquad .$$

The KZDE can be reformulated as first order differential equation in the variable η for the vector $\underset{\sim}{Y}$. I reproduce the general solution without going into any detail of its verification (consult for this purpose references [4] and [5]). The components Y_k turn out to be generalized hypergeometric functions with an Euler contour integral representation of the form

$$Y_k(\eta) = \eta^{k-1+j_1\cdot j_2/\kappa} (\eta-1)^{j_1 j_3/\kappa} \int \prod_{i=1}^{1} dt_i \prod_{j<s} (t_j-t_s)^{1/\kappa}$$
$$\cdot \prod_{i=1}^{k-1} t_i^{-j_2/\kappa-1} (t_i-1)^{-j_3/\kappa} (t_i-\eta)^{-j_1/\kappa-1} \tag{3.2}$$
$$\cdot \prod_{i=k}^{1} t_i^{-j_2/\kappa} (t_i-1)^{-j_3/\kappa-1} (t_i-\eta)^{-j_1/\kappa}, \quad \kappa = \tfrac{1}{2}(k + c_v)$$

The KZDE has, as an $(1+1)$ dimensional matrix differential equation, $(1+1)$ independent vector solutions. Those are found by choosing different contours in Eq. (3.1) around the singularities of the integrand. I will present for later use three linearly equivalent bases of solution, called

$$\{Y_k^{(i)}\}, \qquad \{\hat{Y}_k^{(i)}\}, \qquad \{\hat{\hat{Y}}_k^{(i)}\}, \qquad i = 1,\ldots, 1+1$$

respectively which have the virtue to display the singularity structure at the singular points $\eta = 0, 1, \infty$ in a diagonalized form, each base achieving the diagonalization at one singular point. We shall concentrate on the first components

$$\gamma^{(i)} = \eta^{j_1 j_2/\kappa} (\eta-1)^{j_1 j_3/\kappa} \int_0^{\eta} \prod_{j=1}^{i-1} dt_j \ldots \int_1^{\infty} \prod_{j=1}^{1} dt_j \ldots \tag{3.3}$$
$$\simeq \eta^{\alpha_i}(1+O(\eta)); \quad \alpha_i = -(C_{j_1} + C_{j_2} - C_{j_1+j_2-i+1})/2\kappa + (i-1)$$

$$\hat{\gamma}^{(i)} = \eta^{j_1 j_2/\kappa} (\eta-1)^{j_1 j_3/\kappa} \int_{\eta}^{1} \prod_{j=1}^{i-1} dt_i \int_{-\infty}^{0} \prod_{j=i}^{1} dt_i$$
$$\simeq (\eta-1)^{\beta_i}(1+O(\eta-1)); \quad \beta_i = -(C_{j_1} + C_{j_3} - C_{j_1+j_3-i+1})/2\kappa \tag{3.4}$$

$$\hat{\hat{\gamma}}^{(i)} = \eta^{j_1 j_2/\kappa} (\eta-1)^{j_1 j_3/\kappa} \int_{\eta}^{\infty} \prod_{j=1}^{i-1} dt_j \int_0^{1} \prod_{j=1}^{1} dt_j$$
$$\simeq \left(\frac{1}{\eta}\right)^{\gamma_i}\left(1+O\left(\frac{1}{\eta}\right)\right); \quad \gamma_i = -(C_{j_4} - C_{j_1} - C_{j_4-j_1+i-1})/2\kappa . \tag{3.5}$$

The contours in Eqs. (3.3) - (3.5) connecting the same endpoints are arranged to lie one over the other. The integrals are to be understood as being continued from values of the exponents in the integrand where the integrals are manifestly convergent.

IIIb. The conformal bootstrap

The singular exponents appearing in Eqs. (3.3) - (3.5) have a simple interpretation within the operator product algebra. Choosing the particular configuration of points $z_4 = \infty$, $z_3 = 1$, $z_2 = 0$, $z_1 = \eta$ one projects from the full four point function to its projectively invariant part:

$$\langle \varphi_{j_4}(\infty)\ \varphi_{j_3}(1)\ \varphi_{j_2}(0)\ \varphi_{j_1}(\eta)\rangle \simeq \underset{\sim}{Y}(\eta)$$

The points of coincidence of φ_{j_1} with φ_{j_2}, φ_{j_3} and φ_{j_4} are under these circumstances at $\eta = 0$, $\eta = 1, \eta = \infty$ respectively. The singular exponents in Eqs. (3.3) - (3.5) should be those which are to be expected from the operator product expansion. That this is in fact the case is easily seen. Considering for example the operator product expansion fusing φ_{j_1} and φ_{j_3} to $\varphi_{j_1+j_3-(k-1)}$ one has

$$\varphi_{j_1}(\eta)\ \varphi_{j_3}(1) \sim (\eta-1)^{-\Delta_{j_1}-\Delta_{j_3}+\Delta_{j_1+j_3-(k-1)}}\ \varphi_{j_1+j_3-(k-1)} \ +$$

$$+ \ \text{less singular terms.}$$

The exponent $-(\Delta_{j_1} + \Delta_{j_3}) + \Delta_{j_1+j_3-(k-1)}$ is identical with β_k in (3.4). The singularities of the fundamental systems of solutions (3.3) and (3.5) have an analogous interpretation. (The occurence of κ-independent integers in the exponents α_k is explained through the presence of Kac-Moody fields contributing in the corresponding channel to the leading singular behaviour).

We want to restore the $\bar{z}$-dependence of correlation functions which has been ignored up to now. We assume for simplicity that all operators $\varphi_{j,\bar{j}}$ to be considered are Lorentz scalars, which means that j equals $\bar{j}$. The $\bar{z}$-dependence of the correlation functions is governed by the complex conjugate of the KZDE whose general solution was given in Eq. (3.2). Our task amounts then to superimpose products of special solutions taken from Eqs.(3.3) - (3.5) with suitable complex conjugate solutions so that the resulting expression is a one valued expression in the Euclidean plane. This project is identical (in this context) with the so called conformal bootstrap program.

We choose one specific basis of solutions of the KZDE, say the one represented through Eq. (3.3) and write the general form of the four point function in terms of this basis as

$$<\varphi_{j_1,j_1}\ldots\varphi_{j_4,j_4}>_{1,1} = \prod_{i<j} |z_i - z_j|^{-2\gamma_{ij}} \sum_{s,t} \alpha^{s,t}\, Y^s \overline{Y}^t \; . \quad (3.6)$$

The subscripts (1,1) shall indicate the projection onto the respective singlets of the two global SU(2) symmetries. The strategy of Dotsenko and Fateev to find the correct constants guaranteeing one-valuedness of (3.6) consists in the following: one valuedness around $\eta = 0$ is easily achieved by choosing $\alpha^{s,t}$ in diagonal form,

$$\alpha^{s,t} = \alpha^s \delta^{st} \quad (3.7)$$

Phases of the holomorphic and antiholomorphic parts developing along paths around $\eta = 0$ clearly cancel each other this way. It is possible to show that at values of the isospins j_i, allowed by unitarity (see below), the ansatz (3.7) is mandatory. One-valuedness around the point $\eta = 1$ can similarly be obtained making use of the basis (3.4) for the ansatz

$$<\varphi_{j_1,j_2}\ldots\varphi_{j_4,j_4}> \simeq \sum_{s,t} \beta^s \delta^{st} \hat{Y}^s \overline{\hat{Y}}^t \quad (3.8)$$

We drop here and in the following the, for our purposes, unessential prefactor $\Pi|z_i - z_j|^{-2\gamma_{ij}}$. Suppose we know the linear dependence of one basis, say (3.3), from the other basis, (3.4), given through coefficients γ^{st}:

$$Y^s = \Sigma\, \gamma^{st}\, \hat{Y}^t \quad (3.9)$$

The concrete computation of the coefficients γ^{st} is not too difficult but sufficiently tedious. The compatibility of (3.7) and (3.8) is established if one shows that there are constants α^s satisfying

$$\sum_s \alpha^s \gamma^{st} \gamma^{st'} = 0, \qquad t \neq t' \tag{3.10}$$

(The constants γ^{st} are in fact real in the SU(2)-model).

Dotsenko and Fateev determine the (l+1) coefficients α^s from conditions of the type (3.10) and show through explicit calculations in the concrete models they deal with that an overdetermination of the α^s, which would imply the identical vanishing of the four point functions, does not take place, (note that (3.10) represents $l(l+1)/2$ conditions for the (l+1) constants α^s). A systematic reason for the fact that overdetermination is avoided can be extracted from another representation of solutions of the KZDE given by

$$\begin{aligned} \langle \varphi_{j_1,j_1} \cdots \varphi_{j_4,j_4} \rangle_{1,1} &= \prod_{i<j} |z_i - z_j|^{2\gamma_{ij}} \; |\eta|^{2j_1 j_2/\kappa + 2(k-1)} \\ &\cdot |\eta - 1|^{2j_1 j_3/\kappa} \int \prod_{i=1}^{l} d^2 t_i \prod_{j<s} |t_i - t_j|^{2/\kappa} \prod_{i=1}^{l} |t_i|^{-2j_2/\kappa} \\ &\cdot |t_i - 1|^{2j_3/\kappa - 2} \; |t_i - \eta|^{-2j_1/\kappa} \end{aligned} \tag{3.11}$$

The manipulations needed for a verification of the KZDE, in particular partial integrations, are formally the same for the two-dimensional volume integral representation (3.11) as for contour integrals introduced in the preceding section. One-valuedness of (3.11) is obvious. What is not so obvious - this is the major disadvantage of (3.11) compared to the contour integral representation - is how to make transparent the structure of singularities starting from (3.11). I want therefore to indicate the procedure which allows to transform successively (3.11) back into products of contour integrals. The tool to be applied is clearly Stokes' theorem in the form

$$\int_R dx \wedge dy \frac{\partial f}{\partial \bar{z}} = \frac{1}{2i} \oint_{\partial R} f\, dz \quad , \tag{3.12}$$

or

$$\int_R dx \wedge dy \frac{\partial f}{\partial z} = - \frac{1}{2i} \oint_{\partial R} f\, d\bar{z}, \quad z=x+iy, \quad \bar{z}=x-iy, \tag{3.13}$$

with R denoting a simply connected region. Let us consider for notational convenience instead of the above introduced expressions the function

$$F(z,\bar{z}) = \int \prod_{i=1}^{n} d^2t_i |t_i|^{2\alpha_i^{(1)}} |t_i-1|^{2\alpha_i^{(2)}} |t_i-z|^{2\alpha_i^{(3)}} \cdot \prod_{i<j\leq n} |t_i-t_j|^{2\delta_{ij}} \tag{3.14}$$

which may be rewritten, using the definition

$$g_j(z;\, t_j,\, t_{j+1} \ldots t_n) = t_j^{\alpha_j^{(1)}} (t_j-1)^{\alpha_j^{(2)}} (t_j-z)^{\alpha_j^{(3)}} \cdot \prod_{k>j} (t_j - t_k)^{\delta_{jk}}$$

as

$$F(z,\bar{z}) = \int \prod_i d^2t_i \prod_j g_j \bar{g}_j \tag{3.14'}$$

g_j appearing in the integrand of (3.14') should be regarded as a function which is holomorphic in t_j in a cut plane with cuts from $t_{j+1},\ldots,t_n,0,z$ and 1 to ∞. The picture for g_1 is given in the following figure.

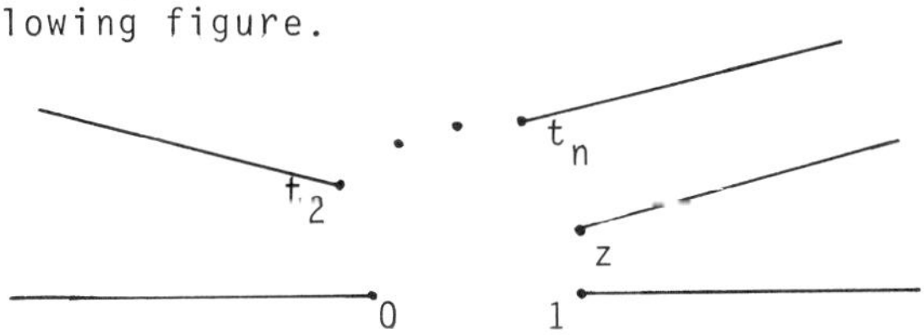

Fig. 1

We want to transform the two-dimensional integral over t_1 into a couple of contour integrals. Introducing for this purpose

$$G_1(z;\, t_1 \ldots t_n) = \int_{t_1}^{t_2} dt\, g_1(z;\, t\, , t_2 \ldots t_n)$$

with a contour of integration not crossing the cuts of fig. 1 one obtains with the help of Eq. (3.13)

$$\int d^2t_1 |g_1|^2 = \int_D d^2t_1\, \partial_{t_1} G_1 \bar{g}_1 = -\frac{1}{2i} \int_{\partial D} d\bar{t}_1\, G_1 \bar{g}_1 \qquad (3.15)$$

D denotes here the cut plane of fig. 1. The integral along the boundary ∂D is given by the discontinuity of $G\cdot\bar{g}$ integrated along the cuts of fig.1. One has

$$\partial_{t_1} \operatorname{disc} G_1 \bar{g}_1 = \operatorname{disc} \partial_{t_1} G_1 \bar{g}_1 = \operatorname{disc} |g_1|^2 = 0$$

It follows that Eq. (3.15) can be transformed into

$$\int d^2t_1 |g_1|^2 = -\frac{1}{2i} \sum_j A_j \int_{C_{t_j}} d\bar{t}_1\, \bar{g}_1$$

$$A_j = \int_{t_1}^{t_2} dt\, g_1(z;\, t, t_2, \ldots, t_n); \qquad t_{n+1}=z,\ t_{n+2}=1,\ t_{n+3}=0.$$

We have so achieved our goal to decompose the two-dimensional integral d^2t_1 into contour integrals. I leave it as a task for the reader to work out the decomposition of the successive integrations $d^2t_2 d^2t_3$ ...

The above indicated connection between the volume integral representation and the contour integral representation allows us to get a systematic understanding why the described strategy of Dotsenko and Fateev leads to non-trivial results for a given configuration of isospins. One has still to extract from the coefficients α^s (cf. e.g. Eq. (3.10) and some normalization integrals the products of expansion coefficients and from those the expansion coefficients themselves. The actual computations, required for this purpose, turn out to be extremely tedious. They do not, as they stand, give any insight into the mechanism responsible for the "miraculous" result (which is of course an a priori consistency condition) that the determination of an expansion coefficient through consideration of different isospin configurations lead to identical expressions.

IIIc. Closed operator algebras

I quote once more for convenience the Kac-Moody algebra relations

$$[j_n^a, j_m^b] = f^{abc} j_{n+m}^c - \frac{k}{2} n \delta^{ab} \delta_{n+m,o} \qquad (3.16)$$

Putting m = n = 0 one obtains in particular the Lie-algebra relations of the global symmetry

$$[j_o^a, j_o^b] = f^{abc} j_o^c$$

There are other finite Lie-algebras contained in (3.16). For example, the combinations

$$P^- = j_{-1}^1 + i j_{-1}^2, \quad P^+ = j_1^1 - i j_1^2, \quad P^3 = k.\mathbb{1} - i j_o^3$$

in the case of SU(2) (a,b,c = 1,2,3) form a second SU(2) Lie-algebra, which has been called "pseudospin" by Gepner and Witten [12]. The highest weight eigenvalues of P^3 are found to be

$$k/2 - m(j) \qquad (3.17)$$

with m(j) denoting the eigenvalues of j_o^3, m(j) =(j,j-1,...,-j). Insisting on unitary representation of pseudo-spin one infers from (3.17):

i) k is a positive integer
ii) the allowed isospins j carried by local operators occuring in a field theory with Osterwalder-Schrader reflection positivity satisfy

$$j \leq k/2 \qquad (3.18)$$

(Field theoretic reflection positivity is supposed to imply necessarily the unitarity of pseudo-spin representations). Fateev and Zamolodchikov [4] and Gepner and Witten [12] argue that the emergence of the selection rule (3.18) should be the consequence of a degeneracy in the highest weight representations of the Kac-Moody algebra, namely the occurence of some states of zero norm in these representations. I sketch here, following ref. [7], how to recover the selection rule through a closer inspection of four point functions.

The family of functions of which different bases are given in Eqs. (3.3) - (3.5),- let us call if F -

$$F = \{Y^{(i)}\} = \{\hat{Y}^{(i)}\} = \{\hat{\hat{Y}}^{(i)}\}$$

has the following properties:

i) F is (1+1)-dimensional

ii) The functions in F have power-type singularities at $\eta = 0,1,\infty$

iii) F is closed under analytic continuation on the Riemann surface consisting of the complex plane with branch points at $\eta = 0,\ 1,\ \infty$.

I want to derive a criterior which supplies a necessary condition for the existence of a subfamily $G \subset F$ satisfying also the above listed conditions (i) - (iii). The criterion is easily inferred from the inspection of the logarithmic derivative $\partial_z \ln W_f \equiv V$ of the Wronski determinant of F. V(z) does not depend on the specially chosen basis of F. It is a meromorphic function in the entire plane which takes at finite points the form

$$V(z) = \frac{\Sigma\, \alpha_n - l(l+1)/2}{z} + \frac{\Sigma\, \beta_n - l(l+1)/2}{z-1} + \Sigma \frac{n_a}{z-z_a} \tag{3.19}$$

with n_a denoting non-negative integers, $z_a \neq 0,\ 1$. V(z) behaves near ∞ as

$$V(z) = -\frac{\Sigma\, \gamma_n + l(l-1)/2}{z} + \mathcal{O}(z^{-2}) \tag{3.20}$$

The residua of the poles at $z = 0,\ 1,\ \infty$ in Eqs. (3.19) and (3.20) are found by evaluating consecutively V with the help of the bases (3.3), (3.4) and (3.5) respectively. The additional terms on the r.h.s. of (3.19) emerge from possible zeros of W_f. Comparison of (3.19) and (3.20) gives the sum rule

$$\sum_{n=1}^{l+1} (\alpha_n + \beta_n + \gamma_n) = l(l+1)/2 - k, \qquad k = \Sigma n_a \qquad . \tag{3.21}$$

The last equation is the necessary condition for closure under analytic continuation we are looking for.

The strategy to be followed in search of closed operator algebras consists then in finding first subsets of complete families of solutions of the KZDE whose singular exponents satisfy for themselves the above criterion. Suppose we find such a subset which is, say, m-dimensional. The second step amounts to finding a factorization in the KZDE such that the subset is the family of solutions of one factor (standing in the product of differential operators on the right side). Cutting down in this way sistematically to subfamilies of solutions wherever is possible, one achieves in fact a restriction which comprises the selection rule (3.18). For more details I refer to ref. [7].

IV. Relations to Other Critical Theories

A general recipe to construct the correlation functions of critical systems with central Virasoro charge $c < 1$ (cf. [15]) has been given by Dotsenko and Fateev [6] by means of a screened Coulomb gas. The four point functions found by these authors have the general form

$$\begin{aligned} F_{m,n}(z_1,\ldots,z_4) &= \int\ldots\int \prod_{i=1}^{m} dt_i \prod_{j=1}^{n} dt'_j \prod_{i,j} (t_i - t'_j)^{-2} f_m \hat{f}_n \\ f_m &= \prod_{i=1}^{m} \prod_{s=1}^{4} (t_i - z_s)^{a_s} \prod_{i<j} (t_i - t_j)^{b} \\ \hat{f}_n &= \prod_{i=1}^{n} \prod_{s=1}^{4} (t'_i - z_s)^{\hat{a}_s} \prod_{i<j} (t'_i - t'_j)^{\hat{b}} \end{aligned} \tag{4.1}$$

where the exponents $a_s(\hat{a}_s)$, $b(\hat{b})$ are functions of c and m(n). One sees, comparing Eqs. (4.1) and (3.2), that the structure of the contour integrals is identical for the particular case $m = 0$ or $n = 0$ in (4.1). Those special situations ($m = 0$ or

$n= 0$) correspond to closed operator subalgebras in the $c < 1$ models.

The functions $F_{n,m}$ represent the general solution of a linear differential equation of $(n+1)(m+1)$-th order, which can be derived from degeneracy relations of the Virasoro algebra. Special solutions are again given through specific choices of contours in (4.1). One has to know how various special solutions are related to each other under analytic continuation in order to be able to go through the bootstrap program, as it has been sketched in section IIIb. One needs in other words information on the monodromy of the system of functions represented by the contour integral (4.1). The problem is facilitated through the observation of Dotsenko and Fateev [6] that the monodromy group factorizes on f_m and f_n in the integrand on the r.h.s. of (4.1). It means that the monodromy groups of $F_{o,n}$ and $F_{m,o}$ determine completely the monodromy of $F_{m,n}$. One is so in this respect lead to the consideration of two SU(2)-like situations. The analysis of closed operator in the $c < 1$ systems can in fact be deduced step by step from the corresponding analysis in the SU(2) model.

The Dotsenko-Fateev-Coulomb gas construction is easily supersymmetrized by addition of a free Majorana field. The unitary $N = 1$ supersymmetric theories with $c < \frac{3}{2}$ have been worked out along these lines in ref. [16]. The previous remark on monodromy groups and connections to the SU(2) model applies also in this case.

Two other links connecting the theory of representations of the SU(2) Kac-Moody algebras with the $c < 1$ systems are to be mentioned:

i) Goddard, Kent and Olive show how to construct all unitary highest weight representations of the Virasoro algebra with $c < 1$ starting from representations of the tensor product of two SU(2) Kac-Moody algebras with suitably chosen central charges and factorizing out the diagonal subgroup.

ii) An observation due to Gepner [22] which has further been worked out and exploited by Cappelli, Itzykson and Zuber, is that the characters of the $c < 1$ highest weight Virasoro algebra representations transform under the modular group in a manner which is neatly related to the transformation behaviour of a product of characters of two SU(2) Kac-Moody algebras.

The points (i) and (ii) and the preceding remarks about monodromy groups are certainly related to each other. The precise conceptual connection however has to be found.

The list of models related to the SU(2) WZW model is prolonged through systems with an underlying parafermion algebra and $N = 2$ supersymmetric models with central Virasoro charge $c < 3$. The parafermion systems are in fact (cf. [19]) isomorphic to the SU(2) WZW models. A representation of the discrete $N = 2$ supersymmetric series of models through parafermion systems was found in [13].

We finally mention some recent results concerning the SU(3) WZW models: Fateev and Zamolodchikov [20] constructed a new discrete series of models whose most characteristic feature is a conserved current of dimension 3. A generalized Goddard-Kent-Olive coset construction involving highest weight representations of the SU(3) Kac-Moody algebra has been worked out by Bais et al. [21]. Comparing the four point correlation functions constructed in [20] and the newly found solutions of the KZDE for the SU(3) WZW ([11]) one concludes that the latter model is related to the discrete series of Fateev and Zamolodchikov in a way which almost copies the relation between the SU(2) WZW model and the $c < 1$ systems.

Does there exist any conformally invariant field theory in two dimensions which is not related to a WZW model ?

Acknowledgement

These lectures were written while I am visiting the L.P.T.H.E. at the University Paris VI. I thank the CNRS for financial support making this visit possible. I thank the members of the L.P.T.H.E. for their hospitality extended to me.

References

[1]. E.Witten, Comm.Math.Phys. 92 (1984) 453.

[2]. W.G.Knizhnik and A.B.Zamolodchikov, Nucl.Phys. B247 (1984) 83.

[3]. R.Dashen and Y.Frishman, Phys.Rev. D11 (1975) 2781.

[4]. V.A.Fateev and A.B.Zamolodchikov, Yad.Fiz. 43 (1986) 75.

[5]. P.Christe and R.Flume, Nucl.Phys. B282 (1987) 466.

[6]. V.J.S.Dotsenko and V.A.Fateev, Nucl.Phys. B240 (1984)312; Nucl.Phys. B251 (1985) 691.

[7]. P.Christe and R.Flume, Phys.Lett. B188 (1987) 219.

[8]. G.F.Dell'Antonio, Y.Frishman and D.Zwanziger, Phys.Rev. D6 (1972) 988.

[9]. A.A.Belavin, A.M.Polyakov and A.B.Zamolodchikov, Nucl.Phys. B241 (1984) 333.

[10]. S.Mukhi, Phys.Lett. 162B (1985) 345.

[11]. R.Flume, to be published.

[12]. D.Gepner and E.Witten, Nucl.Phys. B278 (1986) 493.

[13]. Z.Qiu, Phys.Lett. 188B (1987) 207.

[14]. P.Christe, preprint Bonn-HE-87-16 (1987) to be published in Nucl.Phys. B.

[15]. D.Friedan, Z.Qiu and S.H.Shenker, Phys.Rev.Lett. 52 (1984) 1575.

[16]. M.Bershadsky, V.Knizhnik and M.Teitelman, Phys.Lett. 151B (1985) 31.

[17]. P.Goddard, A.Kent and D.Olive, Comm.Math.Phys. 103 (1986) 195.

[18]. A.Cappelli, C.Itzykson and J.-B.Zuber, Nucl.Phys. B280 (1987) 445.

[19]. A.B.Zamolodchikov and V.A.Fateev, Zh.Eksp.Teor.Fiz. 89 (1985) 380.

[20]. V.A.Fateev and A.B.Zamolodchikov, Nucl.Phys. B280 (1987) 644.

[21]. F.A.Bais, P.Bouwknegt, M.Surridge and K.Schoutens, preprints Amsterdam JTFA 87-12 and JTFA 87-18.

[22]. D.Gepner, Nucl.Phys. B287 (1987) 111.

Physikalisches Institut der Universität Bonn

Nussallee 12, 5300 Bonn 1

KAC-MOODY ALGEBRAS IN 2D CONFORMAL FIELD THEORIES

Peter Bouwknegt
Institute for Theoretical Physics, University of Amsterdam
Valckenierstraat 65, 1018 XE Amsterdam, The Netherlands

Abstract

We review some results in the classification program of 2-dimensional conformal field theories, and point out the relation to the existence of a coset pair (g, g') of affine Kač-Moody algebras.
The coset pair $(A_1^{(1)} \oplus A_1^{(1)}, A_1^{(1)})$ is treated in detail and in such a way that the generalization to other coset pairs becomes suggestive.

1 Introduction

All the examples of two dimensional conformal field theories (CFT's) that have been constructed over the past few years, seem to suggest an intimate relation between a (rational) 2-D CFT and a coset pair $(g, g'), g' \subset g$ of affine Kač-Moody algebras. It is the hope that this relation might eventually evolve in a complete classification of all 2-D CFT's, which is of vital importance for understanding string theories as well as 2-D critical phenomena [1].

The main object of a 2-D CFT is the stress energy tensor $T_{\mu\nu}(\vec{x})$,satisfying the conservation law $\nabla^\mu T_{\mu\nu} = 0$ and the trace condition $T_\mu{}^\mu = 0$ [2]. It is convenient to work in a conformal gauge for the metric, and to introduce complex coordinates $z = x_1 + ix_2$ (we work Euclidean !) The stress energy tensor then splits into two components $T(z)$ and $\bar{T}(\bar{z})$,depending only on z respectively $\bar{z}$. Because both components can be treated similarly we will restrict the discussion to the component $T(z)$.

The most general operator product expansion (OPE) for $T(z)$ can be shown to be

$$T(z)T(w) = \frac{c/2}{(z-w)^4} + \frac{2T(w)}{(z-w)^2} + \frac{\partial T(w)}{z-w} + \cdots \tag{1}$$

where c is a c–number called the central charge and the ... stands for terms regular in the limit $z \to w$.

From (1) it follows that the modes L_m of $T(z) = \sum_{m \in Z} L_m \; z^{-m-2}$ satisfy the infinite dimensional algebra

$$[L_m, L_n] = (m-n)L_{m+n} + \frac{c}{12} m(m^2-1)\delta_{m+n,0}. \tag{2}$$

ISBN 0-12-218100-X

This so-called Virasoro algebra is the quantum version of the algebra of analytic transformations of z, i.e. the conformal group in two dimensions.

The spectrum of a conformal field theory can be assembled into irreducible representations of this Virasoro algebra. The physically interesting representations are the irreducible highest weight modules (HWM's) $L(h,c)$, characterized by the central charge c and the L_0 eigenvalue h of the highest weight vector $\mid h,c>$.
The highest weight vectors are in 1-1 corrsepondence with so-called primary fields $\Phi_h(z)$ of conformal dimension h, i.e. fields satisfying the OPE

$$T(z)\Phi_h(w) = \frac{h\Phi_h(w)}{(z-w)^2} + \frac{\partial\Phi_h(w)}{z-w} + \cdots. \tag{3}$$

The correspondence is given by

$$\mid h,c> = \lim_{z\to 0} \Phi_h(z) \mid 0> \tag{4}$$

where $\mid 0>$ denotes the SL(2,R) vacuum.

An important constraint on the operator content of a 2-D CFT comes from the requirement of unitarity. It has been shown [3,4] that this restricts the possible h and c values of the HWM's $L(h,c)$ to either[1]

$$c \geq 1 \qquad , h \geq 0 \tag{5}$$

or

$$\begin{aligned} c = c(m) = & \quad 1 - \tfrac{6}{m(m+1)} \\ h = h(p,q) = & \quad \tfrac{((m+1)p-mq)((m+1)p-mq+2)}{4m(m+1)} \\ & 0 \leq p \leq m-2 \quad , 0 \leq q \leq m-1 \end{aligned} \tag{6}$$

In terms of the character χ of a HWM $L(h,c)$

$$\chi_h(\tau) = q^{-c/24}\,\mathrm{Tr}_{L(h,c)}\, q^{L_0} \qquad , q = e^{2\pi i\tau} \tag{7}$$

the torus partition function $\mathcal{Z}$ can be written as [1]

$$\mathcal{Z} = \sum_{h,h'} \overline{\chi_h(\tau)}\, \mathcal{N}_{h,h'}\, \chi_{h'}(\tau) \tag{8}$$

where $\mathcal{N}_{h,h'}$ denotes the multiplicity of the HWM $L(h,c) \times L(h',c)$ of the left/right Virasoro algebra.

A second constraint on the operator content comes from the requirement that this partition function is independent of the specific parametrization of the torus, i.e. $\mathcal{Z}$ should be invariant under the modular transformations[5]

$$\tau \to \frac{a\tau+b}{c\tau+d}, \qquad \begin{pmatrix} a & b \\ c & d \end{pmatrix} \in \mathrm{SL}(2,\mathrm{R}). \tag{9}$$

[1] The convention adopted here differs a little from the usual one, but allows for a better comparison to the results of section 4.

For the discrete series (6) all possible modular invariants (8) have been classified [6]. This classification turns out to be closely related to properties of the Kač-Moody algebra $A_1^{(1)}$.

The rest of the paper is organized as follows. In section 2 we will define the notion of a Kač-Moody algebra and discuss the basic elements of its representation theory. We will also point out how to associate a 2-D CFT to a coset pair (g, g') of Kač-Moody algebras. In section 3 we will explain how to use this relation to derive modular invariants for the Virasoro algebra. In particular we will discuss how the representation theory and classification of modular invariants for the discrete series (6), as discussed above, is related to the 'diagonal' coset pair $(A_1^{(1)} \oplus A_1^{(1)}, A_1^{(1)})$. In section 4 we discuss possible generalizations to more general coset pairs (g, g'). In particular we will discuss the diagonal coset pair $(g \oplus g, g)$, where we find an extension of the Virasoro algebra by higher order Casimir operators.

2 Kač-Moody algebras

In this section we will introduce affine Kač-Moody algebras, their representation theory and the relation to 2-D conformal field theories. A more detailed introduction can be found in [7], which also contains the proper historical references. The necessary mathematical background can be found in [8].

2.1 Definitions and representation theory

To every Lie algebra $\bar{g}$, defined by the commutation relations

$$[T^a, T^b] = f^{ab}_{\ \ c}\, T^c \tag{10}$$

where the generators $\{T^a, a = 1, \ldots, \dim(\bar{g})\}$ are normalized to

$$\mathrm{Tr}(T^a T^b) = \gamma^{ab} \tag{11}$$

one can associate an (untwisted) affine Kač-Moody algebra g (also denoted by $\bar{g}^{(1)}$) with generators $\{J^a_m, a = 1, \ldots, \dim(\bar{g}), m \in Z\}$, defined by

$$[J^a_m, J^b_n] = f^{ab}_{\ \ c} J^c_{m+n} + k m \gamma^{ab} \delta_{m+n,0}. \tag{12}$$

The c-number k is called the central charge or level. Notice that the zero modes J^a_0 satisfy the commutation relations of the underlying Lie algebra $\bar{g}$.

In a field theory having a Kač-Moody symmetry, the J^a_m are the Fourier modes of a conserved current

$$J^a(z) = \sum_{m \in Z} J^a_m\, z^{-m-1}, \qquad \bar{\partial} J^a(z) = 0. \tag{13}$$

The commutation relations (12) are then equivalent to the OPE

$$J^a(z) J^b(w) = \frac{k\gamma^{ab}}{(z-w)^2} + f^{ab}_{\ \ c}\, \frac{J^c(w)}{z-w} + \cdots. \tag{14}$$

The relevant representations [9] are the so-called integrable highest weight modules $L(\Lambda)$, characterized by an integral dominant weight of level $k \in Z_{>0}$

$$\Lambda \in P_+^k \equiv \{\lambda \in h^* \,|< \lambda, \alpha_i^v >\in Z_{\geq 0}, i = 0, 1, \ldots, \text{rank}(\bar{g}), \lambda \text{ of level } k\} \qquad (15)$$

where h^* denotes the dual Cartan subalgebra of g and $\{\alpha_i^v\}$ is a system of dual simple roots of g. The projection $\bar{\Lambda}$ of Λ onto the weight lattice of the underlying Lie algebra $\bar{g}$ gives the highest weight of the $\bar{g}$ representation the vacuum is in. At every level k there are only a finite number of integrable HWM's.

2.2 The sugawara construction

The reason that Kač-Moody algebras are so useful is that every 2-D field theory having a Kač-Moody symmetry is automatically a conformal field theory. The reason is that every HWM of a Kač-Moody algebra g can be extended to a HWM of the semi-direct sum $g + Vir$ of a Kač-Moody algebra and a Virasoro algebra. This construction is known as the Sugawara construction [10].

The expression for the field $T(z)$ in terms of the currents $J^a(z)$ is

$$T(z) = \frac{1}{2(k + \text{g})} \gamma_{ab} \, (J^a J^b)(z) \qquad (16)$$

where g is the dual Coxeter number of g [8]. We also adopted the convention to write normal ordering of operators by parentheses. Using the expression in modes or the operator techniques as explained in [22], it is easily shown that

$$T(z) J^a(w) = \frac{J^a(w)}{(z - w)^2} + \frac{\partial J^a(w)}{z - w} + \cdots \qquad (17)$$

proving that $J^a(z)$ is a primary field of conformal dimension 1 with respect to $T(z)$. Using (17) one can verify that $T(z)$ indeed satisfies the Virasoro OPE (1), with a central charge

$$c = c(g, k) = \frac{k(\text{dim}\bar{g})}{k + \text{g}}. \qquad (18)$$

This proves that every HWM $L(\Lambda)$ of g extends to a HWM of $g + Vir$. The L_0-eigenvalue (i.e. conformal dimension) of the highest weight vector $| \Lambda >$ is given by

$$h_\Lambda = \frac{(\bar{\Lambda}, \bar{\Lambda} + 2\bar{\rho})}{2(k + \text{g})}. \qquad (19)$$

The numerator is the eigenvalue of the second order Casimir in the representation $\bar{\Lambda}$ of $\bar{g}$. If we consider the HWM $L(\Lambda)$ as a representation of the Virasoro algebra only, then the HWM will in general not be irreducible. The irreducible components will have conformal weights which differ by integers from those of (19).

To summarize; we have constructed HWM's of the Virasoro algebra from those of the Kač-Moody algebra by means of the Sugawara construction. However, it can be shown that the c-value (18), satisfies the inequality $c(g, k) \geq \text{rank}(\bar{g})$. In particular, it is not possible to realize the $c < 1$ region in this way. In the next section we will give a construction which can produce c-values in the domain $c < 1$.

2.3 The coset construction

The Goddard-Kent-Olive coset construction [4] is the construction of yet another Virasoro algebra. It amounts to taking a Kač-Moody subalgebra g' of g (both g and g' not necessarily simple). One then constructs the Sugawara operators $T(z)$ and $T'(z)$ satisfying Virasoro algebras of central charge respectively $c(g,k)$ and $c(g',k')$ as in (16) and (18).

The value k' is determined by the embedding $g' \subset g$ through $k' = jk$, where j is the so-called Dynkin index of the embedding [11].

It is easily shown that the difference

$$\tilde{T}(z) = T(z) - T'(z) \tag{20}$$

also generates a Virasoro algebra. The central charge $c(g,g',k)$ of this coset-Virasoro algebra $\widetilde{Vir}$ is

$$0 \leq c(g,g',k) = c(g,k) - c(g',k'). \tag{21}$$

A very important property of this coset-Virasoro algebra $\widetilde{Vir}$ is that it commutes with the subalgebra g'. This allows us to write every HWM $L(\Lambda)$ of g as a sum over irreducible HWM's of the direct sum $g' \bigoplus \widetilde{Vir}$. In general these branching rules are not finite. In fact one can show that a HWM $L(\Lambda)$ of g is finitely reducible into HWM's of $g' \bigoplus \widetilde{Vir}$ *if and only if* $c(g,g',k) < 1$ [12,13,14,15,16].

If the condition of this finite reducibility theorem is fulfilled we can write the character of the HWM $L(\lambda)$ of g at level k, as a finite sum of characters of $g' \bigoplus \widetilde{Vir}$ at level $k' = jk$ and coset central charge $c(g,g',k)$

$$\chi_\lambda^{g,k} = \sum_{\Lambda,h} a_{\lambda;\Lambda,h}\, \chi_\Lambda^{g',k'}\, \chi_h^{\widetilde{Vir},c(g,g',k)}. \tag{22}$$

(For a yet unknown reason it seems that the branching coefficients $a_{\lambda;\Lambda,h} \in Z$ are restricted to be either zero or one.)

In the next section we will show how to use this coset construction to derive the representation theory of the Virasoro algebra.

3 Construction of modular invariant partition functions

3.1 The construction

In this section we will show how the finite reducibility theorem can be applied to construct (finite) modular invariant sesquilinear combinations of characters of the Virasoro algebra.

Suppose we are given a modular invariant combination

$$\mathcal{Z} = \sum_{\lambda,\lambda'} \overline{\chi_\lambda}\, \mathcal{N}^g_{\lambda,\lambda'}\, \chi_{\lambda'} \tag{23}$$

of characters χ_λ of a Kač-Moody algebra g at level k. (In this the $\mathcal{N}^g_{\lambda,\lambda'}$ are positive integers which give the multiplicity of the primary field $\Phi_{\lambda,\lambda'}$.) Assume furthermore we are given a Kač-Moody subalgebra g' of g such that $c(g,g',k)<1$. Then we may rewrite $\mathcal{Z}$ as a finite combination of characters $\chi^{g'}_\Lambda \chi^{\widetilde{Vir}}_h$ of the direct sum $g' \oplus \widetilde{Vir}$ at level $k'=jk$ and coset charge $c(g,g',k)$

$$\mathcal{Z} = \sum_{\Lambda,\Lambda',h,h'} \overline{(\chi^{g'}_\Lambda \chi^{\widetilde{Vir}}_h)} \mathcal{N}^{g'\oplus\widetilde{Vir}}_{\Lambda,h;\Lambda',h'} (\chi^{g'}_{\Lambda'} \chi^{\widetilde{Vir}}_{h'}) \tag{24}$$

where the invariant tensor $\mathcal{N}^{g'\oplus\widetilde{Vir}}_{\Lambda,h;\Lambda',h'}$ is determined by the branching rule (22)

$$\mathcal{N}^{g'\oplus\widetilde{Vir}}_{\Lambda,h;\Lambda',h'} = \sum_{\lambda,\lambda'} \mathcal{N}^g_{\lambda,\lambda'}\, a_{\lambda;\Lambda,h}\, a_{\lambda';\Lambda',h'} \tag{25}$$

As (24) is a simple rewriting of (23) this may not seem very deep.
New interesting results are obtained by the observation that contractions of modular invariant tensors again produce modular invariant tensors. If, for example, we are given an invariant tensor $\mathcal{N}^{g'}_{\Lambda,\Lambda'}$ of g' at level k' we may construct the invariant tensor

$$\mathcal{N}^{\widetilde{Vir}}_{h,h'} = \sum_{\Lambda,\Lambda'} \mathcal{N}^{g'}_{\Lambda,\Lambda'} \mathcal{N}^{g'\oplus\widetilde{Vir}}_{\Lambda,h;\Lambda',h'} \tag{26}$$

for the Virasoro algebra.

It schould be stressed that from the above presentation it is by no means clear that *all* possible sesquilinear modular invariants $\mathcal{N}_{h,h'}$ can be obtained in this way.

3.2 Modular invariant combinations of Virasoro characters for $c<1$

Let us apply the techniques developed in the previous sections to find modular invariant sesquilinear combinations of characters in the discrete series (6). Consider thereto the Kač-Moody algebra $g = A^{(1)}_1 \oplus A^{(1)}_1$ at level $(1, m-2)$ where $m \geq 3$ is arbitrary.
The diagonal Kač-Moody subalgebra $g' = A^{(1)}_1$ has a central charge $m-1$. Using the dual Coxeter number $\mathrm{g}(A^{(1)}_1)=2$, equation (18) and (21) we have a coset charge [4]

$$\begin{aligned} c(g,g',(1,m-2)) &= c(A^{(1)}_1,1) + c(A^{(1)}_1,m-2) - c(A^{(1)}_1,m-1) \\ &= \tfrac{1.3}{1+2} + \tfrac{(m-2).3}{(m-2)+2} - \tfrac{(m-1).3}{(m-1)+2} = 1 - \tfrac{6}{m(m+1)} \equiv c(m). \end{aligned} \tag{27}$$

This embedding thus satisfies the condition for finite reducibility. In fact one has been able to prove [4]

$$\chi^{A^{(1)}_1,1}_r \chi^{A^{(1)}_1,m-2}_p = \sum_{\substack{q=0,\dots,m-1 \\ r+p-q\in 2Z}} \chi^{A^{(1)}_1,m-1}_q \chi^{\widetilde{Vir},c(m)}_{h(p,q)} \qquad \text{for } r=0,1 \quad p=0,1,\dots,m-2 \tag{28}$$

where the $A_1^{(1)}$-characters χ_n denote the character of the HWM with $SU(2)$-spin $\frac{n}{2}$ (i.e. dimension $n+1$) repesentation for the ground state.
The conformal dimensions $h(p,q)$ occurring in (28) are precisely all those allowed by unitarity (see (6)).

The branching coefficients (22) are thus given by

$$a_{r,s;t,h} = \delta_{p,s}\delta_{t,q}\frac{1}{2}(1+(-1)^{r+s-t}) \qquad , h = h(p,q) \tag{29}$$

Now taking arbitrary modular invariants $\mathcal{N}_{r,r'}^{A_1^{(1)},1} = \delta_{r,r'}$, $\mathcal{N}_{s,s'}^{A_1^{(1)},m-2}$ and $\mathcal{N}_{t,t'}^{A_1^{(1)},m-1}$ of $A_1^{(1)}$ at level $1, m-2$ and $m-1$ respectively, we obtain by (24) and (25) the Virasoro modular invariant at $c = c(m)$

$$\begin{aligned}\mathcal{N}_{p,q;p',q'}^{Vir,c(m)} &= \sum_{r,r'} \mathcal{N}_{r,r'}^{A_1^{(1)},1} \mathcal{N}_{s,s'}^{A_1^{(1)},m-2} \mathcal{N}_{t,t'}^{A_1^{(1)},m-1} a_{r,s;t,p,q}\; a_{r',s';t',p',q'} \\ &= \mathcal{N}_{p,p'}^{A_1^{(1)},m-2} \mathcal{N}_{q,q'}^{A_1^{(1)},m-1},\end{aligned} \tag{30}$$

where we explicitly used $p+p' \in 2Z$, $q+q' \in 2Z$, which is necessary for $\mathcal{N}_{p,p'}^{A_1^{(1)},m-2}$ and $\mathcal{N}_{q,q'}^{A_1^{(1)},m-1}$ to be modular invariant tensors.
So we obtain the result that modular invariant tensors for the Virasoro algebra can be made out of two modular invariants for the $A_1^{(1)}$ Kač-Moody algebra[17,12,16] by (30). One can even prove that all modular invariants can be otained in this way [6].

3.3 Modular invariant combinations of $A_1^{(1)}$ characters

In the previous section we explained the result that modular invariant tensors for the Virasoro algebra with central charge $c = c(m) < 1$ are given by two modular invariants for the $A_1^{(1)}$ -Kač-Moody algebra. Though it lies somewhat outside the scope of this paper we would like to show in this section how the techniques of the previous sections can also be applied to find all modular invariants for $A_1^{(1)}$ [18,12,19]. It turns out that the modular invariants for $A_1^{(1)}$ are classified by (again) the simply-laced algebras $\bar{g} = A_n, D_n$ and E_n [6]. We denote the invariants by $\mathcal{N}(\bar{g})$.

A combination which is invariant for every untwisted affine Kač-Moody algebra g' is the diagonal combination, i.e. the sum of holomorphic squares of all characters at the given level k', this one is called $\mathcal{N}(A_{k'+1})$.

Further modular invariants can be constructed by embedding g' into a bigger affine Kač-Moody algebra g such that $c(g, g', k) = 0$. This is called a conformal embedding. When this is possible one obtains new combinations for g' at level $k' = jk$ by using known (for instance the diagonal) combinations for g by (24), because for $c(g, g', k) = 0$ the coset Virasoro representation can only be the trivial one.

For $A_1^{(1)}$ there are three possibilities [20].

- $A_1^{(1)} \subset A_2^{(1)}, (k=1, j=4)$, resulting in $\mathcal{N}(D_4)$
- $A_1^{(1)} \subset C_2^{(1)}, (k=1, j=10)$, resulting in $\mathcal{N}(E_6)$

- $A_1^{(1)} \subset G_2^{(1)}, (k = 1, j = 28)$, resulting in $\mathcal{N}(E_8)$

The other invariants can be obtained by embeddings of the form $A_1^{(1)} \oplus g' \subset g$ such that $c(g, g' \oplus A_1^{(1)}, k) = 0$, and projecting out with an invariant of g'.

Though one can obtain all possible modular invariants for $A_1^{(1)}$, the construction has a few disadvantages. It is by no means clear that we should obtain *all* modular invariants and the embedding that can be used to obtain an invariant is not unique. Perhaps the most important drawback is that the relation between the classification of modular invariants and the classification of simply-laced Lie algebras is not explained. It should be said that Pasquier [21] has constructed lattice models based on the Dynkin diagrams of simply-laced Lie algebras giving rise to the above partition functions, but I think it would be fair to say that we still need a deeper insight in this correspondence.

4 Extended Virasoro algebras

4.1 Motivation

We have seen that the classification of (rational) 2-D conformal field theories with central charge $c < 1$ is intimately related to the existence of the coset pair $(A_1^{(1)} \oplus A_1^{(1)}, A_1^{(1)})$ of affine Kač-Moody algebras. We would like to obtain the classification of conformal field theories with $c \geq 1$ by extending this relation to general coset pairs (g, g'). However, the methods we used were based on the validity of the finite reducibility theorem, i.e. for coset charges $c(g, g', k) < 1$. From the point of view of coset pairs (g, g') there is however no indication that their applicability should be restricted to $c < 1$. So, what singles out the special c-values $c(g, g', k)$ from all those allowed by unitarity requirements?
An obvious answer is that there may be other symmetries in the game, as it is for instance well-known that the N-extended superconformal algebras each have their own discrete series which extend beyond $c < 1$.
Because the final objective is to obtain a generalized coset algebra $\mathcal{W}$ for which the branching rule $g \rightarrow g' \oplus \mathcal{W}$ is finite, we are looking for primary fields of integer conformal dimension, singlets under g', which connect isomorphic representations of g' starting at different grades in the reduction of g to g'. The minimal set of such fields which gives rise to a finite branching rule is what we would like to call the extended coset Virasoro algebra $Vir(g, g', k)$.
By realizing that the Sugawara stress energy tensor is, what could be called, the affinization of the second order Casimir it was suggested in [22,23] that the affinized higher order Casimirs could be of some relevance. This idea will be discussed in more detail in the next section.

4.2 The Casimir algebra

The starting point for the construction of the Casimir algebra is a conserved current $J(z)$, taking values in some Lie algebra $\bar{g}$ of rank l, such that its components $J^a(z)$

with respect to an antihermitean basis $\{T_a, a = 1, \cdots, \dim(\bar{g}), \mathrm{Tr}(T_a T_b) = \gamma_{ab}\}$ satisfy the Kač-Moody operator product expansions (14).
We introduce the operators

$$T^{(\lambda_i)}(z) = \mathcal{A}^{(\lambda_i)}(g,k) d_{abc\ldots}(J^a(J^b(J^c(\cdots)(z) \tag{31}$$

where $\mathcal{A}^{(\lambda_i)}(g,k)$ is some suitable normalization constant and $d_{abc\ldots}$ is the traceless completely symmetric invariant tensor of order $\lambda_i (i = 1, 2, \ldots, l)$ of $\bar{g}$, so that $T^{(\lambda_i)} = d^{abc\ldots} T_a T_b T_c \ldots$ is the λ_i-th order Casimir of $\bar{g}$.
In particular $T(z) = T^{(2)}(z)$ is the usual Sugawara stress energy tensor, satisfying the Virasoro OPE (1) with central charge $c(g,k)$ as in (18), and the $T^{(\lambda_i)}$ are primary fields of conformal dimension λ_i with respect to $T(z)$.

In [22] it was argued that *if and only if* $k = 1$ the operator product algebra of the currents $T^{(\lambda_i)}(z), (i = 1, \ldots, l)$ closes in the sense that in the operator product expansions of the $T^{(\lambda_i)}$ one only obtains the $T^{(\lambda_i)}$, their descendants or normal ordered products thereof.

Moreover, we argued that the diagonal extended coset Virasoro operator product algebra $Vir(g \oplus g, g, (k_1, k_2))$ closes on the coset analogues $\tilde{T}^{(\lambda_i)}(z)$ of the Casimir operators *if and only if* either k_1 or k_2 equals one. By a coset analogue $\tilde{T}^{(\lambda_i)}(z)$ we mean an operator constructed out of the $g \oplus g$-currents, which is a primary field of conformal dimension λ_i under the coset Virasoro algebra, commutes with the diagonal subalgebra and reduces to $T^{(\lambda_i)}$ in the limit that one of the k_i equals infinity.

We will not give an explicit form of the operator product algebra (more details can be found in [22,24]), but only mention that the first nontrivial extension
$Vir(A_2^{(1)} \oplus A_2^{(1)}, A_2^{(1)}, (1,k))$ turns out to be the Zamolodchikov Z_3-algebra [25], which was investigated in detail in [26]. As already mentioned the operator product algebras that one obtains, contain normal ordered products of the coset Casimir operators $\tilde{T}^{(\lambda_i)}$, so the modes of the Casimir operators do not satisfy ordinary Lie algebra commutation relations. Rather, it is a big associative algebra living in the universal enveloping algebra of a Lie algebra! However, this is perfectly legitimate from a physical point of view because the states created by these composite operators are present in the physical spectrum (i.e. Verma module) anyhow. The fundamental issue involved is really associativity of the OPE, but because this holds for the Kač-Moody currents we started from, it evidently holds for these algebras too.

The representation theory of the algebras $Vir(g \oplus g, g, (1,k))$, for general simply laced g, turns out to be a beautiful generalization of the $g = A_1^{(1)}$ results mentioned in section 1 [22,24,27].
There exists unitary HWM's $L(p,q)$ of the extended coset algebras $Vir(g \oplus g, g)$ for

$$c = c(m) = l\left(1 - \frac{\mathrm{g}(\mathrm{g}+1)}{m(m+1)}\right) \qquad , m = \mathrm{g}+1, \mathrm{g}+2, \cdots \tag{32}$$

parametrized by two dominant integral g-weights p and q of level $m-\mathrm{g}$ and $m-\mathrm{g}+1$

respectively. The conformal dimensions of the highest weight vectors are given by

$$h^{(m)}(p,q) = \frac{((m+1)\bar{p} - m\bar{q}, (m+1)\bar{p} - m\bar{q} + 2\bar{\rho})}{2m(m+1)} \qquad \begin{array}{l} p \in P_+^{m-g} \\ q \in P_+^{m-g+1} \end{array} \tag{33}$$

where $\bar{p}, \bar{q}$ denote the projections of the g-weights onto the weight lattice of the underlying Lie algebra $\bar{g}$, and

$$\bar{\rho} = \frac{1}{2} \sum_{\alpha \in \bar{\Delta}_+} \alpha \tag{34}$$

i.e. half the sum of positive roots of $\bar{g}$.

In [22] evidence was provided for the conjecture that in terms of this extended algebra, say $\mathcal{W}$, there exists a finite reducibility $g \oplus g \rightarrow g \oplus \mathcal{W}$. The branching rule was argued to be

$$\chi_r^{g,1} \chi_p^{g,m-g} = \sum_{\substack{q \in P_+^{m-g+1} \\ \bar{r} + \bar{p} - \bar{q} \in \Lambda_R(\bar{g})}} \chi_q^{g,m-g+1} \chi_{L(p,q)}^{\mathcal{W},c(m)} \tag{35}$$

for $r \in P_+^1, p \in P_+^{m-g}$. The methods of section 3 were used to construct modular invariants for $\mathcal{W}$ using this branching rule [22].

It should be stressed that the results mentioned in this section have only been proved partially. However, a lot of consistency tests have been made which make these conjectures highly plausable.

4.3 Discussion and outlook

In this review we have tried to illustrate how Kač-Moody algebras can be used in the study of 2-D conformal field theories. We have shown in detail how the classification of 2-D CFT's with central charge $c < 1$ is related to the coset pair $(A_1^{(1)} \oplus A_1^{(1)}, A_1^{(1)})$ of Kač-Moody algebras. In the previous section we have motivated that the generalization of the $A_1^{(1)}$ results to more general coset pairs (g, g') would involve studying properly extended coset Virasoro algebras. It is not yet solved how to construct the 'minimal' coset algebra which gives rise to finite reducibility for a general coset pair (g, g'), but some examples are known (cf. [28]) for a possible approach).

We discussed the diagonal coset pair $g \subset g \oplus g$ at level $(1, k)$ and argued that in this case the proper coset algebra is based on the Casimir operators of $\bar{g}$. A generalization of the Feign–Fuchs construction also provided the representation theory of this extended Virasoro algebra. Also for the diagonal pair $A_1^{(1)} \subset A_1^{(1)} \oplus A_1^{(1)}$ at a general level (k_1, k_2) there has been some recent investigation of the Feign–Fuchs construction, suggesting that for $k_1 \geq 2$ the coset algebra is generated by the stress energy tensor and a dimension $\frac{k_1+4}{k_1+2}$ current [29]. From the motivation of obtaining finite branching rules it seems however that it suffices to consider extensions by integer dimension currents only. It would be interesting to see if the results of [29] can be reformulated in terms of an extended Virasoro algebra using integer dimension currents only.

Apart from the hope that there exists a 1-1 correspondence between 2-D CFT's and coset pairs (g, g') of Kač-Moody algebras the results so far seem to support an additional claim. Namely; *there exists a basis of rational conformal field theories for which the partition function is just given by a finite sum of holomorphic squares of irreducible characters of properly extended conformal algebras.* (A similar statement holds presumably for correlation functions.) In the language of Friedan and Shenker [30] this would mean that it suffices to consider only vector bundles over moduli space which have the trivial metric $\mathcal{N}_{h,h'} = \delta_{h,\bar{h}}$.

This work was financially supported by the 'Stichting voor Fundamenteel Onderzoek der Materie' (F.O.M.).

References

[1] see for instance the other contributions in this proceedings.

[2] A.A.Belavin, A.M.Polyakov, A.B.Zamolodchikov, Nucl.Phys.B241(1984)333

[3] D.Friedan, Z.Qiu, S.Shenker, Phys.Rev.Lett.52(1984)1575; *Vertex Operators in Mathematics and Physics*,eds. J.Lepowsky et al. p.419; Comm.Math.Phys.107-(1986)535

[4] P.Goddard, A.Kent, D.I.Olive, Phys.Lett.152B(1985)88, Comm.Math.Phys.103-(1986)105

[5] J.L.Cardy, Nucl.Phys.B270[FS16](1986)186

[6] A.Cappelli, C.Itzykson, J.B.Zuber, Nucl.Phys.B270[FS18] (1987)445; Saclay preprint PhT.87/59; A.Kato, Tokyo preprint UT-509 (may '87)

[7] P.Goddard, D.I.Olive, Int.J.Mod.Phys.A1(1986)303

[8] V.G.Kač, D.H.Peterson, Adv.Math.53(1984)125; V.G.Kač, *Infinite dimensional Lie algebras*, Cambridge Univ. Press, 1985

[9] D.Gepner, E.Witten, Nucl.Phys.B278(1986)493

[10] H.Sugawara, Phys.Rev.170(1968)1659; V.G.Knihnik, A.B.Zamolodchikov, Nucl.-Phys.B247(1984)83; P.Goddard, D.I.Olive, Nucl.Phys.B257[FS14](1985)226, I.T.Todorov, Phys. Lett.153B(1985)77

[11] F.A.Bais, F.Englert, A.Taormina, P.Zizzi, Nucl.Phys.B279(1987)529

[12] P.Bouwknegt, Nucl.Phys.B290[FS20](1987)507

[13] V.G.Kač, D.H.Peterson, Proc.Math.Acad.Sci.USA.78(1981) 3308

[14] P.Goddard, W.Nahm, D.I.Olive, Phys.Lett.160B(1985)111

[15] V.G.Kač, M.Wakimoto, MIT preprint '86

[16] P.Bowcock, P.Goddard, Nucl.Phys.B285[FS19](1987)651

[17] D.Gepner, Nucl.Phys.B287(1987)111

[18] P.Bouwknegt, W.Nahm, Phys.Lett.184B(1987)359

[19] W.Nahm, Stony Brook preprint ITP-SB-87-7

[20] P.Bouwknegt, F.A.Bais, Nucl.Phys.B279(1987)561; A.Schellekens, N.Warner, Phys.Rev.D34(1986)3092

[21] V.Pasquier, Nucl.Phys.B285[FS19](1987)162, Saclay preprint PhT-87-14

[22] F.A.Bais, P.Bouwknegt, K.Schoutens, M.Surridge, Amsterdam/Utrecht preprints ITFA 87-12/THU 87-18 (aug.'87), ITFA 87-18/THU 87-21 (sep.'87), to be published in Nucl.Phys.B and ITFA 87-22/THU 87-29 (dec'87), to be published in the proceedings of the Copenhagen meeting 'Perspectives in String Theory', sep.'87

[23] T.Hayashi, thesis Nagoya Univ.; J.Thierry-Mieg, Cambridge preprint, DAMTP-87-29 (nov.'87)

[24] V.A.Fateev, S.L.Lykyanov, Landau preprint '87

[25] A.B.Zamolodchikov, Theor.Math.Phys.65(1986)1205

[26] V.A.Fateev, A.B.Zamolodchikov, Nucl.Phys.B280[FS18] (1987)644

[27] V.Pasquier, Saclay preprint PhT 87-125 (sep.'87)

[28] M.R.Douglas, Caltech preprint CALT-68-1453 (sep.'87)

[29] J.Bagger, D.Nemeschansky, S.Yankielowicz, Harvard preprint HUTP-87/A073 (sep.'87); D.Kastor, E.Martinec, Z.Qiu, Chicago preprint EFI-87-58 (aug.'87); F.Ravanini, NORDITA preprint 87/56 (oct.'87)

[30] D.Friedan, S.H.Shenker, Nucl.Phys.B281(1987)509

CONFORMAL ALGEBRAS AND DEFECTS IN THE QUANTUM ISING CHAIN

Malte Henkel

Fachbereich Physik, Universität Essen
Postfach 103764, D-4300 Essen 1, Germany

Abstract: At the bulk critical point, the quantum Ising chain with defects is studied. The exact diagonal form of the Hamiltonian for an infinite number of sites is found from a numerical study of the spectrum. The spectrum-generating algebra, which in case of one or two defects is a U(1) Kac-Moody-Virasoro algebra of central charge $c=1$, is constructed and used to compute all critical exponents. The behaviour of the partition function under modular transformation is investigated.

1. INTRODUCTION

A statistical model at the critical point of a second order phase transition is scale-invariant. In addition, many models do also exhibit *conformal* invariance. In two dimensions, conformal invariance plays a central role in constructing a systematic theory of critical correlations, as first recognised by Belavin, Polyakov and Zamolodchikov [1] (for reviews, see [2]). In particular, the critical exponents describing the powerlike decay of the correlation functions at the critical point can be exactly calculated from the representation theory of the Virasoro algebra.

It was shown by Cardy [3] that a model defined in the (half-) infinite two-dimensional plane can be mapped onto an infinitely long strip of finite width L via a conformal

ISBN 0-12-218100-X

transformation, with periodic (free) boundary conditions, respectively (see Fig.1).This particular strip geometry also arises naturally in the Hamiltonian limit of the transfer matrix (see [4]).Taking this limit amounts to taking the continuum limit in one direction,while one retains a finite lattice in the orthogonal directions,of width $L=Na$,where a is the lattice constant.The Hamiltonian H ,describing a two-dimensional model,is then defined on a quantum chain of N sites.

If the system is conformally invariant,H is proportional to the generator L_0 of the Virasoro algebra,of central charge c

$$[L_n, L_m] = (n-m) L_{n+m} + \frac{c}{12} (n^3-n)\delta_{n+m,0} \tag{1.1}$$

and its spectrum is classified according to the irreducible representations of the Virasoro algebra,characterized by the lowes eigenvalues of L_0 (which correspond to the scaling dimensions of the primary fields of conformal theory)[2].At the critical point,one has for the correlation length ξ_i of a scaling field ϕ_i (in case of periodic boundary conditions) [3]

$$\xi_i^{-1} = E_i - E_0 = 2\pi x_i N^{-1} \tag{1.2}$$

where E_i is an energy level of H, E_0 is the ground state energy and the critical exponent x_i describes the correlation function in the two-dimensional plane

$$\langle \phi_i(r) \phi_i(0) \rangle = r^{-2x_i} \tag{1.3}$$

Using finite-size scaling techniques,Gehlen and Rittenberg showed how to identify the primary fields and to obtain *all* critical exponents for specific models, including the Ising, Potts, Ashkin-Teller and D_6 -symmetric quantum chains[5], for various boundary conditions.

Here, we want to study a more general situation. Consider the upper half plane (which corresponds to a quantum chain with free boundary conditions). It is separated from the lower halfplane by an inpenetrable wall, which is lying on the real axis. Let the wall become partly penetrable, that is, consider an *infinite two-dimensional plane with an infinite line of defects* [6,7] (see Fig.1). In particular, we will ask whether one still has conformal invariance in the presence of defects.

We are going to examine the simplest case study: the Ising model with a defect line. From the general classification [1,2] of conformally invariant models, valid if no defects are present, one knows the central charge $c=\frac{1}{2}$ for the Ising model. This implies that the critical exponents take certain fixed rational values [8]. On the other hand, the critcal exponents (which in this case are generalized surface exponents) of the 2D Ising model with defects do depend continously on the defect strength κ [9]. Since it is known that for all unitary models (for which H is hermitian) the critical exponents can only take values from a discrete set of rational numbers if $c<1$ [8], one should expect $c\geq 1$ for the 2D Ising model with defects, if conformal invariance is valid.

Indeed, we shall explicitly contruct the spectrum-generating Virasoro algebra, with $c=1$, for the quantum Ising chain with defects. The construction of this conformal algebra shows that at least for the Ising model with defects, conformal invaiance is still valid. In addition, the conformal algebra conains a U(1) Kac-Moody algebra. At the critical point, we thus have a conserved current, which is not obvious in the original statistical mechanics formulation.

In section 2, we give the quantum chain formulation of the

Ising model with a defect line. From the spectrum of H, several features are found which are different from the spectrum of the defectless Ising model. In section 3, the spectrum generating conformal algebra is explicitly constructed, and in section 4, these results are used to explain the observations which were made in the spectrum in section 2. The behaviour of the partition function under modular transformations is investigated. In section 5, we mention some further lines of development.

2. QUANTUM ISING CHAIN WITH DEFECTS

2.1. General formulation

We now write down the Hamiltonian for the quantum Ising chain with two defects, described by defect strengths κ_1 and κ_2, at the bulk critical point

$$H = -\frac{1}{2} \sum_{n=1}^{N} [\sigma^z(n) + \sigma^x(n)\sigma^x(n+1)]$$

$$+ \frac{1}{2} \sum_{i=1}^{2} (1-\kappa_i)\ \sigma^x(\frac{2i-1}{4} N)\sigma^x(\frac{2i-1}{4} N + 1) \qquad (2.1)$$

where σ^x and σ^z are Pauli matrices and periodic boundary conditions are assumed. If one has $\kappa_1=1$, for example, one has no defect at the site $\frac{N}{4}$, while for $\kappa_1=0$, a free surface is obtained. Since H commutes with the operator

$$Q = \frac{1}{2} (1 - \prod_{n=1}^{N} \sigma^z(n)) \qquad (2.2)$$

the spectrum of H can be decomposed into sectors 0 and 1 corresponding to the eigenvalues of Q.

Exact results are available for the case $\kappa=\kappa_1=\kappa_2$. The critical exponent $x_\sigma(\kappa)=\frac{1}{2}\eta_{\|}(\kappa)$ for the magnetization correlation function is given by [9]

$$x_\sigma(\kappa) = \frac{1}{2} \left(\frac{2}{\pi} \arctan \left(\frac{1}{\kappa}\right) \right)^2 \tag{2.3}$$

while on the other hand the energy density exponent $x_\epsilon(\kappa)=1$.

Under the assumption of conformal invariance, Turban [10] showed that the critical exponents can still be computed from the spectrum of the Hamiltonian, although x_i does depend on κ_1 and κ_2

$$\xi_i^{-1} = E_i - E_0 = 2\pi\, x_i(\kappa_1,\kappa_2)\, N^{-1} \tag{2.4}$$

This relation is of central importance in the sequel, once having established the conformal invariance of the model. This can be done by two ways. First, one can establish [6] a one-to-one correspondence with the analysis of Kadanoff [24] of the Ising model with an infinite defect line in the infinite 2D plane. Second, as will be done here, one can explicitly construct the conformal spectrum-generating algebra.

2.2. Diagonalization of the Hamiltonian

The diagonalization of H is done in two steps, following a technique of Lieb, Schultz and Mattis [11]. By a Jordan-Wigner transformation, H is written in terms of fermionic operators C_n

$$H = - \sum_{n,m=1}^{N} \left[C_n^+ A_{nm} C_m + \frac{1}{2} \left(C_n^+ B_{nm} C_m^+ + h.c. \right) \right] \tag{2.5}$$

where A is a symmetric and B an antisymmetric matrix. By a

Bogoljubov transformation, H is written in diagonal form

$$H = \frac{2\pi}{N} \sum_{k=1}^{N} \Lambda_k \quad \eta_k^+ \eta_k \tag{2.6}$$

where the one-fermion energies Λ_k are determined from

$$(A + B) (A - B) \psi_k = (\frac{2\pi}{N} \Lambda_k)^2 \psi_k \tag{2.7}$$

This formulation has the advantage that one has, for N sites, only to deal with $N \times N$ matrices as compared with $2^N \times 2^N$ matrices if the Hamiltonian is diagonalized directly. Consequently, lattices with $N \approx 200$ can be studied easily, instead of $N \approx 10$ via the direct approach.

The extrapolation of the finite-size data towards the limit $N \to \infty$ is done using the algorithm of Bulirsch and Stoer [12]

$$\begin{aligned} &T_{-1}^{(N)} = 0 \qquad , \qquad T_0^{(N)} = T(h_N) \quad , \\ &T_m^{(N)} = T_{m-1}^{(N)} + \left[T_{m-1}^{(N+1)} - T_{m-1}^{(N)} \right] \cdot \\ &\quad \left[\left[\frac{h_N}{h_{N+m}}\right]^{\omega} \left[1-(T_{m-1}^{(N+1)}-T_{m-1}^{(N)})/(T_{m-1}^{(N+1)}-T_{m-2}^{(N+1)}) \right] - 1 \right]^{-1} \end{aligned} \tag{2.8}$$

where $T(h_N)$ are the finite-size data, $h_N=1/N$ and ω is a free parameter. One expects that the sequence $T_1^{(N)}$ converges faster than $T_0^{(N)}$ and so on. A comparison with other frequently used extrapolation algorithms shows that the Bulirsch-Stoer method yields in general more reliable results [13].

N	Λ_1	Λ_2	Λ_3	Λ_4
10	0.38448159	0.58270842	1.30841399	1.50073493
30	0.39347004	0.59620667	1.37805910	1.58020757
50	0.39514801	0.59873483	1.38718258	1.59056066
70	0.39585401	0.59980033	1.39056188	1.59440226
90	0.39624277	0.60038758	1.39229942	1.59638031
110	0.39648886	0.60075953	1.39335273	1.59758068
130	0.39665863	0.60101624	1.39405796	1.59838501
150	0.39678283	0.60120407	1.39456263	1.59896094
170	0.39687762	0.60134747	1.39494140	1.59939341
190	0.39695234	0.60146054	1.39523604	1.59972995
210	0.39701277	0.60155197	1.39547172	1.59999924
∞	0.3975836	0.6024164	1.3975835	1.602416

Table 1: Finite-size data for the lowest four one-fermion energies in the sector Q=0 for $\kappa_1=\frac{1}{2}$ and $\kappa_2=1$. The limit $N\to\infty$ was obtained form eq.(2.8), with $\omega=1$. The expected accuracy is two units in the last given digit.

In table 1, we give the lowest four one-fermion energies, for the case $\kappa_1=\frac{1}{2}, \kappa_2=1$ and the sector Q=0. In contrast to the defectless case, there is no exact formula for the Λ's available. We observe the following, which holds for arbitrary values of κ_1 and κ_2 :

1. All the higher one-fermion energies can be written as the sum of either Λ_1 or Λ_2 and a positive integer.
2. $\Lambda_1+\Lambda_2 = 1$.

These observations suggest the following parametrization of H in the limit $N\to\infty$ [6]

$$H = \frac{2\pi}{N} \sum_{r=0}^{\infty} [(r + \frac{1}{2} - \Delta)\, a_r^+ a_r + (r + \frac{1}{2} + \Delta)\, b_r^+ b_r\,] - \frac{\pi\zeta}{6N} \quad (2.9)$$

where a_r and b_r are fermionic annihilation operators. In writing this, we drop the bulk and surface contributions. While this form accounts for the observations made above, the numeical data also yield a conjecture for a relation between Δ and κ_1 and κ_2 [7] (which reproduces, for example, the data of table 1 to at least 6 digits)

$$\Delta(\kappa_1, \kappa_2) = \left[\begin{array}{ll} D(\kappa_1) - D(\kappa_2) & \textit{if } Q=0 \\ 1/2 - D(\kappa_1) - D(\kappa_2) & \textit{if } Q=1 \end{array} \right. \quad (2.10)$$

where

$$D(\kappa) = \left| \frac{1}{\pi} \arctan\left(\frac{1}{\kappa}\right) - \frac{1}{4} \right| \quad (2.11)$$

For negative values of κ, one has to use the relation $arctan(-1/\kappa)=\pi-arctan(1/\kappa)$. Since $x_\sigma(\kappa)$ (see eq. (2.3)) is computed from the lowest gap in the sector Q=1, we reobtain the known exact result. Finally, the parameter ζ describing the leading finite-size correction to the ground state energy is given by

$$\zeta = \frac{1}{2} - 6\,\Delta^2(\kappa_1, \kappa_2\ ;\ Q=0\ ) \quad (2.12)$$

In case of a defectless quantum chain with periodic boundary conditions, $\zeta=c$ is the central charge of the spectrum generating Virasoro algebra [14].

Before we are going to exploit the ansatz (2.9) for the con struction of the conformal algebra (see section 3), we present the full spectrum of the Ising quantum chain with defects.

2.3. Examination of the spectrum

Consider the spectrum of H, for the case $\kappa_2=1$ in the sector Q=0 (Fig. 2). In case of periodic ($\kappa_1=1$) and antiperiodic ($\kappa_1=-1$) boundary conditions, the spectrum is known [5,15] to be generated from a pair of $c=1/2$ Virasoro algebras, the representations being (0,0)+(1/2,1/2) for periodic and (1/16,1/16) for antiperiodic boundary conditions. For free boundary conditions ($\kappa_1=0$), the spectrum is given by the (0) representation of a single c=1/2 Virasoro algebra [16]. In these cases, one has a *finite* number of primary fields, given by the mentioned irreducible representations of the Virasoro algebra. The spectrum is then given by the *conformal towers* of the primary fields. Thus, the critical exponents of the excited states can be written as the sum of the exponent of a primary field and a positive integer (half-integer for $\kappa_1=0$).

For the other values of κ_1, one interpolates between these cases. We observe the following :

1. Beginning from the lowest level x_0, we find excited states, whose values for $x_i(\kappa)$ are given by x_0 plus a positive integer. Moreover, the degeneracy d_R of the R'th excited state in this tower is given by the generating function

$$P(q) = \sum_{R=0}^{\infty} d_R\, q^R = \prod_{n=1}^{\infty} (1-q^n)^{-1} \qquad (2.13)$$

   that is, d_R is the number of partitions of R.

2. There are additional levels, which have a noninteger difference in the values of their critical exponent with

the ground state.These levels also have excited states whose degeneracies are given by eq. (2.13).

To summarize: we find evidence for new primary fields (in fact,we shall see that there are infinitely many of them) with conformal towers of excited states described by eq. (2.13).This pattern of degeneracies does not appear for $c<1$ [17],rather,it is characteristic for $c\geq 1$.

These observations are going to be explained in section 4.1.

3. VIRASORO AND KAC-MOODY ALGEBRA

In this section,we explicitly construct the spectrum generating conformal algebra.Let η_α be fermionic operators such that $\eta_\alpha^+=\eta_{-\alpha}$ and $\{\eta_\alpha,\eta_\beta\}=\delta_{\alpha+\beta,0}$ .Consider the generators [7]

$$L_n(\Delta) = \frac{1}{2}\sum_{p\in Z+\frac{1}{2}} (p-\Delta) : \eta_{n-p+\Delta}\, \eta_{p-\Delta} : \tag{3.1}$$

where the dots denote normal ordering.These generators generalise the well-known realization of the $c=1/2$ Virasoro algebra in terms of free fermions [18],where $\Delta=1/2$ corresponds to periodic (Ramond [19]) and $\Delta=0$ to antiperiodic (Neveu-Schwarz [20]) boundary conditions.Comparing with (2.9), the Hamiltonian will be built from two generators $L_n(\Delta)$ (see below).

From now on,we take $\Delta\neq 0,1/2$ .Then,the generators do not satisfy the unitarity condition $L_n^+=L_{-n}$,but instead we have

$$L_n^+ (\Delta) = L_{-n} (-\Delta) \tag{3.2}$$

The $L_n(\Delta)$ satisfy the commutation relations

$$[L_n(\Delta) , L_m(\Delta)] = \frac{n-m}{2} L_{n+m}(\Delta) - \frac{1}{24} (n^3-n) \delta_{n+m,0} + \frac{n}{4} (\Delta^2 - \frac{1}{4}) \delta_{n+m,0} \tag{3.3}$$

To bring this into the canonical form of a Virasoro algebra, define

$$\tilde{L}_n (\Delta) = 2 L_n (\Delta) + \frac{1}{2} (\Delta^2 - \frac{1}{4}) \delta_{n,0} \tag{3.4}$$

and the $\tilde{L}_n(\Delta)$ satisfy a Virasoro algebra with a *negative* central charge c=-2.This illustrates the general fact that there are no unitary representations for a Virasoro algebra with negative central charge [8].

The commutator of L with L^+ yields

$$[L_n(\Delta) , L_m(-\Delta)] = \frac{n-m}{2} L_{n+m}(-\Delta) - \frac{n^2}{2} W_{n+m}(-\Delta) + \frac{1}{4} (\frac{n^3}{3} + 2n^2\Delta + n(\Delta^2-\frac{1}{12})) \delta_{n+m,0} \tag{3.5}$$

where a new unitary field $W_n(\Delta)$ is introduced

$$W_n(\Delta) = \frac{1}{2} \sum_{p\in Z+½} : \eta_{n-p+\Delta} \eta_{p-\Delta} : - \frac{\Delta}{2} \delta_{n,0} = - W_n(-\Delta) \tag{3.6}$$

$$W_n^+ (\Delta) = W_{-n} (\Delta) \tag{3.7}$$

which satifies a U(1) Kac-Moody algebra

$$[W_n (\Delta) , W_m (\Delta)] = \frac{n}{4} \delta_{n+m,0} \tag{3.8}$$

and closes with the L's

$$[\ L_n(\ \epsilon\Delta\ )\ ,\ W_m\ (\ \Delta\ )\ ] = -\ \frac{m}{2}\ W_{n+m}(\ \Delta\ )\ +\ \epsilon\ \frac{n^2}{2}\ \delta_{n+m,0} \tag{3.9}$$

with $\epsilon=\pm1$.

To write the Hamiltonian in terms of the L's,we need one $L_n(\Delta)$ and one $L_n(-\Delta)$.Define new unitary generators $K_n(\Delta)$

$$K_n\ (\ \Delta\ ) = L_n\ (\ \Delta\ ) + L_n\ (\ -\Delta\ ) + \frac{1}{2}\ \Delta^2\ \delta_{n,0} \tag{3.10}$$

and if one identifies in eq. (2.9)

$$a_r\ =\ \eta_{r+1/2-\Delta}\quad ,\quad b_r\ =\ \eta_{r+1/2+\Delta} \tag{3.11}$$

one can relate H to the generator $K_0(\Delta)$:

$$H = \frac{2\pi}{N}\ (\ K_0\ (\Delta)\ -\ \frac{1}{24}\ ) \tag{3.12}$$

where the constant term comes from the regularised sum of the zero point energies of the fermionic oscillators in eq. (2.9).Comparing (3.12) with (2.9),we obtain

$$\zeta\ =\ \frac{1}{2}\ -\ 6\ \Delta^2 \tag{3.13}$$

in agreement with the earlier obervation eq. (2.12).

Finally,we write the *unitary* conformal algebra :

$$[\ K_n(\Delta)\ ,\ K_m(\Delta)\ ] = (n-m)\ K_{n+m}(\Delta)\ +\ \frac{1}{12}\ (n^3-n)\ \delta_{n+m,0} \tag{3.14}$$

$$[\ K_n(\Delta)\ ,\ W_m(\Delta)\ ] = -m\ W_{n+m}(\Delta) \tag{3.15}$$

$$[\ W_n(\Delta)\ ,\ W_m(\Delta)\ ] = \frac{n}{4}\ \delta_{n+m,0} \tag{3.16}$$

which is a U(1) Kac-Moody-Virasoro algebra with central

charge c=1. From the fact that H is part of this conformal algebra, we conclude that the Ising quantum chain with defects at the bulk critical point is conformally invariant.

4. CONSEQUENCES OF THE CONFORMAL ALGEBRA

Having established the conformal algebra, we now collect the consequences for the spectrum of the Hamiltonian, which explain the observations made in section 2.3. Then, we compute the partition function on a torus and study its properties under modular transformation.

4.1. Spectrum in terms of irreducible reprensentations

We now can explain the earlier observations made in section 2.3 about the spectrum of H (see Fig.2). Because of c=1, there is no need for fixed critical exponents, in contrast to the defectless case, where $c=1/2$. There are infinitely many primary fields, which are labelled by the eigenvalues of $W_0(\Lambda)$. Their critical exponents are in the sector Q=0 ($\Lambda_0=\Lambda(Q=0)$ see (2.4)) [7]

$$x_{0,p}^{\pm} = 2p(p \pm \Lambda_0) \quad , \quad w_0 = \pm p - \frac{1}{2}\Lambda_0 \quad , \quad p\geq 1$$
$$x_{0,0} = 0 \quad , \quad w_0 = \quad -\frac{1}{2}\Lambda_0 \tag{4.1}$$

and in the sector Q=1 ($\Lambda_1=\Lambda(Q=1)$)

$$x_{1,p}^{+} = 2p(p+1\pm\Lambda_1)+\frac{1}{2}\pm\Lambda_1 + \frac{1}{2}(\Lambda_1^2-\Lambda_0^2) \quad , \quad w_0 = \pm(p+\frac{1}{2}) - \frac{1}{2}\Lambda_1 \tag{4.2}$$

and Λ_0 and Λ_1 are taken from eq. (2.10).

From the presence of the U(1) Kac-Moody current, it follows that the degeneracies d_R of *all* excited states are given by the generating function P(q) of eq. (2.13)[7]. This is *not* the case if one only has a $c=1$ Virasoro algebra without a U(1) Kac-Moody algebra.

Finally, for $\Delta=\frac{1}{4}$ the central charge is reduced from $c=1$ to $c=\frac{1}{2}$:

$$K_n\ (\tfrac{1}{4}) = \frac{1}{2}\ L_{2n}\ (0) + \frac{1}{32}\ \delta_{n,0} \tag{4.3}$$

(recall that $L_n(0)$ satisfies a $c=\frac{1}{2}$ Virasoro algebra). This is realized, in the case of a quantum chain with free boundary conditions $\kappa_1=0, \kappa_2=1$, but it also occurs, for example, in the case $\kappa_1=\kappa_2=\sqrt{2} - 1$.

4.2. Partition function

We want to compute the partition function on a torus of length N' and width N. This can be done using a well-known formula [15,21]. By eq. (2.4) the energy differences E_l-E_0 are related to some exponents x_l. We have in the limit $N,N' \to\infty$ but $\delta=N'/N$ fixed

$$Z = \text{Tr}\ e^{-H\ N'} = \exp(\frac{\pi\zeta}{6N}) \sum_M d_M\ e^{-2\pi\ x_M \delta} \tag{4.4}$$

In writing this, we drop the nonuniversal bulk and surface contributions. Using the results of section 4.1, Z is readily evaluated

$$Z = q^{-\frac{1}{24}}\ P(q) \sum_{p\in Z} \left\{ q^{2(p+\Delta_0/2)^2} + q^{2(p+(1+\Delta_1)/2)^2} \right\} \tag{4.5}$$

where $q = e^{-2\pi\delta}$ and $P(q)$ is from eq.(2.13).

We now study the behaviour of Z under the modular transformation $\delta \to 1/\delta$, which corresponds to interchanging the roles of N and N' .The transformed partition function Y reads [7]

$$Y = q^{-\frac{1}{24}} P(q) \sum_{k\in Z} q^{k^2/8} \left\{ \cos(k\Delta_0\pi) + \cos(k(1+\Delta_1)\pi) \right\} \quad (4.6)$$

The system described by the partition function Y has the critical exponents of the defectless Ising model,as can be seen by applying eq. (4.4) to Y .This observation is not surprising.In the Ising model with defects,the correlation functions are measured between points which are parallel to the defect line (Fig.2).For the modular transformed model, however,one measures perpendicular to the defect line.

To conclude,although the Ising model with as well as without defects is conformally invariant,the model with defects is *not* modular invariant,even if the defectless model is.

5. FURTHER DEVELOPMENTS

We now briefly mention some further studies about quantum chains with defects.From the discussion of the quantum chain with one or two defects,we have constructed the conformal algebra,if the Hamiltonian can be written,in the limit $N\to\infty$, in terms of a single pair of real free fermions.More fermions could appear if more defects are taken into account.As an example,consider the quantum Ising chain with three defects :

$$H = -\frac{1}{2}\sum_{n=1}^{N} \left[\sigma^z(n) + \sigma^x(n)\sigma^x(n+1) \right] + \frac{1}{2}\sum_{i=1}^{3} (1-\kappa_i)\, \sigma^x\!\left(\frac{2i-1}{6}N\right) \sigma^x\!\left(\frac{2i-1}{6}N + 1\right) \quad (5.1)$$

Using the techniques of section 2, we obtain in the $N\to\infty$ limit [22]

$$H = \frac{6\pi}{N} \sum_{i=1}^{3} \left\{ \sum_{r=0}^{\infty} \left[(r+\tfrac{1}{2}-\Delta_i)\, a_r^{+(2i-1)} a_r^{(2i-1)} + (r+\tfrac{1}{2}+\Delta_i)\, b_r^{+(2i)} b_r^{(2i)} \right] - \frac{1}{12}\left(\frac{1}{2} - 6\,\Delta_i^{\,2} \right) \right\} + \epsilon_{bulk} N + \epsilon_{surface} \tag{5.2}$$

where $\Delta_i = \Delta_i(\kappa_1,\kappa_2,\kappa_3)$ depends on all three defect strengths and also on the sector Q.

Instead of one, we now have three pairs ("generations") of fermions. Constructing the conformal algebra, H is decomposed into three Hamiltonians $H(\Delta_i)$, each of the form treated in section 3. The Hamiltonians $H(\Delta_i)$ each give rise to a c=1 Virasoro algebra (which are mutally commuting) and thus the spectrum of H is generated by a c=3 Virasoro algebra.

Generalizing the construction of section 3 leading to the Kac-Moody algebra, one further obtains [22] a twisted realization of an SO(6) Kac-Moody algebra with central charge $k=1$. This is just the generalization of the well-known [18] fermionic contruction of Kac-Moody algebras to twisted fermions. Moreover, one can even show [22] that this system (5.2) admits a nonlinear realization, in terms of twisted fermions, of the $N=4$ $SO(4) \otimes U(1)$ extended superconformal algebra of Schoutens [23]. Generalizing the techniques of Seiberg and Schwimmer [25], one can also construct an automorphism to the untwisted realization of the $N=4$ superconformal algebra mentioned above.

Acknowlegements

It is a pleasure to thank A. Patkós for a fruitful collaboration. I also thank V. Rittenberg for useful discussions.

REFERENCES

[1] AA Belavin,AM Polyakov and AB Zamolodchikov,Nucl.Phys. B241,333 (1984)

[2] JL Cardy in Domb and Lebowitz(Eds),Phase Transitions and Critical Phenomena ,Vol.11(1987),London(Academic)
C Itzykson Invariance conforme et modeles critiques bidimensionelles,Marseille preprint CPT-86/P1915
V Rittenberg in Barut and Doebner(Eds),Conformal Groups and related symmetries,Springer Lecture Notes in Physics 261(1986),Berlin(Springer)

[3] JL Cardy J.Phys. A17,L385 (1984)

[4] JB Kogut Rev.Mod.Phys. 51,659 (1979)

[5] Gv Gehlen and V Rittenberg J.Phys. A19,L625 (1986)
M Baake,Gv Gehlen and V Rittenberg J.Phys. A20,L479 (1987) and L487
G Schütz J.Phys. A in press
M Henkel J.Phys. A20,995 (1987)
FC Alcaraz,MN Barber and MT Batchelor Phys.Rev.Lett. 58, 771 (1987)

[6] M Henkel and A Patkos J.Phys. A20,2199 (1987)

[7] M Henkel and A Patkos Nucl.Phys. B285[FS19],29 (1987)

[8] D Friedan,Z Qiu and S Shenker Phys.Rev.Lett. 52,1575 (1984)

[9] R Bariev Sov.Phys. JETP 50,613 (1979)
BM Mc Coy and JHH Perk Phys.Rev.Lett. 44,840 (1980)
I Peschel and KD Schotte Z.Phys. B54,305 (1984)

[10] L Turban J.Phys. A18,L325 (1985)

[11] E Lieb,T Schultz and D Mattis Ann.of Phys 16,407 (1961)
LG Guimaraes and JR Drugowich de Felicio J.Phys. A19, L341 (1987)

[12] R Bulirsch and J Stoer Num.Math. 6,413 (1964)

[13] M Henkel and G Schütz Bonn preprint HE-87-18

[14] I Affleck Phys.Rev.Lett. 56,746 (1986)
HW Blöte,JL Cardy and MP Nightingale Phys.Rev.Lett. 56,

742 (1986)
[15] JL Cardy Nucl.Phys B270[FS17],186 (1986)
[16] JL Cardy Nucl.Phys B275[FS18],200 (1986)
Gv Gehlen and V Rittenberg J.Phys A19,L631 (1986)
[17] A Rocha-Caridi in Lepowsky et al(Eds) Vertex Operators in Mathematics and Physics(1985),Berlin(Springer)
[18] P Goddard and D Olive Int.J.Mod.Phys. A1,303 (1986)
[19] P Ramond Phys.Rev. D3,2493 (1971)
[20] A Neveu and JH Schwarz Nucl.Phys B31,86 (1971)
[21] C Itzykson and JB Zuber Nucl.Phys. B275[FS17],580 (1986)
[22] M Henkel and A Patkos Niels Bohr preprint NBI-HE-87-59
[23] K Schoutens Nucl.Phys B , in press
[24] LP Kadanoff Phys.Rev. B24,5382 (1981)
[25] N Seiberg and A Schwimmer Phys.Lett 184B,191 (1987)

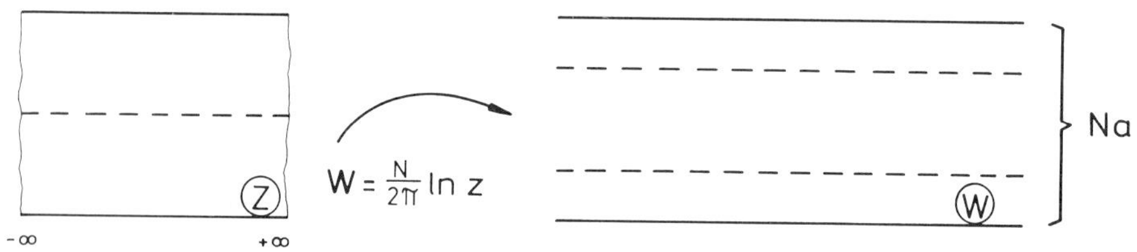

Fig.1 Conformal mapping between the infinite 2D plane with an infinite line of defects (denoted by the broken line) and the infinitely long strip of finite width *Na* with two defects.

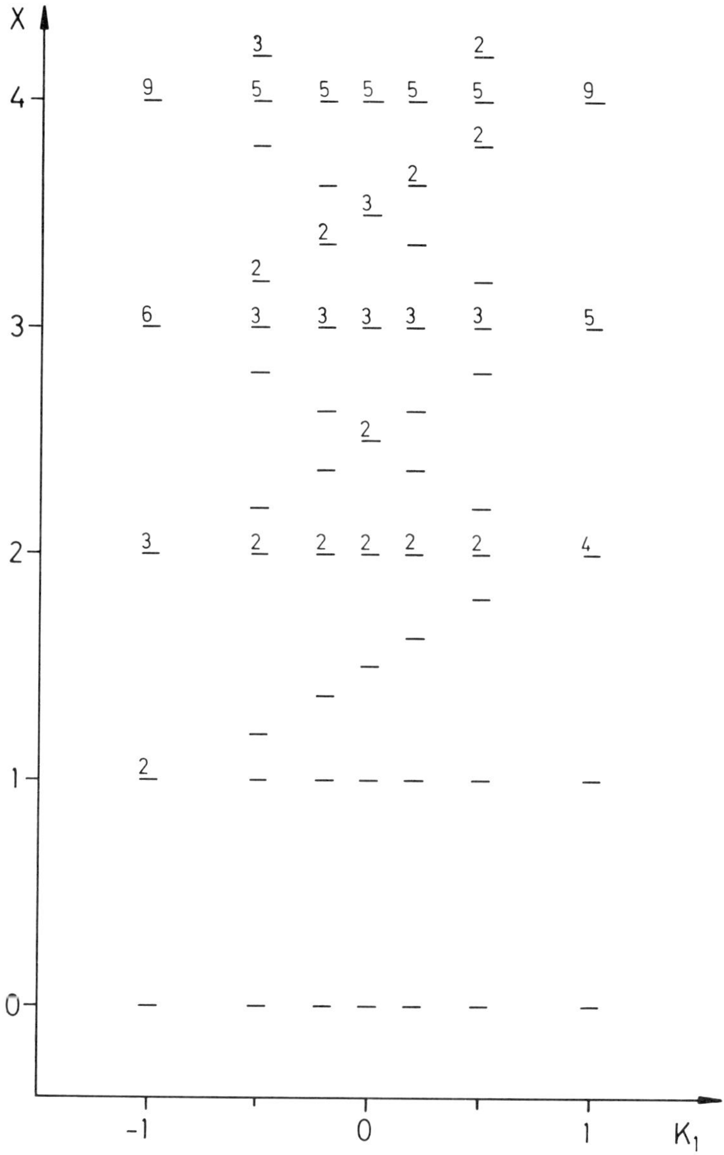

Fig.2 Spectrum in the sector Q=0 for $\kappa_1=\frac{1}{2}$ and $\kappa_2=1$. The numbers denote the degeneracies of the levels.

CONFORMAL QUANTUM FIELD THEORIES AND INTEGRABLE SYSTEMS

M. KAROWSKI

ABSTRACT. The Yang-Baxter algebra method to solve integrable models is reviewed. Statistical systems at second order phase transition points should exhibit conformal invariance for long distances. Therefore they will yield conformal quantum field theories in the scaling limit. Their conformal properties can be analysed using finite size scaling behaviour. For integrable lattice models in two dimensions methods are proposed to calculate from the Bethe ansatz solution the "conformal anomaly" and scaling dimensions. As an application results for the q-state Potts model and modified six-vertex models are presented. It is conjectured that the latter exhaust the class of all unitary conformal quantum field theories with "central charge" $c < 1$.

1. Introduction.

I would like to report on results concerning conformal properties of integrable models in two dimensions. Some of the investigations rely on a collaboration with H. de Vega [1]. Here I will give only a summary of the main results and an introdution to the "Bethe ansatz" method which has been used. The detailed calculations can be found in ref. [2].

Conformal invariance of two dimensional quantum field theories has become a subject of main interest because of its connection to string theory. A string forms a two dimensional world sheet in space time. Reparametrization invariance of this surface causes the importance of conformal concepts in string theories. Statistical mechanical systems at second order phase transition points (i.e. critical systems) should possess conformal invariance[1] [3]. We shall use this fact to obtain information on conformal quantum field theories. It would be interesting to find integrable lattice models which yield these conformal quantum field theories in the scaling limit.

[1]New results on representation theory of conformal invariance explain the old mysterious observation that critical exponents in two dimensions are usually rational numbers.

ISBN 0-12-218100-X

Since a direct calculation of the central charge of the Virasosro algebra and the conformal dimensions for specific models is usually not simple, it seems worthwhile to determine them for integrable models from the Bethe ansatz solution. This will be done for the six-vertex model (the critical eight-vertex model) and for some modified six-vertex models related to Potts models.

The theory of integrable systems has a long history. The technique known as **Yang-Baxter algebra** method, **algebraic Bethe ansatz** or **quantum inverse method** [4] has been applied to a large number of models. The spectrum has been calculated, the states have been constructed, and for some models also form factors and correlation functions have been obtained. The method can be applied to various kinds of models:

(i) quantum chains in one dimension, e.g. Heisenberg models [5]*

(ii) classical statistical models in two dimensions, e.g. Ising [6]* and vertex [7]* models

(iii) quantum field theories in $1+1$-dimensions [8]*:

bosonic models, e.g. nonlinear Schrödinger-, sine-Gordon-, σ- models

fermionic models, e.g. massive Thirring-, Goss-Neveu models.

2. Integrable Systems.

2.1 The six-vertex model. I like to explain the Yang-Baxter algebra method in terms of the six-vertex model, a model of type (ii), a classical statistical lattice model in two dimensions. Its partition function on an $N \times M$ periodic square lattice L is

$$Z = \sum_{conf.} \prod_{x \in L} \omega(x) \tag{1}$$

where the sum extends over all allowed "bond configurations". Each bond can accept one of two states chararterized by $\alpha = 1,2$ or arrows

$$\text{horizontal bonds} \quad \alpha = 1,2 = \leftarrow\,, \ \rightarrow$$
$$\text{vertical bonds} \quad \beta = 1,2 = \ \uparrow\,, \ \downarrow\,.$$

The configuration is allowed if at each vertex x there are two incoming and two out going arrows. The vertex weights $\omega(x)$ are determined by the six bond configurations

*and later works

at the lattice site x. They assume the following values [4,9]

$$\begin{aligned} S^{11}_{11} &= \leftarrow\!\!\overset{\uparrow}{\underset{\uparrow}{|}}\!\!\leftarrow = S^{22}_{22} = \rightarrow\!\!\overset{\downarrow}{\underset{\downarrow}{|}}\!\!\rightarrow &= a = \sin(\gamma - \theta) \\ S^{12}_{12} &= \leftarrow\!\!\overset{\downarrow}{\underset{\downarrow}{|}}\!\!\leftarrow = S^{21}_{21} = \rightarrow\!\!\overset{\uparrow}{\underset{\uparrow}{|}}\!\!\rightarrow &= b = \sin\theta \\ S^{21}_{12} &= \rightarrow\!\!\overset{\uparrow}{\underset{\downarrow}{|}}\!\!\leftarrow = S^{12}_{21} = \leftarrow\!\!\overset{\downarrow}{\underset{\uparrow}{|}}\!\!\rightarrow &= c = \sin\gamma \end{aligned} \tag{2}$$

By this notation the weights have been arranged as a $4 \times 4-$ matrix

$$\begin{matrix} & \delta & \\ \gamma & -\!|\!- & \alpha \\ & \beta & \end{matrix} = S^{\gamma\delta}_{\alpha\beta}(\theta, \gamma) = \begin{pmatrix} a & & & \\ & b & c & \\ & c & b & \\ & & & a \end{pmatrix}. \tag{3}$$

The parametrization in terms of the "spetral parameter" θ and the "coupling" γ has been introduced for later convenience (cf. eq. (11)). A trivial overall factor is set equal to one.

In order to express the partition function in terms of the matrix S we define the **monodromy matrix**

$$T^{\gamma\{\delta\}}_{\alpha\{\beta\}}(\theta, \gamma) = S^{\gamma\ \delta_1}_{\alpha_2\beta_1}(\theta, \gamma) S^{\alpha_2\delta_2}_{\alpha_3\beta_2}(\theta, \gamma) \ldots S^{\alpha_N\delta_N}_{\alpha\ \beta_N}(\theta, \gamma) = \ -|-\,|-\cdots-|-\,. \tag{4}$$

On the r.h.s. we have a matrix product with respect to the "horizontal space" and a tensor product with respect to N "vertical spaces".

We define the **transfer matrix** as a trace in the horizontal space

$$\tau(\theta, \gamma) = \mathrm{tr}_h\ T(\theta, \gamma). \tag{5}$$

Then we can write the partition function (1) as a trace in the tensor product of the vertical spaces

$$Z = \mathrm{tr}_v\, \tau^M. \tag{6}$$

If we are able to solve the eigenvalue problem of the transfer matrix

$$\tau\psi = \lambda\psi \tag{7}$$

we can calculate the partition function in terms of the eigenvalues

$$Z = \sum_i \lambda_i^M. \tag{8}$$

In the thermodynamic limit only the maximal eigenvalue contribute to the free energy per lattice site

$$f = -kT\frac{1}{NM}\log Z \approx \ kT\frac{1}{NM}\log \lambda^M_{max} \qquad \text{for } N, M \to \infty. \tag{10}$$

2.2 Yang-Baxter algebra. The model is integrable which means the eigenvalue problem (7) can be solved since the monodromy matrix (4) generates a Yang-Baxter algebra. This will be shown in the following. For fixed parameter γ the matrix of vertex weights S fulfils the Yang-Baxter equation

$$S_{\alpha'\,\beta'}^{\alpha''\beta''}(\theta-\theta')\;S_{\alpha\,\gamma'}^{\alpha'\gamma''}(\theta)\;S_{\beta\,\gamma}^{\beta'\gamma'}(\theta')=S_{\beta'\,\gamma'}^{\beta''\gamma''}(\theta')\;S_{\alpha'\,\gamma}^{\alpha''\gamma'}(\theta)\;S_{\alpha\,\beta}^{\alpha'\beta'}(\theta-\theta'). \tag{11}$$

This identity represents a set of relations for the vertex weights a, b and c which follow from the parametrization (2) and trigonometric identities. The Yang-Baxter equation (11) for the matrix S implies by means of iterations the Yang-Baxter relation for the monodromy matrix T

$$S_{\alpha'\,\beta'}^{\alpha''\beta''}(\theta-\theta')\;T_{\alpha\,\{\gamma'\}}^{\alpha'\{\gamma''\}}(\theta)\;T_{\beta\,\{\gamma\}}^{\beta'\{\gamma'\}}(\theta')=T_{\beta'\,\{\gamma'\}}^{\beta''\{\gamma''\}}(\theta')\;T_{\alpha'\,\{\gamma\}}^{\alpha''\{\gamma'\}}(\theta)\;S_{\alpha\,\beta}^{\alpha'\beta'}(\theta-\theta'). \tag{12}$$

This will be the basic formula in the following. It reflects the integrability of the model. Note that it implies the commutativity of the transfer matrix for different spectral parameters

$$[\tau(\theta),\tau(\theta')]=0 \tag{13}$$

which represents (an infinite number of) conservation laws. We write the monodromy matrix T defined by eq. (4) explicitly as a matrix in the two dimensional horizontal space

$$T=\begin{pmatrix} A & B \\ C & D \end{pmatrix} \tag{14}$$

where A, B, C and D are operators in the tensor product of the vertical spaces. From the Yang-Baxter relation (12) we obtain the commutation rules

$$\begin{aligned} A(\theta)B(\theta') &= \frac{a(\theta'-\theta)}{b(\theta'-\theta)}B(\theta')A(\theta)-\frac{c(\theta'-\theta)}{b(\theta'-\theta)}B(\theta)A(\theta') \\ D(\theta)B(\theta') &= \frac{a(\theta-\theta')}{b(\theta-\theta')}B(\theta')D(\theta)-\frac{c(\theta-\theta')}{b(\theta-\theta')}B(\theta)D(\theta') \end{aligned} \tag{15}$$

We do not need the other formulas in the following.

2.3 Algebraic Bethe ansatz. The resolution of the six-vertex model has been traced back to the eigenvalue problem of the transfer matrix

$$\tau(\theta)\psi=(A(\theta)+D(\theta))\psi=\lambda\psi \tag{16}$$

which will be solved by the **algebraic Bethe ansatz** method. This method is quite analogous to the algebraic procedure used in quantum mechanics to solve eigenvalue problems (as e.g. for the angular momentum) or more general to find irreducible representations of Lie-algebras. The Yang-Baxter relations (12) correspond to the Lie-algebra structure relations. The diagonal elements A and D of T play the rôle of the "weights". The off diagonal elements B and C are creation and annihilation operators of "pseudoparticles", respectively.

The "pseudovacuum" ϕ is defined by

$$C(\theta)\ \phi = \rightarrow\!\big|\!-\cdots-\!\big|\!\leftarrow\, = 0 \qquad \forall\, \theta\ . \tag{17}$$

From eqs. (2, 4 and 14) we find the solution of this equation

$$\phi^{\{\beta\}} = \prod_{i=1}^{N} \delta_{\beta_i 1} = \uparrow \ldots \uparrow\ . \tag{18}$$

It is an eigenstate of A and D, fulfilling

$$\begin{aligned} A(\theta)\ \phi &= \ \leftarrow\!\updownarrow\!\leftarrow \cdots \leftarrow\!\updownarrow\!\leftarrow\, = a^N(\theta)\ \phi \\ D(\theta)\ \phi &= \ \rightarrow\!\updownarrow\!\rightarrow \cdots \rightarrow\!\updownarrow\!\rightarrow\, = b^N(\theta)\ \phi\ . \end{aligned} \tag{19}$$

However, applying B to ϕ one gets something new

$$B(\theta)\ \phi = \leftarrow\!\big|\!-\cdots-\!\big|\!\rightarrow\, \neq 0, \phi. \tag{20}$$

This fact motivates us to write an eigenstate of the transfer matrix as an **algebraic Bethe ansatz**

$$\psi = B(\theta_1)\ldots B(\theta_n)\ \phi \tag{21}$$

where the parameters θ_i are to be determined.

We follow the general strategy [4] of the algebraic Bethe ansatz method to solve eq. (16). We apply $A(\theta)$ and $D(\theta)$ to ψ and push them through all the B's using the commutation rules (15). From the first terms on the r.h.s. of eq. (15) we get the "wanted" contributions

$$(A(\theta) + D(\theta))\ \psi = (\lambda_A(\theta) + \lambda_D(\theta))\ \psi\ +\ \text{"unwanted terms"} \tag{22}$$

where from eqs. (15 and 19) we find

$$\lambda_A(\theta) = a^N(\theta) \prod_{i=1}^{n} \frac{a(\theta_i - \theta)}{b(\theta_i - \theta)} \qquad \lambda_D(\theta) = b^N(\theta) \prod_{i=1}^{n} \frac{a(\theta - \theta_i)}{b(\theta - \theta_i)}. \tag{23}$$

The "unwanted terms" arise from the second terms in eqs. (15). They contain a $B(\theta)$ and so they can never give a vector proportional to ψ. The unwanted terms vanish if the **Bethe ansatz equations** hold

$$\left(\frac{b(\theta_j)}{a(\theta_j)}\right)^N \prod_{i=1}^{n} \frac{a(\theta_j - \theta_i)}{b(\theta_j - \theta_i)} \frac{b(\theta_i - \theta_j)}{a(\theta_i - \theta_j)} = -1 \qquad j = 1, \ldots, n. \tag{24}$$

Obviously these equations are nessesary for ψ to be an eigenstate of τ. This is because the eigenvalue

$$\lambda(\theta) = \lambda_A(\theta) + \lambda_D(\theta) \tag{25}$$

cannot have poles at $\theta = \theta_i$ as eqs. (2 and 23) would suspect. The cancelation condition of these poles is equivalent to the Bethe ansatz equations (24). It can be shown [4] that they also imply the cancelation of the unwanted terms. We have transformed the eigenvalue problem of the transfer matrix (16) to sets of trancendental equations (24) for the parameters $\theta_1, \ldots, \theta_n$, $n = 0, \ldots, \frac{1}{2}N$ (Neven).

Solutions of Bethe ansatz equations are available for small or very large N and n. We are interested in the latter case. I will explain some more technical details. Introducing a "rapidity" q, a "momentum" p and a "scattering phase shift" Φ for the pseudoparticles by

$$\theta = \tfrac{1}{2}(iq + \gamma) \qquad e^{ip(q)} = \frac{\sinh \frac{1}{2}(q - i\gamma)}{\sinh \frac{1}{2}(q + i\gamma)} \qquad e^{i\Phi(q)} = \frac{\sinh \frac{1}{2}(q + 2i\gamma)}{\sinh \frac{1}{2}(q - 2i\gamma)}. \tag{26}$$

we can write the Bethe ansatz equations (24) as

$$\exp(iNp(q_j)) \prod_{i=1}^{n} \exp(\Phi(q_j - q_i)) = -1$$
$$j = 1, \ldots, n \qquad 0 \le n \le N/2. \tag{27}$$

The "physical" interpretation of this equations is, that the total phase factor around one period $1, \ldots, N$ has to be equal to one. This phase factor consists of the free part $\exp(iNp(q_j))$ and the scattering parts with respect to all other pseudoparticles $\exp(i\Phi(q_j - q_i))$. Each set of roots of eq. (27) $\{q_1, \ldots, q_n\}$ defines an eigenstate of the transfer matrix by eq. (21) and an eigenvalue λ given by eqs. (23, 25 and 26)

$$\log \lambda_A(\theta, \gamma) = N \log a - \sum_{i=1}^{n} e(q_i) \tag{28}$$
$$e(q) = i(p(q + 2i\theta) - \pi)$$

We will consider the region $0 < \theta < \gamma/2$ where $|\lambda_D| < |\lambda_A|$. So λ_D will not contribute to the partition function for large M.

There exist detailed investigations of the Bethe ansatz equations (27) (see e.g. ref. [10]). The analysis simplifies for the discrete values of the "coupling"

$$\gamma = \pi/\nu \qquad \nu = 2, 3, 4, \ldots. \tag{29}$$

We are interested in these values. So we assume eq. (29) in the following.

2.4 Solutions of Bethe ansatz equations: real roots. If we define the phase function

$$z(q) = Np(q) + \sum_{i=1}^{n} \Phi(q - q_i). \tag{30}$$

the logarithm of eq. (27) yields

$$2\pi I_j = z(q_j) \qquad j = 1, \ldots, n. \tag{31}$$

We observe that sets of roots $\{q_1, \ldots, q_n\}$ are given by sets of numbers $\{I_1, \ldots, I_n\} \subset \mathbf{Z} + \frac{1}{2} \cap \frac{1}{2\pi}]z(-\infty), z(\infty)[$ where eqs. (26 and 30) yield

$$\begin{aligned} z(-\infty) &= 2\pi n + 2\gamma\Sigma \\ z(\infty) &= 2\pi N - 2\gamma\Sigma \end{aligned} \tag{32}$$

and $\Sigma = \frac{1}{2}N - n$ is the "spin" of the Bethe state.

We introduce the density of roots

$$\rho(q) = 2\pi \sum_{i=1}^{n} \delta(q - q_i) \tag{33}$$

and the "density of holes"

$$\varphi(q) = \frac{dz(q)}{dq} - \rho(q). \tag{34}$$

The latter name is motivated by the fact that

$$\int_Q^{Q'} \frac{dq}{2\pi} \varphi(q) = I(Q') - I(Q) - \#\text{roots} \in [Q, Q']$$

is the number of possible minus the number of actual roots in an interval $[Q, Q']$. But note that only in the limit $N, n \to \infty$, where the roots become dense on the real line, the function φ is of the form

$$\varphi(q) = 2\pi \sum_{holes} \delta(q - q_h). \tag{35}$$

Taking the derivative d/dq of eq. (30) we obtain with eqs. (33 and 34)

$$z' = \rho + \varphi = Np' + \Phi' * \rho \tag{36}$$

where the convolution is defined by $(f * g)(q) = \int (dq/2\pi) f(q - q')g(q')$. Hence the density of roots of the Bethe ansatz equations (corresponding to eigenstates $\psi(q_1, \ldots, q_n)$ of the transfer matrix) can be written in terms of the "density of holes"

$$\rho = \rho_0 - R * \varphi \tag{37}$$

where $R = (1 - \Phi')^{-1}$ and $\rho_0 = N\, R * p'$ can be calculated by means of the Fourier transformation $f(q) = \int dx e^{-iqx} \tilde{f}(x)$

$$\tilde{R}(x) = \frac{\sinh \pi x}{2 \cosh \gamma x \, \sinh(\pi - \gamma)x} \qquad \tilde{p}'(x) = \frac{\sinh(\pi - \gamma)x}{\sinh \pi x} \tag{38}$$

$$\tilde{\rho}_0(x) = \frac{N}{2 \cosh \gamma x}. \tag{39}$$

The eigenvalue of the transfer matrix is given by

$$\begin{aligned} \log \lambda_A(\theta, \gamma) &= N \log a - \int \frac{dq}{2\pi} \rho(q) e(q) \\ &= N \log a - \int dx \tilde{\rho}(-x) \tilde{e}(x). \end{aligned} \tag{40}$$

In the thermodynamic limit $N \to \infty$ the problem is solved by eqs. (37-40) since φ is of the form (35).

(i) Ground state: There is a "Dirac-sea" of real roots. Since $\Re e(q) < 0$ for real q the eigenvalue is maximal if all possible real roots are occupied. From eq. (32) follows $n = \frac{1}{2}N$ and $\Sigma = 0$ which means that the ground state is a singlet. Eqs. (38-40) imply

$$\rho(q) = \rho_0(q) = \frac{N\pi}{2\gamma} \left(\cosh \frac{\pi q}{2\gamma} \right)^{-1} \tag{41}$$

$$\log \lambda_A(\theta, \gamma) = \log \lambda_{A,0}(\theta, \gamma) = N \log a + N \int_{-\infty}^{\infty} \frac{dx}{x} \frac{\sinh(\pi - \gamma)x \, \sinh 2\theta x}{2 \cosh \gamma x \, \sinh \pi x}. \tag{42}$$

(ii) Holes: For excitations given by holes at q_h we find from eqs. (35, 37, 38 and 40)

$$\rho(q) = \rho_0(q) - \sum_{holes} R(q - q_h) \tag{43}$$

$$\lambda_A(\theta, \gamma) = \lambda_{A,0}(\theta, \gamma) \prod_{holes} \tan \frac{\pi}{\gamma} \left(iq_h + \tfrac{1}{4}(\gamma - 2\theta) \right). \tag{44}$$

2.5 Solutions of Bethe ansatz equations: complex roots. For "excited states" there may be, in addition to holes, complex roots q: the "k-strings" (one could say "bound states" of k pseudoparticles)

$$q = \text{real} + i\gamma(-k+1, -k+3, \ldots, k-1) \qquad 1 < k \leq \nu - 1 \tag{45}$$

and the "antipseudoparticle"

$$q = \text{real} + i\pi. \tag{46}$$

The density of the real roots and the eigenvalue λ, if complex roots are present, can by computed analogously to eqs. (43 and 44) (see e.g. refs. [10,1]).

3. Conformal Invariance.

3.1 Virasoro algebra. In two dimensions conformal invariance is a rather strong restriction since the transformation group is infinite dimensional, in contrast to higher dimensions. It is related to the Virasoro algebra [11]

$$\begin{aligned} [L_n, L_m] &= (n-m)L_{n+m} + \tfrac{1}{12}c(n^3 - n)\delta_{n,-m} \\ c &= \text{"central charge"} \end{aligned} \tag{47}$$

for which a well developed theory exists [12]. Belavin, Polyakov and Zamolodchikov [13] classified conformal quantum field theories in terms of the central charge

$$c = 1 - \frac{6}{\nu(\nu-1)} \tag{48}$$

where ν is a rational number if there exists only a finite number of "primary fields", i.e. fields transforming under conformal transformations

$$\begin{aligned} z &= x + iy \to w(z) \\ \bar{z} &= x - iy \to \bar{w}(\bar{z}) \end{aligned} \tag{49}$$

like

$$A(z, \bar{z}) \to \left(\frac{dw}{dz}\right)^{\Delta} \left(\frac{d\bar{w}}{d\bar{z}}\right)^{\bar{\Delta}} A(w, \bar{w}). \tag{50}$$

The conformal dimensions can assume values given by the Kac-formula [12]

$$\begin{aligned} \Delta, \bar{\Delta} = h_{pq} &= \frac{(\nu p - (\nu-1)q)^2 - 1}{4\nu(\nu-1)} \\ 1 \leq q \leq p < \nu - 1 & \qquad p, q = \text{integer}. \end{aligned} \tag{51}$$

Friedan, Qiu and Shenker [14] have shown that unitarity restricts the number ν to

$$\nu = \text{integer} \geq 2 \quad \text{or} \quad c \geq 1. \tag{52}$$

It would be interesting to find integrable lattice models which yield in the scaling limit conformal quantum field theories correspondig to these unitary representations of the Virasoro algebra. Therefore it is useful to calculate the central charge and the conformal dimensions for integrable critical statistical models. This can be done by analysing finite size effects.

3.2 Finite size behaviour. For a statistical system on a periodic $N \times M$ square lattice one introduces a transfer matrix τ as explained in sect. 2. The partition function can be written as

$$Z = \operatorname{tr} \tau^M = \sum_i \lambda_i^M$$

where the sum extends over all eigenvalues of the transfer matrix. Cardy [15] has shown that conformal invariance implies for the maximal eigenvalue

$$\lambda_{max} \approx \exp(-Nf + \frac{1}{N}\frac{\pi}{6}c) \qquad \text{for} \quad N \to \infty \tag{53}$$

where f is the free energy per site (in the thermodynamic limit) and c is the central charge of the Virasoro algebra (cf. eq. (47)).

Introducing in addition the "crossed" transfer matrix $\hat{\tau}$ (defined on an M-chain) we obtain for $1 \ll N \ll M$ two expressions for the partition function

$$Z \approx \exp(-NMf) \sum_i (\hat{\lambda}_i / \hat{\lambda}_{max})^M \tag{54a}$$

$$Z \approx \exp(-NMf + \frac{M}{N}\frac{\pi}{6}c) \tag{54b}$$

where only the "low energy excitations" with $\hat{\lambda}_i \approx \hat{\lambda}_{max}$ contribute to the sum in eq. (54a). Hence two methods to investigate the partition function on a strip $1 \ll N \ll M \to \infty$ are availble:

(a) one considers the transfer matrix $\hat{\tau}$ for an infinite chain ($M \to \infty$) and calculates the large N corrections to the sum in eq. (54a) or

(b) one considers the transfer matrix τ on an N-sites chain and calculates the large N corrections to the maximal eigenvalue λ_{max} according to eq. (54b).

By both of these procedures one obtains the central charge of the model.

In addition the conformal dimensions of operators of the model can be determined if one looks for excited states with energy $E = -\Re \log \lambda$ and momentum $P = -\Im \log \lambda$. As is argued in ref. [6] one has for $N \to \infty$

$$\begin{aligned} E_i - E_0 &\approx \frac{2\pi}{N} d_i \\ P_i - P_0 &\approx \frac{2\pi}{N} s_i \end{aligned} \tag{55}$$

where $d_i = \Delta + \bar{\Delta}$ ($s_i = \Delta - \bar{\Delta}$) is the "scaling dimension" ("spin") of the operator A associated to the excitation ψ_i, i.e. $\langle 0|A|\psi_i\rangle \neq 0$.

Since a direct calculation of the central charge and the conformal dimensions for specific models is usually not simple, it seems worthwhile to determine them for integrable models from the Bethe ansatz solution. This will be done in the following for the six-vertex model and for some modified six-vertex models related to Potts models and other models of interest. Method (a) has been developed in ref. [1], method (b) is applied in sect. 4 and ref. [2].

4. Results for Six-Vertex and Potts Models.

The Potts models are a generalizations of the Ising model where the spin at each site can accept q values instead of two. For $q = 2, 3, 4$ the models possess a critical point which we consider in the following. The partition function on a cylinder [9] can be written in terms of a six-vertex model with a "seam" from the bottom to the top of the cylinder. This seam carries extra Boltzmann weights which produce an extra term $-i\gamma m$ on the r.h.s. of eq. (28) and extra terms $2\gamma m$ on the r.h.s. of eqs. (30 and 32) with $m = 1$[2]. The number q ($q = 2, 3, 4$) is related to the six-vertex coupling $\gamma = \pi/\nu$ ($\nu = 4, 6, \infty$) by

$$q = 4\cos^2 \gamma. \tag{56}$$

In the following the partition function Z_m and other formulas apply for $m = 0$ to the ordinary six-vertex model and for $m = 1$ to the Potts model

$$Z_0 = Z^{6-vertex} \qquad Z_1 = Z^{Potts}. \tag{57}$$

4.1 Ground state. In refs. [16,17,1,2] techniques were developed to calculate finite size corrections for a large chain of length N. Using these methods we find [1,2] for the "ground state"

$$\log \lambda_{max} \approx Nf + \frac{1}{N}\frac{\pi}{6}\left(1 - \frac{6m^2}{\nu(\nu-1)}\right) \sin \nu\theta. \tag{58}$$

[2]The variable m has been introduced for later convenience.

The vertex weights (2) fulfil $a = b$ for $\nu\theta = \pi/2$ where we expect isotropy and conformal invariance in the scaling limit. Comparing eqs. (58) and (53) we conclude that the central charges of the Virasoro algebra of the six-vertex and the q-state Potts models are

$$c^{6-vertex} = 1 \tag{59a}$$

$$c^{Potts} = 1 - \frac{6}{\nu(\nu-1)}. \tag{59b}$$

4.2 Excitations. As already mentioned in sect. 2 the ground state correspond to the maximal set of real roots $q_1, \ldots, q_{N/2}$. Excitations possess "holes" and complex roots. We obtain [2] for these states taking eqs. (55) into account[3]

$$\log \lambda_i - \log \lambda_{max} \approx -\frac{2\pi}{N}\{(\Delta + \bar{\Delta})\sin\nu\theta + i(\Delta - \bar{\Delta})\cos\nu\theta\}. \tag{60}$$

The general expressions for the conformal dimensions Δ and $\bar{\Delta}$ are rather complicated. But the primary fields correspond to low energy excitations with a small number of holes $h_\pm$ "near to the last roots" $q_\pm$ which approach $\pm\infty$ for $N \to \infty$. In addition there may be a number $n_\pm^{(k)}$ of k-strings "near to $q_\pm$"[4]. For these configurations of roots we obtain [2]

$$\Delta = \frac{\left(\nu\left(h_+ - 2\sum_k n_+^{(k)}(k-1)\right) - \Sigma + m\right)^2 - m^2}{4\nu(\nu-1)} + \text{integer} \tag{61a}$$

$$\bar{\Delta} = \frac{\left(\nu\left(h_- - 2\sum_k n_-^{(k)}(k-1)\right) - \Sigma - m\right)^2 - m^2}{4\nu(\nu-1)} + \text{integer} \tag{61b}$$

where $\Sigma = \frac{1}{2}h - \sum_k n^{(k)}(k-1)$ is the "spin"[5] of the Bethe state and h ($n^{(k)}$) is the total number of holes (k-strings). We have $\Sigma = 0$ since we consider fixed boundary conditions at the top and the bottom of the cylinder. It turns out [2] that we get minimal values for Δ and $\bar{\Delta}$ (i.e. the integers on the r.h.s. of eqs. (61) vanish) if some conditions are fulfilled: There are no holes inside $[q_-, q_+]$, $n_+^{(k)} = 0$ for all k, $n_-^{(k)} = 1$ for only one k and the number of holes have to fulfil $h_+ = h_- = k - 1$. For the six-vertex model we obtain the conformal dimensions from eqs. (61) with

[3]Note that the imaginary term correspond to the momentum since $\tau(\theta = 0) = \exp(-iP)$.

[4]The k-strings which do not approach $\pm\infty$ like the last real roots $q_\pm$ as $N \to \infty$ will not affect Δ and $\bar{\Delta}$.

[5]Σ must not be confused with the "spin" s of the field operator (cf. eq. (55)).

$m = 0$. For the q-state Potts model the extra weights on the seam imply [2] that for the minimal Δ and $\bar{\Delta}$ only the k-strings with

$$k = 2, 3, \ldots, \nu - m - 1 \tag{62}$$

contribute (i.e. for $m = 1$ the $\nu - 1$-string and the "antipseudoparticle" do not appear).

Comparing eqs. (61) for $m = 1$ and $\Sigma = 0$ with the Kac-formula (51) we find for the Potts model the conformal dimensions

$$\Delta = \bar{\Delta} = h_{k,1} \qquad k = 1, 2, \ldots, \nu - 2. \tag{63}$$

These correspond to the "energy operators" of the Potts model (cf. [14,8,15]). The equivalence of the Potts model and a six-vertex model with "seam" for $m = 1$ holds for the partition functions (cf. eq. (57)) but not for the operator content. The order parameter of the Potts model, for example, appears in a six-vertex model with a "seam" for $m = \nu/2$ (cf. [15]).

If we consider a partition function of a generalized model

$$Z = Z_1 + \sum Z_m \tag{64}$$

where the sum extends over some values of $m \in [2, \nu - 1]$ we obtain the dimensions corresponding to Bethe states with a k-string

$$\Delta = \bar{\Delta} = h_{k+m-1,m} \qquad k = 1, \ldots, \nu - m - 1 \quad 1 \leq m \leq \nu - 1. \tag{65}$$

Note that the ground state always belongs to Z_1. The "magnetization" operators of the $q = 3$-state Potts model ($\nu = 6$) correspond to $m = \nu/2 = 3$ (cf. [19]). Their dimensions are therefore (in agreement with numerical estimates [19] and conjectures [18,15])

$$\Delta = \bar{\Delta} = h_{k+2,3} \qquad k = 1, 2. \tag{66}$$

The $1/q$-para fermion operators [20] correspond to "boundary conditions" $m = \nu/q = 2$ for $q = 2, 3$ [20,19]. The Bethe states have spin $\Sigma = 1$ and a symmetric hole configuration $h_+ = h_- = 1$. For $\Sigma \neq 0$ the integers on the r.h.s. of eqs. (61) only vanish, if there are no higher strings. Hence we obtain from eq. (61) the scaling dimensions and spin

$$\begin{aligned} d &= \Delta + \bar{\Delta} = h_{21} + h_{23} = \frac{\nu - 1}{2\nu} + \frac{\nu^2 - q^2}{2q^2\nu(\nu - 1)} \\ s &= \Delta - \bar{\Delta} = h_{21} - h_{23} = 1/q \end{aligned} \tag{67}$$

in agreement with ref. [20,19]. Finaly let us look for the "chiral" operators of the 3-state Potts model (cf. [18,15]). Their conformal dimensions are given by the "boundary conditions" $m = \nu/2 = 3$ and Bethe states with $\Sigma = 2$ or four holes. For $h_+ = 0, 1, 2, 3$ and $h_- = 4 - h_+$ one finds from eq. (61), respctively

$$\Delta = h_{11}, h_{21}, h_{31}, h_{41} = 0, \tfrac{2}{5}, \tfrac{7}{5}, 3$$

$$\bar{\Delta} = h_{15}, h_{25}, h_{35}, h_{45} = 3, \tfrac{7}{5}, \tfrac{2}{5}, 0.$$

At the end let us consider a partition function given by eq (64) where the sum extends over all $m = 2, \ldots, \nu - 1$. The dimension formula (65) obviously describes the dimensional spectrum of the RSOS models [21], in the language of ref. [22] the $(A_\nu, A_{\nu-1})$ series of conformal quantum field theories. For this interpretation of eq. (64) the number m is not connected to extra weights on a "seam" or to "boundary conditions", but it is related [2] to a "Θ-vaccum" like angle appearing in the construction of Bethe states [4]. The RSOS models of ref. [21] can be generalized [23] to models corresponding to all modular invariant conformal quantum field theories with $c < 1$ [22]. All these models are related to modified six-vertex models. Results for these models will be published elsewhere [24].

References.

[1] H.J. de Vega and M. Karowski, Nucl. Phys. B285 [FS19] (1987) 619;
M. Karowski, in 'String Theory Quantum Cosmology and Quantum Gravity Integrable and Conformal Invariant Theories', Proceedings of the Paris-Meudon Colloquium 22-26 sept. 1986, ed. H.J. de Vega et al., World Scientific Pub. Co (1987);
M. Karowski, to be published in the Proceedings of the 'XXVI Internationale Universitätswochen für Kernphysik', Schladming, Feb. 1987.

[2] M. Karowski, to be published.

[3] A.M. Polyakov, ZhETF 55 (1969) 1026;
A.A. Migdal, ZhETF 55 (1969) 1964;
K. Wilson, Phys. Rev. 179 (1969) 1499.

[4] L.A. Takhtadzhan and L.D. Faddeev, Russian Math. Surveys 34 (1979) 186.

[5] H. Bethe, Z. Phys. 71 (1931) 205.

[6] L. Onsager, Phys. Rev. 65 (1944) 117.

[7] R.J. Baxter, Ann. Phys. 70 (1972) 193.

[8] C.N. Yang, Phys. Rev. Lett. 19 (1967) 1312.

[9] R. J. Baxter, Exactly solvable models (Academic Press, 1982).

[10] M. Takahashi and M. Suzuki, Prog. Theor. Phys. 48 (1972) 2187.

[11] M.A. Virasoro, Phys. Rev. D10 (1970) 2933.

[12] V.G. Kac, Lecture Notes in Physics 94 (1979) 441;
B.L. Feigin and D.B. Fuchs, Fuct. Anal. Appl. 16 (1982) 114;
A. Rocha-Caridi, in 'Vertex operators in mathematics and physics', ed. J. Lepowsky et al. (Springer, NY, 1984) p. 451.

[13] A.A. Belavin, A.M. Polyakov and A. B. Zamolodchikov, Nucl. Phys. B241 (1984) 333.

[14] D. Friedan, Z. Qiu and S. Shenker, Phys. Rev. Lett. 52 (1984) 1575.

[15] J. L. Cardy, Nucl. Phys. B270 [FS16] (1986) 186.

[16] H. J. de Vega and F. Woynarovich, Nucl. Phys. B251 (1985) 439;
F. Woynarovich and H. P. Eckle, J. Phys. A20 (1987) L97.

[17] C. J. Hamer, J. Phys. A19 (1986) L3335;
C. J. Hamer, J. Stat. Phys. 47 (1987) 331;
L. V. Ardeev and B. D. Dörfel, J. Phys. A19 (1986) L13;
H. J. Hamer, C. R. W. Quispel and M. T. Batchelor, Canberra, preprint 9-1987.

[18] V. S. Dotsenko, Nucl. Phys. B235 [FS11] (1984) 54.

[19] F. C. Alcaraz, M. N. Barber and M. T. Batchelor, Phys. Rev. Lett. 58 (1987) 771.

[20] B. Nienhuis and H. J. F. Knops, Phys. Rev. B32 (1985) 1872.

[21] G. E. Andrews, R. J. Baxter and P. J. Forester, J. Stat. Phys. 35 (1984) 193.

[22] A. Cappelli, C. Itzykson and J. B. Zuber, Nucl. Phys. B280 [FS18] (1987) 445.

[23] V. Pasquier, Nucl. Phys. B285 [FS19] (1987) 162.

[24] M. Karowski, to be published.

M. Karowski
Theoretical Physics
ETH-Hönggerberg
CH-8093 Zürich, Switzerland

II. STRING THEORY AND RELATED TOPICS

STRING FIELD THEORY

I.Ya. Aref'eva

1. Introduction

1.1. Motivations

In the last years an essential change of point of view took place in elementary particle physics, the role played by the notion of particle being taken over by the notion of string. This perestrojka touches the main notions of theoretical physics and the situation can be compared with the creation of General Relativity and Quantum Mechanics.

ISBN 0-12-218100-X

String theory has a long history. It started as a theory of hadrons in the 70's [1], but the most impressive success was the realization that string theory might provide a natural framework for describing all particle interactions, including gravity; a remarkable step in this direction was made by Green and Schwarz [2].

The main idea of string theory is quite simple: the fundamental object of the theory is a curve instead of a point (as is usual in local field theory). However the construction of a string theory is a rather difficult task. For example, the dynamics of a point particle is described by the Newton equation. What is the natural analogue of this equation for the strings ? We shall discuss this question below.

It is well known that in local field theory all the properties may be derived from a Lagrangian [3]. In string theory the situation is not so simple: for a long time the theory was only a set of rules (which were not derived from a general principle), and which allowed the prediction of interesting physical properties. For example when mixed with the idea of supersymmetry, string theory predicted the following features:

- space time dimension $D = 10$.
- (super)gravity and (super)Yang-Mills theory as low energy limits.
- absence of divergencies.
- prediction of the gauge group ($E_8 \times E_8$, $SO(32)$).
- compatibility with the known phenomenology.

There are strong arguments in favour of these properties of (super)string theory,but a profound understanding of them can be achieved only by a non-perturbative investigation. However this would be possible only in a second quantized or string field theory approach.

In the usual scheme of the quantum field theory (QFT) one starts with a set of fields and a Lagrangian is obtained by imposing a gauge invariance principle. This allows us to develop a perturbative theory or to get non-perturbative effects.

Since some limit cases of string theories are gauge theories it seems natural to believe that some gauge invariant principle lies at the basis of string theory. This invariance principle will be analyzed below. So, we shall see that there are three types of gauge invariant theories in physics: Yang-Mills theory, general relativity and string field theory. It is well known that the mathematical background of the general relativity is classical differential geometry, while that of Yang-Mills theory is the geometry of fibre bundles. I shall argue that analysis on non-associative superalgebras may be considered as the mathematical background of string field theory.

Let me stress the main motivations for the development of SFT: i) to understand the gauge principle behind string theory, ii) to investigate non-perturbative questions such as dynamical symmetry breaking and spontaneous compactification, iii) to get unitarity "automatically".

1.2. Basic facts about operator quantization of strings

A string can be viewed as a surface embedded in the Minkowski space-time M^D, which is described parametrically by a set of functions $X^\mu = X^\mu(\sigma,\tau)$, where σ is a space-like coordinate, $0 \leqq \sigma \leqq \pi$, and τ is a time-like coordinate on the surface.

By analogy with the point particle the action is given by the total area of the world surface

$$S = \frac{1}{4\pi\alpha'} \int_0^\pi d\sigma \int d\tau \quad [(\dot{X}X')^2 - \dot{X}^2 X'^2]^{1/2} \quad ,$$

$$\dot{X}^\mu = \frac{\partial X^\mu}{\partial\tau} \, , \quad X'^\mu = \frac{\partial X^\mu}{\partial\sigma}$$

where α' is the universal Regge slope parameter, and in the following we shall choose the units such that $\alpha' = 1$.

This form of the action is due to Nambu and Goto. A more convenient form of the action is [6]:

$$S = -\frac{1}{4\pi} \int d\sigma d\tau \sqrt{-g} \quad g^{\alpha\beta} \, \partial_\alpha X^\mu \partial_\beta X_\mu \tag{1.1}$$

where $g_{\alpha\beta}$ is the metric on the surface, and plays the role of a Lagrangian multiplier. Note that the metric g in the action (1.1) has no dynamics. A string model for which the 2-dimensional metric satisfies dynamical equations is considered in [7].

Using the reparametrization invariance the equation of motion for the action (1.1) can be reduced in the conformal gauge to the following form

$$\left(\frac{\partial^2}{\partial\tau^2} - \frac{\partial^2}{\partial\sigma^2}\right) X^\mu(\tau,\sigma) = 0 \tag{1.2}$$

together with the constraints

$$(\dot{X}^\mu \pm X'^\mu)^2 = 0 \tag{1.3}$$

Two types of boundary conditions will be used: $X'^\mu(\tau,0) = X'^\mu(\tau,\pi) = 0$ for open strings and $X^\mu(\tau,0) = X^\mu(\tau,\pi)$ for closed strings. We shall restrict ourselves to open strings for which the most general solution of Eq. (1.2.) is given by

$$X^\mu(\tau,\sigma) = x^\mu + p^\mu\tau + i \sum_{n\neq 0} \frac{1}{n} a_n^\mu e^{-in\tau}\cos n\sigma$$

where x^μ is the center of mass position.

The operator quantization amounts to consider a_n^μ as operators satisfying the following commutation relations

$$\begin{aligned} [a_n^\mu, a_m^\nu] &= n\delta_{n+m}\eta^{\mu\nu} \\ [x^\mu, p^\mu] &= i\eta^{\mu\nu} \end{aligned} \tag{1.4}$$

where $\eta^{\mu\nu}$ is the Minkowski metric. The a_n with positive (respectively negative) indices are considered as annihilation (respectively creation) operators.

The left hand side of the constraint (1.3) can be written (at $\tau = 0$) as

$$\frac{1}{2}(\dot{X}^{\mu} \pm X'^{\mu})^2 \Big|_{\tau=0} = \sum_{n=-\infty}^{\infty} L_n \, e^{\mp i n \sigma}$$

where L_n are the operators

$$L_n = \frac{1}{2} \sum_{k=-\infty}^{\infty} :a^{\mu}_{n+k} \, a^{\mu}_{-k}: \quad .$$

These operators satisfy the Virasoro algebra

$$[L_n, L_m] = (n-m)L_{n+m} + \frac{D}{12} n(n^2 - 1)\delta_{n+m} \tag{1.5}$$

in which the central charge depends on D, the dimension of space-time.

Because of the Lorentz covariance of the commutation relations (1.4) the space spanned by the oscillators contains negative norm states as in the Gupta-Bleuler quantization of quantum electrodynamics. The space of the on-shell physical states is a subspace of the entire space. It is characterized by the following conditions [1]:

$$(L_0 - 1)\Phi = 0 \tag{1.6a}$$

$$L_n\Phi = 0 \quad , \qquad n > 0 \quad . \tag{1.6b}$$

L_0 contains the string mass operator,

$$2(L_0 - 1) = p^2 + 2(\sum_{n>0} a^{\mu}_{-n} a^{\mu}_{n} - 1) = \Box + M^2 \quad ,$$

so Eq. (1.6a) gives the equation of motion of physical states and (1.6b) can be considered as a constraint.

To understand the physical content of the theory it is useful to expand Φ in eigenstates of the mass operator M^2. Let $|0\rangle$ be the state annihilated by all of the a_n, $n \geq 1$, i.e. $a_n|0\rangle = 0$, $n \geq 1$. A basis for the Hilbert space of states is formed by applying a_{-n} to $|0\rangle$. An arbitrary state $|\Phi\rangle$ may be expanded as

$$|\Phi\rangle = [\varphi(x) - iA_\mu(x)a^\mu_{-1} - \frac{1}{2}h_{\mu\nu}(x)a^\mu_{-1}a^\nu_{-1} + iB_\mu(x)a^\mu_{-2}+\ldots]|0\rangle \tag{1.7}$$

Here the coefficient functions depend on the center-of-mass position x. This vector state looks as an infinite component field theory. Each term in (1.7) is an eigenvector of the mass operator: $M^2\varphi(x)|0\rangle = -2\varphi(x)|0\rangle$, $M^2A_\mu a^\mu_{-1}|0\rangle = 0$, etc. So, $\varphi(x)$ is a tachyon, A_μ is a massless vector field, $h_{\mu\nu}$ is a massive tensor field,etc.

The state (1.7) corresponds to some functional $\Phi[X(\sigma)]$. The connection between the state $|\Phi\rangle$ given in the oscillator representation and the functional $\Phi[X(\sigma)]$ is the same as for the usual oscillator case:

$$\Phi[X(\sigma)] = \langle X(\sigma)|\Phi\rangle \quad ,$$

where the state $|X(\sigma)\rangle$ is given by the following formula $(X^\mu(\sigma) \equiv x^\mu + \sqrt{2}\sum_{n=1}^{\infty} x^\mu_n \cos n\sigma,\ a_{-n} = a^+_n)$

$$|X(\sigma)\rangle = \prod_{n\geq 1}(1/\pi)^{1/4}\exp\{-\frac{x_n^2}{2} - \frac{1}{2n}a^+_n a^+_n + i(2/n)^{1/2}a^+_n x_n\}|0\rangle .$$

The bra and ket in these formulae mean that we calculate the usual Fock-space scalar product. In the following, I shall omit the bra notations.

So, in SFT the main object is the functional $\Phi[X(\sigma)]$ or the corresponding vector in the infinite product of Fock spaces.

Let us consider another functional

$$\Omega[X(\sigma)] = \{\omega(x) - i\omega_\mu(x)\ a^\mu_{-1} + \ldots\}|0\rangle$$

and the following transformation $\Phi \to \Phi + \delta_\Omega\Phi$, where $\delta_\Omega\Phi = L_{-1}\Omega$. Since

$$L_{-1}\Omega = \{-i\partial^\mu\omega(x)a^\mu_{-1} + \ldots\}|0\rangle \tag{1.8}$$

from Eqs. (1.7), (1.8) we get $\delta_\Omega A^\mu(x) = \partial^\mu\omega(x)$, i.e. the linearized Yang-Mills gauge transformation. Thus, this transformation can be considered as a string gauge transformation. A general string gauge transformation has the form

$$\delta\Phi = \sum_{n\geq 1} L_{-n}\,\Omega_n \quad . \tag{1.9}$$

2. String Field Theory

2.1. Free bosonic action

In the SFT the main object has to be the functional $\Phi[X(\sigma)]$ or the corresponding vector in the infinite product of Fock spaces. To develop the SFT we need an action. This action has to satisfy three conditions: i) it should give us Eqs. (1.6a), (1.6b); ii) it must be gauge invariant under (1.9); iii) for low level fields it has to reproduce the usual actions.

A simple way to get a convenient action is to introduce ghost fields c_n and to consider an action

$$S = \langle\Phi, Q\Phi\rangle \quad , \tag{2.1}$$

where $Q = \sum_{n=-\infty}^{\infty} c_{-n}(L_n - \delta_{n,0})$. If c_{-n}, $n \geq 0$ (c_n, $n > 0$) are realized as creation (annihilation) operators and if $|\Phi\rangle$ has no ghost excitations, then the equation $Q|\Phi\rangle = 0$ implies

$$[c_0(L_0 - 1) + \sum_{n\geq 1} c_{-n}L_n]|\Phi\rangle = 0 \quad . \tag{2.2}$$

Since $c_{-n}|\Phi\rangle$ are linearly independent, Eq.(2.2) yields Eqs.(1.6). Therefore the single equation $Q|\Phi\rangle = 0$ is equivalent to the set of Eqs. (1.6).

However the action (2.1) with such a Q has no gauge invariance. Fortunately, one can modify the operator Q by adding some extra terms up to getting a nilpotent operator Q

$$Q = \sum_{n=-\infty}^{\infty} c_{-n}(L_n + \tfrac{1}{2} L_n^{gh}) \quad . \tag{2.3}$$

Here $L_n^{gh} = \Sigma(n+m)b_{n-m}c_m$, c_n and b_n are anticommuting ghosts

$$\{c_m, c_n\} = 0 = \{b_n, b_m\} , \{c_m, b_n\} = \delta_{n+m} , \qquad (2.4)$$

which satisfy the relations $c_n^+ = c_{-n}$, $b_n^+ = b_{-n}$. This operator Q satisfies the nilpotency relation

$$Q^2 = 0 \qquad (2.5)$$

only for D = 26, and it is nothing but the BRST-charge connected with the Virasoro algebra.

Due to the nilpotency condition (2.5) and hermiticity property of Q, the action (2.1) is invariant under gauge transformations $\delta\Phi = Q\Lambda$. Notice that the new Q is a sum of terms each of them containing at least one annihilation operator. Thus, if Φ has no ghost excitations, then $Q|\Phi\rangle = 0$ implies again the infinite set of constraints (1.6).

Let me note that the BRST-charge connected with an algebra with generators L_n

$$[L_n, L_m] = f_{nm}^k L_k, \quad k = 1,\ldots,N ,$$

is given by the formula [35]

$$Q = \Sigma \; c_{-n}L_n - \frac{1}{2} \Sigma f_{mn}^k \; c_{-m}c_{-n}b_k$$

where c_n and b_n are anticommuting ghosts

$$\{c_n, c_m\} = 0 = \{b_n, b_m\} , \quad \{c_n, b_m\} = \delta_{n+m} .$$

The action (2.1) for the free open bosonic string field theory was proposed in Refs. [8-10]. For a more complete list of references on the field theory free action for the bosonic string and its quantized version see [11]. One can show that for a general string field expanded as:

$$|\Phi\rangle = [\varphi(x)+iA_\mu(x)a^\mu_{-1} + B(x)b_{-1}\, c_0+\ldots]|0\rangle \quad , \qquad (2.6)$$

and with Q as in (2.3), one gets:

$$\langle\Phi,Q\Phi\rangle = \int dx\{-\tfrac{1}{2}\varphi(x)(\Box-1)\varphi(x) + \tfrac{1}{2}A_\mu\Box A_\mu + B^2(x) + 2B(x)\partial^\mu A_\mu + \ldots\} \qquad (2.7)$$

After elimination of the auxiliary field B(x) this action reproduces the tachyonic and Maxwell actions.

2.2. String field algebra (interaction)

Let me consider a set of string fields $\mathcal{B}$. By string fields we mean functionals of string coordinates as well as of ghost string coordinates. Within our usual understanding of local gauge invariance the parameters of gauge transformations are objects of the same nature as the fields undergoing the gauge transformations. For example the usual local Yang-Mills field is a Lie algebra valued vector field and the gauge parameter is also a Lie algebra valued scalar field. So, let us suppose that the parameters of gauge transformations are from some subset K of the set of string fields $\mathcal{B}$.

If we deal with gauge theory then it is natural to suppose that if we perform subsequently two infinitesimal gauge transformations and compare the result with the result of the same gauge transformations performed in the opposite order then the difference between them has to be represented also as a gauge transformation with a parameter which is a multiplication of two initial gauge parameters. So, let us postulate the fundamental relation expressing the closure of the algebra of gauge transformation

$$\delta_{\Omega_1}\delta_{\Omega_2} - \delta_{\Omega_2}\delta_{\Omega_1} = \delta_{\Omega_1\circ\,\Omega_2} \quad . \qquad (2.8)$$

Here δ_Ω is an infinitesimal gauge transformation, depending upon the string field Ω and δ_Ω acts on some string functional. One can think about the string field Ω as a parameter of a gauge transformation. $\Omega_1 \circ \Omega_2$ which appears in the right hand side of Eq. (2.8) is a multiplication of two string fields Ω_1 and Ω_2. By multiplication we mean as usual that to any given pair of elements Ω_1, Ω_2 one associates a third one in a bilinear way. One can see from Eq. (2.8) that the following relation holds

$$\Omega_1 \circ \Omega_2 = - \Omega_2 \circ \Omega_1 \tag{2.9}$$

Therefore for the subset $K \subset B$ which consists of parameters of gauge transformations the multiplication appears in a natural way, so K is an algebra with multiplication law satisfying the anticommutation relation (2.9). After that, it is natural to think the set of all string fields B as an algebra. As I mentioned, the main lesson from the field theory of free bosonic string is that the fields are functionals of bosonic coordinates as well as of anticommuting ghost string coordinates. So, we have a $\mathbb{Z}_2$ grading on B from the very beginning, i.e. $B = B_0 \oplus B_1$ and an element $\Phi \in B$ has a degree $|\Phi|$ which is 0 (even element) or 1 (odd element) depending upon whether Φ lies in B_0 or in B_1. For further considerations it is important to distinguish two different cases. In the first case the algebra of gauge transformations consists of even elements of B, i.e. $K = B_0$, in the second one it consists of odd elements, i.e. $K = B_1$.

For the first possibility, B^I, the multiplication law $\circ$ must be an even operation, i.e. the degree of the product $\Phi \circ \Psi$ is $|\Phi| + |\Psi|$. It is natural to believe that there is a multiplication law for arbitrary elements of B^I with the following property

$$\Phi \circ \Psi + (-1)^{|\Phi||\Psi|} \Psi \circ \Phi = 0 \tag{2.10}$$

which is a generalization of Eq. (2.9).

The second possibility, $\mathcal{B}^{II}$, is to consider the gauge parameters as odd elements, i.e. $K = \mathcal{B}_1^{II}$. In this case the multiplication law has to be odd, i.e. for two elements Φ, Ψ the degree of the product $\Phi \circ \Psi$ is $|\Phi|+|\Psi| + 1$. It is natural to consider the multiplication law $\circ$ for arbitrary elements of $\mathcal{B}^{II}$ and to demand that the multiplication law be commutative in the sense that the following identity holds

$$\Phi \circ \Psi - (-1)^{|\Phi||\Psi|} \Psi \circ \Phi = 0 \tag{2.11}$$

which is another generalization of (2.9), i.e. the even elements commute and the odd elements anticommute. Such an algebra $\mathcal{B}^{II}$ is called a nonassociative commutative superalgebra.

In order to justify further assumptions on the two different multiplication laws we have to specialize the gauge transformation. There are two ways to do this. Namely, to consider the natural generalization of the usual gauge transformation law for matter fields or for gauge fields. To deal with a transformation of gauge string fields we need a differential on the algebra $\mathcal{B}$. However in order to write the gauge transformation for matter string field the multiplication is sufficient. So, let me consider the matter-like transformation law, i.e.

$$\delta_\Omega \Psi = \Psi \circ \Omega \tag{2.12}$$

Here we assume that the parameter Ω of the gauge transformation and the matter string field Ψ are objects of the same algebra and moreover that the transformation law can be written as a product in this algebra. Substituting the gauge transformation (2.12) in the Eq. (2.8) one gets the relation

$$(\Psi \circ \Omega_1) \circ \Omega_2 - (\Psi \circ \Omega_2) \circ \Omega_1 = \Psi \circ (\Omega_1 \circ \Omega_2) \quad . \tag{2.13}$$

In Eq. (2.13) the Jacobi identity is easily recognized. Starting from (2.10) and assuming natural linear properties under multiplication of Grassmann numbers one can deduce from (2.13) the Jacobi identity for arbitrary homogeneous elements of $\mathcal{B}$

$$(-1)^{|\Phi||\chi|}\,\Phi\circ(\Psi\circ\chi)+(-1)^{|\chi||\Psi|}\,\chi\circ(\Phi\circ\Psi)+ \\ +(-1)^{|\Psi||\Phi|}\,\Psi\circ(\chi\circ\Phi)=0 \tag{2.14}$$

Eqs. (2.10) and (2.14) give us on the whole $\mathcal{B}^{I}$ a structure of Lie super-algebra.

The string field algebra $\mathcal{B}^{II}$ is a nonassociative commutative superalgebra with a multiplication law satisfying the Jacobi identity [12]. So, from physical reasons we get (after some additional postulates) only two minimal schemes [13]:

i) $\mathcal{B}^{I}$ is a Lie superalgebra;

ii) $\mathcal{B}^{II}$ is a commutative superalgebra with Jacobi identity.

In some particular cases $\mathcal{B}^{I}$ can be realized with the help of an associative $*$ product

$$\Phi\circ\Psi=\Phi*\Psi-(-1)^{|\Phi||\Psi|}\,\Psi*\Phi \quad .$$

By analogy with the Yang-Mills theory one can take the nonabelian gauge transformation of the string gauge field to be of the form

$$\delta_{\Omega}\Phi \quad = \quad Q\Omega+\Phi\circ\Omega \quad , \tag{2.15}$$

where Q is a linear map of the algebra $\mathcal{B}$ into itself and Q^2 is zero. In this case Eq. (2.8) gives in addition to identity (2.13) the Leibnitz rule for the derivative Q of the algebra $\mathcal{B}$. Q can be considered as an exterior derivative on string fields (string differential forms [9]).

There are two possible string realizations of an algebra $\mathcal{B}$ [16]. The first of them uses the Witten vertex [10, 18], the second uses the light-cone-like vertex proposed by Hata et al [15] and by Neveu and West [14].

To construct an action one can also proceed in two ways. In the first one, proposed by Witten, one deals with an integral on the algebra, in the second one with an inner product on the algebra $\mathcal{B}$. So, the main ingredients of the Witten construction are: a product $*$, an exterior derivative Q and an integral $\int$. In other words, in this case one deals with a Witten-Connes cycle, $W = (*, Q, \int)$.

The main ingredients of the construction proposed by Volovich and myself [16], are: a product (on $\mathcal{B}^{I}$ or $\mathcal{B}^{II}$), an exterior derivative Q and an inner product $(\cdot,\cdot)$.

The action for the gauge string field Φ which is an arbitrary even (odd) element of $\mathcal{B}^{I}$ ($\mathcal{B}^{II}$) has the form

$$S_{\Phi} = (\Phi, Q\Phi) + \frac{1}{3}(\Phi \circ \Phi, \Phi) = \int \Phi * Q\Phi + \frac{2}{3}\int \Phi * \Phi * \Phi \tag{2.16}$$

This action is invariant under the gauge transformation (2.15).

We conclude this section with the following table which summarizes the relationship between Yang-Mills theory and interacting open bosonic string field theory.

	Yang-Mills Theory	String Field Theory
Gauge field	$A = A_{\mu}\, dx^{\mu}$	$\Phi[X(\sigma), c(\sigma), b(\sigma)]$
Gauge transformation	$\delta A = d\lambda + [A, \lambda]$	$\delta\Phi = Q\Lambda + \Phi \circ \Lambda$
Differential	d	Q
Multiplication	[,]	$\circ$ or $*$
Action	$\int tr(F_{\mu\nu}^{2})$	$(\Phi, Q\Phi) + \frac{1}{3}(\Phi \circ \Phi, \Phi) = \int \Phi * Q\Phi + \frac{2}{3}\int \Phi * \Phi * \Phi$

Note that in the last year the interesting suggestion to consider the pure cubic action has been proposed (for more details see [31]).

3. Superstring Field Theory (SSFT)

3.1. Free action

In SSFT we will glue together the ideas just tested for bosonic strings with general ideas of supersymmetric field models. The motivation for developing a field theoretic approach to superstrings is the same as for bosonic string. But there are some additional reasons, namely: i) to investigate non-perturbative questions such as SUSY breaking and spontaneous compactification; ii) to have unitarity "automatically". (In the first quantization approach there is a problem with spin structures).

The main object of SSFT will be $\Phi[X(z), \Psi(z), \text{ghosts}]$, where $X(z)$ and $\Psi(z)$ are world sheet bosonic and fermionic coordinates. Different boundary conditions for Ψ correspond to different sectors - The Neveu-Schwarz (NS) sector and the Ramond (R) sector:

$$\text{NS:} \quad \Psi^{\mu}(z) = \sum_{r} H_r^{\mu} z^{-r} \quad , \quad r \text{ is half integer}$$

$$\text{R:} \quad \Psi^{\mu}(z) = \sum_{n} \Gamma_n^{\mu} z^{-n} \quad , \quad n \text{ is integer} \quad ,$$

$$\{H_r^{\mu}, H_s^{\mu}\} = -2\eta^{\mu\nu} \delta_{r+s} \quad , \quad \{\Gamma_n^{\mu}, \Gamma_m^{\nu}\} = -2\eta^{\mu\nu} \delta_{n+m} \quad .$$

The fermionic ghost representation for the NS and R strings has the following content: anticommuting (anti)ghosts(b)c corresponding to the bosonic coordinates X and commuting (anti)ghosts $(\bar{t})t$ corresponding to world sheet spinors Ψ.

$\Phi[X(\sigma), \Psi(\sigma)]$ as a vector from the Hilbert space of the first quantization approach satisfies the following constraints:

$$L_n \Phi[X(\sigma), \Psi(\sigma)] = 0 \quad , \quad n \geq 0$$

$$\text{NS: } G_r \Phi[X(z), \Psi(z)] = 0, \; r \geq \tfrac{1}{2}; \quad \text{R: } F_n \Phi[X(z), \Psi(z)] = 0, \; n \geq 0$$

where L_n and $G_r(F_n)$ are the generators of the super Virasoro algebra

$$[L_m, L_n] = (m-n)L_{m+n} + \frac{D}{8} m(m^2 - 1)\delta_{m+n,0}$$

NS sector: R sector:

$$[L_m,G_r]=(\tfrac{1}{2}m-r)G_{m+r} \quad ;[L_m,F_n]= (\tfrac{1}{2}m-n)F_{m+n}$$

$$\{G_r,G_s\}=2L_{r+s}+ \frac{D}{2}(r^2-\tfrac{1}{4})\delta_{r+s};\{F_m,F_n\}=+2L_{m+n}+ \frac{D}{2} m^2\delta_{m+n}$$

The main observation for the bosonic string is that if one takes the BRST-charge Q connected with the Virasoro algebra then the equation $Q\Phi = 0$ gives us the correct set of constraints. Moreover, one can get this equation from the action.

One can expect a similar situation for fermionic strings and this does work for the NS sector but not for the R sector. This is due to two reasons [19-21]: i) the appearence of new zero modes of the commuting superconformal ghosts; ii) the presence of two kinetic energy operators in the BRST operator connected with the R super Virasoro algebra (see Appendix):

$$Q_R = c_0L_R + \overline{c}_0(M_R + t_0^2) + t_0F + \overline{t}_0K + \tilde{Q}_R \quad . \tag{3.1}$$

Of course we would like to have the Dirac equation for the zero mass-level field $\chi^\alpha(x)$ of the R string $\Phi[X,\Psi] = \chi^\alpha(x)|0>_\alpha+\ldots$

Up to now there have been three versions [19-25] of a free R-string action:

i) The Witten construction [19] is based on an inverse picture changing operator Y(z).

ii) The Kazawa, Neveu, Nicolai, West [20] and the Terao and Uehara [22] construction which uses a notion of "consistently truncated fields".

iii) The construction due to Medvedeev and myself based on truncated BRST-charge and the truncated scalar product. (Note that the notion of the truncated BRST-charge was also used by Banks et al [21] in the context of string differential forms).

I will start with the third possibility. Following the paper [23] I shall show that the appropriate action for the R-string can be derived from general algebraic considerations. I shall discuss the other proposals later. As for the bosonic string, a field theory action for the R-string has to satisfy the three requirements pointed out in Sect. 2. Due to the relation $F_0^2 = L_0$ we may drop out the mode c_0 which corresponds to the operator L_0. But if one drops out one mode from the BRST-operator then the corresponding truncated operator

$$Q^{tr} = \sum_{n\neq n_0} c_{-n}L_n - \frac{1}{2}\sum_{m,k,n\neq n_0} f^k_{mn}\, b_k c_{-n} c_{-m}$$

is no more nilpotent on the whole space. However there is a subspace H^{tr} specified by the constraint

$$H^{tr} = \{\varphi;\ (\sum_{n,m\neq n_0} c_{-n}c_{-m}f^{n_0}_{mn})\varphi = 0\} \quad ,$$

where the condition $(Q^{tr})^2\varphi = 0$ holds. So, $Q_R^{tr} = t_0 F + \bar{t}_0 K + \tilde{Q}_R$ is nilpotent on vectors satisfying the constraint $(t_0^2 + M_R)\Psi = 0$. Instead of solving directly this equation note that in order to have gauge invariance it is sufficient to have a "weak" nilpotency, i.e. only an average of $(Q_R^{tr})^2$ should be zero. This justifies the introduction of a new scalar product $(\cdot,\cdot)_{tr}$ for which the relation $(\Psi_1, (t_0^2 + M_R)\Psi)_{tr} = 0$ holds. Hence, let me introduce the scalar product by the formula $(\Psi_1,\Psi_2)_{tr} = (\Psi_1, Y\Psi_2)$ where the kernel Y is given by the Yamron [24] δ-function

$$Y = \delta(t_0^2 + M_R) \quad . \tag{3.2}$$

With this definition the equality

$$(\Psi_1, Y(t_0^2 + M_R)\Psi_2) = 0 \tag{3.3}$$

is expected to hold for all Ψ. It is possible to give a precise meaning to the last two equations [23].

Now we shall describe the space H_R. The nonzero physical and ghost modes are in ordinary Fock representation $\breve{H}$ where negative frequency operators are realized as creation operators, so the total vacuum $|\breve{0}>$ is the tensor product of the vacua for each mode. Because the zero modes of the R fermionic oscillators act like gamma matrices, the vacuum and the fields for the R-string also carry SO(9,1) Dirac spinor indices, which we suppress as usual.

The bosonic ghost zero modes t_o, $\bar{t}_o$ satisfy the commutation relation $[\bar{t}_o, t_o] = 1$ and the hermiticity properties $t_o = t_o^+$, $\bar{t}_o = -\bar{t}_o^+$. These operators cannot be realized in the standard way. Let $|0>$, respectively $|0)$ be states such that $\bar{t}_o|0> = 0$, $t_o|0) = 0$ and define $|m> = t_o^m|0>$, $|n) = \bar{t}_o^n|0)$. It is easily seen that

$$(|n>, |m)) \equiv <n|m) = m!\delta_{mn}$$

$$(|m), |n>) \equiv (m|n> = m!\delta_{mn}$$

under the normalization assumption $(0|0> = 1$. Also $(n|m) = <n|m> = 0$. Let $h_{|0>} = \{\Sigma\, a_n t_o^n|0> \equiv \Sigma\, a_n|n>\}$ **and** $h_{|0)} = \{\Sigma\, a_n \bar{t}_o^n|0) \equiv \Sigma\, a_n|n)\}$ be the corresponding "Fock" spaces. The space for the R-string states will be $H_R = H_{|0>} \oplus H_{|0)}$ where $H_{|0>} = \breve{H} \otimes h_{|0>}$ and $H_{|0)} = \breve{H} \otimes h_{|0)}$ can be considered as the Fock spaces corresponding to the total vacua $|\breve{0}> \otimes |0>$ and $|\breve{0}> \otimes |0)$. In what follows we shall omit, as a rule, all the vacuum structures except the t_o structure.

Now we are ready to give the precise definition for $\delta(t_o^2 + M_R)$:

$$\delta(t_o^2+M_R) = \sum_{n=0}^{\infty} \sum_{k=0}^{2n+1} \frac{(-M_R)^n}{(2n+1-k)!k!} |2n + 1 - k)(k| \quad .$$

It is easy to check the desired relation

$$\delta(t_o^2 + M_R)(t_o^2 + M_R)|\psi> = 0 \quad .$$

To obtain from $Q_R^{tr}\psi = 0$ the Dirac-like equation $F\Psi = 0$ we have to define ψ as a vector from the $H_{|0>}$-space, denoting it by $|\Psi>$.

Let me introduce the important notation of truncated field [23]. The truncated field $|\psi^{tr}>$ is a state constructed from $|\psi> = \sum_{n\geq 0} A_n|n>$ as follows:

$$|\psi^{tr}> = \sum_{n\geq 0}(-M_R)^n A_{2n}|0> + \sum_{n\geq 0}(-M_R)^n A_{2n+1}|1> \quad ,$$

where A_n are arbitrary polynomials in nonzero modes. It is obvious that

$$\delta(t_o^2 + M_R)(|\psi> - |\psi^{tr}>) = 0 \quad .$$

so $|\psi>$ and $|\psi^{tr}>$ are indeed equivalent with respect to the metric $Y = \delta(t_o^2 + M_R)$. The scalar product for $H_{|0>}$ vectors can be expressed as follows:

$$(\psi,\chi)^{tr} = (|\psi>,|\chi>)^{tr} = -<\psi^{tr},\delta'(t_o)\chi^{tr}> \quad ,$$

where $\delta'(t_o) = -|0)(1| - |1)(0|$. It follows from [24] that Q_R is a hermitian operator with respect to $(\cdot,\cdot)^{tr}$.

The action for the R-string

$$I_R = <\psi, YQ_R^{tr}\psi> \tag{3.4a}$$

can naturally be defined in terms of $(\cdot,\cdot)^{tr}$. Moreover, due to the relation $(Q_R^{tr}|\psi>)^{tr} = (Q_R^{tr}|\psi^{tr}>)^{tr}$ we can write the R-action (3.4a) through the components of the truncated field $|\psi^{tr}> = \Psi_o|0> + \Psi_1|1>$,

$$I_R = (\Psi,\sigma\,\mathbb{Q}_R\Psi) \tag{3.4b}$$

where

$$\Psi = \begin{pmatrix}\psi_0\\ \psi_1\end{pmatrix} \quad , \quad \mathbb{Q}_R = \begin{pmatrix}\tilde{Q}_R & Z\\ F & \tilde{Q}_R\end{pmatrix} \quad , \quad \sigma = \begin{pmatrix}0 & 1\\ 1 & 0\end{pmatrix} \quad ,$$

$$Z = -\tfrac{1}{2}(FM_R + M_RF)$$

or

$$I_R = (\Psi_0, F\Psi_0) + (\Psi_0, \tilde{Q}_R \Psi_1) + (\Psi_1, \tilde{Q}_R \Psi_0) + (\Psi_1, Z\Psi_1) \quad , \qquad (3.4b')$$

The scalar products in the last line are scalar products in the $\check{H}$ space.

The form (3.4a) of the action coincides with that proposed by Yamron [24], while the component formula (3.4b') is the same as the formula obtained for S_R in the papers by Kazawa, Neveu, Nicolai, West [20] and by Terao and Uehara [22].

Note that the action (3.4a) is minimal as concerns the number of independent field components and the structure of BRST charge, and this makes it suitable for the purpose of studying the interacting string.

Recently [28] it has been shown that all the R-string actions [19-25] possess in fact the following structure

$$S_R = \langle \psi, G\ Q\ \psi \rangle \quad , \qquad (3.5a)$$

where the metric G admits the following decomposition

$$G = T_G^+\ g\ T_G \quad , \qquad (3.6)$$

T_G being a projector on a two dimensional subspace and g a metric in a two-dimensional space (we have in mind the dimension of world-sheet supersymmetric ghost zero modes space). A significant feature of G is its BRST invariance

$$[G, Q] = 0 \qquad (3.7)$$

that guarantees the gauge invariance of the action, and allows us to establish that (3.5a) depends only on G-truncated fields $\psi^{tr} = T_G\ \psi$, i.e.

$$S_R = \langle \psi^{tr}, g(\mathbb{Q}_G\ \psi^{tr})^{tr} \rangle = \langle \psi^{tr}, g\ \mathbb{Q}_G\ \psi^{tr} \rangle \qquad (3.5b)$$

The $\mathbb{Q}_G$ is the modified BRST charge which is weakly nilpotent with respect to the G-metric, i.e. $\langle \psi, G(\mathbb{Q}_G)^2 \psi \rangle = 0$.

So, an inverse picture-changing operator playing the rôle of a metric in a space of ghost modes is the origin of the appearence of a truncated string field as well as of a truncated BRST charge.

Note that the representation (3.5a) with the decomposition (3.6) for the action (3.4a) has just been demonstrated.

The proof that the Witten action $S_R^{(w)}$,

$$S_R^{(w)} = \langle\psi, Y(z)\ Q\ \psi\rangle \quad , \tag{3.8}$$

where $Y(z) = c(z)\delta'(t(z))$, $z = \pi/2$, is also of the form (3.5b), is less trivial. The following formula holds

$$Y(z) = \mathcal{T}_w^+ \left(|0)(1| + |1)(0| - 2\check{t}(z)|1)(1| \right)\mathcal{T}_w \tag{3.9}$$

where

$$\mathcal{T}_w = \mathcal{T}_w^c \mathcal{T}_w^t \quad ;$$

$$\mathcal{T}_w^c = ((0|c(z) \; ; \quad \mathcal{T}_w^c(\varphi_0 + c_0\varphi_1)|0)) = \varphi_0 - \check{c}(z)\varphi_1 \quad ,$$

$$\mathcal{T}_w^t(\sum_n \psi_n |n\rangle) = (\sum_n (-1)^{n+1}(n-1)(\check{t})^n \psi_n)|0\rangle + (\sum_n (-1)^n n(\check{t})^{n-1}\psi_n)|1\rangle \quad ,$$

$$c(z) = c_0 + \check{c}(z), \quad t(z) = t_0 + \check{t}(z) \quad , \quad b_0|0)) = 0.$$

It is important to note that in close analogy to $Y=\delta(t_0^2+M)$, $Y(z)$ has a nontrivial kernel, i.e. $Y(z)|\psi\rangle = 0$, consisting of the following vectors: $|\psi\rangle = c(z)|A\rangle + t^2(z)|B\rangle$ with $|A\rangle$ and $|B\rangle$ being "smooth" Fock vectors, i.e. vectors which do not provide the Schwinger terms in OPE around the point z. We denote by $\mathcal{M}$ the space of these "smooth" vectors. This fact can be really checked by means of OPE if we define

$$Y(z)|\psi(z)\rangle = \lim_{z'\to z} Y(z')|\psi(z)\rangle \quad .$$

Denoting the components of a truncated field as ρ_0 and ρ_1, i.e. $\mathcal{T}_w|\psi\rangle = \rho_0|0\rangle + \rho_1|1\rangle$ it is a matter of simple calculation to check that the change of variables

$$\rho_0 = \Psi_0 + \check{c}\, F\, \Psi_1$$
$$\rho_1 = \Psi_1 \qquad (3.10)$$

casts (3.8) into the form (3.4b') that has been found in [26].

3.2. Supersymmetry of the free NSR action

It is convenient to write down the action for open superstrings using a single string field Φ in the form

$$I = (\Phi,\ Q\ \Phi),$$

where $\Phi = \begin{pmatrix} a \\ \Psi \end{pmatrix}$, $Q = \begin{pmatrix} Q_{NS} & 0 \\ 0 & Q_R \end{pmatrix}$. The action is invariant under the linear transformation

$$\delta\Phi = V\Phi$$

if

$$[V,\ Q] = 0$$

if we choose the matrix V to be off-diagonal,

$$V = \begin{pmatrix} 0 & V_{R,NS} \\ V_{NS,R} & 0 \end{pmatrix}$$

i.e. the variation of the bosonic NS string field is expressed through fermionic R-field,and reciprocally the equation $[V,\ Q] = 0$ means $Q_{NS}\ V_{R,NS} = V_{R,NS}\ Q_R$; $Q_R V_{NS,R} = V_{NS,R}\ Q_{NS}$.

The last equations show that $V_{R,NS}$ and $V_{NS,R}$ are intertwining operators for the BRST-charges.

In fact, things are not so simple, since the scalar product for the R-sector is highly degenerated

$$S = (\Phi,\ J\ Q\ \Phi), \qquad J = \begin{pmatrix} -1 & 0 \\ 0 & G \end{pmatrix} \qquad (3.11)$$

The action I is invariant under $\delta\Phi=V\Phi$ if the following condition

$$V^{+}J\,Q + J\,Q\,V = 0 \tag{3.12}$$

holds.

The structure of the matrix V has to be specified. Namely, we put

$$V_{R,NS} = \mathcal{V}_{R,NS}\,\mathcal{T} \qquad \text{and} \quad V_{NS,R} = -\,(\delta')^{-1}\mathcal{V}_{NS,R}$$

$$(\delta'(t_0))^{-1} = -|0><1| - |1><0|\,; \; (\delta')^{-1}\delta' = P_{|01>},\, \delta'(\delta')^{-1} = P_{|01)},$$

$$P_{|01>} = |0>(0| + |1>(1| \;; \quad P_{|01)} = |0)<0| + |1)<1| \quad .$$

So,

$$V = \begin{pmatrix} 0 & \mathcal{V}^{+}\mathcal{T} \\ (-\delta')^{-1}\mathcal{V} & 0 \end{pmatrix} \quad , \qquad V^{+} = \begin{pmatrix} 0 & \mathcal{V}^{+}(-\delta')^{-1} \\ \mathcal{T}^{+}\mathcal{V} & 0 \end{pmatrix}$$

Eq. (3.12) reduces to the single equation

$$-P_{|01)}\,\mathcal{V}Q_{NS} + \delta'\,\mathcal{T}\,Q_R^{tr}(\delta')^{-1}\mathcal{V} = 0$$

It is possible to find $\mathcal{V}$ in the form $\mathcal{V} = \mathcal{U}W$, where W is an intertwining operator between Q_{NS} and Q_R, i.e.$WQ_{NS} = Q_RW$, and $\mathcal{U}$ is an intertwining operator (on a subspace only) between Q_R and Q_Y:

$$P_{|01)}(-\mathcal{U}Q_R + \mathbb{Q}_Y\mathcal{U}) = 0 \quad .$$

W is given by the total fermion emission vertex [19-23] (spin operator in the context of the conformal field theory [36]).

It is not difficult to guess the answer for $\mathcal{U}$[23]:

$$\mathcal{U} = ((0| - |1)<0|\;F\;((0|c_0 .$$

So, the explicit form of SUSY transformations is

$$\delta a = [W^{+} - W^{+}c_0F|0>(1|]\psi^{tr}|0)) \tag{3.13a}$$

$$\delta\psi^{tr} = (-\delta')^{-1} ((0|[W - |1)<0|Fc_0W]a \tag{3.13b}$$

Let me discuss the connection between the SUSY transformations proposed by Witten [19] and (3.13). The Witten SUSY transformations are of the form

$$\delta a = W^+\psi \tag{3.14a}$$

$$\delta\psi = XWa \tag{3.14b}$$

where $X(z)$ is the picture changing operator [36]. As soon as the form (3.5b) of the action is established Eq. (3.14a) seems rather puzzling. Indeed, the transformation law for the NS field a includes the total field ψ, although the action depends only on $\psi_W^{tr} = T_W\psi$ and thereby the transformation must be of the form $\delta a = z\psi_W^{tr} = zT_W\psi$. So we have to search for an operator which is able to transform ψ_W^{tr} into ψ for all smooth vectors. Such an operator does exist and is nothing else than the well-known "unity" $X(z)Y(z)$ multiplied by the factor $|0))$. In fact, we find that

$$\lim_{z'\to z}\lim_{z''\to z} X(z')(Y(z'')|0))\,T_W\psi) = \psi \quad , \quad \psi \in M \; . \tag{3.15}$$

The standard calculation yielding $X(z)Y(z) = 1$ is really correct when we suppose that the operator $Y(z)$ acts on regular vectors (finite Fock vectors for example). But in singular cases we have to take into account the nonassociativity of OPE and we have to define the action of operators carefully:

$$X(z)Y(z)\psi(z) \equiv \lim_{z''\to z}\lim_{z'\to z} X(z'')(Y(z')\psi(z)) \tag{3.16}$$

The Eq. (3.15) holds only for the states ψ from the smooth space. For the states from the kernel of $Y(z)$, for example for $\psi = t^2(z)|0>$, the formula (3.15) does not hold because the l.h.s. is identically zero: $T_Wt^2(z)\psi = 0$. A proof of Eq. (3.15) can be found in [28].

The above discussion leads to the following modification of the transformation law (3.14a):

$$\delta a = W^{+}X\ Y|0))\ \psi_W^{tr} \tag{3.14a'}$$

so we conclude that $W^{+}XY|0))$ is a proper operator for SUSY transformation in the truncated space. The r.h.s. of (3.14a') coincides with the r.h.s. of (3.14a) for $\psi \in M$.

Let me compare the transformation laws (3.14) and (3.13) supposing that the components of ψ_W^{tr} are smooth vectors. The state vector ψ_W^{tr} in the r.h.s. of (3.14a') can be parametrized in terms of Ψ_0 and Ψ_1 ($\Psi_0, \Psi_1 \in M$) as in Eqs. (3.10), $\psi_W^{tr} = (\Psi_0 + \check{c}(z)F\Psi_1)|0> + \Psi_1|1>$ that gives

$$\delta a = W^{+}XY|0))\ \psi_W^{tr} = W^{+}XY[(\Psi_0 + \check{c}(z)F\Psi_1)|0> + \Psi_1|1>]|0)) =$$

$$= W^{+}(\Psi_0|0> + \Psi_1|1>)|0)) - W^{+}c_0F\Psi_1|0>|0)) \quad ,$$

which coincides with the Eq. (3.13a).

To compare Eqs. (3.13b) and (3.14b) we note [28] that, due to the structure (3.5b) of the Witten action, it is sufficient to look at $(\delta\psi)^{tr} = T_W(XWa)$. The r.h.s. of this formula can be calculated using the following representation (compare with Eq. (B.2) in ref. [23]):

$$T_W = (|0><1| + |1><0| + 2\check{t}(z)|0><0|)Y(z)$$

that yields

$$(\delta\psi)^{tr} = ((0|[|0><1| + |1><0| - 2\check{t}(z)|0><0|]\ Wa \quad .$$

Exploiting the commutation relation $\{\check{c}(z), F\} = -2\check{t}(z)$, we find for the components $\delta\Psi_0$ and $\delta\Psi_1$ the same transformation rule as in Eq. (3.13b). So the equivalence of SUSY transformations (3.14) and (3.13) is established. But this does not give the on shell closure of the algebra (3.13) in spite of closure of the algebra of the Witten transformations [19].

3.3. Interacting open superstrings

It is convenient to write down the action for the open superstring using a single string field in the form

$$I = (\Phi, GQ\Phi) + \frac{1}{3}(\Phi, G(\Phi \circ \Phi))$$

where $\Phi = \begin{pmatrix} a \\ \Psi \end{pmatrix}$, a and Ψ are functionals from supersymmetric sectors of the NS and R string and

$$\Phi_1 \circ \Phi_2 = \begin{pmatrix} a_1 \\ \Psi_1 \end{pmatrix} \circ \begin{pmatrix} a_2 \\ \Psi_2 \end{pmatrix} = \begin{pmatrix} a_1 \circ a_2 + \Psi_1 \circ \Psi_2 \\ a_1 \circ \Psi_2 + \Psi_1 \circ a_2 \end{pmatrix}$$

So, to define $\Phi_1 \circ \Phi_2$ it is necessary to define three kinds of multiplications.

As it was demonstrated by Medvedev and myself [23] it is possible to define the product of the NS and the R states as well as the product of two R states by requiring supersymmetry. Naively:

$$V(\Phi \circ \Phi) + V\Phi \circ \Phi + \Phi \circ V\Phi = 0$$

$$\Psi_1 \circ \Psi_2 \sim (V_{R,NS}\Psi_1) \circ (V_{R,NS}\Psi_2)$$

i.e. one can expect that the product of two R states is determined in the following way. First using the operator $V_{R,NS}$ the R states are converted into NS states and these ones are multiplied as usual. But we have to take into account the degeneracy of the metric G. In this case SUSY invariance of the interaction implies:

$$V^{+}G(\Phi \circ \Phi) + G(V\Phi \circ \Phi + \Phi \circ V\Phi) = 0 .$$

Apart from the formulas defining the NS-R and R-R vertices, the last equation puts constraints on the NS-NS vertex. To complete the construction one needs an explicit form of the NS-NS vertex.

4. Toward the Closed String Field Theory

As it was seen from the previous lectures, recently, important results have been obtained in the gauge invariant field theory of free and interacting open bosonic strings as well as open superstrings. Closed strings are also considered in some papers, however, the field theory of closed strings is now in a rather unsatisfactory status. In particular, closed string theory pretending to describe the general relativity starts from a linearized approximation.

In this section I discuss the general principles which are behind the theory of closed strings and in particular the fundamental general covariance principle in string field theory. I also consider a new formulation of string field theory in terms of vielbein and spin connection string fields [29,30].

4.1. General relativity invariance and string field theory

As it is well known, the closed string theories in some sense generalize the general relativity theory. Therefore, it is natural to think that at the base of closed string field theories lies a gauge invariance principle.

In general relativity, we deal with general covariance principle. Note that the gauge groups of general relativity and Yang-Mills theory have essentially different features. The gauge transformations in gravity are generated by the Lie derivative $L_\xi = i_\xi d + di_\xi$ and they are homogeneous, in contrast to the non-homogeneous transformations in Yang-Mills theory, which are given by the covariant derivative $D = d + A$.

There is also another reason why the gauge group for string theory with Yang-Mills type invariance and the general coordinate gauge group should have different features. The string field can be considered as an infinite set of usual local fields. Let us suppose that we have the set of local gauge fields A_μ, $A_{\mu\nu}$, ... If the gauge transformations of these fields are described by the set of gauge parameters λ, λ_μ, ... , namely, $\delta A_\mu = \partial_\mu \lambda$, $\delta A_{\mu\nu} = \partial_{[\mu}\lambda_{\nu]}$, ... , then we will name gauge invariance the Yang-Mills type one. In particular, in string field theory the gauge parameter is a string field, to which the set of local fields λ, λ_μ, ... corresponds.

A different picture arises when we deal with general transformations. In this case for any given set of fields on a manifold M (for example, on $\mathbb{R}^D$) coordinate transformations are generated by only one vector-function, connected with the change of coordinates, $\delta x^\mu = \xi^\mu(x)$. This fact is closely connected with the universality of gravity and it also concerns the string theory. A general coordinate transformation in string field theory $\Phi[x(\sigma)] \to \Phi[x(\sigma) + \xi(x(\sigma))]$ will not be specified by a string functional but only by one usual vector field ξ^μ. So, we see the fundamental difference between a gauge group of Yang-Mills type and the group of general coordinate transformations in string field theories.

This means that a scalar functional $\Phi[x(\sigma)]$ transforms under general coordinate transformations as

$$\delta_G \Phi[x(\sigma)] = L_\xi \Phi[x(\sigma)] = \int \xi^\mu(x(\sigma)) \frac{\delta\Phi[x(\sigma)]}{\delta x^\mu(\sigma)} d\sigma \tag{4.1}$$

This is nothing but the string analogue of the Lie derivative for scalar fields. It can be expressed also in the following form:

$$L_\xi = \int :\xi^\mu(x\ (\sigma))P_\mu(\sigma): d\sigma$$

where P_μ is the canonical conjugate momenta.

For the string Lie derivative the usual relations hold up to the possible anomalies which can, in principle, appear caused by a normal ordering operation.

4.2. String vielbein formalism

Let me note that there is a reason for the search of an essentially new formulation of field theory of interacting closed strings which would not be the naive generalization of the theory of open strings. It is well known that closed superstring in the limit $\alpha' \to 0$ have to describe the supergravity.

Recall that the main object in the supergravity is not the metric itself but the Cartan variables, the vielbein e_μ^a and the spin connection ω_μ^{ab}. These local fields cannot be described as zero-mass-level components of some scalar string field.

Let me consider the string vielbein fields E_a and E^a and a string spin connection field Ω_b^a depending on the string position $X^\mu(\sigma)$ as well as on the ghost coordinates. The indices $a,b = 1, \ldots D$ are indices referring to the Lorentz frame, and the functional Ω_b^a takes values in the Lie algebra of the Lorentz group. The local frame one-form $e^a = e_\mu^a dx^\mu$ and the spin connection $\omega_b^a = \omega_{b\mu}^a dx^\mu$ transform under local Lorentz transformations as

$$\begin{aligned} \delta_G e_\mu^a &= L_\xi e_\mu^a = \xi^\nu \partial_\nu e_\mu^a + \partial_\mu \xi^\nu e_\nu^a \quad , \\ \delta_G \omega_{b\mu}^a &= L_\xi \omega_{b\mu}^a = \xi^\nu \partial_\nu \omega_{b\mu}^a + \partial_\mu \xi^\nu \omega_{b\nu}^a \quad . \end{aligned} \tag{4.2}$$

To generalize the transformations (4.2) to string functionals E_a, E^a and Ω_b^a it is natural to suppose that $\delta_L E^a = \Lambda_b^a \circ E^b$,

$$\delta_L E_a = E_b \circ \Lambda_a^b, \quad \delta_L \Omega_b^a = Q\Lambda_b^a + (\Omega \circ \Lambda)_b^a \quad . \tag{4.3}$$

Hence we suppose that the string fields E_a, E^a transform under the string counterpart of the usual Lorentz gauge transformation as the matter in accordance with the homogeneous transformation (4.2) and the string field Ω_b^a transforms as a connection, i.e. as a usual Lie algebra valued string field. Then we suppose that the action of the general coordinate transformation on string Cartan variables is given by the Lie derivative

$$\delta_G E^a = L_\xi E^a, \qquad \delta_G \Omega_b^a = L_\xi \Omega_b^a \quad , \tag{4.4}$$

where L_ξ is given by the formula (4.1) (if we work in conformal gauge we have to add in r.h.s. of (4.1) some additional terms).

Let us introduce the string curvature

$$R^a_b = Q\Omega^a_b + \frac{1}{2}(\Omega \circ \Omega)^a_b \quad ,$$

which transforms homogeneously under string gauge transformations as well as under general coordinate transformations

$$\delta_G R^a_b = L_\Xi R^a_b = (Q\Xi)\ R^a_b$$

We believe that it is natural to introduce the following string analogue of the Einstein action

$$S = (R^a_b, E_a \circ E^b) \tag{4.5}$$

Here $(\cdot,\cdot)$ is the natural scalar product of string fields. It is easy to verify that under natural assumptions on the multiplication $\circ$, the action (4.5) is gauge invariant and also invariant under the transformation (4.3) and (4.4).

The usual vielbein and spin connection appear at the first mass level of the expansions

$$E^a(\alpha) = e^a_\mu \alpha^\mu_1 + \ldots \ , \ E_a(\bar{\alpha}) = e^a_\mu \bar{\alpha}^\mu_1 + \ldots \quad ,$$

$$\Omega^b_a(\alpha,\bar{\alpha}) = \omega^{ab}_\mu(\alpha^\mu_1 + \bar{\alpha}^\mu_1) + \ldots \quad .$$

5. p-adic Strings

In string theory, as in other modern physical theories, the usual ideas, going back to Euclid, Riemann and Einstein about space-time as a manifold with a metric, have been unquestioningly accepted. However, the superstring theory deals with very small distances of Planck order and it is very questionable that the standard structure of space-time, based on the notion of real numbers and the usual geometry, could indeed be applicable here. It seems also that a proper understanding of the problems of gravitational collapse and the cosmological singularity cannot be obtained in terms of the standard concept of space-time.

Recently, Volovich has suggested [32] a new theory, which abandons the usual concept of space-time as well as the usual concept of quantum strings. It was argued there that at Planck distances a non-Archimedian geometry might emerge which could be realized on the p-adic numbers, and therefore in the corresponding physical models one should use p-adic numbers instead of real ones. The usual description in terms of real numbers would emerge only in the macroscopic limit.

A possible rôle of the non-Archimedian p-adic space in physics was noted in Ref. [33].

In this section I shall briefly review the Volovich papers [32, 34].

5.1. Sub-Planckian physics and non-Archimedian geometry

Certainly, on distances of the Planck order, the effects of quantum gravity have to play a crucial rôle. It is known that gravitational measurement cannot be localized in a sub-Planckian domain. More exactly it was shown in quantum gravity that in contrast to quantum electrodynamics essential limitations on field measurements arise which follow from the equivalence principle. In particular one has the inequality $\Delta\Gamma \gtrsim \ell_{P\ell}^2/\ell^3$ where $\Delta\Gamma$ denotes the uncertainty in determination of a connection coefficient, ℓ is the length scale, $\ell_{P\ell} = (hGc^{-3})^{\frac{1}{2}}$ is the Planck length and G is Newton's constant.

Then there is an absolute limitation on length measurements $\ell \gtrsim \ell_{P\ell}$, so the Planck length is the smallest possible distance that can in principle be measured. The reason is the following. If we want to locate an elementary particle or string we need to have energy greater than the Planck mass $m_{P\ell} = (hcG^{-1})^{\frac{1}{2}}$. But in such a case the corresponding gravitational field will have a horizon at $r = 2Gm_{P\ell}c^{-2} = 2\ell_{P\ell}$, shielding whatever happens inside the Schwarzschild radius. So no information of geometry in a sub-Planckian region is available. We see that in some sense a smallest quantum of space and time exists and it is of the order of the Planck scale. This means that we need a new theory instead of the usual quantum gravity or string theory in this high energy region. We have to understand what quantum space-time means at the Planck scale. Note that a string is a curve in the usual space and therefore the string theory cannot help us here.

Indeed, what do we mean by usual space ? Our intuitive understanding of properties of space is expressed in the axioms of elementary geometry. Let us consider elementary geometry. According to the Archimedian axiom any given large segment on a straight line can be surpassed by successive addition of small segments along the same line. In fact, this is a physical axiom which concerns the process of measurement. Two different scales are compared in this axiom. If one imagines that the small segment has a size less than the Planck length and the large one has a macroscopic or even cosmological size, then the application of this axiom becomes questionable. As we just discussed, the Planck length is the smallest possible distance that can in principle be measured.

In the context of the sub-Planckian physics Volovich has suggested to abandon the Archimedian axiom and to construct a physical theory using p-adic numbers on which a non-Archimedian geometry can be realized. The p-adic numbers (see next section) look rather natural from a mathematical point of view, they are no more puzzling than the real numbers. In Ref. [34] it was argued that a wide range of applications is possible for them, not only in the theory of elementary particles but also in other fields where one has to deal with complex or fractal behaviour, for example, in turbulence theory, statistical mechanics, dynamical systems, etc.

5.2. p-adic numbers

Let me recall the basic algebraic notions which are needed for the construction of a physical theory. A field K is a set together with two operations, addition and multiplication, such that K is a commutative group under addition, $K\setminus\{0\}$ is a commutative group under multiplication and the distributive law holds. Examples of a field are the field of real numbers, the field of complex numbers and the field $\mathbb{Q}$ of rational numbers.

They are infinite fields. The simplest example of a finite field is the set of integers modulo a prime p, i.e., the field F_p of numbers 0,1,2,...,p-1 modulo p for some prime number p, $F_p = \mathbb{Z}/p\mathbb{Z}$, where $\mathbb{Z}$ are integers. This is a finite Galois field. The field F_2 (or $\mathbb{Z}_2$) is well known in supersymmetry theory.

If K is a field, it contains either the field of rational numbers $\mathbb{Q}$ as a subfield (in this case K is said to have characteristic 0) or the field F_p for some prime number p (in this case K is said to have characteristic p). So, the simplest fields are $\mathbb{Q}$ and F_p.

Some fields have an additional property of being "ordered" according to a relationship of "greater than" or "less than". For example the field of rational numbers and the field of real numbers are ordered. Note that the notion of the arrow of time is closely connected with the order of the field of real numbers. Usually, one considers time as the real axis.

To develop mathematical analysis we need a norm. A norm on the field K is a map, denoted $|\cdot|$, from K to the non-negative real numbers such that $|x+y| \leq |x| + |y|$ (triangle inequality), $|xy| = |x|\cdot|y|$ and $|x| = 0$ if and only if $x = 0$. A basic example of a norm on the field $\mathbb{Q}$ of rational numbers is the usual absolute value. It is known(Ostrowski theorem) that every non-trivial norm on $\mathbb{Q}$ is equivalent to the usual absolute value or to the p-adic norm for some prime p. The p-adic norm is defined as follows. Let p be any prime number, p = 2,3,5,...,137, ... Any non-zero rational number $x = M/N$ where M and N are integers can be represented in the form $x = p^f m/n$ where m and n are integers which are not divisible by p and f is an integer. Such a representation is possible because each integer greater than 1 can be written as a product of primes. The p-adic norm is $|x|_p = |p^f m/n|_p = p^{-f}$. Examples: $|9|_3 = 9^{-1}$, $|9^{-1}|_3 = 9$, $|10|_3 = 1$. We have $|m|_p \leq 1$ for any integer m. The p-adic norm is a norm on $\mathbb{Q}$. This norm satisfies not only the triangle inequality but also a stronger inequality $|x + y|_p \leq \max(|x|_p, |y|_p)$. A norm is called non-Archimedian if this inequality holds. Thus, the p-adic norm is a non-Archimedian norm on $\mathbb{Q}$.

The completion of $\mathbb{Q}$ with the usual absolute value is the field of real numbers. The completion of $\mathbb{Q}$ with the p-adic norm is the p-adic number field $\mathbb{Q}_p$. Any p-adic number $x \in \mathbb{Q}_p$ can be represented as a series $x = p^f(a_0 + a_1 p + a_2 p^2 + \dots)$. Here a_i are integers, $0 \leq a_i \leq p-1$. This series is convergent in the p-adic norm because $|a_n p^n|_p = p^{-n}$. This series can be compared to a decimal which represents a real number: $1/10 \leftrightarrow p$. The p-adic numbers can be defined by such series. Addition and multiplication are defined in an obvious way.

Our intuition about distance is based on the Archimedian metric. Some properties of the non-Archimedian metric seem very strange at first. Here are some examples.

1) In non-archimedian geometry all triangles are isosceles. If $|x|_p < |y|_p$, then $|y|_p = |y-x|_p$.
2) Any point in a disc is its centre. Define a disc of radius r centered at a: $D_a(r) = \{x \in \mathbb{Q}_p : |x-a|_p \leq r\}$. Then if $b \in D_a(r)$, we have $D_a(r) = D_b(r)$.
3) If we consider two balls then either a ball is contained inside the other one or they do not intersect. Such balls remind us the elementary particles.

5.3. p-adic dual amplitudes

In the construction of the p-adic mathematics and the associated physics one can follow two ways. In the first one complex-valued functions depending on variables belonging to $\mathbb{Q}_p$ are considered. Here the results of representation theory on local compact fields are useful. In the second way one has to do with functions taking values in the field $\mathbb{Q}_p$.

Both of these ways were considered in Refs. [32, 34]. In order to develop a p-adic physical theory it was proposed in Ref. [32] to start with the Veneziano amplitude which is the core of the contemporary string theory.

As it is well known, the Veneziano amplitude can be represented in terms of gamma functions or as an integral:

$$A(s,t) = \frac{\Gamma(-\alpha(s))\Gamma(-\alpha(t))}{\Gamma(-\alpha(s)-\alpha(t))} = \int_0^1 dx\, x^{-\alpha(s)-1}(1-x)^{-\alpha(t)-1} \tag{5.1}$$

where $\alpha(s) = \alpha_0 + \alpha's$. Using the p-adic interpolation of the gamma function we obtain all the ingredients of the theory, namely the momenta and the amplitude itself, as p-adic-valued quantities:

$$A_p(s,t) = \frac{\Gamma_p(-\alpha(s))\Gamma_p(-\alpha(t))}{\Gamma_p(-\alpha(s) - \alpha(t))} \tag{5.2}$$

where Γ_p is the p-adic gamma function. Here $\alpha(s) = \alpha_0 + \alpha's$, α_0 and α' are p-adic numbers and $s = \langle k_1+k_2, k_1+k_2\rangle$, $\langle\cdot,\cdot\rangle$ is a bilinear form on $\mathbb{Q}_p^D$ and $k_i \in \mathbb{Q}_p^D$. For an ingeter n the p-adic gamma function is defined in the following way

$$\Gamma_p(n) = (-1)^n \prod_{\substack{j<n \\ p\nmid j}} j \tag{5.3}$$

Here $p \nmid j$ means that j is not divisible by p.

The following correspondence principle was noticed [32]:

$$\lim_{p\to\infty} A_p(m,n) = A(m,n) \quad , \tag{5.4}$$

where m and n are integers.

In this approach there is an analogy with the standard approach in which all the ingredients of the theory are based on real numbers. In particular in such theories it is possible to consider general transformations of coordinates. Therefore this approach presents obvious advantages but of course there are questions concerning the physical interpretation of such amplitudes (see [34] for a discussion).

Another generalization of the Veneziano amplitude uses the interpretation of the amplitude as a convolution of two characters [32]. The generalized Veneziano amplitude is

$$A(\gamma_a,\gamma_b) = \int_K \gamma_a(x)\gamma_b(1-x)dx \quad . \tag{5.5}$$

Here K is a field, $\gamma_a(x)$ is a multiplicative character on K, dx is a measure on K (for example, the Haar measure if K is a locally compact field). The field K describes the internal structure of the string, the dual object $\hat{K}$ (which is a set of characters on K) being related with the space-time structure. Choosing K in an appropriate way, one can obtain the usual Veneziano amplitude (for $K = \mathbb{R}$, $\gamma_a(x) = |x|^a$), the Virasoro-Shapiro amplitude (for $K = \mathbb{C}$), the p-adic amplitudes and the finite Galois field amplitude.

It was suggested [32] that for closed p-adic strings the parameters take values in quadratic extensions of $\mathbb{Q}_p$, because the usual closed string amplitude is obtained with the help of a quadratic extension of the field of real numbers. The fact that there are three different quadratic extensions of $\mathbb{Q}_p$ indicates that there are three types of closed p-adic strings. This suggests the hypothesis that the graviton can have an additional quantum number (p-adic colour).

Appendix

The fermionic ghost representations for the NS and R strings have the following content: anticommuting (anti)ghosts $(\bar{s})s$ corresponding to the bosonic coordinates X_μ and commuting (anti)ghosts $(\bar{\lambda})\lambda$ or $(\bar{t})t$ corresponding to world-sheet spinors H_μ or Γ_μ of the NS and R strings respectively. Ghost mode expansions have the form

$$\lambda(z) = \Sigma\lambda_r z^{-r}, \quad \bar{\lambda}(z) = \Sigma\,\bar{\lambda}_r z^{-r}, \quad r: \qquad \text{half-integer}$$

$$t(z) = \Sigma t_n z^{-n}, \quad \bar{t}(z) = \Sigma\bar{t}_n z^{-n}, \quad n: \qquad \text{integer}$$

$$[\lambda_r, \bar{\lambda}_s] = \delta_{r+s,o} \quad , \qquad [t_n, \bar{t}_m] = \delta_{m+n,o} \quad .$$

Ghost modes hermiticity properties are $\lambda_r^+ = \lambda_{-r}$, $\bar{\lambda}_r^+ = -\bar{\lambda}_{-r}$, $t_m^+ = t_{-m}$, $\bar{t}_m^+ = -\bar{t}_{-m}$. The zero modes Γ_o^μ have the algebra and the conjugation properties of ten-dimensional gamma matrices.

The BRST charges in the NS and R sectors have the form

$$Q_{NS} = Q_B + \sum_n (L_n^H - L_n^\lambda)s_{-n} + \sum_r G_r\lambda_{-r} + \sum_n M_{NSn}\bar{s}_{-n} \qquad ,$$

$$Q_R = Q_B + \sum_n (L_n^\Gamma + L_n^t)s_{-n} + \sum_n F_n t_{-n} + \sum_n M_{Rn}\bar{s}_{-n} \quad ,$$

where

$$L_n^H = \frac{1}{8\pi i}\oint dz z^{n+1} :\frac{H(z)}{\sqrt{z}} \frac{d}{dz} \frac{H(z)}{\sqrt{z}}: = -\frac{1}{4}\sum_s (s+\tfrac{1}{2}) :H_{n-s}H_s: \quad ,$$

$$L_n^\Gamma = \frac{1}{8\pi i}\oint dz z^{n+1} (:\frac{\Gamma(z)}{\sqrt{z}} \frac{d}{dz} \frac{\Gamma(z)}{\sqrt{z}}: + \frac{5}{2z^2}) =$$

$$- \frac{1}{4}\sum_m (m+\tfrac{1}{2}) :\Gamma_{n-m}\Gamma_m: + \frac{5}{8}\delta_{n,o} \qquad ,$$

$$L_n^\lambda = -\frac{1}{4\pi i}\oint dz z^{n+1} (:\sqrt{z}\lambda(z) \frac{d}{dz} \frac{\bar{\lambda}(z)}{z^{3/2}} + 3\frac{\bar{\lambda}(z)}{z^{3/2}} \frac{d}{dz}(\sqrt{z}\lambda(z)): - \frac{1}{z^2})$$

$$= \sum_r (\frac{n}{2} + r) : \bar{\lambda}_{n-r}\lambda_r: + \frac{1}{2}\delta_{n,o} \qquad ,$$

$$L_n^t = -\frac{1}{4\pi i}\oint dz z^{n+1} (:\sqrt{z}t(z)\frac{d}{dz} \frac{\bar{t}(z)}{z^{3/2}} + 3\frac{\bar{t}(z)}{z^{3/2}} \frac{d}{dz}(\sqrt{z}t(z)): - \frac{3}{4z^2})$$

$$= \sum_m (\frac{n}{2} + m) :\bar{t}_{n-m}t_m: + \frac{3}{2}\delta_{n,o} \qquad ,$$

$$G_r = -\frac{1}{2\pi}\oint dzz^{r-1}\ H(z)P(z) = -i\sum_m \frac{1}{\sqrt{2}}\alpha_{-m}H_{m+r} \quad ,$$

$$F_n = \frac{1}{2\pi i}\oint dzz^{n-1}\Gamma(z)P(z) = \sum_m \frac{1}{\sqrt{2}}\alpha_{-m}\Gamma_{n+m} \quad ,$$

where α_m are the modes of the first-quantized bosonic coordinate X_μ ,

$$P^\mu(z) = \frac{1}{2}\, z\frac{d}{dz}\, X^\mu(z) = \frac{1}{\sqrt{2}}\sum_n \alpha_n^\mu z^{-n} \quad ,$$

$$M_{NSn} = -\frac{1}{2\pi i}\oint dzz^{n-1}(\lambda(z))^2 = \sum_r \lambda_{-r}\lambda_{r+n} \quad ,$$

$$M_{Rn} = \frac{1}{2\pi i}\oint dzz^{n-1}\ (t(z))^2 = \sum_m t_{-m}t_{m-n} \quad .$$

It is useful to expand Q_{NS} and Q_R in ghost zero modes

$$Q_{NS} = s_o L_{NS} + \bar{s}_o M_{NS} + \tilde{Q}_{NS}$$

$$Q_R = s_o L_R + \bar{s}_o(M_R + t_o^2) + t_o F + \bar{t}_o K + \tilde{Q}_R$$

here $M_{NS} = M_{NSo}$, $M_R = M_{Ro} - t_o^2$,

$$F = F_o + F_o^{gh} = F_o + \frac{d}{dt}\,\Sigma(L_n^t s_{-n} + M_n^t \bar{s}_{-n}) \quad .$$

References

[1]. G.Veneziano, Phys.Rep. 9C(1974) 199.
S.Mandelstam, Phys.Rep. 13C (1974) 259.
J.Scherk, Rev.Mod.Phys. 47 (1975) 123.
C.Rebbi, Phys.Rep. 12C (1974) 1.
J.H.Schwarz, Phys.Rep. 8C (1973) 269.
V.Alessandrini, D.Amati, M.Le Bellac, and D.Olive, Phys.Rep. 1C (1971) 269.

[2]. M.B.Green and J.H.Schwarz, Phys.Lett. 149B (1984) 117.
M.B.Green, J.H.Schwarz, E.Witten, *Superstring Theory*, Cambridge University Press, Cambridge, UK (1987).

[3]. N.N.Bogolubov, D.V.Shirkov, *Introduction to the Theory of Quantized Fields*, Interscience, N.Y. (1959).

[4]. A.N.Schellekens, Lecture in this volume.

[5]. M.J.Duff, T.Inami, C.N.Pope, E.Sezgin, K.S.Stelle, preprint IC/87/74, CERN-Th.4737/87.

[6]. L.Brink. P.Di Vecchia, P.S.Howe, Phys.Lett. 65B (1976)471.
S.Deser, B.Zumino, Phys.Lett. 65B (1976) 369.

[7]. M.O.Katanajev, I.V.Volovich, Phys.Lett.B175 (1986) 413.

[8]. A.Neveu, P.West, Phys.Lett. 165B (1985) 63.

[9]. T.Banks, M.Peskin, Nucl.Phys. B264 (1986) 513.

[10]. E.Witten, Nucl.Phys. B268 (1986) 253.

[11]. G.Thorn, Nucl.Phys. B287 (1987) 61.

[12]. I.Ya.Aref'eva, I.V.Volovich, in: *Fields and Geometry*, Proceedings of XXII Winter Karpacz School, World Scientific, Singapore (1986).

[13]. I.Ya.Aref'eva, I.V.Volovich, Phys.Lett. B182 (1986) 312.

[14]. A.Neveu, P.West, Phys.Lett. B168 (1986) 192.

[15]. H.Hata, K.Itoh, T.Kugo, H.Kunitomo, K.Ogawa, Phys.Lett. B172 (1986) 196.

[16]. I.Ya.Aref'eva, I.V.Volovich, Theor.Math.Phys. 67 (1986) 460; Phys.Lett. 182B (1986) 163.

[17]. H.Hata, K.Itoh, T.Kugo, H.Kunitomo, K.Ogawa, Phys.Rev. D34 (1986) 2360; D35 (1986) 1356.

[18]. D.J.Gross, A.Jevicki, Nucl.Phys. B283 (1987) 1; Nucl.Phys. B287 (1987) 225.

[19]. E.Witten, Nucl.Phys. B276 (1986) 291.

[20]. Y.Kazama, A.Neveu, H.Nicolai, P.West, Nucl.Phys. B276 (1986) 366.

[21]. T.Banks, M.E.Peskin, C.R.Preitshopf, D.Friedan,E.Martinec, Nucl.Phys.B274 (1986) 71.

[22]. H.Terao, S.Uehara, Phys.Lett. 173B (1986) 134.

[23]. I.Ya.Aref'eva, P.B.Medvedev, Preprint ITEP 22 (1987)

[24]. I.T.Yamron, Phys.Lett. 174B (1986) 69; Preprint UCH-PTH-86/35.

[25]. H.Ooguri, Phys.Lett. 172B (1986) 204.
S.P.de Alwis, N.Ohta, Phys.Lett. 174B (1986) 573.
W.Siegel, Maryland preprint UMDEPP 87-60 (1986).
G.D.Daté, M.Günaydin, M.Pernici, K.Pilch, P.van Nieuwenhuizen, Phys.Lett. 171B(1986) 82.
A.Le Clair, J.Distler, Nucl.Phys. B273 (1986) 552.

[26]. B.Sazdović, Preprint Brown-HET-611.

[27]. D.J.Gross, A.Jevicki, Preprint Brown-HET-603 (1987).

[28]. I.Ya.Aref'eva, P.B.Medveded, Phys.Lett.B (in press).

[29]. J.-L.Gervais, Preprint LPTENS 86/1 (1986); LPTENS 86/29 (1986).

[30]. I.Ya.Aref'eva, I.V.Volovich, Trieste preprint, IC/87/61.

[31]. G.I.Horowitz, Preprint UCS-BTH-87-38(1987).

[32]. I.V.Volovich, Theor.Math.Phys. 71 (1987) 337; Class.and Quant.Gravity 4 (1987) L83.

[33]. V.S.Vladimirov, I.V.Volovich, Theor.Math.Phys.59 (1984)3.

[34]. I.V.Volovich, Preprint CERN-Th-4781/87 (1987).

[35]. E.S.Fradkin, G.A.Vilkovisky, Phys.Lett. 55B (1975) 224.
I.A.Batalin, G.A.Vilkovisky, Phys.Lett. 69B (1978) 309.

[36]. D.Friedan, E.Martinec, S.Shenker, Nucl.Phys. B271 (1986) 93.

Steklov Mathematical Institute
Ul. Vavilova 42, Moscow 117 333

LIOUVILLE SUPERSTRINGS

Jean-Loup GERVAIS

Physique Théorique, Ecole Normale Supérieure
24 rue Lhomond - 75231 PARIS cédex 05 FRANCE

I. INTRODUCTION

Soon after Polyakov's pioneering work on the integration measure over surfaces [1,2], I started an extensive program, in collaboration with A. Neveu, in order to solve the corresponding quantum Liouville theory [3-8]. Besides its key role in the building of the Liouville string theories, our work has unravelled rather interesting structures. Although they are suggested by the quantum Liouville theory, they go well beyond this particular dynamics. They give a new insight into the general problem of conformally invariant field theories [5,8], and suggest the existence of a whole class of new critical models in two dimensions [6,7].

In a series of more recent papers [9-13], A. Bilal and I studied the associated string dynamics. We showed that, in the case of the purely bosonic strings, as well as for the Neveu-Schwarz-Ramond (NSR) strings, there exist two new critical values for the number D of space-time dimensions where the string theories, that include the quantum Liouville dynamics, following refs [3-8], are completely consistent and exhibit the same remarkable properties as the standard string models. In particular, the new NSR strings are space-time supersymmetric and may be formulated by means of new types of light-cone superstring formalism, similar to the famous one of Green and Schwarz, where supersymmetry is manifest. Remarkably, these critical dimensions are two out of the three values predicted, in my earlier work with Neveu, purely from the Liouville world-sheet dynamics [9].

ISBN 0-12-218100-X

A preliminary remark is in order, concerning the numerology, of these values. As we shall argue, the new string theories in D dimensions have D-1 physical degrees of freedom. Altogether the number of such physical degrees of freedom is thus three times two raised to the power n (resp two to the power n) for the bosonic (resp NSR) models, with n=1,2, for the new models, and n=3 for the standard one. In each case, the new values are a natural extension of the old one. Moreover, for the NSR model, these numbers coincide with the real dimensions of the algebras of complex numbers, quaternions and octonions, respectively. Besides the real numbers, these are the only known division algebras. The existence of a division algebra is crucial in the building of superstrings field as we shall see.

II THE BASIC FEATURES OF THE BOSONIC LIOUVLLE STRING DYNAMICS

II-1 The relevance of the Liouville dynamics for string theories

In this section we briefly review the arguments of Polyakov [1]. Following Brink, Di Vecchia, and Howe [14], one takes the string position X^{μ} and the world-sheet metric tensor g_{ab} as independent degrees of freedom. The action is

$$\mathcal{S} = \frac{1}{2\pi} \int d\sigma d\tau \sqrt{-\det g}\; g^{ab}\, \partial_a X^{\mu}\, \partial_b X^{\mu} \quad ; \quad g^{ab} = (g^{-1})_{ab} \tag{2.1}$$

σ and τ are the parameters of the surface generated by the string motion. This action is invariant under reparametrization of this world-sheet and under the Weyl transformations

$$X \Rightarrow X \quad ; \quad g_{ab} \Rightarrow f(\sigma,\tau)\, g_{ab} \tag{2.2}$$

where f is an arbitrary function of σ and τ. In the partition function

$$Z = \int \mathcal{D}X\, \mathcal{D}g\; e^{-\mathcal{S}} \tag{2.3}$$

we first integrate over X. Consider

$$Z_0(g) = \int \mathcal{D}X\; e^{-\mathcal{S}} \tag{2.4}$$

The well-known formula [15]

$$g^{ab} \frac{\delta Z_0}{\delta g^{ab}} = g^{ab} < T_{ab} > = \frac{D}{24\pi} (R + cste) \tag{2.5}$$

relates the trace anomaly of the energy-momentum tensor T to the world sheet scalar curvature R. This trace anomaly breaks the Weyl invariance but preserves the reparametrization invariance. We shall fix the parametrization by imposing

$$g_{ab} = e^{\phi} \begin{vmatrix} -1 & 0 \\ 0 & 1 \end{vmatrix} \tag{2.6}$$

In this case, equation (2.5) leads to

$$Z_0(\phi) = \exp\left(- \frac{iD}{48\pi} \int d\sigma \, d\tau \, [(\partial_\tau\phi)^2/2 - (\partial_\sigma\phi)^2/2 + e^{\phi}]\right) \tag{2.7}$$

On the other hand [1], the measure of integration over ϕ involves the Jacobian (Faddeev-Popov determinant) which is equal to

$$\Delta(\phi) = \exp\left(\frac{i26}{48\pi} \int d\sigma \, d\tau \, [(\partial_\tau\phi)^2/2 - (\partial_\sigma\phi)^2/2 + e^{\phi}]\right) \tag{2.8}$$

The coefficient of $\exp(\phi)$ may be freely changed by adding an irrelevant constant to ϕ. We took advantage of this in order to get rid of the arbitrary constant in formula (2.5) and set this coefficient equal to one.

Remarkably, equations (2.7) and (2.8) both involve the same Liouville action and the coefficients in front are opposite at D=26. If one assumes that the integral which is common to eqs. (2.7) and (2.8) remains finite at D=26, it follows that the Liouville dynamics disappears at the critical dimension. This common wisdom, which leads to the so-called Polyakov measure, assumes, however, that ϕ is a background field. Indeed, when the dimension D tends to 26, the Liouville theory behaves, as if its Planck constant was going to infinity. The functional integral over ϕ will thus be dominated by very irregular configurations, unless it is a priori restricted to continuous functions. If ϕ is considered as a full-fledged quantum field, one cannot draw any definite conclusion about the behavior at D=26, from the above discussion, since the kinetic term of (2.7,8) is unbounded in the limit. One must first solve the quantum

Louville theory. This is the program which I started to follow with A. Neveu in 1982 [3-8]. We studied the quantum dynamics by operator methods in order to avoid any ambiguity. A brief account of the key features of our work follows next.

II-2 Highlights of the Liouville dynamics

As a starting point, we shall take the action

$$S = \frac{1}{16\pi\hbar} \int d\sigma \, d\tau \, [(\partial_\tau\phi)^2/2 - (\partial_\sigma\phi)^2/2 + e^{\phi}] \tag{2.9}$$

This particular rewriting of the factor in front is convenient since the parameter $\hbar$ so defined will indeed play the role of Planck's constant. In view of formulae (2.7), (2.8), Polyakov's argument which, we just outlined, gives

$$\hbar_P = 3 / (26 - D) \tag{2.10}$$

As we shall see, however, this formula will be modified by finite renormalization effects, and it is better to leave $\hbar$ as a free parameter to begin with.

Consider, first, the classical Louville dynamics [3]. The field equation

$$4\partial^2_{uv}\phi = e^{\phi} \;;\quad u = \tau + \sigma \;;\quad v = \tau - \sigma \tag{2.11}$$

can be easily used to derive the relations

$$\partial_u \left[e^{\phi/2}\, \partial^2_{vv}\, e^{-\phi/2}\right] = \partial_v \left[e^{\phi/2}\, \partial^2_{uu}\, e^{-\phi/2}\right] = 0 \tag{2.12}$$

It follows that we can write equations of the form

$$[\,-\partial^2_{uu} + U^{(L)}(u)]\, e^{-\phi/2} = [\,-\partial^2_{vv} + U^{(R)}(v)]\, e^{-\phi/2} = 0 \tag{2.13}$$

where $U^{(L)}$ and $U^{(R)}$ are functions of a single variable. This allows to separate the right-movers and the left-movers in a natural way, since the general solution of the last partial differential equations is

$$e^{-\phi/2} = [\Psi^{(R)}_1(v)\, \Psi^{(L)}_2(u) - \Psi^{(R)}_2(v)\, \Psi^{(L)}_1(u)]/(2\sqrt{2}) \tag{2.14}$$

where the psi's are functions of one variable, which obey the Schrödinger equations

$$[-\frac{d^2}{dx^2} + U^{(L)}(x)]\ \Psi^{(L)}_{1,2}(x) = [-\frac{d^2}{dx^2} + U^{(R)}(x)]\ \Psi^{(R)}_{1,2}(x) = 0 \tag{2.15}$$

Let us next display the basic features of the canonical formalism and of the quantization following ref. [4]. Since this part deals with the left and right movers separately and in exactly the same way we drop all R and L superscripts.

At the classical level, it is convenient to define

$$\lambda(x) = \frac{1}{\Psi_1(x)}\frac{d\Psi_1(x)}{dx} \quad ; \quad \tilde{\lambda}(x) = \frac{1}{\Psi_2(x)}\frac{d\Psi_2(x)}{dx} \tag{2.16}$$

The boundary conditions are always such that U is periodic with a period which we choose to be 2π. Generically, the two solutions of (2.15) may be selected in such a way that λ and $\tilde{\lambda}$ are periodic. We then let

$$\lambda(x) = \sum_n \lambda_n\, e^{-inx} \quad ; \quad \tilde{\lambda}(x) = \sum_n \tilde{\lambda}_n\, e^{-inx} \tag{2.17}$$

Standard Wronskian arguments show that $\lambda_0 + \tilde{\lambda}_0 = 0$. As is well-known, these logarithmic derivatives satisfy the Riccati equations

$$U = (\lambda)^2 + d\lambda/dx = (\tilde{\lambda})^2 + d\tilde{\lambda}/dx \tag{2.18}$$

One can show [4] that the canonical Poisson-brackets associated with (2.9) are equivalent to

$$i\left\{\lambda(x_1)\,,\ \lambda(x_2)\right\} = 2\pi\hbar\ \delta(x_1 - x_2) \tag{2.19}$$

or to the same relation with λ replaced by $\tilde{\lambda}$. This Poisson -bracket structure is such that if we let

$$U(x) = 2\hbar \sum_n \Lambda_n\, e^{-inx} \tag{2.20}$$

the operators Λ_m satisfy the classical Poisson-bracket Virasoro algebra

$$i\left\{\Lambda_m\,,\ \Lambda_n\right\} = (m-n)\ \Lambda_{m+n} + \frac{n^3}{4\hbar}\ \delta_{m+n,0} \tag{2.21}$$

One may verify that the psi fields transform as conformal fields ([16], [17]) through Poisson-brackets. Their conformal weights is 1/2, and equation (2.14) shows that $\exp(-\phi/2)$ has conformal weights $\delta = \bar{\delta} = 1/2$. This is consistent since it follows that the product $\exp(\phi)dudv$ which appears in the action (2.9) is invariant under conformal transformations.

Let us now move on to the quantum case. As shown in refs [4,5], the structure summarized above extends with surprisingly small modifications. In eq. (2.19), the Poisson bracket is replaced by a commutator. Equation (2.18) becomes

$$U = N[(\lambda)^2] + d\lambda/dx = \tilde{N}[(\tilde{\lambda})^2] + d\tilde{\lambda}/dx \tag{2.22}$$

where N and $\tilde{N}$ denote normal ordering with respect to the modes of λ and $\tilde{\lambda}$ respectively. The key property, derived in ref.[4] and further discussed in ref. [8], is that, generically, equation (2.22), together with the condition $\lambda_0 + \tilde{\lambda}_0 = 0$, establishes a quantum correspondence between λ and $\tilde{\lambda}$ which is such that, if λ_m satisfies the quantum analog of (2.19), the same is true for $\tilde{\lambda}_m$. One thus has

$$[\lambda_m , \lambda_n] = [\tilde{\lambda}_m , \tilde{\lambda}_n] = \hbar \, m \, \delta_{m+n,0} \tag{2.23}$$

The quantum version of (2.16) is

$$\Psi_1(x) = N[\exp(i\eta \sum_n \lambda_n e^{-inx}/n)] \quad ; \quad \Psi_2(x) = \tilde{N}[\exp(i\eta \sum_n \tilde{\lambda}_n e^{-inx}/n)] \tag{2.24}$$

Equation (2.23) shows that the two sets of operators λ_m and $\tilde{\lambda}_m$ separately satisfy standard free field commutation relations, and Ψ_1 and Ψ_2 are given by expressions which are both identical to a standard string vertex-operator for a single string component. The above expression is slightly abbreviated. The zero-modes are to be handled in the usual way (see [5] for details). The factor η, which is a pure number, exhibits a quantum modification with respect to (2.16). Its value is determined as follows. Ψ_1 and Ψ_2 are conformal fields ([16], [17]) with the same conformal weight. This can be verified from the formulae we just recalled, together with the quantum version of (2.20) which gives

$$\Lambda_n = \frac{1}{2\hbar}N[\sum_r \lambda_r \lambda_{n-r} - in\lambda_n + \frac{1}{4}\delta_{n,0}] = \frac{1}{2\hbar}\tilde{N}[\sum_r \tilde{\lambda}_r \tilde{\lambda}_{n-r} - in\tilde{\lambda}_n + \frac{1}{4}\delta_{n,0}] \qquad (2.25)$$

An easy computation shows that the Virasoro algebra is satisfied with central charge

$$c = 1 + \frac{3}{\hbar} \qquad (2.26)$$

From the short distance expansion of products of ψ_1 and ψ_2 one next generates a conformal family following [4], [6], [8]. In the same way as in the classical case, some of its members must have conformal weight equal to one, since one should be able to build a conformally invariant action with an interaction term that is a marginal operator. Classically ψ_1 and ψ_2 raised to the power minus two do have unit conformal weights. Imposing the same condition at the quantum level leads to the equation

$$2 \hbar \eta^2 - \eta + 1 = 0 \qquad (2.27)$$

Another motivation for this choice of η is that the psi's so defined satisfy a second order differential equations that is the quantum analogue of (2.15). (see [5], [6],and [8]). It plays a key role, but we shall not consider it explicitly. This differential equation was independently derived in ref. [17] and applied to two-dimensional critical systems.

Equation (2.27) has the two solutions

$$\eta_\pm = \frac{1}{4\hbar}(1 \pm \sqrt{1 - 8\hbar}) \qquad (2.28)$$

Substituting them into (2.24) we get four operator psi's instead of the two we had at the quantum level. This is not surprising since for $\hbar \to 0$, η_+ blows up. Using formula (2.28) with η_+ leads to quantum operators which have no smooth classical limit. Nevertheless, the conformal family generated by performing operator products with both solutions of (2.27) has a remarkable property [6], [8]. Indeed, the corresponding set of conformal weights is given by

$$\epsilon(m,n) = \frac{1}{48}\,[(13-c)(m^2+n^2) - 24mn - (1-c) + (n^2-m^2)\sqrt{(c-1)(c-25)}] \qquad (2.29)$$

where m and n are arbitrary positive or negative integers. This generalizes Kac's formula [18]. Indeed, if m and n are restricted to have the same sign, (2.29) is precisely equal to the highest weight which is such that a zero of a Kac determinant occurs at the level m times n. It is crucial that the conformal family derived from the discussion summarized above is more general than Kac's formula. Indeed, for c>1, the null states only exist for negative highest weights and are unphysical (the corresponding representation may never be unitary). In general, the conformal weights (2.29) are not all positive and only a subset may appear in the spectrum of the Louville theory.

Since the string has a finite length, the Liouville field theory is considered with a finite extension in σ and boundary conditions must be specified. We only consider the free string case where the world-sheet parameters vary over a strip parallel to the τ axis, and begin with the boundary conditions relevant for open strings. By convention, σ is taken to run from 0 to π. Let us go back, for a short while to the classical case. The conditions are, [3,13],

$$\partial_\sigma e^{-\phi/2} = \frac{1}{\sqrt{2}} \text{ at } \sigma=0 \ ; \ \partial_\sigma e^{-\phi/2} = -\frac{1}{\sqrt{2}} \text{ at } \sigma=\pi \tag{2.30}$$

This may seem complicated, but one easily sees from equation (2.14) that these conditions only involve the Wronskian of the solutions of (2.15) and are easily implemented. One finds that $U^{(R)}$ and $U^{(L)}$ are equal and periodic with period 2π. Equations (2.15) then show that $\psi^{(L)}_{1,2}$ and $\psi^{(R)}_{1,2}$ are linearly related and one ends up with a formula of the form, [3,13],

$$e^{-\phi/2} = \sum_{m,n=1,2} \alpha_{mn} \Psi_m[u]\Psi_n[v] \tag{2.31}$$

where the coefficients are determined up to one unknown constant that reflects the possibility of an arbitrary time translation. For closed string one would expect that ϕ should be simply taken as periodic. Although it is conceptually simpler to begin with, this case turned out to be much more delicate than the boundary condition relevant for open string; and we shall not discuss it. The closed strings will come out later from the one-loop open-string diagrams

As shown in [6] and [7], the spectrum of the quantum Liouville theory may be very simply summarized from formula (2.29), since the physical conformal weights are obtained by choosing all values of m and n that give positive highest weights. We next discuss the Liouvillle spectrum by applying this simple rule. In this connection, the square-root of formula (2.29) plays a key role. Clearly it is real only if $c<1$ or if $c>25$. The first case is the standard region of two-dimensionnal critical systems. We shall not discuss it since it is excluded by condition (2.26) --assuming of course that $\hbar$ is positive. In the region $c>25$, one has $\hbar<1/8$, and the solutions of (2.27) are both real. This is the weak coupling regime of the Liouville theory, that includes its semi-classical limit $\hbar \to 0$. The quantum solution is thus constructed using η_- only. This allows to extend (2.31) to the quantum case [5] in such a way that locality in σ is satisfied. The requirement that the operators so defined can be restricted to the subset of positive highest weights of (2.29), completely determines the quantum solution [5]. Moreover this is really completely consistent [5] when c takes the discrete set of values

$$c = 1 + \frac{6(N+1)^2}{N} \quad ; \quad N \text{ positive integer} \tag{2.32}$$

The set of physical conformal weights is then given by (ν is a positive integer)

$$\epsilon(1\ ,\ 2m-N) \quad ; \quad 0 \leq m \leq \nu \quad ; \quad N = 2\nu+1 \tag{2.33}$$

$$\epsilon(1\ ,\ 2m-1-N) \quad ; \quad 1 \leq m \leq \nu \quad ; \quad N = 2\nu \tag{2.34}$$

Remarkably one finds that c must be such that the square-root of formula (2.29) is rational. This is similar to the series put forward by Belavin Polyakov and Zamolodchikov [17] in the region $c<1$. Such a rationality requirement is natural since, if it is not satisfied, the set of values taken by

(2.29) for m and n arbitrary positive or negative is dense over the whole real line.

There remains to discuss the region $1<c<25$. There the square-root of eq.(2.29) is pure imaginary. One gets real conformal weights only if m and n are equal or opposite. The first possibility is excluded since it only gives negative or vanishing values. One finds [7] the spectrum

$$\epsilon(m\ ,\ -m) \equiv \frac{1}{24}\ [(m^2(25\ -\ c)\ +\ (c\ -\ 1)] \tag{2.35}$$

The quantum Liouville theory exists [7] only if c takes the values 7 , 13 , and 19. This discrete set comes out from the condition that it must be possible to build an algebra of local observables restricted to the subset of values (2.35) that are real and positive. These local fields involve both solutions of (2.27) which are on the same footing in this case since they are complex conjugate.

In general, the infinite conformal family built up from products of the psi operators play the role of building blocks which are to be used in order to construct the physical conformal operators. These should generate a conformal family that is closed over conformal fields with positive conformal weights. This conformal bootstrap requirement is so stringent that it can be fulfilled only for the special values of c indicated above. As we next discuss, this fixes the values of the space-time dimension where the string may be embedded.

II-3 The Liouville string theories

The Virasoro generators L_m of the X modes commute with the Λ_m operators. The consistency of the string theory requires two conditions. First the total central charge must be 26. The Liouville coupling constant is thus determined by

$$c\ =\ 26\ -\ D\ \ ;\ \ \hbar\ =\ \frac{3}{25\ -\ D} \tag{2.36}$$

this value of $\hbar$ differs from (2.10). This is due to the fact that formula (2.26) involves a quantum contribution besides the classical term $3/\hbar$. Second, the open string spectrum is to be determined from the mass-shell condition

$$(L_0 + \Lambda_0 - 1)| \text{ phys.} > = 0 \qquad (2.37)$$

These two conditions ensure, in particular, that there exists a BRST nilpotent operator, so that two degrees of freedom are unphysical.

From condition (2.36), one sees that the weak coupling regime of Liouville only corresponds to D<1. The associated string theories may be given a meaning only after continuation in D. One obtains

$$D = 1 - \frac{6(N-1)^2}{N} \quad ; \quad N \text{ positive integer} \qquad (2.38)$$

Although they are unphysical these cases exhibit an interesting structure [5].

For physical string building, we are however only interested in the values of D such that 1<D<25. This precisely selects the region 1<c<25 where the square-root of (2.29) is pure imaginary. From the allowed values c=7, 13, 19, we conclude that Liouville string theories may exist only if D=7 ;13 ; 19. Condition (2.37) leads to the mass spectrum

$$M^2_{m,\mathcal{N}} = (m^2 - 1)\,\frac{D-1}{12} + 2\,\mathcal{N} \qquad (2.39)$$

where m is the integer that determines the highest weight of the Liouville theory (see formula (2.35), and where the integer $\mathcal{N}$ is determined by the occupation numbers describing the harmonic excitations both of the Liouville mode and of the X modes.

III THE FIVE-DIMENSIONAL LIOUVILLE SUPERSTRING THEORY

From now on, we shall study the Neveu-Schwarz (NS) Ramond (R) Liouville string theories that incorporate the 2D supersymmetric Liouville (Sliouville) theory, following the original suggestion of Polyakov [2]. The general structure of these theories is very similar to that of the purely bosonic theories which we just reviewed. The values of D predicted by the Sliouville dynamics are 3, 5, 7 (see ref.[9]) The properties of the purely bosonic case which we just summarized may be straightforwardly extended to the present case, and we shall not elaborate upon them. In the present section, we discuss the characteristic features of the five-dimensonal string theory.

III-1 The equality between Neveu-Schwarz and Ramond partition functions

The distinctive feature of the standard superstrings, as compared with the purely bosonic ones, is that they are space-time supersymmetric in 10 dimensions. The first hint in this direction was obtained by Gliozzi, Scherk, and Olive [20] who proved that the Neveu-Schwarz (NS) and Ramond (R) partition functions are equal. In this section we study the NSR Sliouville strings from this viewpoint, and show that they satisfy similar equalities. The present section is a summary of ref. [9]

The weak coupling solution of refs.[3-5] has been extended to the Sliouville theory by Arvis [21] and Babelon [22]. The associated representation of the conformal algebra reads

$$[\Lambda_m , \Lambda_n] = (m-n)\, \Lambda_{m+n} + \frac{c}{8}\,(m^3-\text{"}n\text{"}) \quad ; \quad c = 1 + \frac{2}{\hbar} \tag{3.1}$$

In this formula, and in the following, quantities between quotation marks are only present in the NS sector. In the R sector they are to be replaced by zero. The weak coupling regime is now for c>9. As shown in ref. [21] and [22], the Sliouville field theory can be solved through a complicated transformation to free fields similar to λ and $\tilde{\lambda}$. These fields have the same structure as one component of the standard space-time NSR fields. In particular, there is a NS Sliouville field and a R Sliouville field. The Sliouville theory has its own NS and R sectors. Formula (2.29) becomes

$$\epsilon(m,n) = \frac{1}{32}\,[(5-c)(m^2+n^2) - 8mn - "(1-c)" + (n^2-m^2)\sqrt{(c-1)(c-8)}] \tag{3.2}$$

The string theory will be consistent if c + D = 10, and if the spectrum is determined from the mass shell condition

$$(L_0 + \Lambda_0 - "1/2")|\,\text{phys.}> = 0 \tag{3.3}$$

These conditions are equivalent to the existence of a nilpotent BRST operator in each sector. From (3.2), one sees that, in the same way as in the bosonic case, the interesting region for string theory building coincides with the interval where the square-root of formula (3.2) is pure imaginary. The physical spectrum corresponds to m+n=0 as before. The mass spectrum is thus given by:

$$M^2(m, N_1, N_2) = \frac{(D-1)}{8}\,(m^2-"1") + 2N_1 + 2N_2 \tag{3.4}$$

N_1 (N_2) are the contributions of the non-zero 2D bosonic (fermionic modes) to Λ_0 and L_0. It is easy to derive the partition functions of the physical string states:

$$Z = \sum_{m,\,N_1,\,N_2} q^{M^2(m,\,N_1,\,N_2)/2}\; Z_X \tag{3.5}$$

Using formula (3.4), one obtains

$$Z^{\pm}_{NS} = \Big(\sum_m q^{(m^2-1)(D-1)/8}\Big)\,\frac{1}{2}\Big[\prod_{n>0}(1 + q^{n-1/2})^{D-1} \pm \prod_{n>0}(1 - q^{n-1/2})^{D-1}\Big]\,Z_X \tag{3.6}$$

$$Z_R = d_s \sum_m q^{m^2(D-1)/8} \prod_{n>0}\Big(1 + q^n\Big)^{D-1} Z_X \tag{3.7}$$

where the common factor Z_X comes from the transverse components of X and from the Liouville harmonic bosonic modes:

$$Z_X = f(q)^{D-1} \quad ; \quad f(q) = \prod_{n>0} (1 - q^n)^{-1} \tag{3.8}$$

The string theory is built up in such a way that the ghost -killing mechanism removes two degrees of freedom. In D dimensions the total number of degrees of freedom is D+1, and hence the number of physical degrees of freedom is D-1. This is why the partition functions of the non-zero modes are standard harmonic oscillator partition functions raised to the power D-1. The factor d_S represents the number of Dirac physical components of the zero-mode spinor of Ramond. It will be determined below. In the NS sector, we already separated the two G-parity sectors, since this was a key point in the work of G.O.S. In practice, the equality between Z_{NS} and Z_R is only possible if they can be expressed as Jacobi theta functions of q or q^2. This is true for any values of D for all factors in (3.6) and (3.7) apart from the contribution of the zero-modes of the Sliouville field theory. For 1<D<9, this latter factor takes a suitable form only for the new critical dimensions D=3 and 5.

For D=5, the basic identities are

$$2Z^{(0)} \prod_{n>0} \left(1+q^n\right)^4 = q^{-1/4}\, Z^{(1/2)}\, \frac{1}{2}\Big[\prod_{n>0} (1+q^{n-1/2})^4 + \prod_{n>0} (1-q^{n-1/2})^4\Big] \tag{3.9}$$

$$2Z^{(1/2)} \prod_{n>0} \left(1+q^n\right)^4 = q^{-1/4}\, Z^{(0)}\, \frac{1}{2}\Big[\prod_{n>0} (1+q^{n-1/2})^4 - \prod_{n>0} (1-q^{n-1/2})^4\Big] \tag{3.10}$$

We have introduced the Sliouville zero-mode partition functions

$$Z^{(0)} = \sum_{p=-\infty}^{+\infty} q^{p^2} = \Theta_3(q) \quad ; \quad Z^{(1/2)} = \sum_{p=-\infty}^{+\infty} q^{(p+1/2)^2} = \Theta_2(q) \tag{3.11}$$

which are equal to standard theta functions [19] with vanishing first argument. This particular rewriting will be convenient below. The proof of equalities (3.9) and (3.10) follows from standard properties of the theta functions. Ultimately they are equivalent to

$$\sqrt{\theta_3^4(q) - \theta_4^4(q)} = \pm\theta_2^2(q) \tag{3.12}$$

These last relations are the square-roots of the one that gives the GSO [20] formula for the standard superstring. Indeed, by equating the products of the left and of the right members of equations (3.9) and (3.10), one obtains the famous GSO relation

$$8\prod_{n>0}\left(1+q^n\right)^8 = q^{-1/2}\,\frac{1}{2}\Big[\prod_{n>0}(1+q^{n-1/2})^8 - \prod_{n>0}(1-q^{n-1/2})^8\Big] \tag{3.13}$$

It is now our task to connect (3.9) and (3.10) with the string state partition functions. This is done by selecting appropriate values of m in (3.6,7). Since Z_X is a common factor, we need not discuss it. Apart from an overall factor, the left-hand side of eq. (3.9) (resp. (3.10)) coincides with (3.7) if the summation over m is restricted to the even (resp odd) values. Similarly, the right-hand side of (3.9) (resp (3.10)) corresponds to formula (3.6) with the plus (resp minus) sign if the summation is restricted to the odd (resp even) values of m. Consider, now, the factor d_S. The zero-mode Sliouville R field anticommutes with the D zero-mode Dirac-matrices of the space-time R field. It follows that altogether we have a realization of the Dirac algebra with D+1=6 Dirac matrices. The Dirac spinors thus have 8 complex components. Imposing the Weyl condition and the Dirac equation leaves 4 real degrees of freedom and we conclude that d_S is equal to this number. Both sides of equations (3.9) and (3.10) must be multiplied by 2. For the R sector this gives the right factor d_S, but, in the NS sector, we are led to assume that each value of m is doubly degenerate.

One sees that, contrary to the ten-dimensional superstring, both G-parities lead to supersymmetric theories. For the even G-parity NS sector, the lightest particles are massless scalars and spin-one-half particles. For the odd case, the lightest bosonic particle has spin one as in standard superstrings. It is massive, however. Expanding the right-hand side of (3.10), one finds that its mass is equal to 1/2.

III-2 Group triality

Another key feature of the Green-Schwarz superstring is the remarkable triality property of O(8) that appears as the group of transverse rotation in the light-cone formalism. Summarizing ref.[11], we shall next show that, for the D=5 Sliouville string theory, a different group emerges, in a dynamical way, with triality properties similar to the ones of O(8). They are the basis of the superstring formalism which will be the subject of the following section. (The D=3 model is discussed along the same line in ref. [12]. We shall consider this case later on).

We shall hereafter make use of a six-dimensional notation that combines the space-time fields X_μ, and Γ_μ whose indices run from 1 to 5, together with the Sliouville free fields denoted X_6 and Γ_6 . One may visualize the present D=5 Sliouville string theory as a six-dimensional one where the Lorentz invariance is broken down to the D=5 Lorentz group. This breaking is rather subtle, and cannot be reduced to a trivial compactification of one space-dimension. It is important to stress that the theory is strictly five-dimensional even if we use a six-dimensional language. In particular, the equality between partition functions which we discussed above is for the five-dimensional masses. From now on, we make use of the light-cone formulation of string theories. A space-time vector has components

$$V_\pm = (V_1 \pm V_2)/\sqrt{2} \quad , \text{and } V_i \quad \text{with} \quad i=3,4,5 \tag{3.14}$$

By convention, lower case latin indices run from 3 to 5 and refer to the true transverse coordinates. When the Sliouville fields are included, it is convenient to use an O(4) notation and to introduce two complex components:

$$V^3 = \frac{(V_3 + iV_4)}{\sqrt{2}} \; ; \; V^{\bar{3}}=(V^3)^+ \quad ; \quad V^5 = \frac{(V_5 + iV_6)}{\sqrt{2}} \; ; \; V^{\bar{5}}=(V^5)^+ \tag{3.15}$$

Lower and upper indices describe real and complex components respectively. The lower index 6 refers to the Sliouville degrees of freedom.

In the light-cone formulation, the plus and minus components of X and Γ are eliminated. Let us next bosonize the four remaining NSR fields. They are equivalent to two bosonic modes α^1 and α^2. Define

$$\Omega^{\pm\,\pm}_{1,2}(z) = \sqrt{z}\; K_1^{\pm 1} K_2^{\pm 1}\; z^{(\pm\alpha_0^1 \pm \alpha_0^2)/\sqrt{2}}\; N\Big(\exp\Big[-\sum_{n\neq 0} \frac{z^{-n}}{n} \frac{(\pm\alpha_n^1 \pm \alpha_n^2)}{\sqrt{2}}\Big]\Big) \tag{3.16}$$

The indices 1 and 2 of Ω are for later convenience. The operators K are the usual zero-mode shift operators:

$$[\alpha_0^1 \, , \, K_1] = K_1/\sqrt{2} \quad ; \quad [\alpha_0^2 \, , \, K_2] = K_2/\sqrt{2} \tag{3.17}$$

The four NSR fields are given by

$$\Gamma^3(z) \sim \Omega^{+\,+}_{1,2}(z); \quad \Gamma^{\bar{3}}(z) \sim \Omega^{-\,-}_{1,2}(z); \quad \Gamma^5(z) \sim \Omega^{+\,-}_{1,2}(z); \quad \Gamma^{\bar{5}}(z) \sim \Omega^{-\,+}_{1,2}(z) \tag{3.18}$$

For compactness, we leave the well-known cocycle factors aside. By standard computations one may verify that the fields so defined satisfy the anticommutation relations

$$[\Gamma_I(e^{i\sigma_1}) \, , \, \Gamma_J(e^{i\sigma_2})]_+ = \delta(\sigma_1 - \sigma_2)\, \delta_{I,J} \; ; \quad I,J = 3,4,5,6 \tag{3.19}$$

The two bosonic modes we just introduced generate both the space-time and Sliouville Fermi Fock spaces. By the standard quark-current construction [23], one derives the generators of the $\hat{O}(4)$ Kac-Moody algebra, which is the direct product of two $\widehat{SU}(2)$ Kac-Moody algebras. The first $\widehat{SU}(2)$ generators read (double dots mean fermionic normal ordering):

$$I_+^{(1)} = :\Gamma^5\Gamma^3: \quad ; \quad I_-^{(1)} = :\Gamma^{\bar{3}}\Gamma^{\bar{5}}: \quad ; \quad I_3^{(1)} = (:\Gamma^3\Gamma^{\bar{3}}: + :\Gamma^5\Gamma^{\bar{5}}:)/2 \tag{3.20}$$

The other $\widehat{SU}(2)$ generators are deduced from these relations by permuting 5 and $\bar{5}$. In bosonic terms, both operators take the Frenkel-Kac form [23] (s=1, 2):

$$I_\pm^{(s)}(z) \sim z(K_s)^{\pm 2} \, z^{\pm\alpha_0^s\sqrt{2}} \, N\left[\exp\left[\mp \sum_{n\neq 0} \frac{z^{-n}}{n}\sqrt{2}\alpha_n^s\right]\right]; \quad I_3^{(s)}(z) = \sum_n \alpha_n^s z^{-n}/\sqrt{2} \tag{3.21}$$

Eqs (3.17) show that the eigenvalues of the zero modes α_0^1 and α_0^2 are discrete. The corresponding two-dimensional lattice is of course the weight-lattice of O(4), which is the direct sum of two SU(2) weight-lattices. Its points may be separated into four cosets. One is the set of points with coordinates that are integer multiples of $\sqrt{2}$. Denoted (0,0) it corresponds to products of two integer-spin representations of SU(2). Two more cosets are obtained by shifting the preceding one by $1/\sqrt{2}$ along the two SU(2)-axes successively. Denoted (0,1/2) and (1/2,0) they describe representations that are are products of an integer-spin with a half-integer-spin representation. Finally, the set (1/2,1/2) which is the product of two half-integer-spin representations is obtained after shifting the first coset in both directions simultaneously by $1/\sqrt{2}$. The group generators do not mix the cosets. The NSR fields mix (0 , 0) and (1/2 , 1/2) on the one hand (this is the NS sector), and (0 , 1/2) and (1/2 , 0) on the other hand (this is the R sector). (0 , 0) and (1/2 , 1/2) are the odd and even G-parity sectors of NS respectively. The other two cosets correspond to the two chirality projections of the R sector.

The construction of the NSR fields and of the group generators which we just carried out is not specific to O(4). The O(8) NSR fields are constructed in a similar way [23].

The distinctive feature of this last case is that there are additional lattice-vectors of unit length, besides those that give the NSR fields, that can be used to construct the light-cone superfields of Green and Schwarz. This of course reflects the triality of O(8) [23]. In the present case the only unit vectors have coordinates $(\pm 1/\sqrt{2}, \pm 1/\sqrt{2})$. They all appear in formula (3.18), and there is no way to build additional spinor fields besides the NSR ones. This is not surprising, since the Dynkin diagram of O(4) has no triality. There is thus no way to achieve triality from the NSR fields alone even including the Sliouville NSR field as we did above. We shall next show, however that triality is recovered if one further includes the Sliouville bosonic mode . This agrees with its key role in the equality between the partition functions of the space-time bosons and fermions which we derived above.

The spectrum of highest weights may be rewritten as

$$\epsilon(2r, -2r) = \frac{(r\sqrt{2})^2}{2} + \text{"}\frac{1}{4}\text{"} \; ; \text{ or } \epsilon(2r+1, -2r-1) = \frac{(1/\sqrt{2}+r\sqrt{2})^2}{2} + \text{"}\frac{1}{4}\text{"}; \tag{3.22}$$

where we separated the even and odd values. From the way we wrote these expressions it is clear that, apart from the last constant term, the bosonic Sliouville zero mode spectrum precisely corresponds to the one-dimensional weight-lattice of SU(2) that just appeared above in the $\widehat{O}(4)$ bosonization. Since the bosonic Sliouville spectrum is made up with the harmonic excitations of those ground states, one may carry out the Frenkel-Kac construction in this Hilbert space, thereby generating the level-one representation of $\widehat{SU}(2)$. We thus have another $\widehat{SU}(2)$ at our disposal which comes entirely from the Sliouville bosonic modes, and is purely dynamical. It describes internal quantum numbers in the present physical five-dimensional world. The trick will be to treat it on the same footing as the generators we built up --see formula (3.21)-- from the NSR fields. We shall denote all quantities relevant to this third $\widehat{SU}(2)$ by the same symbols as for the first two, apart from the value of the upper indices which is set equal to 3. In particular, α^3 denotes the harmonic oscillators of the Sliouville bosonic modes and the third $\widehat{SU}(2)$ generators are given by formula (3.21) with s=3. Altogether with have an $SU(2)\otimes SU(2)\otimes SU(2)$ lattice and triality is due to the exact mathematical equivalence between these three SU(2)'s. The complete weight lattice is now tridimensional and cubic. There are eight cosets, corresponding to the two choices of integer or half -integer spin representations for the three axes. The NS and R sectors are each made up with two disconnected pieces corresponding to the two cosets of the third SU(2). Concerning each SU(2) Kac-Moody algebra, the two cosets made up with the integer and half-integer spin representations respectively are the irreducible components of the level-one representations [23]. The corresponding zero-mode partition functions already appeared in formula (3.11).

Let us now apply this analysis to rederive the equality between partition functions which we discussed in the preceding section. The mass-shell conditions (3.3) becomes, after bosonization,

$$\left(\frac{M^2}{2} - \Xi_0\right) |\ \rangle = 0; \quad \Xi_0 = \frac{1}{2}\sum_{s=1}^{3}\sum_{n}\left(\alpha_n^s\,\alpha_{-n}^s\right) - \frac{1}{4} + \frac{1}{2}\sum_{n=1}^{\infty}\Pi_{i,-n}\Pi_{i,\,n} \tag{3.23}$$

where the last term is the contribution of the transverse X modes. The partition function (3.5) is thus given by the partition function of the operator Ξ_0 which is completely symmetric in the three α modes.

It now becomes very easy to understand the equalities between partition functions. First it is clear that, for each of the two cosets (1/2,0,0) and (0,0,1/2) the partition function is equal to

$$q^{-1/4}\ Z^{(1/2)}\ (Z^{(0)})^2\ f^3(q) \tag{3.24}$$

The first coset describes an O(4) chiral R sector, while the second is the odd G-parity NS sector. The equality between space-time boson and fermion partition functions is a direct consequence of the equivalence between the three SU(2), which therefore plays the same role as the O(8) triality of the Green-Schwarz superstring. The identity (3.10) is now immediately rederived by writing the partition functions of both cosets of formula (3.24) in the NSR language.

The second case of equality between partition functions, is treated in exactly the same way. The two cosets are (1/2,1/2,0), and (0,1/2,1/2). Both partition functions are equal to

$$q^{-1/4}\ Z^{[0]}\ (Z^{[1/2]})^2\ f^3(q) \tag{3.25}$$

The first coset describes the even G-parity sector of NS, while the second is a O(4) chiral sector of R.

One sees that the O(8) triality has been replaced by the $SU(2)^3$ triality. Remarkably the physics of the D=5 Sliouville string theory is precisely such that this is possible. It is worth noting that the $SU(2)^3$ triality was investigated, prior to ref [11], in order to perform the Frenkel-Kac construction for non-simply-laced groups [24].

Another general point is in order before we terminate this paragraph. The O(8)-superstring makes an essential use of the special properties of the O(8)-gamma-matrices which are themselves a consequence of the properties of octonions. In the present case, there are four NSR fields and quaternions play a key role. Let us define

$$\sigma_I\ ;\ I = 3,4,5,6\ ;\quad \sigma_3 = \begin{vmatrix}0&1\\1&0\end{vmatrix}\ ;\ \sigma_4 = \begin{vmatrix}0&-i\\i&0\end{vmatrix}\ ;\ \sigma_5 = \begin{vmatrix}1&0\\0&-1\end{vmatrix}\ ;\ \sigma_6 = \begin{vmatrix}i&0\\0&i\end{vmatrix} \tag{3.26}$$

In the quaternionic language, the NSR fields are represented by

$$\Gamma = \Gamma_I\ \sigma_I = \vec{\Gamma}\ \vec{\sigma} + i\ \Gamma_6 \tag{3.27}$$

where we introduced a convenient vector notation in the true transverse coordinate space. The transformation laws of Γ

under O(4) are

$$[\,\Gamma\,,\,I^{(1)}_r] = -\Gamma\,\frac{\sigma_r}{2} \quad ; \quad [\,\Gamma\,,\,I^{(2)}_r] = \frac{\sigma_r}{2}\,\Gamma \tag{3.28}$$

It is convenient, next to discuss the graded algebra satisfied by the fermionic operators: (primes denote derivatives with respect to ln(z))

$$\mathcal{G}^{(1)}_{\alpha\beta} = (\,\Pi^{+}\Gamma + \Gamma^{/}\sqrt{2})/2 \quad ; \quad \mathcal{G}^{(2)}_{\alpha\beta} = (\,\Pi\,\Gamma^{+} - \Gamma^{+/}\sqrt{2})/2\,. \tag{3.29}$$

They are such that their traces are equal to the total superconformal generators:

$$\mathcal{G}_n = \oint\frac{dz}{2\pi iz}\, z^n\,(\Pi_I\Gamma_I + i\sqrt{2}\,\Gamma_6^{\,/}) \quad ; \quad \Pi_I = iX_I^{/} \tag{3.30}$$

It is straightforward to compute the commutators of the operators without multiplying the 2x2 matrices. This is a common practice in 2-d integrable systems, and we use the same notation. One obtains, in particular,

$$[G^{(s)}_m \overset{\otimes}{,} G^{(s)}_n]_+ = -\sigma_I\tau_I\,\Big(\,\mathcal{L}_{m+n} + 2\delta_{m,-n}(m^2-\text{"}1/4\text{"})\Big)/2$$

$$+\,(m-n)\Big(\sigma_I\tau_I\,\vec{\tau}\,\vec{I}^{(s)}_{m+n} - \sqrt{2}\,\vec{\sigma}\,\vec{\Pi}_{m+n}\,\sigma_I\tau_I\Big)/2 \tag{3.31}$$

where the $\mathcal{L}$'s are the total Virasoro generators

$$\mathcal{L}_n = \oint\frac{dz}{2\pi iz}\, z^n\,\Big(\,N(\Pi_I\Pi_I) + :\Gamma_I^{/}\Gamma_I: + i\sqrt{2}\,\Pi_6^{\,/}\Big)/2 + \text{"}1/4\text{"}\,\delta_{n,0} \tag{3.32}$$

The quaternions which appear in the first and second operators of the left-hand side are noted σ and τ respectively and are to be considered as acting in different spaces. The algebra of the zero-mode components is simply:

$$[G^{(s)}_0 \overset{\otimes}{,} G^{(s)}_0]_+ = -\sigma_I\tau_I\,L_0/2 \tag{3.33}$$

The linear terms of eqs. (3.29) are needed in order to recover the full Virasoro operator (3.32) in (3.31). They are responsible for the contribution to the commutation relation that is proportional to the Π field. Apart from this, formula (3.31) coincides with the N=4 superconformal algebra commutation relation [25]. This is a novel application of this mathematical structure, where the index of the fermionic generators partly corresponds to the transverse components of the X_i and Π_i space-time fields.

Next we remark that these two sets of N=4 superconformal generators have non-trivial commutation relations with one another. One obtains

$$[\tilde{\mathcal{G}}^{(1)} \overset{\otimes}{,} \tilde{\mathcal{G}}^{(2)}]_+ = -\vec{\sigma}\vec{I}^{(1)} \quad \vec{\tau}\vec{I}^{(2)} + \frac{1}{4}\vec{\sigma}\vec{\tau}(\Pi_6^2 - \vec{\Pi}^2) + \frac{1}{2}(\vec{\Pi}\vec{\sigma})(\vec{\Pi}\vec{\tau}) - \frac{1}{2}\Pi_6 \epsilon_{ijk}\Pi_i \sigma_j \tau_k$$

where, in order to simplify, we removed all linear terms and subtracted the traces from the generators (this is why we put wiggles above them). This algebra is intriguing. There appear new operators of conformal weight 2 on the right-hand side. Commuting them leads to enlarge the algebra further, and this process does not seem to stop. This structure remains to be investigated.

III-3 The light-cone superstring formalism

The superstring fields are obtained by exchanging the role of the three $\widehat{SU}(2)$. They are given by formulae similar to (3.18). One obtains two new sets of fermionic fields

$$S^3(z) \sim \Omega^{+\,+}_{1,3}(z);\; S^{\bar{3}}(z) \sim \Omega^{-\,-}_{1,3}(z);\; S^5(z) \sim \Omega^{+\,-}_{1,3}(z);\; S^{\bar{5}}(z) \sim \Omega^{-\,+}_{1,3}(z)$$

$$T^3(z) \sim \Omega^{+\,+}_{2,3}(z);\; T^{\bar{3}}(z) \sim \Omega^{-\,-}_{2,3}(z);\; T^5(z) \sim \Omega^{+\,-}_{2,3}(z);\; T^{\bar{5}}(z) \sim \Omega^{-\,+}_{2,3}(z) \qquad (3.34)$$

where the new Ω fields are defined by formulae similar to (3.16). The fiels S and T commute with the G parity and the chirality projections. They are indeed the equivalent of the Green-Schwarz fields and are to be used in order to construct the supersymmetry generators. We shall, forinstance, discuss the S fields. There are two types of light-cone supersymmetry generators. As usual, the first type is essentially proportional to the zero mode of the superstring field. Indeed, one has

$$Q_+ \sim \sqrt{2p_+}\, \underset{\sim}{S}_0 \; ; \quad \underset{\sim}{S} = \begin{pmatrix} S^{\bar{5}} \\ S^3 \end{pmatrix} \tag{3.35}$$

The other supersymmetry generator is proportional to the zero mode component of the fermionic generator $\mathcal{S}$ obtained from $\mathcal{G}$ by permuting the axes 2 and 3, in the lattice. Let us define (the linear term of eq. (3.30) is irrelevant here)

$$\hat{\underset{\sim}{\mathcal{S}}} = (\vec{\pi}\,\vec{\sigma} + i\,\alpha^2)\, \hat{\underset{\sim}{S}} \tag{3.36}$$

The other supersymmetry generator is given by

$$Q_- \sim \sqrt{\frac{2}{p^+}}\, \hat{\underset{\sim}{\mathcal{S}}}_o \; ; \quad \hat{\underset{\sim}{\mathcal{S}}} = \begin{pmatrix} \hat{\underset{\sim}{\mathcal{S}}}^{\bar{5}} \\ \hat{\underset{\sim}{\mathcal{S}}}^{3} \end{pmatrix} \tag{3.37}$$

Of course, both Q_+ and Q_- are two component spinors. They satisfy the "light-cone supersymmetry algebra":

$$[Q^a_+ \, , \, \bar{Q}^b_+]_+ = 2\; p^+\, \delta^{ab}; \quad [Q^{\dot{a}}{}_- \, , \, \bar{Q}^{\dot{b}}{}_-]_+ = \frac{2}{p^+}\, \mathcal{L}_o \quad \delta^{\dot{a}\dot{b}}$$

$$[Q^{\dot{a}}{}_- \, , \, \bar{Q}^b{}_+]_+ = \sqrt{2}\; p_I\; \sigma^{\dot{a}b}_I; \; [Q^a{}_+ \, , \, \bar{Q}^{\dot{b}}{}_-]_+ = \sqrt{2} p_I (\sigma^+_I)^{a\dot{b}} \tag{3.38}$$

The reason for the above quotation marks is that, in the second line, there appears a quantity which we called P_6 since it is contracted with σ_6 . It is, however given by

$$P_6 = \sqrt{2}\; (I^{(2)}_3)_0 \tag{3.39}$$

The Lorentz properties are thus very unusual, since this "sixth" component is not a five-dimensional Lorentz scalar.

The theory is Lorentz invariant in five dimensions, since one can verify [11] the closure of the Lorentz algebra. This is basically due to the fact that we carefully arranged everything in such a way that the central charge of the superVirasoro generators (3.30) is equal to 8. The interplay between Lorentz invariance and supersymmetry is rather novel, however, and remains to be investigated.

IV THREE-DIMENSIONAL LIOUVILLE SUPERSTRING

In this last section, we review the particular features of the three-dimensional Liouville superstring, following ref. [12]. The general structure is similar to the one of the five-dimensional model we just discussed. We shall skip the details, and focus on the following two novel points.

First it has been conjectured by Polyakov [1,2] and worked out in more detail by Dotsenko [26] (see also ref. [27]) that the three-dimensional Ising model should be equivalent to a fermionic string theory at D=3. This string theory should, most probably, be the three-dimensional Liouville superstring. This equivalence would provide an alternative way to compute e.g. the critical exponents of the Ising model, and we shall discuss preliminary results in this direction.

Second, the supersymmetry generators will turn out not to obey any graded Lie algebra commutation relation. In this connection, let us note that, in the work of Goddard, Nahm, Olive and Schwimmer [24] which we quoted above about the D=5 case, the discussion of the magic square of Freudenthal suggests that, in the same way as the octonions and quaternions are related to the groups with triality SO(8) and SU(2)xSU(2)xSU(2), the complex numbers and hence the D=3 NSR string might be related to the abelian subgroup U(1)xU(1) of SU(3). This scheme does not work, however, and the way supersymmetry is realized here is quite different from the five-dimensional case. Indeed, the operators relating the space-time fermions to the space-time bosons of this theory are vertex-operators associated with lattice vectors of length squared one half. Hence their operator product expansion contains square-root singularities. These operators are neither commuting nor anticommuting but of a more general nature.

IV.1 The NSR formulation

Let us first briefly summarize the basic feature of the light-cone formulation. In three dimensions, the components of a vector like Γ_μ are

$$\Gamma_+ = (\Gamma_1 + \Gamma_2)/\sqrt{2}\,, \quad \Gamma_- = (\Gamma_1 - \Gamma_2)/\sqrt{2} \quad \text{and } \Gamma_3 \tag{4.1}$$

We will unify the transverse field Γ_3 and the Liouville field Γ_L into a complex field:

$$\Gamma = (\Gamma_3 + i\,\Gamma_L)/\sqrt{2} \tag{4.2}$$

The super-Virasoro generators in the light-cone are

$$G_n = \oint \frac{dz}{4\pi i z}\, z^n\, (\Pi_3\Gamma_3 + \Pi_L\Gamma_L + i\sqrt{3}\,\Gamma_L{}') \qquad ; \qquad \Pi = iX'$$

$$L_n = \oint \frac{dz}{2\pi i z} z^n \left[N(\Pi_3\Pi_3 + \Pi_L\Pi_L) + :(\Gamma_3'\Gamma_3 + \Gamma_L'\Gamma_L): + i\sqrt{3}\Pi_L' \right] + \text{"}\tfrac{3}{8}\text{"}\delta_{n,0} \tag{4.3}$$

where primes mean $z\,\partial/\partial z$ and "3/8" = 3/8 for NS and 0 for R. The generators (4.3) obey the standard superVirasoro algebra with a central charge equal to 8, the transverse fields contributing 1 while the Liouville fields contribute 7. Hence, in a covariant formalism, together with the + and − fields the total central charge is 10 cancelling exactly the ghost contribution.

The zero-modes of the world-sheet bosonic Liouville field Π_L are [9] (see formula (3.2) with m+n=0):

$$(\Pi_L)_0 = \sqrt{\frac{D-1}{8}}\, m, \qquad m \in \mathbb{Z} \tag{4.4}$$

and for D=3 we see that these zero-modes are integers or

half-integers. Hence the eigenvalues $\mathcal{E}$ of the Virasoro operator $\mathcal{L}_0$ are of the form

$$\epsilon_{NS} = \frac{k^2}{2} + \frac{m^2}{8} + N + \frac{3}{8} \qquad m\,,\ 2N \in \mathbb{Z}$$

$$\epsilon_{R} = \frac{k^2}{2} + \frac{m^2}{8} + N \qquad m\,,\ N \in \mathbb{Z} \tag{4.5}$$

for the Neveu-Schwarz and Ramond sector, respectively. After this brief review let us turn to the 3D Ising model.

IV.2 The Critical Exponents of the 3D Ising model

At the present time the connection between the 3D Ising model and a fermionic string theory has only been established on a lattice [26,27]. Since the continuum limit of the latter theory has not been taken so far in a satisfactory way, it cannot be directly used to compute critical exponents of the Ising model. Following Polyakov [2], we shall assume that the corresponding string theory in the continuum is the Sliouville fermionic string in three dimensions which we presently discuss.

If the two models are in the same equivalence class, the power law behaviour of the Ising Green functions at large distance should be derivable from the infrared behavior of the Sliouville string theory. The latter is governed by the dynamics of the zero-mass string states which should appear in the spectrum. Indeed the correspondence would otherwise be inconsistent, all string Green functions decreasing exponentially at large distances. The Liouville superstring meets this requirement since it possesses massless spin-zero and spin-one-half particles in its spectrum [9]. On the other hand, it is fully world-sheet interacting and there is no reason to expect that the power law behavior of Green functions should be governed by intermediate states involving only one such massless particle, since the interaction need not vanish in the infrared limit. It is therefore difficult to derive quantitative conclusions from the known properties of the Liouville superstring theory in its three-dimensional

(3D) aspect.

On the other hand, for statistical mechanics in two dimensions, the critical exponents are directly given by the corresponding eigenvalues of the Virasoro operator $\mathcal{L}_0$ [28]. Hence we know the power law behavior of the string Green functions as functions of the distance in the parameter space of the world-sheet from the properties of the Liouville dynamics which we just recalled. Earlier attempts [29] to directly connect the two-dimensional (2D) and the 3D behaviors have been unsuccessful. In our opinion, the basic point is that, due to quantum fluctuations of the string position operators X^μ, the corresponding 2D world-sheet "surface" embedded in the 3D space is certainly very irregular for the most important string configurations. Hence we must at least take account of the possibility that its Haussdorf dimension be non-trivial in the large distance limit. In this case the relationship between the 3D distance and the parameter-space distance is of the general form

$$< \sum_{\mu=1}^{3} [X^\mu(\vec{\xi}_1) - X^\mu(\vec{\xi}_2)]^2 > \sim \left[(\vec{\xi}_1 - \vec{\xi}_2)^2\right]^\rho \tag{4.6}$$

where $\vec{\xi}$ are world-sheet parameters lumped into 2D euclidean vectors. On the left- and right-hand sides of eq. (4.6), appear the squares of the usual euclidean norms in 3 and 2 dimensions, respectively. In the following they will be simply denoted by vertical bars. The bracket on the left hand side means that some quantum average over the X field is understood. Similar scaling laws have already met with great experimental success in other condensed matter problems, as is well-known.

The critical features of the 3D Ising model may be described from a ϕ^4- field theory in 3 dimensions, that is, $\phi = \phi(X^\mu)$ where X^μ is a given point of the three-dimensional euclidean space. A typical large distance behavior is

$$< \phi^P(X_1)\ \phi^P(X_2) > \sim |\ X_1 - X_2\ |^{-2d} \tag{4.7}$$

In this formula, p is an integer and d the dimension of ϕ^P. On the right-hand side, there of course appears the 3D euclidean distance. The 3D NSR Liouville string may be regarded, at least formally, as defining an imbedding of the 2D world-sheet into the 3D space

$$\vec{\xi} \rightarrow X^\mu(\vec{\xi})$$

The idea is that through this mapping, scaling operators of the three-dimensional theory may become primary fields of the corresponding 2D world-sheet theory. If this is true, we also have

$$< \phi^P(X^\mu(\vec{\xi}_1))\ \phi^P(X^\mu(\vec{\xi}_2)) > \sim |\ \vec{\xi}_1 - \vec{\xi}_2\ |^{-2\delta} \tag{4.8}$$

where, by definition, δ is one of the scaling dimensions of the world-sheet conformal field theory. Combining the above three equations, we find the simple relation

$$d = \rho^{-1}\ \delta \tag{4.9}$$

Since the 2D Green functions only involve operators which are local on the world-sheet it is natural to assume that δ does not depend upon the particular choice of operator (more about this point below). The 2D and 3D scaling dimensions are proportional and we shall thus look at appropriate ratios.

For this purpose, we want to make use of formula (4.5). Two preliminary points need be straightened out. First the k^2 term leads, a priori, to a continuous set of values. This is unacceptable and we shall assume that only operators with k=0 should be considered. Second, we have to decide when we use the NS or the R formula. In the ϕ^4 – field theory describing the Ising model, the operators fall into two classes according to whether they are even or odd under the replacement of ϕ by $-\phi$. Products of operators of the same type are even, while products of operators of opposite type are odd. This is similar to the classification of the 2D scaling fields according to whether they belong to the NS and R sectors, and it seems natural to identify even (odd) operators of the 3D Ising model with operators of the NS (R) sectors of the string theory.

Let us now apply these ideas to the simplest even operators. The 2D counterparts of ϕ^2 and ϕ^4 are of the NS type; they have no conformal spin and we have

$$\delta = \epsilon + \bar{\epsilon} = 2\,\epsilon \tag{4.10}$$

Then we get

$$\frac{d_{\phi^4}}{d_{\phi^2}} = \frac{\delta_{\phi^4}}{\delta_{\phi^2}} = \frac{\epsilon_{\phi^4}}{\epsilon_{\phi^2}} \tag{4.11}$$

By the standard renormalization group argument [30] the dimensions of ϕ^2 and ϕ^4 are related to the critical exponents ν and ω by

$$d_{\phi^2} = D - \frac{1}{\nu} \quad , \quad d_{\phi^4} = D + \omega \tag{4.12}$$

Using the values ν =0.6310 ± 0.0015, ω=0.81 ± 0.04 [31], and letting D=3, we find

$$\frac{d_{\phi^4}}{d_{\phi^2}} = 2.692 \pm 0.036 \tag{4.13}$$

On the other hand, using eq. (4.5) we see that in the NS sector

$$\frac{\epsilon^{N=1/2}_{m=1}}{\epsilon^{N=0}_{m=0}} = \frac{8}{3} \tag{4.14}$$

It seems natural that ϕ^2 corresponds to the lowest lying NS state, m=0, N=0. But, at present, we see no really convincing argument why ϕ^4 should be associated with the NS state N=1/2, m=1. The value 8/3, "predicted" by the Liouville string theory is in very good agreement with the renormalization group value given by formula (4.13). One may next compute the value of ρ , since the absolute magnitude of the 2D scaling dimension is specified once the quantum numbers have been determined. One finds

$$\rho = 0.530 \pm 0.001$$

One sees that ρ is close to one-half and the 3D distance increases much more slowly than the 2D distance. This indicates that indeed the world-sheet surface is highly irregular.

We also tried to derive a relation involving the other exponent η, by considering the operator ϕ. The 2D counterpart is in the Ramond sector. We did not find any quantum number assignment in formula (4.5) leading to numerical agreement for the ratios of d_ϕ with d_{ϕ^2} or d_{ϕ^4}, unless we used values of m and n, that are too large to be significant. One way around this is to assume that ρ may take different values in the NS and R sectors. From this viewpoint, the above result $\rho = 0.530 \pm .001$ is valid in the NS sector, while the R value is unknown at present.

Our preceding discussion of the relation between the Ising critical exponents and the Liouville string spectrum is at the level of conjectures, and at best an oversimplified approximation of the real relation which is probably rather complicated. However, the equality between the expressions (4.13) and (4.14) is encouraging and perhaps more than just a numerical coincidence. Clearly further studies are needed, but we may note that it is the first time, to our knowledge, that any numerical relation is established between the 3D Ising model and a string theory, that is in perfect agreement with the numerical values of the critical exponents.

IV.3 The supersymmetry generators

Let us leave aside the discussion of the Ising model, and study the structure of the three-dimensional Liouville NSR string in more detail.

By the fermion-boson equivalence [23] the two world-sheet fermi fields Γ_3 and Γ_L (or Γ and $\bar{\Gamma} = \Gamma^\dagger$, see eq. (4.2)) are equivalent to one world-sheet bosonic field, called $\alpha^1(z)$:

$$\Gamma(z) = \sqrt{z}\, K\, z^{\alpha_0^1}\, N(\exp(-\sum_{n\neq 0} \frac{z^{-n}}{n}\, \alpha_n^1\,)) = \sqrt{z}\, N(\exp(iQ^1(z)))$$

$$[\ \alpha_0^1\ , K\ ] = K \qquad (4.15)$$

The operator K shifts the eigenvalues of the zero-mode by +1. The relevant set of zero-modes are the integers plus the half-integers. When α_0^1 takes an integer eigenvalue, $\Gamma(e^{2\pi i}z) = -\Gamma(z)$, and the corresponding state belongs to the NS sector. If on the other hand, α_0^1 takes a half-integer eigenvalue, $\Gamma(e^{2\pi i}z) = +\Gamma(z)$, and the corresponding state belongs to the Ramond sector.

Let us recall that, according to eq. (4.4), in three dimensions, the zero-modes of the world-sheet bosonic Sliouville field Π_L are integers or half-integers. Hence this bosonic Liouville field Π_L is mathematically exactly equivalent to the field α^1 we just introduced. In order to stress this fact we will write

$\Pi_L(z) \equiv \alpha^2(z)$.

In this completely bosonized form, the equality between NS and R partition function becomes obvious, in the same way as for the D=5 model. The N=4 SUSY generators we had in section III are replaced by N=2 generators. We shall not elaborate upon these facts which are straightforward to verify (see ref. [12] for details).

We are now going to construct operators that relate the space-time bosons of the NS sector and the space-time fermions of the R sector. This is most easily done in the bosonized formalism using vertex operators:

$$V^{\vec{u}}(z) = z^{\frac{1}{2}\vec{u}^2}\, N(\exp(\ iu_1Q^1 + iu_2Q^2\ )): \quad , \qquad \alpha^s = iz\partial_z Q^s \quad , \ s=1,2 \qquad (4.16)$$

It is easy to see that vertex operators shifting α_0^1 or α_0^2 by one unit, i.e. $\vec{u}=(\pm 1,0)$ or $(0,\pm 1)$, as is the case e.g. for Γ and $\overline{\Gamma}$ in eq.(4.15), only connect states of a given sector with themselves. These vertex operators are associated with lattice vectors $\vec{u}$ of unit length, and, hence, are fermionic fields. These are the only fermionic fields we can construct from $\alpha^1(z)$ and $\alpha^2(z)$.

Vertex operators connecting NS and R states are, instead, given by eq. (4.16) with $\vec{u} = (\pm 1/2, \pm 1/2)$. They are thus associated with lattice vectors of length squared $\vec{u}^2 = 1/2$. They are neither bosonic nor fermionic but of a more general nature. These four vertex operators can be split into two pairs:

$$S(z) = z^{\frac{1}{4}} N[\exp(\tfrac{i}{2} (Q^1+Q^2))] \quad , \quad \bar{S}(z) = z^{\frac{1}{4}} N[\exp(- \tfrac{i}{2} (Q^1+Q^2))] \tag{4.17a}$$

and

$$T(z) = z^{\frac{1}{4}} N[\exp(\tfrac{i}{2} (Q^1-Q^2))] \quad , \quad \bar{T}(z) = z^{\frac{1}{4}} N[\exp(-\tfrac{i}{2} (Q^1-Q^2))] \tag{4.17b}$$

Evidently, the S and T fields mutually commute. S and $\bar{S}$ act along one diagonal of the $(\alpha_0^1 , \alpha_0^2)$ plane while T and $\bar{T}$ act along the other diagonal. The following discussion is analogous for the S and the T fields, and we will concentrate on the former.

The operator product expansions of the S fields contain square-root singularities

$$S(x)\ \bar{S}(y) = \frac{1}{(x-y)^{1/2}}\ N(S(x)\ \bar{S}(y)) \quad , \quad |x| > |y| \tag{4.18}$$

which make impossible the usual contour deformation argument to obtain commutation or anticommutation relations for the Fourier modes. Nevertheless, $S(z)$ or $\bar{S}(z)$ applied on any state of the bosonic Fock space with finite occupation number yields an analytic function except for the zero-mode term which is

$$z^{\frac{1}{4} + \frac{1}{2}(\alpha_0^1 + \alpha_0^2)} \quad \text{for } S(z) \qquad \text{and} \qquad z^{\frac{1}{4} - \frac{1}{2}(\alpha_0^1 + \alpha_0^2)} \quad \text{for } \bar{S}(z) .$$

Thus we can define Fourier modes

$$S_n = \oint \frac{dz}{2\pi iz} z^n S(z) \quad , \quad \bar{S}_n = \oint \frac{dz}{2\pi iz} z^n \bar{S}(z) \tag{4.19}$$

with n integer for S whenever $\alpha_0^1 + \alpha_0^2 + \frac{1}{2}$ is even and with n half-integer for S whenever $\alpha_0^1 + \alpha_0^2 + \frac{1}{2}$ is odd, and vice versa for S.

In particular the zero-modes of S and $\bar{S}$ transform the four massless states into themselves:

$$S_0 \mid 0 , -\tfrac{1}{2} > = \mid \tfrac{1}{2} , 0 > \quad , \quad S_0 \mid -\tfrac{1}{2} , 0 > = \mid 0 , \tfrac{1}{2} > \quad ,$$

$$\bar{S}_0 \mid \tfrac{1}{2} , 0 > = \mid 0 , -\tfrac{1}{2} > \quad , \quad \bar{S}_0 \mid 0 , \tfrac{1}{2} > = \mid -\tfrac{1}{2} , 0 > \tag{4.20}$$

where $|1/2,0\rangle$ is shorthand for $|\alpha_0^1=1/2, \alpha_0^2=0\rangle$ etc. We see that S_0 and $\bar{S}_0$ indeed transform NS states into R states of the same mass level, and vice versa. They act as space-time supersymmetry generators. This is true at all mass levels: It is clear that S_0 and $\bar{S}_0$ acting on space-time bosons yield space-time fermions and vice versa. On the other hand, the Fourier modes of a vertex operator, V_m, have the usual commutation relations with the Virasoro operator $\mathcal{L}_0$ (which contains no linear term)

$$[\mathcal{L}_0 , V_n] = - n V_n . \tag{4.21}$$

S_0 and $\bar{S}_0$ being the zero-modes of vertex-operators, we see that they commute with the Hamiltonian $\mathcal{L}_0$ and do not change the mass level.

It is also easily seen that the integer modes of S and $\bar{S}$ do not connect the different G-parity subsectors. This is not true for the half-integer modes but they do not appear in the space-time supersymmetry operators which are made up with the integer modes of S and $\bar{S}$ only. This proves the supersymmetry between each of the subsectors separately.

A second set of space-time supersymmetry operators is readily constructed following the same mathod as for the D=5 case. Let us introduce

$$\tilde{\Pi}_3 = \Pi_3 \quad , \quad \tilde{\Pi}_L = \frac{1}{\sqrt{2}}(\ \alpha^2 - \alpha^1\ ) \quad . \tag{4.22}$$

This definition of $\tilde{\Pi}_L$ ensures that it commutes with S and $\bar{S}$. Let now

$$\mathcal{S} = 2\ \tilde{\Pi}^{+}\ S + \sqrt{6}\ S' \quad ; \quad \tilde{\Pi} = (\tilde{\Pi}_3 + i\tilde{\Pi}_L)/\sqrt{2} \tag{4.23}$$

By eq. (4.21) the zero-modes of $\mathcal{S}$ and $\mathcal{S}^+$ again commute with the Hamiltonian $\mathcal{L}_0$ and are space-time supersymmetry generators. Using a reference frame where $(\Pi_3)_0 = 0$, their action on the massless states is

$$\mathcal{S}_0\ |\ 0\ ,\ -\tfrac{1}{2} > \ = \tfrac{i}{2}\ |\ \tfrac{1}{2}\ ,\ 0 > \quad , \quad \mathcal{S}_0\ |\ -\tfrac{1}{2}\ ,\ 0 > \ = -\tfrac{i}{2}\ |\ 0\ ,\ \tfrac{1}{2} > \quad ,$$

$$\mathcal{S}_0^+\ |\ \tfrac{1}{2}\ ,\ 0 > \ = -\tfrac{i}{2}|\ 0\ ,\ -\tfrac{1}{2} > \quad , \quad \mathcal{S}_0^+\ |\ 0\ ,\ \tfrac{1}{2} > \ = \tfrac{i}{2}\ |\ -\tfrac{1}{2}\ ,\ 0 > \tag{4.24}$$

Although S_0, $\bar{S}_0$, $\mathcal{S}_0$ and $\mathcal{S}_0^+$ are true space-time supersymmetry generators, let us stress again that they are not simply commuting or anticommuting due to the square-root singularity in their operator product expansion (4.18). Asking e.g. about the anticommutator of any two of them simply makes no sense.

The relevant operators of the theory are the NSR fields Γ and $\bar{\Gamma}$, defined in eq. (4.15), the supersymmetry operators S and $\bar{S}$ (see eq. (4.17a)) and the generators of a SU(2) Kac-Moody algebra

$$I_3(z) = \tfrac{1}{2}\ (\ \alpha^1(z) + \alpha^2(z)\ ) \quad , \quad I_{\pm}(z) = z\, N\exp\left[i\left(Q^1(z)+Q^2(z)\right)\right] \tag{4.25}$$

The operator product expansions of the S operators are

$$S(x)\ S(y) \sim \frac{(x-y)^{1/2}}{(xy)^{1/4}}\ I_+(y) \quad ; \qquad \bar{S}(x)\ \bar{S}(y) \sim \frac{(x-y)^{1/2}}{(xy)^{1/4}}\ I_-(y)$$

$$S(x)\ \bar{S}(y) \sim \frac{(xy)^{1/4}}{(x-y)^{1/2}} + \frac{(x-y)^{1/2}}{(xy)^{1/4}}\ I_3(y)$$

$$\bar{S}(x)\ S(y) \sim \frac{(xy)^{1/4}}{(x-y)^{1/2}} - \frac{(x-y)^{1/2}}{(xy)^{1/4}}\ I_3(y) \tag{4.26}$$

Hence we see that the operators S, $\bar{S}$, I_3 and $I_\pm$ form an algebra that closes under short distance expansion of the product of any two of them (at least to the leading orders). Thus, although we cannot define a Lie algebra for the Fourier modes, these operators nevertheless satisfy a well defined algebraic structure that underlies the supersymmetry of the theory. The other operator product expansions take the standard form (see ref. [12] for details.

REFERENCES

[1] A.M.Polyakov, Phys. Lett. 103B (1981) 207.

[2] A.M.Pokyakov, Phys. Lett. 103B (198&) 211.

[3] J.-L.Gervais, A.Neveu, Nucl.Phys. B199 (1982) 59; B209 (1982) 125.

[4] J.-L.Gervais, A.Neveu, Nucl.Phys. B224 (1983) 329.

[5] J.-L.Gervais, A.Neveu, Nucl.Phys. B238 (1984) 125; 396

[6] J.-L.Gervais, A.Neveu, Nucl.Phys. B257 [FS14] (1985) 59.

[7] J.-L.Gervais, A.Neveu, Phys.Lett. 151B (1985) 271.

[8] J.-L.Gervais, A Neveu, Com.Math. Phys. 100 (1985) 15; Nucl. Phys. B264 (1986) 557

[9] A.Bilal, J.-L.Gervais, Nucl.Phys. B 284 (1987) 397.

[10] A.Bilal, J.-L.Gervais, Phys.Lett. 187B (1987) 39.

[11] A.Bilal, J.-L.Gervais, Nucl.Phys. B 293 (1987) 1.

[12] A.Bilal, J.-L.Gervais, Liouville superstring and Ising model in three dimensions; preprint LPTENS 87/26, Nucl. Phys. to be published.

[13] A.Bilal, J.-L.Gervais, Exact quantum three-point function of three Liouville highest-weight states; preprint LPTENS 87/33, submited to Nucl. Phys.B

[14] L.Brink, P.di Vecchia, P.Howe Phys. Lett. 65B (1976) 471

[15] see, e.g. , A.S.Schwartz Com. Math. Phys. 64 (1979) 233.

[16] J.-L.Gervais, B.Sakita, Nucl. Phys. B34 (1971) 477.

[17] A.Belavin, A.Polyakov, A.Zamolodchikov, Nucl. Phy. B241 (1980) 333.

[18] V.G.Kac, Proc. Int. Congress of Mathematicians, Helsinki (1978); Lecture Notes in Physics, vol. 94 (Springer, New York, 1979) p. 441.

[19] Higher Transcendental Functions, A.Erdélyi, W.Magnus,

F.Oberhettinger, F.Tricomi; H.Bateman Project, Vol.2, McGraw-Hill 1955.

[20] F.Gliozzi, D.Olive, J.Scherk, Nucl.Phys. B122 (1977) 253.

[21] J.F.Arvis, Nucl. Phys. B212 (1983) 151; B218 (1983) 309.

[22] O.Babelon, Phys. Lett. 141B (1984) 353; Nucl. Phys. B258 (1985) 680.

[23] For a pedagogical review, see P.Goddard, D.Olive, Int. J. Mod. Phys.A1 (1986) 303.

[24] P.Goddard, W.Nahm, D.Olive, A. Schwimmer, Comm. Math. Phys. 107(1986) 179; A.Schwimmer in Proceedings of the 1985 Bonn Firenze Johns Hopkins meeting, World Scientif.; P.Goddard, D.Olive, A.Schwimmer, Phys. Lett. 157B (1985) 393

[25] M.Ademollo, L.Brink, A.D'Adda, R.D'Auria, E. Napolitano, S.Sciuto, E. Del Giudice, P. Di Vecchia, S.Ferrara, F.Gliozzi, R.Musto, R.Pettorino, Nucl. Phys. B114 (1976) 297.

[26] Vl.S.Dotsenko, thesis, Landau Inst. (1981); Nucl. Phys. B 285 [FS19] (1987) 45.

[27] A.R.Kavalov, A.G.Sedrakyan, Nucl. Phys. B285 [FS19](1987) 264.; A. Casher, D. Foerster, P. Windey, Nucl. Phys. B251 [FS13] (1985) 29.

[28] D.Friedan, Z.Qiu, S. Shenker, in Vertex Operators in Mathematics and Physics, Springer, Heidelberg, 1984

[29] Y.Y. Goldsmith, Phys. Lett. 112B (1982 359.

[30] E.Brézin, J.C. Le Guillou, J.Zinn-Justin, in Phase Transitions and Critical Phenomena, ed. by C. Domb and M.S. Green, Academic, New York, 1976, vol. 6.

[31] J.C.Le Guillou, J.Zinn-Justin, J. Physique Lett.46 (1985) 137, J. Physique 48 (1987) 19.

GEOMETRY OF THE KALUZA-KLEIN THEORIES

Richard KERNER
LPTPE, Université Pierre et Marie Curie

1. Introduction

In this series of lectures, devoted to the theories of Kaluza-Klein type, we shall insist on the geometrical aspect of these models, and especially on new possibilities of generalization of such theories.

The idea of the existence of other spacelike dimensions than the three ones accessible to our experimentation, is one of the deepest among the theoretical bases of twentieth-century physics. Although it is often associated with the unification of space and time introduced by Albert Einstein, it is quite different in nature. The pseudo-Euclidean or pseudo-Riemannian character of the space-time metric makes it possible to distinguish locally the time-like direction from the space-like directions. On the other hand, in the absence of particular forces which could break the symmetry, it is impossssible to distinguish the three space-like directions, which are equivalent. The fact that extra space dimensions remain unaccessible to our direct perception needs an explanation, of course.

The deep dependence between the conservation laws and the symmetry groups which govern the essential content of the field theory has been best perceived and stated by Emmy Noether in her famous theorems.

Thus, the conservation laws for energy-momentum tensor are derived from the Lorentz invariance of any field theory. This fact is independent of the representation of the Lorentz group : it may act on vectors, tensors (like $F_{\mu\nu}$), or some Hilbert space of quantum states. In Special Relativity, the linear space on which a real, four-dimensional matrix representation of the Lorentz group acts, is identified with the experimentally perceivable space-time.

In some particular cases the symmetry group of the lagrangian describing the system is bigger than the Lorentz group. For example, the quantum mechanical description of a hydrogen atom displays the O(4)-symmetry, which can not be included in the Lorentz group. Although the group O(4) may be presented as group of rotations of a four-dimensional Euclidean space, it

ISBN 0-12-218100-X

does not mean that the real components of the hydrogen atom evolve in four space-like dimensions. This conclusion is easy to draw, because the particular symmetry O(4) does not apply to other, more complicated systems. The conservation laws which might be interpreted as a consequence of symmetries in more than three space dimensions have to be universal and apply to all physical systems known to us.

At the time when the first five-dimensional theory has been proposed by Kaluza, only one such universal conservation law was known for sure : it was the conservation of the electric charge. Moreover, the electric charge was known to be quantized universally in units e. This quantization is even more restrictive than the quantization of the energy already discovered by Planck : the energy is quantized in portions $\varepsilon = h\nu$, but ν may take on any value ; the electric charge appears in elementary units $\pm e$ regardless of the system or physical situation under consideration.

If one wanted to interpret the charge conservation as a conservation law derived from the symmetry group acting in more than four space-time dimensions, the simplest thing to imagine was that it corresponds to an extra space component of the energy-momentum. On the other hand, it does not participate in the Lorentz transformations of the usual components of $T_{\mu\nu}$: this suggests that the translations in fifth dimension commute with ordinary Lorentz transformations restricted to the four dimensional space-time. This leads to the structure of the full group being a Cartesian product $\mathcal{L}_4 \times G$. In order to explain the conservation of one real quantity only, it is enough to choose a one-parameter Lie group G.

The quantization of the fifth component of the momentum suggests that the fifth direction is compact : as a matter of fact, the eigenvalues of the operator $-i\hbar\, \partial/\partial x^5$ will be automatically discrete if it acts on a circle $x^5 \sim \theta \in [0, 2\pi]$. Of course, the same result can be obtained on $\mathbb{R}^1$ by imposing an infinite potential well : $V = \infty$ for $x^5 < a$ or $x^5 > b$, $V = 0$ for $x^5 \in [a,b]$, but then we must explain where this potential well comes from. The compactness of the fifth dimension is a much more elegant and geometrical way to introduce the charge quantization.

The symmetry group acting effectively and transitively on the circle is U(1), and is isomorphic with the circle itself. This is rather an exception : in the case of higher dimensional symmetry the symmetry group can act on spaces having lower dimension and different topology than the groups themselves. For example, the symmetry group SO(3) can act transitively and effectively on a two-dimensional sphere S^2, etc.

Here we encounter one of the dillemmas of the Kaluza-Klein theories : supposing that we are able to discover the underlying structural group due to the character of the conservation laws, how to be sure what is the dimension and topology of space on which this group is supposed to act ? One of the simplest choices is to identify this space with the group itself, as dictated in the simplest case of $U(1) = S^1$; but such a choice is by no means compulsory. Some hints will be given, as we shall see later, by the fermionic content of the full theory.

The compactness of the extra space dimensions has to be implemented by the possibility of their dimensional comparison with the scales observed in the space dimensions accessible to our experimentation directly. As a matter of fact, a compact space will have a finite volume; this volume, as well as the eventual displacements along the extra dimensions, have to be extremely small in order to avoid their perception in everyday life. Until now, the most convincing is the adoption of Planck scale, because it is related to the most universal interaction : the gravitational one. The Planck scale fixes the order of magnitude of extra dimension at the level of 10^{-33} cm.

In its original version the theory proposed by Kaluza in order to unify the electromagnetic and gravitational phenomena, the space-time was enlarged to five dimensions:one time-like, four space-like dimensions, with the topology of $V_4 \times S^1$. Then, the Einstein-Hilbert lagrangian has been taken, for the metric g_{AB} having the properties

$$\partial_5 g_{AB} = 0, \qquad g_{55} = 1$$

$A, B = 1,2,\ldots,5$; $\mu, \nu = 0,1,2,3$. Denoting $g_{5\mu} = g_{\mu 5} = A_\mu$, Kaluza obtained his famous ansatz

$$g_{AB} = \begin{pmatrix} g_{\mu\nu} + A_\mu A_\nu & A_\nu \\ A_\mu & 1 \end{pmatrix}$$

It turned out that the scalar curvature of this five-dimensional metric was equal to

$$\overset{5}{R} = \overset{4}{R}(g_{\mu\nu}) - \frac{1}{4} F_{\mu\nu} F^{\mu\nu}$$

with $F_{\mu\nu} = \partial_\mu A_\nu - \partial_\nu A_\mu$. Also $\det(g_{AB}) = \det(g_{\mu\nu}).g_{55}$ so that the density $\overset{5}{R} \sqrt{\det(g_{AB})}$ could be reinterpreted, when integrated first over x^5 (which yields the overall factor $2\pi R$, R being the radius of S^1) as a variational principle in four-dimensional space-time leading to the Maxwell equations in the curved space-time and Einstein's equations with the energy-momentum tensor of the electromagnetic field.

At this point, two original sins were committed, one of them of minor nature, and which has been repaired in next versions of this theory, proposed by Jordan and Thiry; it has been pointed out already by Klein and Einstein and Bergmann. The other original sin seems now even more important, and it is quite amazing that it needed almost half a century to become visible. Let us develop briefly on these two points.

The first error is the inconsistency in interpretation. If we take seriously both the idea of the existence of extra dimensions and the validity of the variational principle on the whole space, then it leads to three distinct systems of equations :

$$\overset{4}{R}_{\mu\nu} - \frac{1}{2}\overset{4}{g}_{\mu\nu}\overset{4}{R} \sim - T_{\mu\nu}(F_{\lambda\rho})$$

$$R_{\mu 5} - \frac{1}{2} g_{\mu 5}\overset{5}{R} = 0 \text{ equivalent to } \nabla^\mu F_{\mu\nu} = 0$$

and

$$R_{55} - \frac{1}{2} g_{55}\overset{5}{R} = 0 \text{ equivalent to } \overset{4}{R} = \frac{1}{4} F_{\mu\nu} F^{\mu\nu}$$

The last equation, corresponding to the variation with respect to g^{55}, is an algebraic constraint on $F_{\mu\nu}$, which should be absent in theory of the electromagnetic field. Obviously, the full system is overdetermined : one field degree of freedom is obviously missing. The situation is easily repaired by the introduction of the scalar field ϕ in the following way : the new ansatz is given by

$$g_{AB} = e^{-\phi/3} \begin{pmatrix} g_{\mu\nu} + e^\phi A_\mu A_\nu & e^\phi A_\mu \\ e^\phi A_\nu & e^\phi \end{pmatrix}$$

$$g^{AB} = e^{\phi/3} \begin{pmatrix} g^{\mu\nu} & - e^{-\phi} A^\nu \\ - e^{-\phi} A^\mu & e^{-\phi} + g^{\lambda\rho} A_\lambda A_\rho \end{pmatrix}$$

Now $\overset{5}{R}\sqrt{|\det g_{AB}|}$ turns out to be

$$\sqrt{|\overset{4}{g}_{\mu\nu}|}\,[\overset{4}{R} - \frac{1}{6} g^{\lambda\rho}\partial_{\lambda\phi}\partial_{\rho\phi} - \frac{1}{4} e^{\phi}F_{\mu\nu}\,F^{\mu\nu}]$$

If we neglect the electromagnetic field, this theory is known as scalar-tensor theory of gravitation of Brans and Dicke. The experimental evidence is strongly against this theory.

The second error is of subtler nature and concerns the choice of the variational principle. After the spectacular success of the General Relativity it was taken for granted that the natural extension of this theory to higher dimensions should be based on a similar lagrangian density, i.e. $R\sqrt{|g|}$ + eventually a cosmological term $\Lambda\sqrt{|g|}$. This variational principle, introduced by Einstein and Hilbert, seemed to be unique for the following reasons : it is invariant with respect to the general coordinate transformations and leads to <u>second-order</u> differential equations. There is of course plenty of invariants non-linear in the components of Riemann's tensor, e.g. R^2, R^3, or $R_{AB}R^{AB}$, etc.., but they all lead to the differential equations of the order higher than two. However, among the quadratic invariants of the Riemann tensor there is <u>one and only one</u> that under variation leads to the second-order equations :

$$I_2 = (R_{ABCD}R^{ABCD} - 4R_{AB}R^{AB} + R^2)\sqrt{|g|}$$

But it turns out that in four dimensions I_2 is a full divergence and does not contribute to the Euler-Lagrange equations. The third power combination whose variation would lead to second-order differential equations vanishes identically in four dimensions. That is why the Einstein-Hilbert lagrangian is <u>unique in four dimensions</u>. This is no more true in dimensions higher than four; that is why restraining the Kaluza-Klein theory to the Einstein-Hilbert lagrangian is as unnatural as restraining the General Relativity in four or three dimensions to the truncated variational principle containing the cosmological term $\Lambda\sqrt{|g|}$ only.

Such a situation has occured many times in the history of Science : the fact that Ptolemaean or Aristotelean systems of thought lasted so long enhancing the further development can be explained by the tremendous success they represented at the time of their creation.

The existence of higher-power invariants of the Riemann tensor, all of which are of generic form

$$I_m = \varepsilon^{i_1 i_2 \dots i_{2m}}_{j_1 j_2 \dots j_{2m}} R^{j_1 j_2}_{i_1 i_2} \dots R^{j_{2m-1} j_{2m}}_{i_{2m-1} i_{2m}}$$

and reduce to a full divergence in the dimension 2m suggests that any Kaluza-Klein theory, in order to reveal its full meaning, has to be based on the variational principle of the following form :

$$\sqrt{|g|}\ (\Lambda + \alpha_1 R + \alpha_2 I_2 + \dots + \alpha_m I_m)$$

The series $\alpha_k I_k$ ends up naturally in dimension 4 + N with the term $[\frac{4+N-1}{2}]$, [] meaning the integer part. The coefficients α_k are not dimensionless : the dimension of the whole expression should be cm^{-1}, d = 4 + N; the dimension of R is cm^{-2}, of I_2 cm^{-4}, etc; that is why $\dim[\alpha_{k+1}] = [\dim \alpha_k] - 2$. All these constants can be reduced to the corresponding powers of one universal coupling constant having dimension of length and deducible from the volume of the internal space; the dimensionless coefficients remain to be determined by other considerations.

If we start from the generalized variational principle in 4 + N = d dimensional pseudo-Riemannian space, we would like to find stable solutions displaying the property of spontaneous compactification, i.e. with the topology $V_4 \times B$, B being the "internal" manifold. The existence of the spontaneous compactification has been analyzed with rather negative conclusions, in the framework of the classical theory based on the truncated lagrangian. Many conclusions have to be modified if we take into account the general form of the theory.

Another direction of generalization, pursued actively by many authors during last decade consists of considering internal spaces which have subtler topological properties than a semi-simple Lie group or a homogeneous Riemannian space.

These lectures are essentially devoted to the modifications along these lines brought recently to the Kaluza-Klein theories. We shall start with mathematical preliminaries, making them as short as possible and giving examples of calculations in order to visualise abstract definitions, and to introduce the notation.

2. Geometrical framework : a fibre bundle

Let M be a differential manifold of dimension n, and G a Lie group of dimension N. A principal fibre bundle P(M,G) with a structural group G and basis space M is a manifold P of dimension $N_p = n + N$, which satisfies the following properties :

a) G acts on P on the right, i.e. the following mapping

$$(p,g) \in P \times G \longrightarrow pg \in P$$

is a diffeomorphism of P x G onto P. Often we shall use the notation $pg = R_g p$. The action of G on P is supposed to be effective (i.e., whatever $g \in G$ we take, the points $p \in P$ move; the only element leaving p unchanged is the unit element of G).

b) M is isomorphic with the quotient space of P by the equivalence relation induced by the action of G on P; the canonical projection so obtained,

$$\pi : P \longrightarrow M$$

is differentiable. The equivalence relation in P is defined by

$$p_1 \sim p_2 \quad \text{if there exists } g \in G \text{ such that } \quad p_1 = p_2 g$$

c) P is locally trivial, i.e. each point $x \in M$ has a neighborhood $U \ni x$ such that the set of all points in P that project on U, $\pi^{-1}(U)$, is diffeomorphic with U x G. The set $\pi^{-1}(x)$ is called a fibre over x.

The notion of fibre bundle can be extended to similar structures whose fibres are not isomorphic with the structural group G, but rather with some manifold V on which an action of G is defined via some (linear or non-linear) representation. These manifolds are called associated fibre bundles.

Let P(M,G) be a principal fibre bundle over M with the structural group G; let V be a differentiable manifold on which G acts on the left. (In most cases V is a linear space of some representation of G). We have

$$\nu \in V, \; g \in G, \quad g : V \longrightarrow V, \quad \nu \longrightarrow g\nu \in V$$

Now we can define a right action of G on the product manifold P(M,G)xV as follows :

$$\tilde{R}_g : P(M,G)xV \longrightarrow P(M,G)xV$$

for $(p,\nu) \in P(M,G)xV$ we define $\tilde{R}_g(p,\nu) = (R_g p, g^{-1}\nu)$. The action $\tilde{R}_g$ defines an equivalence relation in P(M,G)xV. An associated fibre bundle E(M,G;P,V) is defined as the quotient of P(M,G)xV by the equivalence relation induced by the action $\tilde{R}_g$ of G ;

$$E(M,G;P,V) = \frac{P(M,G)xV}{\tilde{R}_G}$$

Obviously, dim E = dim M + dim V.

Examples

a) Let G be a Lie group and H its closed subgroup. Let H act on G on the right via right group translations.

$$h \in H, \quad g \in G : \qquad R_h g = gh^{-1}$$

The set of all equivalence classes in G under this action is called the quotient manifold G/H. It is easy to see that G(G/H, H) is a principal fibre bundle over G/H with the structure group H.

Take, for example, the group G of triangular 2x2 real matrices :

$$G \ni g = \begin{pmatrix} a^1 & a^2 \\ 0 & a^3 \end{pmatrix} \quad , \text{ with } a^1 a^3 \neq 0$$

and let H be its subgroup given by

$$H \ni h = \begin{pmatrix} h^1 & 0 \\ 0 & h^3 \end{pmatrix} \quad , \; h^1 h^3 \neq 0$$

The division by the equivalence relation yields G/H = R^1, i.e. the invariant parameter a^2 which is unaffected by the right action of H on G

If we take H to be

$$H \ni h = \begin{pmatrix} h^1 & h^2 \\ 0 & 1 \end{pmatrix},$$

then G/H becomes equal to $R^1 - \{0\}$, i.e. the invariant parameter a^3. The projections are obvious.

One of the most important principal bundles is the <u>bundle of frames</u>. This bundle can be defined given any differentiable manifold M. Let $m \in M$, and let $\{x^i\}$, $i = 1,2,\ldots d = \dim M$, be a local coordinate system in a open neighborhood containing m. A natural basis in the tangent space at m, $T_m(M)$, is given by the set of d derivations $(\partial/\partial x^i)$; under a local change of the coordinate system, $x^i \rightarrow y^j(x^i)$, the basis transforms as $\frac{\partial}{\partial x^i} \rightarrow (\frac{\partial x^k}{\partial y^j})\frac{\partial}{\partial x^k} = (\frac{\partial}{\partial y^j})$. Consider now <u>all</u> possible local frames defined as

$e_i = X^k_i(x)\, \frac{\partial}{\partial x^k}$. These frames can be non-integrable (non-holonomic), if

$\partial_j X^k_i \neq \partial_i X^k_j$. The matrix X^k_i has to be non-singular; therefore it belongs to the group GL(d, R). There are as many local frames as there are elements of GL(d, R); moreover, the group GL(d, R) acts naturally on the right by the group multiplication of the matrices. So, the set of frames has a natural structure of the principal fibre bundle over M. Its dimension is equal to $d + d^2 = \dim M + \dim(GL(d, R))$.

In the same way one can define the principal bundle of <u>orthonormal</u> frames, if one restrict the frames X^i_k to orthogonal matrices only; then the group is O(d, R), and the dimension of the bundle is $d + \frac{d(d-1)}{2} = \frac{d(d+1)}{2}$.

It has to be underlined that a fibre bundle is not always <u>globally</u> isomorphic with the Cartesian product of the basis space and the typical fibre. To visualise a non-trivial case, it is enough to consider a simple discrete group, say Z_2, containing just two elements, identity and an element whose square is equal to identity. If one takes a circle S^1 as a basis space, then a fibre bundle $P(S^1,Z_2)$ can be either a Cartesian product of S^1 with a two-point set (Fig. 1), or a connected set (Fig. 2)

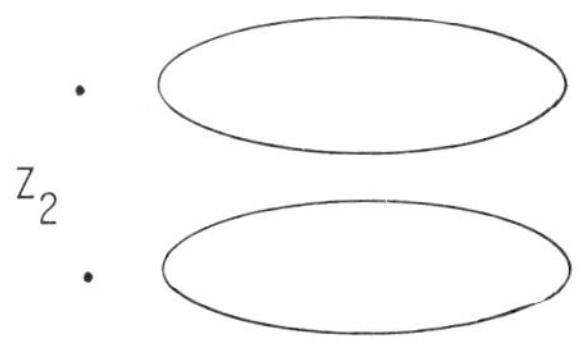

$S^1 \times Z_2$

Fig. 1

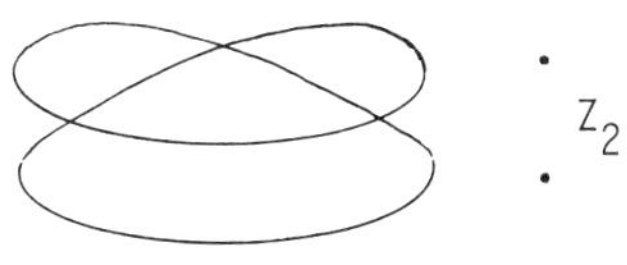

$P(S^1,Z_2)$

Fig. 2

Consider now the action of Z_2 on the manifold isomorphic with $S^1 = \{e^{i\lambda}\}$. The non-trivial element of Z_2 acts as an inversion : $Z_2 \ni a_1 : e^{i\lambda} \to e^{(2\pi i - i\lambda)} = e^{-i\lambda}$. Taking this S^1 as a typical fibre, one can construct an associated fibre bundle in two different ways, with the principal fibre bundle chosen from the previous example :

$$E_1 = \frac{(S^1 \times Z_2) \times S^1}{Z_2} \quad \text{or} \quad E_2 = \frac{P(S^1, Z_2) \times S^1}{Z_2}$$

In the first case the resulting manifold is a torus, while in the second case it is the so-called Klein bottle (fig. 3,4)

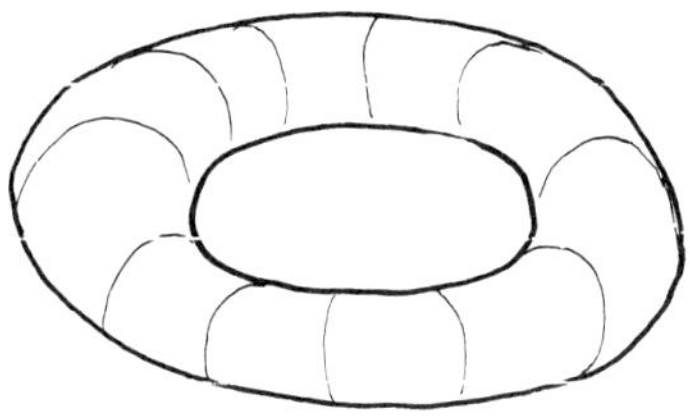

Fig. 3

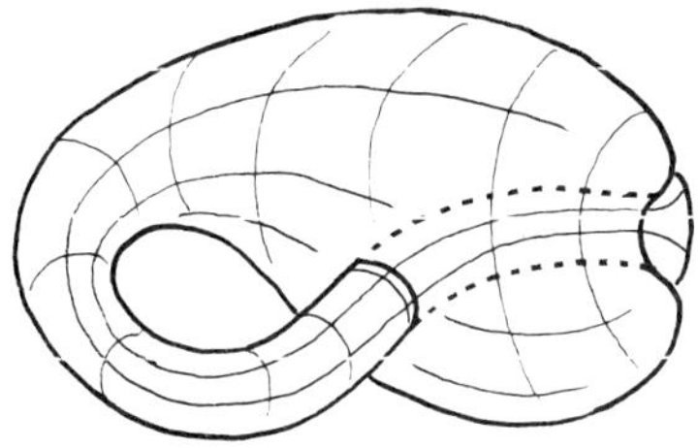

Fig. 4

3 Invariant fields and forms on bundles

First, let us recall some basic properties of the Lie groups. The action of a Lie group on itself (on the right, on the left, or an adjoint action $h \to ghg^{-1}$) is smooth, and represents a differomorphism of the Lie group on itself. If the acting element is very close to the unit element, the induced "motion" is very small, and locally generates a vector field. Through the unit element one can imagine passing of the one-parameter subgroups. The differentiations of functions on the group with respect to these one-parameter actions generate N tangent vectors at the unit element. Let us show on an example how these vectors are obtained.

Let the group G be that of 2 x 2 real triangular matrices :

$$g = \begin{pmatrix} g^1 & g^2 \\ 0 & g^3 \end{pmatrix} \qquad \text{with } g^1 g^3 \neq 0$$

The three independent one-parameter subgroups passing through the unit element $e = \begin{pmatrix} 1 & 0 \\ 0 & 1 \end{pmatrix}$ are

$$\overset{(1)}{\phi}_t = \begin{pmatrix} e^t & 0 \\ 0 & 1 \end{pmatrix}, \quad \overset{(2)}{\phi}_t = \begin{pmatrix} 1 & t \\ 0 & 1 \end{pmatrix}, \quad \overset{(3)}{\phi}_t = \begin{pmatrix} 1 & 0 \\ 0 & e^t \end{pmatrix}$$

Any function on G is a real function of the three variables g^1, g^2, g^3; if we derive with respect to t, we get :

$$\frac{d}{dt} f(g(t), g(t), g(t)) = \frac{\partial f}{\partial g^1}\frac{dg^1}{dt} + \frac{\partial f}{\partial g^2}\frac{dg^2}{dt} + \frac{\partial f}{\partial g^3}\frac{dg^3}{dt}$$

With our choice we obtain the three vectors as

$$(X_1 f)(e) = \frac{\partial f}{\partial g_1}(e), \ (X_2 f)(e) = \frac{\partial f}{\partial g_2}(e), \ (X_3 f)(e) = \frac{\partial f}{\partial g_3}(e) \quad .$$

A global left-invariant vector field on G can be obtained out of each of these vectors by left action of the group. It is sufficient to shift our 1-parameter subgroups by a left action of any group element : take, e.g.

$$g \overset{(1)}{\phi}_t = \begin{pmatrix} g_1 & g_2 \\ 0 & g_3 \end{pmatrix}\begin{pmatrix} e^t & 0 \\ 0 & 1 \end{pmatrix} = \begin{pmatrix} e^t g_1 & g_2 \\ 0 & g_3 \end{pmatrix} ;$$

then derive $f(g \overset{(1)}{\phi}_t)$ with respect to t at t = 0 :

$$\text{then } (X_1 f)(g) = \frac{d}{dt} f(g \overset{(1)}{\phi}_t)\Big|_{t=0} - g_1 \frac{\partial f}{\partial g_1}$$

Similarly, we obtain $(X_2 f)(g) = g_1 \dfrac{\partial f}{\partial g_2}$ and $(X_3 f)(g) = g_2 \dfrac{\partial f}{\partial g_2} + g_3 \dfrac{\partial f}{\partial g_3}$.

The commutators of these vector fields defined globally on the group manifold close to form the <u>Lie algebra</u> of G : $[X_i, X_j] = C^k_{ij} X_k$, where the constants $C^k_{ij} - - C^k_{ji}$ are called the structure constants.

The vector fields X_i defined above are <u>left-invariant</u>, which means that if one acts on the left on the element g with any other element $h \in G$, $g \to hg = b$ then the expressions for X_i in the variables b^k are exactly the same as they were in the variables g^i. This can be verified by applying the group multiplication rule for hg expressed in coordinates, and using the composed derivation rule $\frac{\partial}{\partial(hg)^k} = \frac{\partial g^i}{\partial(hg)^k}\frac{\partial}{\partial g^i}$

The structure constants C^k_{ij} can be interpreted as N matrices N x N : $C^k_{(i)j}$. They provide us with the adjoint representation of the Lie algebra :

$$C^k_{(i)j}C^j_{(m)\ell} - C^k_{(m)j}C^j_{(i)\ell} = C^j_{im}C^k_{(j)\ell}$$

which follows from the Jacobi identity for commutators.

In the case of the Lie group considered above, we have

$$[X_1,X_2] = X_2, \quad [X_2,X_3] = X_2, \quad [X_3,X_1] = 0$$

The adjoint representation of the Lie algebra can be also obtained as a differential at g = e of the adjoint representation of the group : $h \to ad_h$, so that $ad_h(g) = hgh^{-1}$. The matrices

$$\left.\frac{\partial(ad_h(g))^k}{\partial g^\ell}\right|_{g=e} = (ad_h)^k{}_\ell = D^k_\ell(h)$$

give the adjoint representation of the group, i.e.

$$D^k_\ell(h_1)D^\ell_m(h_2) = D^k_m(h_1h_2).$$

Differentiation of $D^k_\ell(h)$ with respect to the group parameters h^j at h=e yields, once again, the structure constants $C^k_{(j)\ell}$.

Given the N global left-invariant fields on the group, one can construct N left-invariant one-forms <u>dual</u> to these fields, i.e.

$$\theta^m = \theta^m_i\, dg^i \quad \text{so that} \quad \theta^m(X_k) = \delta^m_k.$$

For our example, the 1-forms dual to the vector fields X_k are given by :

$$\theta^1 = \frac{1}{g^1}\, dg^1, \quad \theta^2 = \frac{1}{g_1}\, dg^2 - \frac{g^2}{g_1 g_3}\, dg_3, \quad \theta^3 = \frac{1}{g_3}\, dg_3$$

The direct consequence of the commutation relations for the left invariant fields X_i is the Maurer-Cartan identity :

$$d\theta + \frac{1}{2}[\theta,\theta] = 0$$

or, in coordinates,

$$\partial_i \theta^m_k - \partial_k \theta^m_i + C^m_{\ell n} \theta^\ell_i \theta^n_k = 0$$

The action of the Lie group on a fibre bundle generates also the left-invariant vector fields on the bundle, called the vertical fields. They are everywhere tangent to the fibres, and satisfy the same commutation relations as the elements of the Lie algebra of structural group. Their canonical projection on the tangent basis space vanishes identically. We shall denote by σ the canonical imbedding that attributes to each element of the Lie algebra $\mathcal{A}_G$ the corresponding vertical field on the bundle :

$$\sigma : \mathcal{A}_G \longrightarrow T(P)$$

so that

$$\sigma([L_i, L_k]_{\mathcal{A}_G}) = [\sigma(L_i),\ \sigma(L_k)]$$

(Lie bracket of vector fields)

The Lie brackets of the elements of $\mathcal{A}_G$ are transformed into the Lie brackets (commutators) of their images as vector fields.

4. Connections

At each point of a principal bundle $P(M,G)$ we have constructed $N = \dim G$ vectors which cover all the bundle space and span at each point an N-dimensional subspace of $TP(M,G)$ which is called the vertical subspace. The choice of the completing $n = \dim M$ dimensional subspace at each point remains unspecified. A connection ω on $P(M,G)$ is some special choice of a horizontal subspace of the tangent space at each point $p \in P(M,G)$:

1) At each $p \in P$, $T_pP = V_p \oplus H_p$, a direct sum of vertical and horizontal subspaces

2) $H_{pg} = R_g H_p$,

where R_g is the differential of the right translation induced by the element g, $R_g(p) = pg$

3) H_p depends differentiably on p (in other words, H_p is a differentiable distribution of subspaces).

It is easier to operate with an alternative definition of a connection, which is much more useful in calculus. As $\dim H_p = \dim M$, $\dim V_p = \dim G$, in order to determine $n = (\dim M)$- dimensional subspace of an $n + N$ dimensional vector space, it is sufficient to give N linear equations imposed on the $n + N$ dimensional tangent vectors. This can be achieved if we define a Lie-algebra valued 1-form ω at any point $p \in P$; then $X \in T_pP$ will be defined as <u>horizontal</u> if

$$\omega_p(X) = 0$$

The covariance under the right translations means that

$$\omega_{pg}(R_g X) = \mathrm{ad}(g^{-1})\omega(X)$$

It is said that the 1-form ω is of ad-type. On the other hand, if a vector field X is one of the left-invariant vector fields (and vertical by definition), then X is an image of some element of Lie algebra :

$$X = \sigma(L);$$

in such a case,

$$\omega(X) = \omega(\sigma(L)) = L$$

The following relations hold :

$$\omega \cdot \sigma = \mathrm{Id}_{\mathcal{A}_G}\ , \qquad d\pi \cdot \sigma = 0$$

One completes these by introducing the <u>horizontal lift</u> τ_p which associates to a vector in T_xM a unique vector in T_pP (with $x = \pi(p)$), which is horizontal.

Then we have also

$$d\pi \cdot \tau = Id_M, \qquad \omega \cdot \tau = 0$$

Any vector tangent to P can be decomposed now into its horizontal and vertical parts :

$$X = \text{hor } X + \text{ver } X = \tau \cdot d\pi(X) + \sigma \cdot \omega(X) = (- \sigma\cdot\omega)(X) + \sigma\cdot\omega(X)$$

The curvature 2-form of the connection ω is then defined as

$$\Omega = D\omega = d\omega \cdot \text{hor, i.e.}$$

$$\Omega(X,Y) = D\ (X,Y) = d\omega\ (\text{hor } X, \text{ hor } Y).$$

It is proved that the following identity holds :

$$d\omega\ (\text{hor } X, \text{ hor } Y) = d\omega\ (X,Y) + \frac{1}{2}\,[\omega(X), \omega(Y)]_{A_G}$$

Usually the connection form ω on P(M,G) is defined by means of a family of 1-forms, each of them defined on an open subset of the base M. Let $\{U_\alpha\}$ be an open covering of M ; let $\phi_\alpha : \pi^{-1}(U_\alpha) \to U_\alpha \times G$ be the corresponding isomorphism of a subset of P(M,G) over U_α and the Cartesian product $U_\alpha \times G$; it is called a <u>local trivialization</u>. Let $\psi_{\beta\alpha} = \phi_\beta \phi_\alpha^{-1}$ be the <u>transition function</u> defined on M having its values in G. It is identified in physics as a <u>change of a gauge</u>. The fact that $\psi_{\beta\alpha}$ depends only on $x \in M$ comes from the following property of ϕ_α : being an isomorphism, it has to satisfy

$$\phi_\alpha\ (pg) = \phi_\alpha(p)g \text{ ; therefore}$$

$$\psi_{\beta\alpha}(pg) = \phi_\beta(pg)\phi_\alpha^{-1}(pg) = \phi_\beta(p)g.g^{-1}\phi_\alpha^{-1}(p) = \phi_\beta(p)\phi_\alpha^{-1}(p)$$

so that

$$\psi_{\beta\alpha}(pg) - \psi_{\beta\alpha}(p), \text{ i.e. } \psi_{\beta\alpha} = \psi_{\beta\alpha}(x = \pi(p))$$

A cross-section is defined as the mapping

$$\mu_\alpha : U_\alpha \longrightarrow P(M,G), \qquad \mu_\alpha(x) = \phi_\alpha^{-1}(x,e)$$

with $e \in G$ being the unit element of G.

By definition, $\mu_\beta(x) = \phi_\beta^{-1}(x,e) = \phi_\alpha^{-1}(x,e)\phi_\alpha(x,e)\phi_\beta^{-1}(x,e)$
$= \mu_\alpha(x)\ \psi_{\alpha\beta}(x).$

This is also another way of defining the gauge transformation. The mappings of the differential manifolds define also the corresponding mappings of their tangent spaces; for example, the differential $d\mu_\alpha$ maps TM onto TM x $\mathcal{A}_G$.

Let θ be the left-invariant, $\mathcal{A}_G$-valued canonical 1-form on G (the Maurer-Cartan form). For any two open subsets of M, $U_\alpha \cap U_\beta \neq 0$, we define a $\mathcal{A}_G$-valued 1-form on $U_\alpha \cap U_\beta$ by setting

$$\theta_{\alpha\beta} = \overset{*}{\psi}_{\alpha\beta}\theta$$

with $\overset{*}{\psi}_{\alpha\beta}$ the induced mapping of $\psi_{\alpha\beta}$ acting on the 1-forms. In the same way, one defines a 1-form ω_α on U_α by putting

$$\omega_\alpha = \overset{*}{\mu}_\alpha\omega$$

The following important relation holds :

$$\omega_\beta = \mathrm{ad}(\psi^{-1}_{\alpha\beta})\omega_\alpha + \theta_{\alpha\beta}$$

which is a gauge transformation of the 1-form ω_α defined on $U_\alpha \subset M$ under the change of the gauge $\psi_{\alpha\beta}$. This transformation is not homogeneous ; however, the corresponding curvature form Ω_α transforms covariantly :

$$\Omega_\beta = \mathrm{ad}(\psi^{-1}_{\alpha\beta})\Omega_\alpha$$

We shall give an illustration of these formulae on a simple example. Let the structure group of the bundle be the subgroup of 2 x 2 triangular matrices of the form :

$$G \ni h, \qquad h = \begin{pmatrix} h_1 & h_2 \\ 0 & 1 \end{pmatrix}$$

The fibre bundle space on which G acts from the right shall be identified with all 2 x 2 real matrices

$$P(M,G) \ni p = \begin{pmatrix} p_1 & p_2 \\ p_3 & p_4 \end{pmatrix} \quad \text{satisfying } p_1p_4 - p_2p_3 \neq 0,\ p_1p_3 \neq 0.$$

The right action of G on P(M,G) is defined as the matrix multiplication :

$$pg = \begin{pmatrix} p_1 & p_2 \\ p_3 & p_4 \end{pmatrix} \begin{pmatrix} g_1 & g_2 \\ 0 & 1 \end{pmatrix} = \begin{pmatrix} p_1g_1 & p_1g_2 + p_2 \\ p_3g_1 & p_3g_2 + p_4 \end{pmatrix}$$

The basis space $M = P(M,G)/G$ is two-dimensional and should be identified with the invariants of this action of G. It is easy to see that $x_1 = (p_1/p_3)$, $x_2 = p_4 - (p_2p_3)/p_1$ do not change under the action of G ; any other two independent functions of (x_1, x_2) can serve as local coordinates in $M = P(M,G)/G$. The canonical projection π is therefore

$$(p_1, p_2, p_3, p_4) \xrightarrow{\pi} (\frac{p_1}{p_3}, p_4 - \frac{p_2p_3}{p_1}) = (x_1, x_2)$$

The differential $d\pi$ is the Jacobian matrix $\frac{\partial(x_1,x_2)}{\partial(p_1,\dots p_4)}$.

The two left-invariant fields on G are

$$X_1^G = g_1 \, \partial/\partial g_1, \qquad X_2^G = g_1 \, \partial/\partial g_2$$

and satisfy $[X_1^G, X_2^G] = X_2^G$. The Lie algebra $\mathcal{A}_G$ is defined by structure constants $C_{12}^2 = - C_{21}^2 = 1$, others vanishing. The adjoint representation of $\mathcal{A}_G$ is realised therefore by the two matrices

$$C_{(1)} = \begin{pmatrix} 0 & 0 \\ 0 & 1 \end{pmatrix}, \qquad C_{(2)} = \begin{pmatrix} 0 & 0 \\ -1 & 0 \end{pmatrix}; \; C_{(1)}C_{(2)} - C_{(2)}C_{(1)} = C_{(2)}$$

The canonical Maurer-Cartan 1-form on G has to satisfy

$$\theta(X_1^G) = L_1, \qquad \theta(X_2^G) = L_2$$

L_1, L_2 being the two abstract elements of $\mathcal{A}_G$. It is easy to see that if we write

$\theta = \theta^1 L_1 + \theta^2 L_2$, then

$$\theta^1 = \frac{1}{g_1} dg_1, \qquad \theta^2 = \frac{1}{g_1} dg_2 \;;$$

$$d\theta^1 = 0, \qquad d\theta^2 = -\frac{1}{g_1^2} dg_1 \wedge dg_2, \text{ and } \frac{1}{2}[\theta,\theta] = \frac{1}{g_1^2} dg_1 \wedge dg_2,$$

so that the Maurer-Cartan identity is verified.

The adjoint representation of the group G is defined as

$$[\mathrm{ad}(h)]^a_{\ b} = \left.\frac{\partial (hgh^{-1})^a}{\partial g^b}\right|_{g=e}, \quad a,b=1,2,\dots \dim G$$

It is easy to check that in our case

$$\mathrm{ad}(h) = \begin{pmatrix} 1 & h_2/h_1 \\ 0 & 1/h_1 \end{pmatrix}$$

Finally, let us give an example of different cross-sections and transition functions. Let us choose on some open set U_α of M

$$\phi_\alpha(p_1,p_2,p_3,p_4) = \begin{pmatrix} p_1 & p_2 \\ 0 & 1 \end{pmatrix} \in G$$

and on another U_β such that $U_\beta \cap U_\alpha \neq \emptyset$,

$$\phi_\beta(p_1,p_2,p_3,p_4) = \begin{pmatrix} p_3 & p_4 \\ 0 & 1 \end{pmatrix} \in G$$

The corresponding transition function is

$$\psi_{\alpha\beta} = \phi_\alpha \phi_\beta^{-1} = \begin{pmatrix} p_1 & p_2 \\ 0 & 1 \end{pmatrix}\begin{pmatrix} 1/p_3 & -p_4/p_3 \\ 0 & 1 \end{pmatrix} = \begin{pmatrix} p_1/p_3 & -\frac{p_1}{p_3}(p_4 - \frac{p_2 p_3}{p_1}) \\ 0 & 1 \end{pmatrix}$$

and it depends, as it should be, on x_1 and x_2 only.

Also $\psi_{\beta\alpha}(x) = \psi^{-1}_{\alpha\beta}(x) = \begin{pmatrix} 1/x_1 & x_2 \\ 0 & 1 \end{pmatrix}$

The cross-sections corresponding to ϕ_α and ϕ_β are defined as $\mu_\alpha(x) = \phi_\alpha^{-1}(x,e)$. As $\phi_\alpha(p_1,p_2,p_3,p_4) = (\pi(p), \phi_\alpha(p)) = ((x_1,x_2), \begin{pmatrix} p_1 & p_2 \\ 0 & 1 \end{pmatrix})$

then $\mu_\alpha(x_1,x_2)$ is a point p such that $x_1 = p_1/p_3$, $x_2 = p_4 - (p_2p_3)/p_1$, $p_1 = 1$ and $p_2 = 0$; i.e. $\mu_\alpha(x_1,x_2) = (1,0,1/x_1,x_2)$. Similarly, for μ_β one gets $\mu_\beta(x_1,x_2) = (x_1, -x_1x_2,1,0)$.

The image of the canonical 1-form θ obtained by the induced mapping $\psi_{\alpha\beta} = \theta_{\alpha\beta}$ is θ calculated at the value (g_1,g_2) given by $\psi_{\alpha\beta}$, i.e. $g_1 = x_1$, $g_2 = - x_1x_2$; i.e. $dg_1 = dx_1$, $dg_2 = - x_2dx_1 - x_1dx_2$. Then

$$\theta_{\alpha\beta} = (\frac{1}{x_1} dx_1)L_1 - (\frac{x_2}{x_1} dx_1 + dx_2)L_2$$

Finally, let us choose a connection 1-form ω on P(M,G). It is easy to check that ω defined as

$$\omega = (dp_3/p_3)L_1 + (dp_2/p_1)L_2$$

satisfies all the conditions imposed on a connection form. The image of ω on an open set U_α induced by μ_α is

$$\omega_\alpha = -(dx_1/x_1)\ L_1$$

and

$$\omega_\beta = -\left((x_2/x_1)dx_1 + dx_2\right)\ L_2$$

We leave to the reader the check-up of the formula

$$\omega_\beta = ad(\psi_{\alpha\beta}^{-1}(x))\omega_\alpha + \theta_{\alpha\beta}$$

In physical litterature the 1-form ω is denoted by $A = A_\mu dx^\mu = A_\mu^a dx^\mu L_a$ the Lie-algebra valued gauge potential; a cross-section is a mapping S : $M \to G$, and the transformation formula reads as

$$A_\mu \to S^{-1}A_\mu S + S^{-1}\partial_\mu S$$

The curvature 2-form $F_{\mu\nu} = \partial_\mu A_\nu - \partial_\nu A_\mu + [A_\mu, A_\nu]$ transforms homogeneously $F_{\mu\nu} \to S^{-1}F_{\mu\nu}S$.

One of the most important examples of connection and curvature is that of a principal fibre bundle of frames. The Lie algebra of GL(R,n) has an n^2 dimensional basis of matrices E^j_k, j,k = 1,2,...n, the dimension of the basis manifold of the bundle; they satisfy

$$[E^j_k, E^i_\ell] = \delta^i_k E^j_\ell - \delta^j_\ell E^i_k - \delta^j_k E^i_\ell + \delta^i_\ell E^j_k$$

The connection 1-form defined on the basis manifold M can be therefore decomposed as

$$\omega = \omega_i dx^i = \Gamma^k_{i\ell}\ E^\ell_k\ dx^i$$

It is easy to compute that

$$\Omega = D\omega = d\omega + \frac{1}{2}[\omega,\omega] =$$

$$= \frac{1}{2}(\partial_i\Gamma^k_{j\ell} - \partial_j\Gamma^k_{i\ell} + \Gamma^k_{im}\Gamma^m_{j\ell} - \Gamma^k_{jm}\Gamma^m_{i\ell})dx^i \wedge dx^j E^\ell_k =$$

$$= \frac{1}{2} R^k_{ij\ell}\ dx^i \wedge dx^j\ E^\ell_k.$$

$R^k_{ij\ell}$ is called the Riemann tensor, and $R_{j\ell} = R^i_{ij\ell}$ is called the Ricci tensor.

5. Classical Kaluza-Klein theory

Consider a principal fibre bundle $P(V_4,G)$ with the basis space V_4 being a Riemannian manifold endowed with the metric $g_{\mu\nu}$, $\mu,\nu = 0,1,2,3$, and the structural group G endowed with the Killing-Cartan metric g_{ab}, $a,b = 1,2,\ldots,N = \dim G$.

Let ω be the connection 1-form on $P(V_4,G)$, and π the canonical projection $\pi : P(V_4,G) \to V_4$. If e^a is the basis of left-invariant 1-forms on G and L_b the basis of the Lie algebra $\mathcal{A}_G$, then, if a local section $\mu : V_4 \to P(V_4,G)$ is chosen, ω can be written as

$$\omega = (\omega^b_\mu \, dx^\mu + \omega^b_a \, e^a) L_b$$

Let $\sigma_a = \sigma^b_a \, \partial_b$ be the set of left-invariant vector fields on G. The action of G on $P(V_4,G)$ induces the left-invariant fields defined on $P(V_4,G)$: we shall use the same letter σ to denote them :

$$\sigma_a = \sigma^\beta_a \, \partial_\beta \;, \qquad \alpha,\beta = 1,2,\ldots N+4$$

The fields σ_a must satisfy

$$[\sigma_a, \sigma_b] = C^d_{ab} \, \sigma_d$$

One can always choose the local coordinates in such a way, that

$$\sigma^\beta_a = (\sigma^\mu_a, \sigma^b_a) = (0, \sigma^b_a).$$

From now on we shall suppose $\sigma^\mu_a = 0$. Then the bracket $[\sigma_a, \sigma_b]$ reduces to

$$[\sigma_a, \sigma_b] = (\sigma^d_a \, \partial_d \, \sigma^f_b - \sigma^d_b \, \partial_d \, \sigma^f_a)\partial_f = C^g_{ab} \, \sigma^f_g \, \partial_f$$

i.e.
$$\sigma^d_a \, \partial_d \, \sigma^f_b - \sigma^d_b \, \partial_d \, \sigma^f_a = C^g_{ab} \, \sigma^f_g$$

The left-invariant fields σ_a define the mapping from $\mathcal{A}_G$ onto the left-invariant fields on $P(V_4,G)$:

$$\sigma_a = \sigma(L_a), \qquad [\sigma(L_a), \sigma(L_b)] = \sigma([L_a, L_b])$$

The 1-form ω defines the mapping from the tangent space of $P(V_4,G)$ into $\mathcal{A}_G$:

$$X \in TP(V_4,G) \to \omega(X) \in \mathcal{A}_G$$

The condition $\omega \cdot \sigma = Id_{\mathcal{A}_G}$ reduces itself in the coordinate system in which $\sigma^{\mu}_{a} \equiv 0$, to

$$\omega^a_b \, \sigma^b_d = \delta^a_d;$$

therefore, as we deal with two N x N non-singular matrices, also

$$\sigma^a_f \, \omega^f_b = \delta^a_b$$

The Maurer-Cartan equations for ω^a_b follow then directly from the bracket condition on σ_a :

$$\sigma^d_a \, \partial_d \, \sigma^f_b - \sigma^d_b \, \partial_d \, \sigma^f_a = C^g_{ab} \, \sigma^f_g$$

Multiplying by ω and using the fact that

$$(\partial_f \, \omega^a_b)\sigma^b_d = - \omega^a_b \, \partial_f \, \sigma^b_d, \quad (\partial_f \, \sigma^a_g)\omega^g_b = - \sigma^a_g \, \partial_f \, \omega^g_b$$

we arrive to

$$\partial_a \, \omega^d_b - \partial_b \, \omega^d_a = C^d_{gh} \, \omega^g_b \, \omega^h_a$$

which reads in a more elegant way as

$$d\omega + \frac{1}{2} [\omega,\omega] - 0$$

These are the Maurer-Cartan equations in coordinateless form (concerning only the invariant 1-form defined on G or on the typical fibre).

For any tangent vector of $TP(V_4,G)$ we define its horizontal and vertical parts :

$$\text{ver } X = \sigma \cdot \omega(X), \quad \text{hor } X = X - \sigma \cdot \omega(X)$$

It follows that the curvature 2-form Ω defined as

$$\Omega(X,Y) = D\omega(X,Y) = d\omega(\text{hor } X, \text{hor } Y)$$

has the following components in our chosen coordinate system in which $\sigma^{\mu}_{a}=0$:

$$\Omega^{d}_{ab} = \partial_a \omega^{d}_{b} - \partial_b \omega^{d}_{a} + C^{d}_{gh} \omega^{g}_{a} \omega^{h}_{b} \equiv 0$$

$$\Omega^{d}_{\mu b} = -\Omega^{d}_{b\mu} = \partial_\mu \omega^{d}_{b} - \partial_b \omega^{d}_{\mu} + C^{d}_{gh} \omega^{g}_{\mu} \omega^{h}_{b} \equiv 0$$

$$\Omega^{d}_{\mu\nu} = \partial_\mu \omega^{d}_{\nu} - \partial_\nu \omega^{d}_{\mu} + C^{d}_{gh} \omega^{g}_{\mu} \omega^{h}_{\nu}$$

In physics we use the notation A^{b}_{μ} for ω^{b}_{μ} and $F^{b}_{\mu\nu}$ for $\Omega^{b}_{\mu\nu}$. A^{b}_{μ} is called the gauge potential, $F^{b}_{\mu\nu}$ is the gauge field.

In its classical version, the Kaluza-Klein theory is based on the Einstein-Hilbert lagrangian density produced on $P(V_4,G)$ endowed with the canonical metric. This metric is defined in the following way :

if X,Y are tangent vectors to $P(V_4,G)$, then we put

$$g_P(X,Y) = g_{V_4}(d\pi(X),\ d\pi(Y)) + g_G(\omega(X),\ \omega(Y))$$

Locally the coordinates can be chosen in such a way, that

$$d\pi(X) \to X^\mu \ ; \quad \omega(X) \to \omega^{d}_{\beta} X^\beta = \omega^{d}_{\mu} X^\mu + \omega^{d}_{f} X^f$$

Then

$$g_P(X,Y) = \overset{(V4)}{g}_{\mu\nu} X^\mu Y^\nu + \overset{(G)}{g}_{ab}(\omega^{a}_{\mu} X^\mu + \omega^{a}_{g} X^g)(\omega^{b}_{\nu} Y^\nu + \omega^{b}_{h} Y^h)$$

This can be written in a matrix form as

$$\overset{(P)}{g}_{\alpha\beta} = \begin{pmatrix} \overset{(V4)}{g}_{\mu\nu} + \overset{(G)}{g}_{ab} \omega^{a}_{\mu} \omega^{b}_{\nu} & \overset{(G)}{g}_{ab} \omega^{a}_{d} \omega^{b}_{\nu} \\ \overset{(G)}{g}_{ab} \omega^{a}_{\mu} \omega^{b}_{g} & \overset{(G)}{g}_{ab} \omega^{a}_{d} \omega^{b}_{g} \end{pmatrix}$$

The inverse is then

$$\overset{(P)}{g}{}^{\alpha\beta} = \begin{pmatrix} \overset{(V4)}{g}{}^{\mu\nu} & -\overset{(V4)}{g}{}^{\mu\nu} \omega^{d}_{\nu} \sigma^{b}_{d} \\ \overset{(V4)}{g}{}^{\mu\nu} \omega^{d}_{\mu} \sigma^{c}_{d} & \overset{(G)}{g}{}^{cd} + \overset{(V4)}{g}{}^{\mu\nu} \sigma^{b}_{d} \sigma^{c}_{f} \omega^{d}_{\mu} \omega^{f}_{\nu} \end{pmatrix}$$

From now on, we turn to physicists' notation, replacing ω^b_μ by A^b_μ (the gauge potential), and $\Omega^b_{\mu\nu}$ by $F^b_{\mu\nu}$ (gauge field strength tensor).

The Christoffel symbols of our metric are given by

$$\{^{\alpha}_{\beta\gamma}\} = \frac{1}{2} g^{\alpha\delta} (\partial_\beta g_{\delta\gamma} + \partial_\gamma g_{\beta\delta} - \partial_\delta g_{\beta\gamma})$$

and can be computed explicitly, but are quite cumbersome. It is much easier to introduce the non-holomic system of frames which diagonalizes the metric tensor. Let us denote the components of $\overset{P}{g}$ by $g_{\alpha\beta} = (g_{\mu\nu}, g_{\mu b}, g_{bd})$, whereas the components of $\overset{V_4}{g}$ will be denoted by $\overset{0}{g}_{\mu\nu}$, and the components of $\overset{G}{g}$ by g_{ab}. We want to define a local transformation of linear frame such that

$$g_{\alpha\beta} = U^\delta_\alpha U^\gamma_\beta \tilde{g}_{\delta\gamma}, \qquad g^{\alpha\beta} = \overset{-1}{U}{}^\alpha_\delta \overset{-1}{U}{}^\beta_\gamma \tilde{g}^{\delta\gamma}$$

and

$$\tilde{g}_{\delta\gamma} = \begin{pmatrix} \overset{0}{g}_{\mu\nu} & 0 \\ 0 & \overset{0}{g}_{ab} \end{pmatrix}$$

It is easy to see that the matrices U^δ_α and $\overset{-1}{U}{}^\alpha_\beta$ must be of the following form :

$$U^\mu_\nu = \delta^\mu_\nu, \quad U^\mu_a = 0, \quad U^a_\mu = A^a_\mu, \quad U^a_b = \omega^a_b$$

$$\overset{-1}{U}{}^\nu_\lambda = \delta^\nu_\lambda, \quad \overset{-1}{U}{}^\lambda_b = 0, \quad \overset{-1}{U}{}^b_\lambda = -\sigma^b_d A^d_\lambda, \quad \overset{-1}{U}{}^b_d = \sigma^b_d$$

The local frame is then given as

$$\mathcal{D}_\alpha = \overset{-1}{U}{}^\beta_\alpha \partial_\beta,$$

i.e.

$$\mathcal{D}_\mu = \partial_\mu - A^d_\mu \sigma^b_d \partial_b$$

$$\mathcal{D}_b = \sigma^f_b \partial_f$$

Note that

$$[\mathcal{D}_\mu, \mathcal{D}_\nu] = -F^b_{\mu\nu} \mathcal{D}_b$$

$$[\mathcal{D}_a, \mathcal{D}_b] = C^f_{ab} \mathcal{D}_f, \qquad [\mathcal{D}_\mu, \mathcal{D}_a] = 0$$

Under this local transformation of frames the connection coefficients transform according to

$$\widetilde{\{{}^{\alpha}_{\beta\gamma}\}} = \overset{-1}{U}{}^{\beta}_{\beta}\,\overset{-1}{U}{}^{\gamma}_{\gamma}\,U^{\alpha}_{\alpha}\,\{{}^{\alpha}_{\beta\,\gamma}\} + U^{\alpha}_{\alpha}\,\partial_{\beta}\,\overset{-1}{U}{}^{\alpha}_{\gamma}$$

The components of the 1-form ω are particularly simple in this system:

$$\tilde{A}^{h}_{\mu} = \overset{-1}{U}{}^{\nu}_{\mu}\,A^{h}_{\nu} + \overset{-1}{U}{}^{b}_{\mu}\,\omega^{h}_{b} = \delta^{\nu}_{\mu}\,A^{h}_{\nu} - A^{d}_{\mu}\,\sigma^{b}_{d}\,\omega^{h}_{b} =$$

$$= A^{h}_{\mu} - A^{h}_{\mu} = 0,$$

and

$$\tilde{\omega}^{h}_{b} = \overset{-1}{U}{}^{\mu}\,\omega^{h}_{\mu} + \overset{-1}{U}{}^{d}_{b}\,\omega^{h}_{d} = 0 + \sigma^{d}_{b}\,\omega^{h}_{d} = \delta^{h}_{b}$$

The components of the linear connection of $\overset{P}{g}$ expressed in this local frame are

$$\Gamma^{a}_{bd} = \frac{1}{2}\,C^{a}_{bd}, \quad \Gamma^{a}_{\mu\nu} = \frac{1}{2}\,F^{a}_{\nu\mu}, \quad \Gamma^{\nu}_{\mu b} = \Gamma^{\nu}_{b\mu} = \frac{1}{2}\,\overset{o}{g}_{bd}\,\overset{o}{g}{}^{\nu\lambda}\,F^{d}_{\mu\lambda}$$

and $\Gamma^{\mu}_{\nu\lambda} = \{{}^{\mu}_{\nu\lambda}\}$, i.e. Christoffel symbols of $\overset{o}{g}_{\mu\nu}$.

Now we are able to compute the components of the Riemann tensor. It is useful to remind that in a non-holomic system of frames the expression for the Riemann tensor is

$$R_{\alpha\beta}{}^{\gamma}{}_{\delta} = \partial_{\alpha}\Gamma^{\gamma}_{\beta\delta} - \partial_{\beta}\Gamma^{\gamma}_{\alpha\delta} + \Gamma^{\gamma}_{\alpha\eta}\,\Gamma^{\eta}_{\beta\delta} - \Gamma^{\gamma}_{\beta\eta}\,\Gamma^{\eta}_{\alpha\delta} - B^{\eta}_{\alpha\beta}\,\Gamma^{\gamma}_{\eta\delta}$$

where

$$[\mathcal{D}_{\alpha}, \mathcal{D}_{\beta}] = B^{\eta}_{\alpha\beta}\,\mathcal{D}_{\eta}$$

We shall give here the components of $R_{\alpha\beta\gamma\delta} = g_{\gamma\eta}\,R_{\alpha\beta}{}^{\eta}{}_{\delta}$:

$$R_{\mu\nu\lambda\rho} = \overset{(V_4)}{R}_{\mu\nu\lambda\rho} - \frac{1}{4}\,\overset{o}{g}_{ab}(F^{a}_{\mu\lambda}\,F^{b}_{\nu\rho} - F^{a}_{\nu\lambda}\,F^{b}_{\mu\rho} + 2F^{a}_{\mu\nu}\,F^{b}_{\lambda\rho})$$

$$R_{\mu b\lambda\rho} = -\frac{1}{2}\,\overset{o}{g}_{ab}\,D_{\mu}\,F^{a}_{\lambda\rho}$$

$$R_{\mu\nu ab} = \frac{1}{4}\,\overset{o}{g}_{ac}\,\overset{o}{g}_{bd}\,\overset{o}{g}\,(F^{d}_{\mu\rho}\,F^{c}_{\nu\lambda} - F^{d}_{\nu\rho}\,F^{c}_{\mu\lambda}) + \frac{1}{2}\,\overset{o}{g}_{ac}\,F^{d\mu\nu}C^{c}_{db}$$

$$R_{\mu ab\nu} = -\frac{1}{4}\,\overset{o}{g}_{bc}(C^{c}_{ad}\,F^{d}_{\mu\nu} + \overset{o}{g}_{ad}\,\overset{o}{g}{}^{\lambda\rho}\,F^{c}_{\mu\lambda}\,F^{d}_{\nu\rho})$$

$$R_{abcd} = -\frac{1}{4}\,\overset{o}{g}_{cg}\,C^{f}_{ab}\,C^{g}_{fd}$$

Here

$$D_\mu F^a_{\lambda\rho} = \nabla_\mu F^a_{\lambda\rho} + C^a_{bd} A^b_\mu F^d_{\lambda\rho},$$

with ∇_μ denoting the covariant derivative with respect to Christoffel connaction in V_4. Other components ($R_{\mu abc}$) vanish; besides, the Riemann tensor even in the non-holonomic frame has still its symmetry properties :

$$R_{\alpha\beta\,\gamma\delta} = R_{\gamma\delta\,\alpha\beta} \; ; \quad R_{\alpha\beta\,\gamma\delta} = -R_{\alpha\beta\,\delta\gamma}$$

$$R_{\alpha\beta\,\gamma\delta} = -\,R_{\beta\alpha\,\gamma\delta}$$

The Ricci tensor defined by $R_{\alpha\gamma} = R_{\alpha\,\delta\,\gamma}^{\;\;\delta}$ is equal to :

$$R_{\mu\lambda} = \overset{(V_4)}{R}_{\mu\lambda} + \frac{1}{2} \overset{o}{g}_{bd} \overset{o}{g}^{\nu\rho} F^b_{\mu\nu} F^d_{\lambda\rho}$$

$$R_{\mu b} = R_{b\mu} = \frac{1}{2} \overset{o}{g}_{bd} \overset{o}{g}^{\nu\rho} D_\nu F^d_{\rho\mu}$$

$$R_{ab} = -\frac{1}{4} C^f_{ag} C^g_{fb} - \frac{1}{4} \overset{o}{g}_{ad} \overset{o}{g}^{\lambda\rho} \overset{o}{g}^{\mu\nu} F^c_{\mu\lambda} F^d_{\nu\rho} \overset{o}{g}_{bc}$$

Finally, the scalar curvature of $P(V_4,G)$ is equal to

$$\overset{(P)}{R} = \overset{(V4)}{R} - \frac{1}{4} \overset{o}{g}_{ab} \overset{o}{g}^{\nu\rho} \overset{o}{g}^{\mu\lambda} F^a_{\mu\nu} F^b_{\rho\lambda} - \frac{1}{4} \overset{o}{g}^{ab} C^f_{ag} C^g_{fb}$$

The determinant of $\overset{P}{g}$ is equal to the product of the determinants of $\overset{V_4}{g}$ and of $\overset{G}{g}$:

$$\det (g_{\alpha\beta}) = \det (\overset{o}{g}_{\mu\nu}) \times \det (\overset{o}{g}_{ab})$$

The Lagrangian density in $P(V_4,G)$ becomes :

$$|\det \overset{P}{g}| \overset{(P)}{R} = |\det \overset{V4}{g}| \; |\det \overset{G}{g}| (\overset{(4)}{R} - \frac{1}{4} F^a_{\mu\nu} F^{\mu\nu}_a - \frac{1}{4} \overset{o}{g}^{ab} C^f_{ag} C^g_{fb})$$

The last term plays the rôle of cosmological constant, whose value is enormous and can not be justified by any means. If we temporarily forget about this problem, there is still the overdetermination of the Euler-Lagrange system derived from such a variational integrand in $P(V_4,G)$.

As a matter of fact, the system of Einstein equations

$$\overset{(P)}{R}_{\alpha\beta} - \frac{1}{2}\overset{(P)}{g}_{\alpha\beta}\overset{(P)}{R} = \overset{(P)}{g}_{\alpha\beta}$$

with Λ proportional to the last term, $\frac{1}{4} g^{ab} C^{f}_{ag} C^{g}_{fb}$, yields three groups of equations :

$$\overset{(4)}{R}_{\mu\nu} - \frac{1}{2}\overset{o}{g}_{\mu\nu}\overset{(4)}{R} = -\frac{1}{2}(\overset{o}{g}^{\lambda\rho} F^{a}_{\mu\rho} F_{a\nu\lambda} - \frac{1}{4} F^{a}_{\lambda\rho} F^{\lambda\rho}_{a}\overset{o}{g}_{\mu\nu}) + \Lambda\overset{o}{g}_{\mu\nu}$$

$$\overset{(P)}{R}_{\mu b} = 0 \text{ equivalent to } D^{\mu} F^{b}_{\mu\nu} = 0$$

and

$$\overset{(P)}{R}_{ab} - \frac{1}{2}\overset{(P)}{g}_{ab}\overset{(P)}{R} = \Lambda\overset{o}{g}_{ab}$$

This system is implemented by the Bianchi identities

$$D_{\mu} F^{b}_{\lambda\rho} + D_{\rho} F^{b}_{\mu\lambda} + D_{\lambda} F^{b}_{\rho\mu} = 0$$

and

$$\nabla_{\alpha} R_{\beta\gamma\ \varepsilon}^{\ \ \delta} + \nabla_{\gamma} R_{\alpha\beta\ \varepsilon}^{\ \ \delta} + \nabla_{\beta} R_{\gamma\alpha\ \varepsilon}^{\ \ \delta} = 0$$

These identities are satisfied by definition of $F^{b}_{\mu\nu}$ and $R_{\alpha\beta\ \delta}^{\ \ \gamma}$

A "minimal" programme, followed by Kaluza and Klein, consisted in integrating first over the group variables, which in view of the invariance of the expression yields a constant factor, and then apply to the resulting integral in V_4 the variational principle with respect to the independent components of $\overset{(P)}{g}_{\alpha\beta}$, i.e. $\overset{o}{g}_{\mu\nu}$ and $\overset{P}{g}_{\mu b} = A_{b\mu}$. The result is the Einstein-Yang-Mills system in $P(V_4,G)$.

Like in General Relativity, the field equations without sources can be implemented by the equations of motion of point-like particles. The most natural definition, which coincides with the definition of motion in General Relativity when the gauge field $F^{b}_{\mu\nu}$ vanishes, is the geodesic equation :

$$\frac{d^2x^{\alpha}}{ds^2} + \Gamma^{\alpha}_{\beta\gamma}\frac{dx^{\beta}}{ds}\frac{dx^{\gamma}}{ds} = 0$$

with

$$ds^2 = \overset{(P)}{g}_{\alpha\beta}\, dx^{\alpha}\, dx^{\beta}$$

When written explicitely, the geodesic equation separates into two pieces,

$$\frac{d^2x^\mu}{ds^2} + \{^{\mu}_{\nu\lambda}\} \frac{dx^\nu}{ds}\frac{dx^\lambda}{ds} + F^{\mu}_{b\nu} \frac{dx^b}{ds}\frac{dx^\nu}{ds} = 0$$

and

$$\frac{d^2x^b}{ds^2} = 0, \text{ from which follows } \frac{dx^b}{ds} = Q^b = \text{Const.}$$

Therefore,

$$\frac{d^2x^\mu}{ds^2} + \{^{\mu}_{\nu\lambda}\} \frac{dx^\nu}{ds}\frac{dx^\lambda}{ds} + Q_b\, F^{b\mu}_{\ \ \rho} \frac{dx^\rho}{ds} = 0$$

In five-dimensional case Q_5 should be identified with e/m, the ratio of electric charge versus mass of the particle, in order to coincide with the expression for the Lorentz force.

There is another interesting fact, which until now seems to possess mostly aesthetical value. The Yang-Mills equations $D^\mu F^b_{\mu\nu}$ can be deduced from an invariant equation of second order, quite naturally defined on the Riemannian manifold $P(V_4,G)$, and applied to the 1-form ω.

Let us prove the following theorem :

The equations $D^\mu F^b_{\mu\nu} = 0$ are equivalent with the equations

$$(d\delta + \delta d)\, \omega^h = \frac{1}{2} g^{ad} C^f_{ab} C^h_{df}\, \omega^b$$

if the structure constants are traceless : $C^b_{ab} = 0$, and the metric on the fibre is a Killing-Cartan metric g_{ab} with det$(g_{ab}) \neq 0$, i.e. if the group is compact and semi-simple.

The proof is made by explicit calculation. The operator $d\delta + \delta d$ (Laplace-Beltrami operator) is defined via the Hodge duality $*$ as

$$\delta = *^{-1} d *\ ,\quad d\delta + \delta d = d *^{-1} d * + *^{-1} d * d$$

The operator $*$ contains in its definition the metric $g_{\alpha\beta}$, and therefore the connection 1-form ω of the fibre bundle. The explicit expression depends on the degree of the exterior form on which $d\delta + \delta d$ acts. If ϕ is a 0-form, then

$$(d\delta + \delta d)\, \phi = g^{\alpha\beta} \nabla_\alpha \nabla_\beta \phi$$

For a 1-form ω we get, in the component language,

$$[(d\delta + \delta d)\omega]_\gamma = g^{\alpha\beta}[\nabla_\alpha \nabla_\beta \omega_\gamma + (\nabla_\gamma \nabla_\alpha - \nabla_\alpha \nabla_\gamma)\omega_\beta] =$$

$$= g^{\alpha\beta} \nabla_\alpha \nabla_\beta \omega_\gamma + R_\gamma{}^\delta \omega_\delta$$

where $R_\gamma{}^\delta = g^{\delta\beta} R_{\gamma\beta}$ is the Ricci tensor.

For a p-form, this formula can be easily generalized with the matrix R^δ_γ acting subsequently on each index.

Now, as we know all the components in the non-holonomic frame, and recalling that in this frame $\tilde{\omega}^h_\mu = 0$, $\omega^h_d = \delta^h_d$, we obtain easily that

$$(d\delta + \delta d)\omega^h_\mu = (\overset{o}{g}{}^{\rho\lambda} D_\rho F^b_{\lambda\mu}) \omega^h_b$$

under condition that $C^b_{fb} \equiv 0$,

and

$$([d\delta + \delta d]\omega)^h_b = \frac{1}{2} (g^{ad} C^f_{ab} C^g_{df}) \omega^h_g$$

For a simple or semi-simple compact group the last expression is proportional to the identity operator on the group algebra; for the SU(n) groups, in particular, the coefficient is $\frac{N(N-1)}{2}$. In the first equation we can add to the right-hand side any constant multiplied by $\tilde{\omega}^h_\mu = 0$ in this sytem; on the other hand, the whole equation is invariant, and therefore its form will not alter under the change of frames. Therefore, if

$D^\lambda F^b_{\lambda\mu} = 0$, the equation $(d\delta + \delta d)\omega = \Lambda\omega$ is satisfied and vice versa.

All the results obtained until now supposed that the metric on the fibre was an invariant Killing-Cartan metric of the structural group G; in calculations, it meant that $\partial_\mu g_{ab} \equiv 0$. Let us give the general expression for the Kaluza-Klein variational principle when g_{ab} depends on the basis space-time variables. In such a case, the scalar curvature of P(M,G) reads :

$$\overset{P}{R} = \overset{V_4}{R} - \frac{1}{4} g_{ab} g^{\mu\nu} g^{\lambda\rho} F^a_{\mu\lambda} F^b_{\nu\rho} + R_G + D_G$$

where

$$R_G = - g^{ab}(\frac{1}{2} C^c_{ad} C^d_{bc} + \frac{1}{4} g^{df} C_{ch} C^c_{ad} C^b_{bf} + C^d_{ad} C^f_{bf})$$

and

$$D_G = \frac{1}{4} g^{ab} g^{cd} (\mathcal{D}_\mu g_{ac} \mathcal{D}^\mu g_{bd} + \mathcal{D}_\mu g_{ab} \mathcal{D}^\mu g_{cd}) + \nabla_\mu (g^{ab} \mathcal{D}^\mu g_{ab})$$

with

$$\mathcal{D}_\mu g_{ac} = \partial_\mu g_{ac} + C^f_{ad} \omega^d_\mu g_{fc} + C^f_{cd} \omega^d_\mu g_{af}$$

In the case of the unimodular group G, the trace of the structure constants vanishes :

$$C^b_{ab} = C^b_{ba} \equiv 0$$

and

$$R_G = - \frac{1}{2} g^{ab} C^c_{ad} C^d_{bc}$$

This point is also important in the derivation of the expression for the Laplace-Beltrami operator acting on ω : if $C^b_{ab} \neq 0$, the coefficients

$$\frac{1}{2} g^{ad} C^f_{ab} C^g_{df}$$

have to be replaced by the full Ricci tensor, $R_b{}^g{}_p$. When G = U(1), the scalar curvature R contains the contribution of the scalar field $g_{55}(x^\mu) = e^{2\phi/3}$

$$\overset{5}{R} \; |\det g| = |\det g_{\mu\nu}| \; [\overset{4}{R} - \frac{1}{6} g^{\lambda\rho} \partial_\lambda \phi \, \partial_\rho \phi - \frac{1}{4} e^\phi F_{\mu\nu} F^{\mu\nu}]$$

which is the lagrangian of Brans and Dicke.

6. Developments and generalisations

The ultimate ambition of Kaluza-Klein type theories is to explain all known physical fields as natural components of some multidimensional geometrical objects. We have just seen how the four-dimensional metric (gravitational field) and non-abelian fields (gauge potential A^b_μ) can be incorporated into a multi-dimensional metric over a bundle with connection. Enlarging the theory by relieving the inconsistence which follows from the assumption $g_{ab} = 0$, we can introduce as many as N(N+1)/2 (N = dim G) scalar fields, but unfortunately, it is difficult to put them into correspondence with any known fields. On the contrary, we would like to have some explanation of the Higgs scalars, which play such an important role in particle physics, because they give rise to the symmetry-breaking potential and are thus responsible for the observable masses of weak bosons in the Weinberg-Salam model.

The Higgs fields can be introduced quite naturally with the following generalisation of the Kaluza-Klein theory : it is sufficient to suppose that the multi-dimensional space is a fibre bundle whose basis is a fibre bundle, too. Let us take the simplest example : consider the original fibre bundle introduced by Kaluza, $(P_1(M,U(1))$ as a basis of another fibre bundle, with a non-abelian structure group G : $P_2(P_1(M,U_{(1)}), G)$. Let the coordinates in M be x^μ, in U(1), x^5, and y^b in G. Suppose that the connection in P_1 vanishes ; the connection in P_2 will have the components A^b_μ and A^b_5. Its curvature can be decomposed into three types :

$$F^b_{\mu\nu} = \partial_\mu A^b_\nu - \partial_\nu A^b_\mu + C^b_{df} A^d_\mu A^f_\nu$$

$$F^b_{\mu 5} = \partial_\mu A^b_5 - {}_5A^b_\mu + C^b_{df} A^d_\mu A^f_5$$

$$F^b_{55} = 0$$

Denoting A^b_5 by ϕ^b and supposing that A^b_μ is invariant along the U(1)-fibres, i.e. $\partial_5 A^b_\mu = 0$, we have

$$F^b_{\mu 5} = \partial_\mu \phi^b + C^b_{df} A^d_\mu \phi^f = \mathcal{D}_\mu \phi^b$$

The gauge part of the lagrangian becomes then

$$- \frac{1}{4} g_{ab} F^a_{\mu\nu} F^a \quad F^{b\mu\nu} - \frac{1}{2} g_{ab} \mathcal{D}_\mu \phi^a \mathcal{D}^\mu \phi^b$$

and gives the correct minimal interaction of the Higgs scalar multiplet in adjoint representation of G with the Yang-Mills field. There is no symmetry-breaking potential, because $F^b_{55} = 0$. Nevertheless, the above lagrangian yields a solution in a soliton form, identified with the magnetic monopole of Prasad-Sommerfield, if G = SU(2), a,b = 1,2,3; with a spherically symmetric ansatz

$$A^b_0 = 0, \quad A^b_i = \frac{\varepsilon^b_{ik} x^k}{er^2} (1 - K), \quad \phi^b = \frac{\sigma^b}{er^2} H(r)$$

Then the lagrangian density reduces to

$$L = \frac{1}{e^4} \quad K'^2 + \frac{(1 - K^2)^2}{2r^2} + \frac{(rH' - H)^2}{2r^2} + \frac{H^2K^2}{r^2} \quad ,$$

the Euler-Lagrange equations are

$$r^2K'' = K(K^2 + H^2 - 1)$$

$$r^2H'' = 2HK^2$$

and have the solution $K = Cr/\text{sh}\ Cr$, $H = Cr \coth Cr - 1$, C arbitrary constant of dimension cm^{-1}.

The magnetic field B_k can be identified with the combination

$$B_k = \frac{1}{2} E_{k\ell m}(\phi_a F^{a\ell m} - \phi_a \varepsilon^{abc} \mathcal{D}^\ell \phi_b \mathcal{D}^m \phi_c).$$

This solution describes a magnetic monopole whose charge is proportional to the inverse of e, and energy proportional to C.

The Prasad-Sommerfield monopole is a limit case of 't Hooft's monopole, whose lagrangian contains a potential term of the form $V(\phi) = -\frac{\mu^2}{2}\phi_a\phi^a + \frac{\lambda}{4}(\phi_a\phi^a)^2$. The same kind of potential appears in the unified theory of electroweak interactions (the Weinberg-Salam model). This potential can be obtained in a more general scheme of a double fibre bundle with two non-abelian structure groups.

Let us consider such a double bundle $P_2(P_1(M,G_1),G_2)$. Let $a,b = 1,2,\ldots,N_1 = \dim G_1$; $A,B = 1,2,\ldots,N_2 = \dim G_2$. Now the connection 1-form in P_2can be decomposed as A^B_μ, A^B_b. We shall assume that A^B_b has some well-defined invariance properties with respect to the action of G_1, which generalises the former condition $\partial_5 = 0$. As A^B_b is a 1-form on G_1, we shall suppose that it can be decomposed along the Maurer-Cartan forms on G_1 :

$$A^B_b = \phi^B_d(x^\mu)\omega^d_b$$

with ϕ^B_d a matrix of scalar fields depending on x only. Then we can compute explicitly the components F^B_{bd} of the curvature 2-form :

$$F^B_{bd} = \partial_b A^B_d - \partial_d A^B_b + C^B_{EF} A^E_b A^F_d =$$

$$= \phi^B_f(\partial_b\omega^f_d - \partial_d\omega^f_b) + C^B_{EF}\phi^E_g\phi^F_h\omega^g_b\omega^h_d$$

Now we can use the Maurer-Cartan equations :

$$\partial_b\omega^f_d - \partial_d\omega^f_b = -C^f_{gh}\omega^g_b\omega^h_d$$

and obtain

$$F^B_{bd} = \omega^g_b\omega^h_d\ (C^B_{EF}\phi^E_g\phi^F_h - C^f_{gh}\phi^B_f)$$

The part of the lagrangian $-\frac{1}{4} F^B_{bd}F^{bd}_B$ will yield now a quartic expression in the fields ϕ^B_d whose maximum is at $\phi = 0$, and the minimum corresponds to a subset of values of $\phi \neq 0$. For example, if we choose $G_1 = G_2 = SU(2)$, then ϕ^B_d becomes a 3 x 3 matrix; the expression $-\frac{1}{4} F^B_{bd}F^{bd}_B$ becomes proportional to

$$2Tr\,(\phi\phi^T) + Tr\,(\phi\phi^T)^2 - Tr\,(\phi\phi^T\phi\phi^T) - 12\,\det\phi$$

If we choose $\phi^B_d = \phi(x)\delta^B_d$, we get

$$6\phi^2\,(\phi - 1)^2$$

In the Weinberg-Salam theory there are no monopole solutions, because the Higgs field is not in the adjoint representation of SU(2), but a complex doublet of the lowest-dimensional non-trivial representation. Such a doublet can be constructed in a Kaluza-Klein theory on a double fibre bundle if some further assumptions are made concerning the internal symmetry of the bundle. This has been achieved in the construction proposed by N. Manton in 1979.

7. Manton's construction

Consider the six-dimensional space-time with the metric

$$g_{\alpha\beta} = \text{diag}\,(+1,-1,-1,-1,-R^2,-R^2\sin^2\psi)$$

Throughout this paper the greek letters α,β, run from 1 to 6, the greek letters μ, ν... run from 0 to 3, and latin letters a,b,... from 1 to 2, and latin letters i,j from 1 to 3.

The two-dimensional sphere S^2 is parametrized with the two angles ψ and ϕ; its isometries are generated by three independent Killing fields

$$\xi_1 = \xi^b_1\partial_b = \cos\phi\,\frac{\partial}{\partial\psi} - \frac{\cos\psi\,\sin\phi}{\sin\psi}\,\frac{\partial}{\partial\phi}$$

$$\xi_2 = \xi^b_2\partial_b = -\sin\phi\,\frac{\partial}{\partial\psi} - \frac{\cos\psi\,\cos\phi}{\sin\psi}\,\frac{\partial}{\partial\phi}$$

$$\xi_3 = \xi^b_3\partial_b = \frac{\partial}{\partial\phi}$$

whose Lie brackets satisfy the commutation rules of the SO(3)-Lie algebra:

$$[\xi_i,\,\xi_j] = \varepsilon^k_{ij}\xi_k$$

The connection 1-form on $P(M_4 \times S^2, G)$ has the components

$$A^B_\alpha = (A^B_\mu, A^B_b)$$

and the Yang-Mills field tensor

$$F^B_{\alpha\beta} = \partial_\alpha A^B_\beta - \partial_\beta A^B_\alpha + C^B_{EF} A^E_\alpha A^F_\beta$$

where A,B,... run from 1 to N = dim G. The gauge group G is yet unspecified, but supposed to be compact and semi-simple.

The spherical symmetry requirement for the 2-form F means its invariance with respect to SO(3), i.e.

$$\mathcal{L}_{\xi_k} F = 0$$

However, this does not imply that the Lie derivatives of A^B_α with respect to ξ_k must vanish; it is enough to require that

$$\mathcal{L}_{\xi_k} A^B = (\partial_\alpha \xi^\beta_k) A^B_\beta + \xi^\beta_k \partial_\beta A^B_\alpha = \partial_\alpha W^B_k + C^B_{EF} A^E_\alpha W^F_k$$

which means that the Lie derivative of A is a pure gauge. Note that $\xi^\beta_k = (\xi^\mu_k, \xi^b_k) = (0, \xi^b_k)$; W^B_k are three G-Lie-algebra valued scalar fields. The solution found by Manton is the following :

$$W^B_1 = \phi^B_3 \frac{\sin\phi}{\sin\psi}, \qquad W^B_2 = \phi^B_3 \frac{\cos\phi}{\sin\psi}, \qquad W^B_3 = 0$$

From now on we shall drop the Lie algebra indices, denoting by ϕ_3 the corresponding Lie-algebra valued field.

A_α must be of the following form :

$$A_\mu = A_\mu(x^\nu), \quad A_\psi = -\Phi_1(x), \quad A_\phi = \Phi_2(x)\sin\psi - \Phi_3 \cos\psi$$

with Φ_1, Φ_2, Φ_3 satisfying the following constraints :

$$[\Phi_3, A_\mu] = 0, \quad [\Phi_3, \Phi_1] = +\Phi_2, \ [\Phi_3, \Phi_2] = -\Phi_1$$

The brackets are taken in the Lie algebra of G.

If thus constrained Yang-Mills potentials A_α are inserted into the expression $F^A_{\alpha\beta} F^{\alpha\beta}_A$, it shall split into three parts :

$$F^A_{\alpha\beta}F^{\alpha\beta}_A = F^A_{\mu\nu}\ F^{\mu\nu}_A + 2\,\mathcal{D}_\mu\phi^A_k\mathcal{D}^\mu\phi^k_A + (\varepsilon_{k\ell m}\phi^A_m + [\phi_k,\phi_\ell]^A)(\varepsilon^{k\ell m}\phi_{mA} + [\phi^k,\phi^\ell]_A)$$

where we identified $\mathcal{D}_\mu\phi_k = \partial_\mu\phi_k + [A_\mu,\ \phi_k]$

These constraints reduce the gauge group G to the little group of ϕ_3, i.e. the generators commuting with ϕ_3.

The group G has to be chosen in such a way, that the little subgroup of ϕ_3 be exactly SU(2) x U(1); one must have at least two commuting generators in the Lie algebra of G, i.e. it has to be of rank 2. Choosing the basis in the Cartan subalgebra H_1, H_2 such that

$$K(H_a,\ H_b) = \delta_{ab}$$

where K(,) means the Killing product in the Lie algebra, all other generators can be put into one-to-one correspondence with roots. Denoting by

$$h_\gamma = \frac{2\gamma_b H_b}{(\gamma\cdot\gamma)}$$

one can find two eigenvectors x_γ, $x_{-\gamma}$ corresponding to the roots γ and $-\gamma$, so that

$$[h_\gamma,\ x_\gamma] = 2x_\gamma, \qquad [h_\gamma,\ x_{-\gamma}] = -2x_{-\gamma} \quad \text{and} \quad [x_\gamma,\ x_{-\gamma}] = h_\gamma$$

Then the generators

$$\frac{i}{2}\,(x_\gamma + x_{-\gamma}), \qquad \frac{i}{2}\,(x_\gamma - x_{-\gamma}), \qquad \frac{i}{2}\,h_\gamma$$

span an SU(2) subalgebra of G.
One can also choose another element of the Cartan subalgebra, h, so that

$$K(h,h) = k(h_\gamma,\ h_\gamma) = \frac{4}{(\gamma\cdot\gamma)} \text{ and } k(h,\ h_\gamma) = 0$$

Then the Yang-Mills potential dispalying the SU(2) x U(1) gauge symmetry can be chosen as

$$A_\mu = \frac{i}{2}\,(A^-_\mu x_\gamma + A^+_\mu x_{-\gamma} + A^3_\mu h_\gamma + B_\mu h)$$

Later, the rescaled fields A^-_μ and A^+_μ will be identified with the W boson field, whereas the usual combinations of A^3_μ and B_μ will be identified with the Z-boson and the electromagnetic field $A_\mu{}^m$.

In order to have a complex Higgs doublet, one has to choose two other couples of generators corresponding to roots α and $\beta = x - \gamma$, x_α, $x_{-\alpha}$, x_β, $x_{-\beta}$, and with the commutation relations given by

$$[x_\gamma, x_\beta] = x_\alpha, \quad [x_{-\gamma}, x_{-\beta}] = -x_{-\alpha}$$

$$[x_{-\gamma}, x_\alpha] = x_\beta, \quad [x_\gamma, x_{-\alpha}] = -x_\beta$$

and

$$[x_\gamma, x_\alpha] = [x_{-\gamma}, x_{-\alpha}] = [x_{-\gamma}, x_\beta] = [x_\gamma, x_{-\beta}] = 0$$

Then the components of Φ can be chosen as

$$\Phi = \Phi_1 + i\Phi_2 = \phi_1 x_\alpha + \phi_2 x_\beta$$

$$\tilde{\Phi} = \Phi_1 - i\Phi_2 = -\phi_1 x_{-\alpha} - \phi_2 x_{-\beta}$$

and

$$\Phi_3 = \frac{i}{2} Z(h_\alpha + h_\beta), \text{ with } Z = \frac{(\alpha\cdot\alpha)}{2(\alpha\cdot\alpha) - (\gamma\cdot\alpha)}$$

where $h_\alpha = [x_\alpha, x_{-\alpha}]$ and $h_\beta = [x_\beta, x_{-\beta}]$

The roots α, β, γ corresponding to the three Lie groups, SU(3), O(5) and G_2, are displayed on fig. 1.

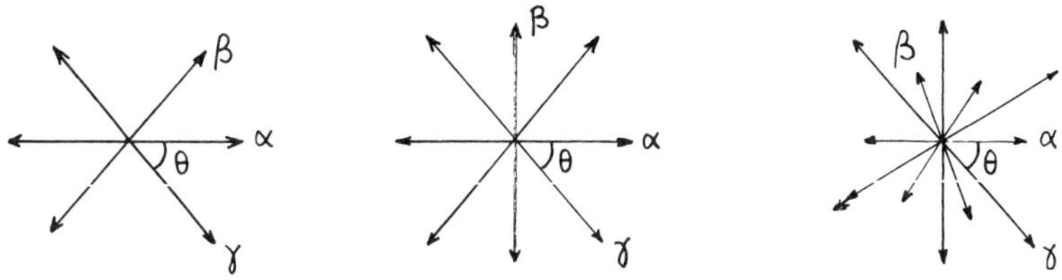

SU(3) O(5) G_2

Fig. 1

With this unique choice of decomposition, one obtains

$$K(F_{\mu\nu}, F^{\mu\nu}) = -\frac{1}{(\gamma\cdot\gamma)} (\partial_\mu A^\ell_\nu - \partial_\nu A^\ell_\mu + \varepsilon^\ell_{mn} A^m_\mu A^n_\nu)^2 - \frac{1}{(\gamma\cdot\gamma)} (\partial_\mu B_\nu - \partial_\nu B_\mu)^2$$

$$K(\mathcal{D}_\mu \Phi^\ell, \mathcal{D}^\mu \Phi_\ell) = K(\mathcal{D}_\mu \Phi, \mathcal{D}^\mu \tilde{\Phi}) =$$

$$-\frac{2}{(\alpha\cdot\alpha)} (\partial_\mu \phi - \frac{i}{2} A^\ell_\mu \sigma_\ell \phi - \frac{i}{2} B_\mu \phi \tan\theta)^+ (\partial_\mu \phi - \frac{i}{2} A^m_\mu \sigma_m \phi - \frac{i}{2} B_\mu \phi \tan\theta)$$

where θ is the angle between the roots γ and α

if one introduces Pauli matrices σ_ℓ and the doublet notation

$$\phi = \begin{pmatrix} \phi_1 \\ \phi_2 \end{pmatrix}, \qquad \phi^+ = (\phi_1^*, \phi_2^*)$$

Finally,

$$K\,(\varepsilon_{mn\ell}\Phi^{\ell} + \frac{i}{2}\,[\Phi_m,\ \Phi_n]\ ,\qquad \varepsilon^{mnj}\Phi_j + \frac{i}{2}\,[\Phi^m,\ \Phi^n]) =$$

$$= K(\Phi_3 + \frac{i}{2}[\Phi,\tilde{\Phi}],\ \Phi_3 + \frac{i}{2}\,[\Phi,\tilde{\Phi}]) = -\frac{2}{(\alpha\cdot\alpha)}\left[2Z - 2\phi^{+}\phi + (\phi^{+}\phi)^{2}\right] = V(\phi),$$

i.e. the symmetry-breaking Higgs potential.
The sum of these three terms, when integrated over S^2, (which yields just the factor $4\pi R^2$, R being the radius of S^2), is the lagrangian of the Weinberg-Salam model in its standard form. The Weinberg angle θ_W can be identified with the angle θ between the roots γ and α, and is equal to 60° for SU(3), 45° for O(5) and 30° for G_2. The masses are related as follows :

$$M_H = M_Z, \qquad \frac{M_Z}{M_W} = \frac{1}{\cos\,\theta}$$

For more details, we send to the original paper by Manton.

8. Higher order Lagrangians; Example : non-linear electrodynamics in Kaluza-Klein theory

As explained already in the Introduction, the lagrangian of the General Relativity extends naturally in more than four dimensions as

$$= \sum \alpha_k I_k,$$

where $I_k = \varepsilon^{i_1 i_2 \dots i_{2k}}_{j_1 j_2 \dots j_{2k}}\, R^{j_1 j_2}_{i_1 i_2} \dots R^{j_{2k-1} j_{2k}}_{i_{2k-1} i_{2k}}$

In the dimension d = 2k, the invariant density I_k turns out to be a closed form, and its integral over a compact manifold yields a topological number (proportional to Euler's characteristic). We have

$$I_0 = \sqrt{|g|}\ , \qquad I_1 = \sqrt{|g|}\ R\ ,$$

$$I_2 = \sqrt{|g|}\ (R_{ABCD}R^{ABCD} - 4R_{AB}R^{AB} + R^2),\ \text{etc...}$$

I_2 is often called the Gauss-Bonnet term.

On the Kaluza-Klein fibre bundle the computation of I_2 yields the following expression (if g_{ab}, the invariant group metric, is constant, which supposes that we neglect the corresponding scalar fields) :

$$I_2 = \overset{(4)}{I_2} + 2\overset{G(4)}{R}\,R + \overset{G}{I_2} + (\overset{G}{R}_{ab} - \frac{1}{2}\overset{G}{R}g_{ab})F^a_{\mu\nu}F^{b\mu\nu} -$$

$$- \frac{1}{2} g_{ab}F^a_{\rho\lambda}\ F^{b\mu\nu}\ (\overset{4}{R}{}^{\rho\lambda}_{\mu\nu} - 4\ \overset{4}{R}{}^{[\rho}_{[\mu}\ \delta^{\lambda]}_{\nu]} + \overset{4}{R}\ \delta^{\rho}_{[\mu}\ \delta^{\lambda}_{\nu]}) +$$

$$+ \frac{1}{16}\,(F^a_{\mu\nu}\ F^{\mu\nu}_a)^2 + \frac{1}{8}\,F_{a\rho\lambda}\ F^{b\rho\lambda}\ F^a_{\mu\nu}\ F^{\mu\nu}_b +$$

$$+ \frac{1}{8}\,F^{a\rho}_{\lambda}\ F^{b\lambda}_{\mu}\ F_{a}\ F^{\mu}_{b\nu} - \frac{1}{2}\,F^{a\rho\mu}\ F_{a\rho\nu}\ F^{b\lambda\nu}\ F_{b\lambda\mu} -$$

$$- \frac{1}{2}\,C_{abc}\ F^c_{\mu\nu}\ F^{a\rho\mu}\ F^{b\nu}_{\rho} + \text{pure divergence}$$

(everything multiplied by $\sqrt{|g|}$).

Recent progress in the string theories have also suggested the presence of these terms in the gravitational sector of the theory (cf. Zwiebach, Deser et al.).

The models exposed above, namely a monopole, Weinberg-Salam model, were based on a Kaluza-Klein picture with the lagrangian $\sqrt{|g|}R$. As the total spaces of these theories are of dimension greater than four, such a truncated lagrangian becomes unnatural. It is beyond the scope of these lectures to give the most recent applications of the full-scale development of the higher order invariants; we shall show only the simplest example; we shall show how a well-defined version of non-linear electrodynamics can be derived form the Kaluza-Klein theory. Although, as we know, the consistent formulation must include all the degrees of freedom possible, i.e. $g_{\mu\nu}$, A_μ and the scalar field ϕ, we shall restrain our analysis to the limit when both ϕ and $g_{\mu\nu}$ are constant, supposing $\phi = 0$ and $g_{\mu\nu} = \eta_{\mu\nu} = (+,-,-,-)$.

The Lorentz invariants of the electromagnetic field are

$$F_{\mu\nu}\ F^{\mu\nu} = -\ 2(\vec{E}^2 - \vec{B}^2) \text{ and } F_{\mu\nu}\ {}^{*}F^{\mu\nu} - 2\vec{E}\cdot\vec{B},$$

the second one being a pseudo-scalar, changing its sign under space-time reflexions. Therefore, the only lagrangian leading to linear equations turns out to be

$$- \frac{1}{4}\,F_{\mu\nu}\ F^{\mu\nu} = \frac{1}{2}\,(\vec{E}^2 - \vec{B}^2)$$

if we want a theory without parity violation. In this sense, Maxwell's theory is unique, and the Kaluza-Klein theory in five dimensions reproduces it faithfully if the variation principle is based on the five-dimension scalar curvature of the bundle $P(V_4, U(1))$.

Non-linear versions of electrodynamics has been proposed by Mie (1912), Born and Infeld (1934). They investigated the theories based on the lagrangians taken as arbitrary functions of the two invariants :

$$L = L(\vec{E}^2 - \vec{B}^2, (\vec{E}\cdot\vec{B})^2)$$

Note that only even powers of $\vec{E}\cdot\vec{B}$ have to appear in order to preserve parity. Let us show that the generalized gravitational lagrangian in five dimensions fixes a unique form of such a lagrangian.

The five-dimensional metric g_{AB} on $P(M_4, U(1))$ is

$$g_{AB} = \begin{pmatrix} \eta_{\mu\nu} - A_\mu A_\nu & - A_\nu \\ - A_\mu & - 1 \end{pmatrix}$$

where we have chosen $g_{55} = -1$, $\eta_{\mu\nu} = \mathrm{diag}(+,---)$.

Discarding the cosmological term, the only lagrangian density leading to the second-order equations is

$$\mathcal{L} = (e^2 R + \gamma I_2)\sqrt{|g|}$$

The scale constant e^2 has the dimension cm^{-2}; $\gamma = \pm 1$, and the sign will be determined later considering the energy whose positivity must be assured.

It is not difficult to compute I_2 expressed by means of $F_{\mu\nu}$:

$$I_2 = \frac{3}{16}\left[(F_{\mu\nu} F^{\mu\nu})^2 - 2 F_{\mu\lambda} F_{\nu\rho} F^{\mu\nu} F^{\lambda\rho} \right]$$

The full lagrangian becomes $(\det(g_{AB}) = 1)$:

$$L = -\frac{1}{4} F_{\mu\nu} F^{\mu\nu} + \frac{3\gamma}{16e^2} (F_{\mu\nu} F^{\mu\nu})^2 - 2 F_{\mu\lambda} F_{\nu\rho} F^{\mu\nu} F^{\lambda\rho} \quad + \text{exact form}$$

The equations are directly derived by variation with respect to :

$$\partial_\lambda \left[F^{\lambda\rho} - \frac{3\gamma}{2e^2} (F_{\mu\nu} F^{\mu\nu}) F^{\lambda\rho} + \frac{3\gamma}{e} F_{\mu\nu} F^{\lambda\mu} F^{\rho\nu} = 0 \right]$$

Note that we have discarded the divergence term in the lagrangian, equal to

$$\partial_\mu F_{\lambda\rho} \partial^\mu F^{\rho\lambda} - 2 \partial^\mu F_{\mu\lambda} \partial_\nu F^{\mu\lambda}$$

which does not alter the Euler-Lagrange equations.

The Bianchi identities corresponding to the two Maxwell equations still hold by definition of $F_{\mu\nu}$:

$$\partial_\lambda F_{\mu\nu} + \partial_\mu F_{\nu\lambda} + \partial_\nu F_{\lambda\mu} = 0$$

In order to discuss the properties of this non-linear electrodynamics, it is much better to express everything in terms of the familiar fields $\vec{E}$ and $\vec{B}$. The lagrangian is then

$$L = \frac{1}{2}(\vec{E}^2 - \vec{B}^2) + \frac{3\gamma}{2e^2}(\vec{E}\cdot\vec{B})^2$$

and the full set of equations is :

$$\text{div}\,\vec{B} = 0, \qquad \overrightarrow{\text{rot}}\,\vec{E} = -\frac{\partial\vec{B}}{\partial t}$$

$$\text{div}\,\vec{E} = -\frac{3\gamma}{e^2}\vec{B}\cdot\overrightarrow{\text{grad}}(\vec{E}\cdot\vec{B}), \quad \overrightarrow{\text{rot}}\,\vec{B} = \frac{\partial\vec{E}}{\partial t} + \frac{3\gamma}{e^2}[\vec{B}\frac{\partial(\vec{E}\cdot\vec{B})}{\partial t} - \vec{E}\times\overrightarrow{\text{grad}}(\vec{E}\cdot\vec{B})]$$

By forming the sum

$$\vec{B}\cdot\frac{\partial\vec{B}}{\partial t} + \vec{E}\cdot\frac{\partial\vec{E}}{\partial t}$$

and using the four equations, we obtain the conservation law :

$$\frac{\partial}{\partial t}[\frac{1}{2}(\vec{E}^2 + \vec{B}^2) + \frac{3\gamma}{2e^2}(\vec{E}\cdot\vec{B})^2] = \text{div}\,(\vec{E}\times\vec{B})$$

in which we easily recognize the energy density and the Poynting vector $\vec{E}\times\vec{B}$. In order to assure the positivity of the energy, we have to set $\gamma = +1$.

The two equations containing div $\vec{E}$ and $\overrightarrow{\text{rot}}\,\vec{B}$ generalize the two Maxwell equations with sources ρ and j, i.e. charge and current densities. Here the field itself acts as a source via non-linear interaction. It is not difficult to check that the induced charge and current densities

$$\rho_{ind} = -\frac{3}{e^2}\vec{B}\cdot\overrightarrow{\text{grad}}\,(\vec{E}\cdot\vec{B})$$

$$\vec{j}_{ind} = \frac{3}{e^2}\quad\vec{B}\cdot\frac{\partial(\vec{E}\cdot\vec{B})}{\partial t} - \vec{E}\times\overrightarrow{\text{grad}}\,(\vec{E}\cdot\vec{B}).$$

satisfy the continuity equation

$$\frac{\partial\rho_{ind}}{\partial t} + \text{div}\,(\vec{j}_{ind}) = 0$$

Whenever the fields $\vec{E}$ and $\vec{B}$ are orthogonal, the equations coincide with Maxwell's system; therefore they admit the linearly polarized wave solutions. However, these waves can no longer be superposed.

It is also tempting to try to find a stationary configuration of fields displaying a non-vanishing total charge and magnetic moment.

In a static case, our equations are :

$$\text{div}\,\vec{B} = 0, \qquad \overrightarrow{\text{rot}}\,\vec{E} = 0$$

$$\text{div}\,\vec{E} = -\frac{3}{e^2}\,\vec{B}\cdot\overrightarrow{\text{grad}}(\vec{E}\cdot\vec{B}), \qquad \overrightarrow{\text{rot}}\,\vec{B} = -\frac{3}{e^2}\,\vec{E}\times\overrightarrow{\text{grad}}\,(\vec{E}\cdot\vec{B})$$

If we expect to find a non-vanishing magnetic moment, the lines of force of $\vec{B}$ must be closed; there must be a chosen axis, say $\vec{z}$, parallel to the magnetic moment $\vec{m}$ resulting from this configuration, and the lines of force of $\vec{B}$ are identical in any plane passing through $\vec{z}$ and parallel to it. The field $\vec{E}$ can also be expected to have this symmetry, at least close to the origin, and approaching the spherical symmetry at great distances, where we expect to see the electric field of a concentrated charge.

The induced current has only azimutal component, like the Poynting vector; the result is an apparent magnetic moment along z and the kinetic angular momentum

$$\vec{v}\times(\vec{E}\times\vec{B}) \sim s\vec{z}$$

There is also a remarkable symmetry which enables one to produce three new solutions out of any particular solution of this type : the equations are invariant under independent reflections $\vec{E}\to-\vec{E}$, $\vec{B}\to-\vec{B}$. The corresponding charges, magnetic momenta and angular momenta of these configurations give the four states whose energy is obviously the same (but we can not be sure if it is finite) :

solution	charge	magnetic moment	spin
$\vec{E},\ \vec{B}$	q	$\vec{\mu}$	$\vec{s}$
$\vec{E},\ -\vec{B}$	q	$-\vec{\mu}$	$-\vec{s}$
$-\vec{E},\ \vec{B}$	-q	$\vec{\mu}$	$-\vec{s}$
$-\vec{E},\ -\vec{B}$	-q	$-\vec{\mu}$	$\vec{s}$

It is nice to see here the four states of a charged fermionic particle : two states of a given charge, with magnetic moment parallel to spin, with spin up or down with respect to $\vec{z}$, and two other states with the opposite charge, magnetic moment anti-parallel to spin, with spin up or down.

Unfortunately, there must be a singularity concentrated on a circle (string) in the ρ-plane, in such a solution. If the space on which the solution is defined is R^3 and all the functions are of C^∞ -class, then the Poincaré lemma says that if $d\theta = 0$, then $\theta = d\chi$. This amounts, in standard 3-dimensional notations, to the following statements :

as $\vec{B}\cdot\overrightarrow{\text{grad}}(\vec{E}\cdot\vec{B}) = \text{div}\,(\vec{B}(\vec{E}\cdot\vec{B}))$, because $\text{div}\,\vec{B} = 0$,
we can write one of the equations as

$$\text{div}\,(\vec{E} + \frac{3}{e^2}\vec{B}(\vec{E}\cdot\vec{B})) = 0,$$

which, after Poincaré's lemma, leads to the existence of the field $\vec{C}$ such that

$$\vec{E} + \frac{3}{e^2}\vec{B}(\vec{E}\cdot\vec{B}) = \overrightarrow{\text{rot}}\,\vec{C}$$

Similarly, because of $\overrightarrow{\text{rot}}\,\vec{E} = 0$, we can write

$$\vec{E}\times\overrightarrow{\text{grad}}\,(\vec{E}\cdot\vec{B}) = -\,\overrightarrow{\text{rot}}\,(\vec{E}(\vec{E}\cdot\vec{B}))$$

and the second of our equations becomes

$$\overrightarrow{\text{rot}}\,(\vec{B} - \frac{3}{e^2}\vec{E}(\vec{E}\cdot\vec{B})) = 0$$

which again, after Poincaré's lemma, leads to the existence of the function ψ such that

$$\vec{B} - \frac{3}{e^2}\vec{E}(\vec{E}\,\vec{B}) = \overrightarrow{\text{grad}}\,\psi$$

Then, multiplying the first identity scalarly by $\vec{E}$ and second by $\vec{B}$, we get

$$\vec{E}^2 + \frac{3}{e^2}(\vec{E}\cdot\vec{B})^2 = \vec{E}\cdot\overrightarrow{\text{rot}}\,\vec{C} = -(\overrightarrow{\text{grad}}\,V)\cdot\overrightarrow{\text{rot}}\,\vec{C} = -\,\text{div}(V\overrightarrow{\text{rot}\vec{C}})$$

and

$$\vec{B}^2 - \frac{3}{e^2}(\vec{E}\cdot\vec{B})^2 = \vec{B}\cdot\overrightarrow{\text{grad}}\,\psi = \text{div}(\psi\vec{B})$$

Therefore, by taking the sum, we obtain a positive-definite expression $\vec{E}^2 + \vec{B}^2$ which is equal to a full divergence :

$$\vec{E}^2 + \vec{B}^2 = \text{div}(\psi\vec{B} - V\,\overrightarrow{\text{rot}}\,\vec{C})$$

Our conditions at infinity make $\vec{E}$ and $\vec{B}$ decrease at least as R^{-2}, if the total charge and energy are to remain finite; then $\psi\vec{B} - V\,\text{rot}\,\vec{C}$ must decrease at least as R^{-3} and its divergence as R^{-4}. Applying now Stokes' theorem to a sphere of radius R, we see that the integral

$$\int_{S_R} (\vec{E}^2 + \vec{B}^2) = \int_{S_R} (\psi\vec{B} - V\vec{rot}\ \vec{C})d\vec{\Sigma} \simeq \frac{1}{R}$$

behaves as 1/R, i.e. vanishes over the whole space.
Therefore, the only solution is $\vec{E} = 0$, $\vec{B} = 0$, or the Poincaré lemma doesn't hold, Ie. there is some singularity between 0 and ∞, which must be a circle because of the imposed symmetry; or the solution has to be found on a space whose topology is $R^3 - S^1$.

It is probable that the solution in a soliton form can be found, like what happened in case of the magnetic monopole, in the non-abelian version of the theory.

References

1. Th. Kaluza, Sitzungsber. d. Berl. Akad., 1921, 966
2. O. Klein, Z. für Physik, 37, 1926, 894
3. A. Einstein, Sitzungsber. preuss. Akad. Wiss., 1927, 23
4. A. Lichnerowicz, Théorie relativiste de la gravitation et de l'électromagnétisme, Masson, Paris, 1955
5. R. Kerner, Ann. Inst. H. Poincaré, 9, 1968, 143
6. A. Trautman, Rep. Math. Phys., 1, 1970, 29
7. C. Brans, R.H. Dicke, Phys. Rev., 124, 1961, 925
8. S. Kobayashi, K. Nomizu, Foundations of differential geometry, Acad. Press, New York, 1963
9. A. Lichnerowicz, Théorie globale des connections et groupes d'holonomie, Dunod, Paris, 1955
10. R. Kerner, Ann. Inst. H. Poincaré, 34, 1981, 437
11. R. Coquereaux, A. Jadczyk, CMP 90, 1983, 79
12. R. Kerner, J. Math. Phys. 1983, 351
13. P. Forgacs, N.S. Manton, CMP 72, 1980, 15
14. J. Harnad, S. Schnider, J. Tafel, LMP 4, 1980, 41
15. N. Manton, Nucl. Phys. B, 1979, 141
16. J. Madore, Phys. Lett. 110A, 1985, 289
17. B. Zwiebach, Phys. Lett. 156B, 1985, 315

DETERMINANTS OF LAPLACIANS ON RIEMANN SURFACES

Yu.I. Manin

1. Notations and Results

Let X be a compact Riemann surface of genus g = 2 with a Hermitian metric γ and L a holomorphic line bundle on X with a Hermitean metric h. We denote by $\Delta = \Delta(L, \gamma, h)$ the Laplacian acting on smooth sections of L (for precise normalization see below, formula (10)). In this note we discuss some modern methods for calculating $\det{}'\Delta = \exp(-\zeta_\Delta'(0))$.

After the work by Quillen [1] and Belavin-Knizhnik [2] it became clear that determinants of Laplacians of this kind can be computed in terms of holomorphic geometry of (X,L): see Bismut-Gillet-Soulé [3] for a very general treatment. Such computations involve the so called determinant line bundles whose explicit description is a problem in algebraic geometry.

In the papers Manin [4] and Beilinson-Manin [5] this problem was solved with the help of technique introduced by Arakelov, Faltings, and Deligne [6], for the case $L = \Omega_X^{\otimes n}$, $n \in \frac{1}{2}\mathbb{Z}$, where Ω_X is the holomorphic cotangent bundle and γ is the metric of constant curvature -1.

Here we return to this problem using instead of Belavin-Knizhnik [2] the more precise results due to Deligne [7]. Combining them with the construction of canonical bases of spaces of holomorphic spinor fields of arbitrary weight from [5] we give our formulas for Laplacians acting on spinor fields. Comparing them with results of D'Hoker and Phong [8] we obtain also expressions for the values of Selberg's zeta-functions at integer points in terms of holomorphic invariants. They are slightly more precise than those given by Beilinson-Manin [9] involving no undetermined constants.

For detailed proofs we refer to [7], [5], and [8].

In order to state main results we need some notation.

ISBN 0-12-218100-X

1. Algebro-geometric data. A holomorphic vector bundle L on X defines a 1-dimensional vector space

$$d(L) = \det H^0(X,L) \det^{-1} H^1(X,L) = \det H^0(X,L) \det H^0(X,L^t).$$

Here $\det V = \wedge^{\dim V}(V)$ and $L^t = L^* \otimes \Omega_X^{-1}$ is the Serre dual bundle. If bases $l = (l_1,\ldots,l_n)$ (resp. $l^t = (l_1^t,\ldots, l_m^t)$) of $H^0(X,L)$ (resp. $H^0(X,L^t)$) are given, they define a basis in d(L):

$$d = d(l)d(l^t) = (l_1 \wedge \ldots \wedge l_n) \otimes (l_1^t \wedge \ldots \wedge l_m^t) \quad .$$

Two holomorphic line bundles L, M on X define a 1-dimensional vector space $\langle L,M\rangle$. Any two meromorphic sections l,m of L, M with disjoint divisors define a basis $\langle l,m\rangle \in \langle L,M\rangle$ such that

$$\langle l,fm\rangle = f(\operatorname{div} l)\langle l,m\rangle \quad , \quad \langle fl,m\rangle = f(\operatorname{div} m) \langle l,m\rangle ,$$

where f is a meromorphic function and $f(\Sigma a_i x_i) = \Pi f(x_i)^{a_i}$, $x_i \in X$, $a_i \in Z$. (This notation should not be confused with that for a Hermitean scalar product which we shall denote by $\langle l|m\rangle$).

The space $\langle L,M\rangle$ bimultiplicatively depends on L,M. Further details see in [6], [7], [5].

2. Hermitean metrics. Suppose now that X, L, M are endowed with Hermitean metrics γ, h, h'. Then both d(L) and $\langle L,M\rangle$ can be endowed by specific Hermitean metric such that length squares d(l) and $\langle l,m\rangle$ are given by explicit formulas in terms of γ, h, h': cf. (8), (9), and (11) below.

Determinants are computed through them via the following Deligne's theorem([7]):

3. Theorem. There is a canonical isomorphism

$$\varphi_L : d(L)^2 d(\Omega_X)^{-2} \overset{\sim}{\to} \langle L,L^t\rangle^{-1} \tag{1}$$

such that for a nonzero $\delta \in d(L)^2 d(\Omega_X)^{-2}$ we have

$$\frac{\det' \Delta(L,\gamma,h)}{\det' \Delta(\Omega_X,\gamma,\gamma^{-1})} = \frac{||\delta||}{||\varphi_L(\delta)||} \quad . \tag{2}$$

4. *Theorem.* (Beilinson-Manin [5]). Let $\Omega^{\frac{1}{2}}$ be an odd theta-characteristic with $\dim H^0(\Omega^{\frac{1}{2}}) = 1$ and put $\Omega^n = (\Omega^{\frac{1}{2}})^{\otimes 2n}$ for $n \in \frac{1}{2}Z$. Then one can construct explicitly bases

$$d_i \in d(\Omega^i) \ , \qquad v \in \langle\Omega^{\frac{1}{2}}, \Omega^{\frac{1}{2}}\rangle$$

in such a way that under the Deligne isomorphism (1) we have

$$\varphi_{\Omega^i}(d_i^2\, d_1^{-2}) = v^{4(i^2-i)} \quad . \tag{3}$$

These bases are described in the following section: see (12), (14), (15) below.

From (2) and (3) we get:

5. *Corollary.* Let $\Delta_i = \Delta(\Omega^i,\gamma,h_i)$ where h_i is induced by γ . Then

$$\det' \Delta_i = \det' \Delta_1 \frac{||d_i||}{||d_1||\,||v||^{2(i^2-i)}} \quad . \tag{4}$$

6. *Selberg's zeta function.* If γ is the metric of constant curvature -1 our Laplacian Δ_n is $\Delta_n^- = -\nabla_z^{n-1}\nabla_n^z$ in the D'Hoker-Phong notation [8]. Hence we have ([8], p.543) for $n \geq 3/2$:

$$\det' \Delta_n = \exp((2g-2)c_{n-1})\, Z_{n-[n]}(X,n) \quad , \tag{5}$$

$$c_n = \sum_{0\leq m<n-\frac{1}{2}} (2n-2m-1)\log(2n-m) - (n+\tfrac{1}{2})^2 +$$

$$+ 2(n-[n])(n+\tfrac{1}{2}) + (n+\tfrac{1}{2})\log 2\pi + 2\zeta'(-1) \ .$$

Here $\zeta(s)$ is the Riemann zeta function and $Z_\varepsilon(X,a)$, $\varepsilon = 0,\frac{1}{2}$ are Selberg's zeta functions

$$Z_\varepsilon(X,s) = \prod_\gamma \prod_{p=0}^{\infty} (1 - \nu(\gamma)^2 e^{-(p+s)l(\gamma)}) \quad ,$$

where γ runs through primitive classes of a discrete subgroup $\Gamma \subset PSL(2, R)$, $X = H/\Gamma$, ν is a character of Γ, defining $\Omega^{\frac{1}{2}}$, $l(\gamma)$ is the length of a geodesic in the class γ.

Moreover ([8], p. 544)

$$\det{}' \Delta_1 = \exp((2g-2)c_o)\, Z_o'(1) \quad . \tag{6}$$

From (4), (5), and (6) we get:

7. Theorem. In the above notation we have for $n \geq 3/2$:

$$Z_{n-[n]}(X,n) = \exp((2g-2)(c_o - c_n)) Z_o'(1) \frac{\|d_n\|}{\|d_1\| \; \|v\|^{2(n^2-n)}} . \tag{7}$$

2. Explicit Formulas

1. Metrics and Laplacians. We start with precise normalization of metrics used in sec. 1, following [7].

Let $\Omega^{oq}(L)$ be the space of smooth (o,q)-forms with values in L, z a local holomorphic coordinate on X. Let metric γ be given by $\langle\partial_z|\partial_z\rangle = \gamma(z,\bar z)$ and let $\langle l|l\rangle_h = h(z,\bar z)$ for a local holomorphic section of L. Then for $l_1, l_2 \in \Omega^{oo}(L)$ we put

$$\langle l_1|l_2\rangle = \frac{i}{2\pi} \int_X \langle l_1|l_2\rangle_h \, \gamma(z,\bar z)\, dz \wedge d\bar z \quad , \tag{8}$$

and

$$\langle l_1 d\bar z|l_2 d\bar z\rangle = \frac{i}{2\pi} \int_X \langle l_1|l_2\rangle_h \, dz \wedge d\bar z \quad . \tag{9}$$

The Laplacian is then given by

$$\Delta: \Omega^{oo}(L) \to \Omega^{oo}(L) \quad ,$$

$$\Delta(fl) = \bar\partial^* \bar\partial(fl) = -h^{-1}\gamma^{-1}\partial_z(\gamma\partial_{\bar z} f)\, l \quad . \tag{10}$$

We have

$$H^0(X,L) = \mathrm{Ker}\bar\partial \subset \Omega^{00}(L) \quad ,$$

$$H^1(X,L) = \mathrm{Coker}\bar\partial^* \subset \Omega^{01}(L) \quad .$$

We endow these spaces by metrics (8) and (9). Corresponding volume forms define the metric on d(L) alluded to in sec. 1.

Finally, on <L,M> we put

$$\log||\langle l,m\rangle|| = \frac{1}{2\pi i}\int d'd''\log||l||\cdot\log||m|| + $$

$$+ \log||l||[\mathrm{div}\ m] + \log||m||[\mathrm{div}\ l] \ , \qquad (11)$$

where

$$f[\sum_i a_i x_i] = \sum_i a_i f(x_i) \ , \quad x_i \in X \ , \quad a_i \in Z$$

and d'd" is understood in the sense of distributions.

2. *Canonical bases.* This part is borrowed from [5]. We start with a symplectic basis $(a_1, \ldots, a_g; b_1, \ldots, b_g)$ in $H_1(X,Z)$ and define the Riemann basis $(\omega_1, \ldots, \omega_g)$ in $H^0(X,\Omega_X)$ by $\int_{a_i}\omega_j = \delta_{ij}$. Then we can introduce the matrix of b-periods $\tau_{ij} = \int_{b_i}\omega_j$ and the corresponding Riemann theta-function θ. If α denotes an odd theta characteristics corresponding to $\Omega_X^{\frac{1}{2}}$, we can define the holomorphic differential φ_0 on X having double zeroes:

$$\varphi_0 = \sum_{j=1}^{g} (\partial/\partial z_j \ \theta[\alpha])(0,\tau)\omega_j \quad .$$

Let $\nu \in H^0(X,\Omega^{\frac{1}{2}})$ be a differential of weight ½ with $\nu^2 = \varphi_0$ and let $\mathrm{div}\nu = \sum_{i=1}^{g-1} P_i$. We shall use ν for several purposes.

a). Firstly, ν defines the canonical element $v \in \langle\Omega^{\frac{1}{2}},\Omega^{\frac{1}{2}}\rangle$ used in Theorem 4, sec. 1:

$$v = \langle\nu,\nu\rangle \quad . \qquad (12)$$

(Since divν is not disjoint with itself some care is needed here: see [5], p. 371).

b). Secondly, ν defines the generator $d_{1/2} = \nu^2$ in $d(\Omega^{\frac{1}{2}}) = \det H^o(X,\Omega^{\frac{1}{2}})^{\otimes 2}$.

c). Thirdly, ν defines exact sequences

$$
\begin{gathered}
0 \to \Omega^i \overset{\nu}{\to} \Omega^{i+\frac{1}{2}} \to \bigoplus_n \Omega_{P_n}^{i+\frac{1}{2}} \to 0 \quad , \\
0 \to H^o(\Omega^i) \overset{\nu}{\to} H^o(\Omega^{i+\frac{1}{2}}) \to \bigoplus_n \Omega_{P_n}^{i+\frac{1}{2}} \to H^1(\Omega^i) \to H^1(\Omega^{i+\frac{1}{2}}) \to 0 \quad .
\end{gathered} \tag{13}
$$

For $i = 3/2$ we get from (13)

$$d(\Omega^{i+\frac{1}{2}}) = d(\Omega^i) \otimes (\bigotimes_{n=1}^{g-1} \Omega_{P_n}^{i+\frac{1}{2}}) \quad . \tag{14}$$

Finally, ν allows us to find explicit bases in fibers $\Omega_{P_n}^{i+\frac{1}{2}}$: if we put $\nu^{\otimes 2} = t_n^2 dt_n$ near P_n, we obtain a distinguished element $(dt_n)^{i+\frac{1}{2}} \bmod (t_n) = \tau_n^{i+\frac{1}{2}}$ in the P_n-fiber. Therefore for $i = 3/2$ we can put

$$d_{i+\frac{1}{2}} = d_i \otimes (\bigotimes_{n=1}^{g-1} \tau_n^{i+\frac{1}{2}}) \quad . \tag{15}$$

A minor change in this construction allows us also to define d_1 and $d_{3/2}$ starting from d_o: see [5], p. 364.

3. *Lengths*. To make (4) and (7) explicit we must now calculate lengths of d_i and ν.

In [5] we suggested to use a normalized basis of $H^o(\Omega^i)$ in such a way that d_i becomes the exterior product of elements of this basis. Then d_i is calculated via (9). Alternatively, one could use (15) directly. The r.h.s. of (15) has a Hermitean metric which is the tensor product of that on $d(\Omega^i)$ and those on fibers $\Omega^{i+\frac{1}{2}}$. It differs from the "right" metric on $d(\Omega^{i+\frac{1}{2}})$ by the multiplier $\prod_{k<1} G(P_k,P_1)^{-1}$ where G is a Green function (cf. [5], p. 372 and [7]). The resulting formulas can serve as a basis for discussing bosonization on Riemann surfaces.

The function $||\langle\nu,\nu\rangle||$ deserves further study. It is closely connected with Noether's theorem for arithmetical surfaces.

References

[1]. Quillen D. Determinants of Cauchy-Riemann operators over a Riemann surface. Funkc.Analiz i ego Pril., 19 (1985), 37-42 (in Russian).

[2]. Belavin A.A., Knizhnik V.A. Algebraic geometry and the geometry of quantum strings. Phys.Lett. 168B (1986), 201-207.

[3]. Bismut J.-M., Gillet H., Soulé C. Analytic torsion, and holomorphic determinant bundles. Preprint T8/87, Univ. Paris-Sud.

[4]. Manin Yu.I. The partition function of the string can be expressed in terms of theta functions. Phys.Lett. B172 (1986), 184-186.

[5]. Beilinson A.A., Manin Yu.I. The Mumford form and the Polyakov measure in string theory. Comm.Math.Phys., 107 (1986) 359-376.

[6]. Deligne P. La formule de dualité globale. In: *Lecture Notes in Mathematics*, Springer, vol. 305 (1970), 481-587.

[7]. Deligne P. Le determinant de la cohomologie. In: *Current Trends in Arithmetical Algebraic Geometry*, Contemp.Math., vol. 67 (1987), AMS, 93-178.

[8]. D'Hoker E., Phong D.H. On determinants of Laplacians on Riemann surfaces. Comm.Math.Phys., 104 (1986), 537-545.

[9]. Beilinson A.A., Manin Yu.I. Values of the Selberg zeta function at integer points. Funkc.Analiz i ego Pril., 21: 1 (1987),68-69 (in Russian).

Steklov Mathematical Institute

Ul. Vavilova 42, Moscow 117 333

REFLECTIONS ON ARITHMETICAL PHYSICS*

Yu.I. Manin

Il en est qui, face à cela, se contentent de hausser les épaules d'un air désabusé et de parier qu'il n'y a rien à tirer de tout cela, sauf des rêves. Ils oublient, ou ignorent, que notre science, et toute science, serait bien peu de choses, si depuis ses origines elle n'avait été nourrie des rêves et des visions de ceux qui s'y adonnent avec passion.

A.*Grothendieck* [0], p.18

The development of theoretical physics in the last quarter of the twentieth century is guided by a very romantic system of values. Aspiring to describe fundamental processes at the Planck scale physicists are bound to lose any direct connection with observable world. In this social context the sophisticated mathematics emerging in the string theory ceases to be only a technical tool needed to calculate some measurable effects and becomes a matter of principle.

Today at least some of us are again nurturing an ancient Platonic feeling that mathematical ideas are somehow predestined to describe physical world, however remote from reality their origins may seem to be.

From this viewpoint one should perversely expect number theory to become the most applicable branch of mathematics.

*) *Dedicated to Alexander Grothendieck on his 60th birthday.*

ISBN 0-12-218100-X

Indeed, one clearly discerns the trend to at least include the number theory into universe of discourse of modern theoretical physics.

The author of these notes was amazed and pleased by the discovery that in order to calculate the Polyakov measure in string theory one can use Faltings' calculation of a specific number theoretical function called "height" (cf. [1], [2],[3]).

Sasha Polyakov told me that after Faltings' talk at the Berkeley International Congress of Mathematicians Ed Witten bought all books in number theory available at a bookshop across the road. (I did not check this with Ed - se non è vero, è ben trovato).

Therefore it is high time to present some musings of a professional number-theorist and an amateur-physicist on the controversial subject of arithmetical physics.

First of all, can one calculate anything physical by unambigously number-theoretical means ? I believe that the answer should be affirmative. Look at one of the beautiful formulas discovered by Euler:

$$\pi^2/6 = \prod_{p \text{ prime}} (1 - p^{-2})^{-1} \qquad (1)$$

The r.h.s. unmistakably belongs to number theory: primes $p = 2,3,5,7,11, \ldots$ are one of its main concerns. I dare say that the l.h.s. involving π is a physical constant although it may need some arguing to convince the reader. In fact, π can be (and had been) measured, just like water boiling temperature or the Earth Equator length. One may say that Euclidean geometry where π appears as a mathematical constant is actually kinematics of ideal solids valid in the macroscopic approximation of flat gravitational vacuum.

In order to understand (1) better it is useful to recall some facts on primes. Classically, a prime number p is defined as a positive integer having no divisors except of 1 and itself. Each integer can be decomposed into a product of primes in a unique way; there is an infinity of primes; they are distributed quite irregularly; the simplest asymptotic formula for the number of primes not exceeding N is N/log N.

This is not however the viewpoint we need here.

A modern explanation of the role of primes is given by Ostrowski's theorem: primes describe all reasonable ways to define a notion of continuity on the set of rational numbers Q (in addition to the traditional one).

To be more precise, define a function $|a|_p$ of $a \in Q$ by $|a|_p = p^{-n}$, if $a = p^n c d^{-1}$ where c,d are integers nondivisible by p. It verifies the usual properties of a norm: $|ab|_p = |a|_p |b|_p$, $|a + b|_p \leq |a|_p + |b|_p$ (in fact, even $\leq \max(|a|_p, |b|_p)$ the so called non-Archimedean traingle inequality). Therefore it defines on Q a topology in which $a_i \to 0$ if $|a_i|_p \to 0$. This topology is called the p-adic one. Addition and multiplication being continuous p-adically, one can define in a standard way the notion of a Cauchy sequence and the set of limits of such sequences which are called p-adic numbers.

The set of p-adic numbers Q_p is a new analog of the set of reals R which can also be constructed in this way with the help of the usual absolute value $|a|$ which is conveniently denoted as $|a|_\infty$. Ostrowski's theorem says that any norm, or valuation, on Q defines the same topology as $|\cdot|_\infty$ or $|\cdot|_p$ for some prime p.

Of course, the properties of Q_p are in many respects very different from those of R. The main reason is that the topology of Q_p is quite unlike that of R: p-adic numbers form a Cantor set, or a "fractal" [4]. Nevertheless, many chapters of classical calculus and geometry were developed over p-adic numbers. For a beautiful introduction see [5].

The physical world as we know it can be described very efficiently by mathematics based upon R (and its further extension C, complex numbers). Recently it was suggested that at the Planck scale a p-adic topology may become more relevant [6]. This however raises a question: why should one particular p be physically marked ? Isn't it more reasonable to believe in a democracy of all available topologies ? (Or at least of all p-adic ones since R is certainly "the first among equals" being defined by the only Archimedean norm).

It appears that Euler's identity (1) and a whole series of similar facts can be explained in a way furnishing a very suggestive picture of such democracy.

Let us start with an almost evident formula $|a|_\infty = \prod_p |a|_p^{-1}$. It means that the knowledge of the ordinary absolute value of a rational number is equivalent to that of all p-adic absolute values simultaneously. Or, totally democratically, $\prod_v |a|_v = 1$, where $v = \infty$ or $p = 2,3,5, \ldots$ Formulas of this kind abound in the number theory and are called product formulas, reciprocity laws etc.

In the last product formula we considered a rational number a in turn as a real, 2-adic, 3-adic ... number. Let us introduce more generally the set of infinite vectors $(a_\infty, a_2, a_3, \ldots)$ where $a_\infty \in R$, $a_p \in Q_p$. Such a vector, with a restriction $|a_p|_p = 1$ for all sufficiently large p, is called an adèle. This term was coined by Claude Chevalley around 1940, together with the word idèle meaning an invertible adèle (i.e. $a_v \neq 0$ for all v, $|a_p|_p = 1$ for large p). Etymology of these words is doubtful; idèle presumably stems from ideal while adèle means additive idèle. Anyway, among number theorists these are now common words.

Let us now imagine the first steps of mathematics in which the notion of a real number is replaced by that of adèle. An adèle $a = (a_v)$ has components, real one a_∞ and p-adic ones a_p for all p's. The set of all adèles is a topological ring A_Q with componentwise addition and multiplication. Its topology combines Archimedean and fractal properties. Rational numbers Q are imbedded into A_Q diagonally: $a \equiv (a, a, a, \ldots)$. A simple but important observation is that $Q \subset A_Q$ is a discrete subgroup, like $Z \subset R$. In other words, a sequence of rational numbers cannot converge in all v-adic topologies simultaneously (e.g. if it converges to zero p-adically for all p's it must consist of infinitely growing integers).

Recalling that $R/Z \simeq U(1)$ is a circle we arrive to the notion of adèlic circle:

Here (3) is established in the same way as (2), the Archimedean part of (4) is proved by a Poisson summation trick, and the p-part of (4) follows from the fact that SL_2 over a finite field of p elements consists of p^3-p^2 points, so that the relative amount of points in $SL_2(Z_p)$ with respect to Z_p^3 is just $1 - p^{-2}$.

Now we can see the following pattern:

- (at least some) essential notions of real and complex calculus and geometry have their adèlic counterparts;
- adèlic objects have a strong tendency to be simpler than their Archimedean components, e.g. the adèlic fundamental domains of arithmetical discrete subgroups of semisimple groups usually have volume 1 (the Siegel-Tamagawa-Weil philosophy, cf. [15]);
- due to this fact and to the "product formulas" like (2) or (3) embodying the idea of democracy for all topologies, an information on the real component of an adèlic object can be read off *either* from this real component *or* from the product of p-adic components for all p's.

With some strain, one can generalize and state the following principle which is the main conjecture of this talk.

On the fundamental level our world is neither real, nor p-adic, it is adèlic. For some reasons, reflecting the physical nature of our kind of living matter (e.g. the fact that we are built of massive particles), we tend to project the adèlic picture onto its real side. We can equally well spiritually project it upon its non-Archimedean side and calculate most important things arithmetically.

The relation between "real" and "arithmetical" pictures of the world is that of complementarity, like the relation between conjugate observables in quantum mechanics.

Of course, one is not obliged to take this metaphysics seriously. A skeptical reader can still use it as a guiding principle in a mathematical study of the structure of string theory.

I shall describe below some recent work which seems promising in this respect.

$$A_Q/Q \simeq (R \times \prod_p Z_p)/Z \quad , \tag{2}$$

where Z_p is the set of p-adic integers ($a \in Z_p \Leftrightarrow |a|_p \leq 1$). We see from (2) that the adèlic circle is a mixture of U(1) and a compact topological totally discontinuous group which can be also described as a "lattice approximation limit" for U(1), i.e. $\varprojlim_n Z/nZ$. Thus we encounter again the Archimedean and fractal properties combined in one object. Fourier analysis based upon A_Q/Q rather than U(1) ties together in a very nice way the usual and the finite Fourier transforms: cf. Tate's thesis [7].

Going one step further we may define the simplest non-commutative adèlic group $SL_2(A_Q)$ which is essentially the set of infinite matrix vectors

$$\{\begin{pmatrix} a_v & b_v \\ c_v & d_v \end{pmatrix} \in SL_2(Q_v) \mid v = \infty, 2, 3, 5, \ldots\}$$

such that (a_v), (b_v), (c_v), (d_v) are adèles. Again, $SL_2(Q)$ is discrete in $SL_2(A_Q)$. With the help of a left invariant differential form on SL_2 and norms $|a|_v$ we can define a left invariant measure $dm = \prod_v dm_v$ on $SL_2(A_Q)$ in the same way as on its classical component $SL_2(R)$. If we normalize dm by the condition $\int_{SL_2(A_Q)/SL_2(Q)} dm = 1$ and then calculate this integral componentwise we get finally a beautiful explanation of (1):

$$1 = \int_{SL_2(A_Q)/SL_2(Q)} dm = \int_{SL_2(R)/SL_2(Z)} dm_\infty \times \prod_p \int_{SL_2(Z_p)} dm_p \ , \tag{3}$$

$$\int_{SL_2(R)/SL_2(Z)} dm_\infty = \pi^2/6 \quad , \quad \int_{SL_2(Z_p)} dm_p = 1 - p^{-2} \quad . \tag{4}$$

To start with, a reinterpretation of the calculation of Polyakov's measure in [1] shows [3] that if we take a point x of the moduli space M_g which has algebraic coordinates, then the density of this measure with respect to a canonical one equals the inverse Archimedean part of a function called height of x.

A remarkable property of the height compatible with our philosophy is that it is defined by a product of factors corresponding to all valuations of the field in which the coordinates of x lie.

I conjecture that one can *define on the space of adèlic points of a universal moduli space an adèlic Polyakov measure whose Archimedean component is the ordinary Polyakov measure. Hopefully, the corresponding total adèlic volume would be computable as in (3), (4), thus giving an arithmetical expression of the string partition function.*

If this hope comes true we shall be able to speak on adèlic strings with some justification. Of course, the main reason for believing in this is the remarkable appearance in the string theory of algebraic varieties (moduli spaces) and measures on them (Mumford forms) which are intrinsically defined over integers and not just over R or C.

In order to explain some details we need an extension of the picture we worked with up to now. Number theory studies not only rational numbers Q but all algebraic numbers $\overline{Q}$ i.e., roots of polynomials with rational coefficients. It is convenient to work with smaller number fields K i.e., with finite-dimensional Q-subspaces of $\overline{Q}$ containing 1 and closed under multiplication. For such a field K one can prove a generalization of the Ostrowski theorem, describing all valuations of K. Since $Q \subset K$, each such valuation w induces on Q a valuation v which is equivalent either to $|\cdot|_\infty$ or to $|\cdot|_p$. We say that w extends,or divides v. Now the following facts are true: a) each valuation of Q extends to a finite number of valuations of K; b) valuations extending $|\cdot|_\infty$ correspond to various embeddings of K into complex numbers. After this theorem has been proved, one can define w-adic numbers K_w, adèles A_K, idèles J_K and all the other objects we considered previously "over Q".

Suppose now that we have a linear space L over K endowed with norms $||\cdot||_w$, one for each valuation w of K, such that $||al||_w = |a|_w ||l||_w$ for $a \in K$, $l \in L$ and $||l||_w = 1$ for almost all w if $l \neq 0$. We can then define the height of $l \in L$ by the formula

$$h(l) = \prod_w ||l||_w \quad .$$

From the product formula $\prod_w |a|_w = 1$ for $a \in K$ one sees that h(l) in fact depends only on the ray Kl in L i.e., the height is a function on the projective space associated with L.

Since the moduli space M_g is in fact defined by algebraic equations with integer coefficients we can embed M_g into such a projective space. If this embedding is constructed with the help of Mumford's determinant vector bundles, we obtain in this way the height connected with the Polyakov measure.

The total height, unlike its Archimedean part, is defined only for points with algebraic coordinates, which, although dense in M_g, do not look very appealing physically. However, in a recent work [8] it was established that just these points appear in a natural way as lattice points in a string lattice approximation scheme.

The situation is as follows. In string lattice approximation one replaces a smooth Riemann surface, i.e., a string world sheet with a metric ds^2, by a triangulated metric surface which is essentially given by combinatorial data listing, say, vertices and lengths of arcs connecting some pairs of vertices. This is a two-dimensional version of the Regge calculus in general relativity.

Consider now only equilaterally triangulated compact oriented surfaces with all arcs of the length 1 (changing this value leaves the surface in the same conformal class). One easily proves that such a surface is endowed with a complex structure compatible with the metric (first delete vertices then extend the obtained complex structure which is possible because the sum of angles at each vertex equals $n\pi/6$ for an integer n). Therefore such a surface defines a point in M_g. The main theorem of [8] foreseen by A.Grothendieck [0] states that in this way we obtain precisely all algebraic points. Thus the number theoretic picture reflects very well a combinatorial-metric picture. An outstanding problem now is establishing further ties between these two descriptions, in particular, calculating height in metric terms.

We now come to the last point of our discussion.

The most extensive generalization of Euler's formula (1) is connected with calculation of adèlic volume of homogeneous spaces of the type $H(A_K)/H(K)$ where H is a semisimple algebraic group and K is an algebraic number field. (We treated briefly the case $H = SL_2$). Similar calculations for other kinds of algebraic varieties, such as M_g, present great difficulties. Why can we then hope to deal arithmetically with M_g ?

A possible way out is remarkably connected with a new approach to another baffling property of the Polyakov partition function, namely, that it is inherently a perturbative expansion. It was suggested by several authors that one should work instead with a sort of Universal moduli space $\hat{M}$ embodying all M_g's (and something else).

In [9], motivated by this idea and operator approach, I conjectured that this $\hat{M}$ should be a homogeneous space with respect to the Virasoro algebra.

This conjecture was recently proved by four groups [10] - [13]. They all use the same basic construction due to Sato and Segal-Wilson. In this construction $\hat{M}$ is an infinite Grassmannian and the "moduli part" of $\hat{M}$ parametrizes triples (X,p,z) where X is a complex Riemann surface, p a point on it and z a local coordinate at this point.

If one succeeds to define the nonperturbative path integral as an integral over $\hat{M}$ it may well become accessible to the arithmetical treatment. To this end we shall need an extension of the Tamagawa-Weil theory to infinite-dimensional groups, like Kac-Moody ones and $GL(\infty)$. (Remark in passing that $vol(SL(n, R)/SL(n,Z)) = \zeta(2) \dots \zeta(n)$ has a well defined limit for $n \to \infty$. Is it possible to obtain it as a volume for $n = \infty$?).

In conclusion I would like to describe very briefly some preoccupations of number theorists which may be relevant to the program of arithmetization of physics.

Number theory has its own Grand Unification Group: it is the so called Galois group $G = Gal(\overline{Q}/Q)$ which consists of all permutations of algebraic numbers among themselves conserving algebraic relations between them with rational coefficients. It is an infinite topological group of "fractal" type whose structure is very complex and in a sense embodies all arithmetics, if one takes into account also some canonical central extensions of its subgroups, the so called Weil groups. To illustrate this assertion, we only note that the maximal abelian quotient G^{ab} is just $\prod_p Z_p^*$ so that primes appear again in a completely unexpected way, essentially as generators of G^{ab}. In studying the representations of G and its closed subgroups a number theorist meets with automorphic and modular functions in much the same way (or, rather, dual one) as a physicist studying the representations of Kac-Moody and Virasoro algebras. A series of deep conjectures due to Langlands [14] connect the representation theory of G with that of groups $H(A_K)$ via modular forms and their Mellin transforms.

I sincerely hope that such remarkable coincidences are not fortuitous.

Finally, I am pleased and proud to dedicate this note to Alexander Grothendieck whose insights immensely influenced mathematics and start to influence physics as well.

References

[0]. Grothendieck A. Esquisse d'un programme. Preprint 1984

[1]. Manin Yu.I. The partition function of the Polyakov string can be expressed in terms of theta functions. Phys.Lett. B172 (1986), 184-186.

[2]. Faltings G. Calculus on arithmetic surfaces. Ann. of Math., 119 (1984), 387-424.

[3]. Smit D.-J. String theory and algebraic geometry of moduli spaces. Preprint ITPh, Utrecht (1987).

[4]. Mandelbrot B. *Fractals*. Freeman, San-Francisco, 1977.

[5]. Koblitz N. *p-adic numbers, p-adic analysis and zeta-functions*. Springer, Heidelberg, 1977.

[6]. Volovich I.V. a) Number theory as the ultimate physical theory. Preprint CERN-TH 4781/87.
b) p-adic space-time and string theory. TMPh, 71:3 (1987), 337-340 (in Russian).

[7]. Tate J. Fourier analysis in number fields and Hecke's zeta-functions. In: *Algebraic Number Theory*, ed. by J.W.S. Cassels, A.Fröhlich. Academic Press, 1967, 305-347.

[8]. Voevodsky V.A., Shabat G.B. Equilateral triangulations of Riemann surfaces and curves over algebraic number fields. Preprint, 1987 (in Russian).

[9]. Manin Yu.I. Critical dimensions of string theories and the dualizing sheaf on moduli spaces of (super)curves. Funkc.Analiz i ego Pril., 20:3 (1986), 88-89 (in Russian).

[10]. Kontsevich M.L. Virasoro algebra and Teichmüller spaces. Funkc.Analiz i ego Pril., 21:2 (1987), 78-79 (in Russian).

[11]. Beilinson A.A., Schechtman V.V. Determinant bundles and Virasoro algebras. Preprint (1987).

[12]. Kawamoto N., Namikawa Yu., Tsuchiya A., Yamada Y. Geometric realization of conformal field theory on Riemann surfaces. Preprint, Nagoya University (1987).

[13]. Arbarello E., De Concini C., Kac V., Procesi C. Moduli spaces of curves and representation theory. Manuscript (1987).

[14]. Langlands R.P. L-functions and automorphic representations. In: *Proc. ICM 1978*, vol.1, Helsinki 1980, 165-176.

[15]. Kneser M. Semi-simple algebraic groups. In: *Algebraic Number Theory*, ed. by J.W.S. Cassels, A. Fröhlich, Academic Press, 1977, 250-265.

Steklov Mathematical Institute
Ul. Vavilova 42, Moscow 117 333

STRINGS ON ORBIFOLDS: AN INTRODUCTION

Hans-Peter Nilles
CERN, Geneva, Switzerland

1. Introduction

These lectures intend to give an elementary introduction to four-dimensional string models as obtained in orbifold compactifications [1], including the basic calculational techniques. To follow the lectures the reader should have some acquaintance with the heterotic string [2]. In Section 2 we start with reviewing basic facts about lattices [3] and θ-functions [4], followed by a discussion of one-loop modular invariance and its implications for heterotic string theory. Section 4 discusses the concept of an orbifold as a geometrical space, while in Section 5 some general properties of strings on orbifolds are discussed. Finally, in Section 6 we explain one example, the Z-orbifold, in detail including the partition function and the constraints from modular invariance.

2. Lattices and theta-functions

Let us discuss briefly the main facts about lattices that will be useful for us [3]. Consider an N-dimensional vector space (V) with basis e_i, $1 \leq i \leq N$. A lattice $\Lambda \in V$ is defined as a set of points of the form

$$\Lambda = \left\{ x = \sum_{i=1}^{N} n_i e_i \,/\, n_i \in \mathbb{Z} \right\} \tag{2.1}$$

The lattice is called <u>integral</u> if $x \cdot y$ is integral for every $x, y \in \Lambda$ ($x \cdot y$ denotes the inner product in V). The lattice is called <u>even</u> if for every $x \in \Lambda$ $x \cdot x$ is even.

For any lattice $\Lambda \subset V$ we can define the <u>dual lattice</u> - denoted Λ^*. It is the set of points $y \in V$ for which $x \cdot y \in Z$ for all $x \in \Lambda$

ISBN 0-12-218100-X

$$\Lambda^* = \{ y \in V \,/\, x \cdot y \in \mathbb{Z} \quad \forall x \in \Lambda \} \tag{2.2}$$

A basis e^{*i}, $i = 1,\dots,N$ satisfies $e_i \cdot e^{*j} = \delta_i^j$. Moreover, if we define

$$g_{ij} = e_i \cdot e_j \tag{2.3}$$

the inverse is given by

$$(g_{ij})^{-1} = g^{ij} = e^{*i} \cdot e^{*j} \tag{2.4}$$

If the lattice is integral, it is contained in its dual. A lattice is called self-dual if $\Lambda = \Lambda^*$. The volume of the lattice is defined as the volume of its fundamental cell $\sqrt{|\det g_{ij}|}$. A lattice is called unimodular, if $|\det g_{ij}| = 1$. Since Λ and Λ^* have reciprocally related numbers of points per unit volume, an integral and unimodular lattice can be shown to be self-dual.

As a simple example let us discuss SU(3)-group lattices. The root lattice is spanned by the two simple roots $e_1 = (\sqrt{2},0)$ and $e_2 = (-1/\sqrt{2},\sqrt{3/2})$.

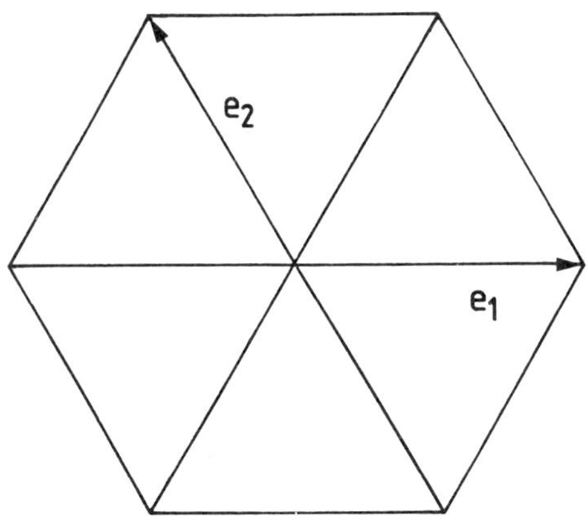

Figure 1

Observe that $g_{ij} = e_i \cdot e_j$ coincides with the Cartan matrix of SU(3). The lattice Γ spanned by e_1 and e_2 is integral and even. Its dual Γ^* is spanned by $e_1^* = (1/\sqrt{2}, 1/\sqrt{6})$ and $e_2^* = (0, \sqrt{2/3})$

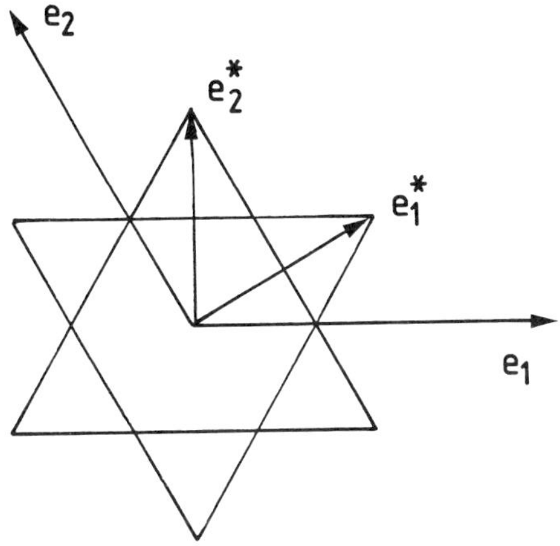

Figure 2

Since Γ is integral, it is contained in Γ^*. Γ^* is the weight lattice of SU(3). The volume of Γ is $\sqrt{3}$ and the volume of Γ^* is $1/\sqrt{3}$.

An even simpler example can be found within the rectangular O(2N) lattices. The root lattice (Λ_R) consists of the points $(n_1, \ldots, n_N)$ with $n_i \in Z$ and $\Sigma\, n_i \in 2Z$, again an even integral lattice. Its dual, the weight lattice of O(N), consists of the points $(n_1, \ldots, n_N)$ where n_i are half integers. It can be decomposed into the following conjugacy classes:

scalar	:	$(n_1, \ldots, n_N)$	Σn_i even
vector	:	$(n_1, \ldots, n_N)$	Σn_i odd
spinor	:	$(n_1 + \frac{1}{2}, \ldots, n_N + \frac{1}{2})$	Σn_i even
antispinor	:	$(n_1 + \frac{1}{2}, \ldots, n_N + \frac{1}{2})$	Σn_i odd.

Take as an example O(10). The states in the scalar conjugacy class with length^2 two correspond to 40 roots. The states in the vector class with $\vec{p}^{\,2} = 1$ to the fundamental (10) of O(10) and the shortest vectors in the spinor (antispinor) class with $p^2 = N/4$ to the $16(\overline{16})$ of O(10).

Let us next discuss partition functions of the lattices. Define $q = e^{2\pi i \tau}$ and

$$Z = \sum_{p \in \Lambda} q^{1/2 p^2} \tag{2.5}$$

and we want to work this out for the O(2N) lattices.

$$Z = \sum_{\{n_i\}} q^{1/2 \sum_{i=1}^{N} n_i^2} \frac{1}{2} \left(1 + (-1)^{\sum n_i}\right) \tag{2.6}$$

corresponds to the root lattice. This can be written as

$$Z = \frac{1}{2} \left[\prod_{i=1}^{N} \sum_{n_i \in \mathbf{Z}} q^{n_i^2/2} + \prod_{i=1}^{N} \sum_{n_i} q^{n_i^2/2} e^{i\pi n_i} \right] \tag{2.7}$$

Jacobi theta-functions are defined as

$$\vartheta\left[{\theta \atop \phi}\right](0|\tau) = \vartheta\left[{\theta \atop \phi}\right] = \sum_{n=-\infty}^{+\infty} q^{\frac{1}{2}(n+\theta)^2} e^{2\pi i (n+\theta)\phi} \tag{2.8}$$

Specializing to $\theta = 0$ we deduce for the expression of (2.7)

$$Z = \frac{1}{2} \left[\vartheta^N \left[{0 \atop 0}\right] + \vartheta^N \left[{0 \atop 1/2}\right] \right] \tag{2.9}$$

while the partition function of the O(2N) weight lattice is given by

$$Z = \vartheta^N \left[{0 \atop 0}\right] + \vartheta^N \left[{1/2 \atop 0}\right] \tag{2.10}$$

Observe that $\theta = \frac{1}{2}$ corresponds to the spinor conjugacy classes.

Very useful is the product representation of θ-functions

$$\frac{\vartheta\left[{\theta \atop \phi}\right]}{\eta} = e^{2\pi i\theta\phi}\, q^{\theta^2/2 - 1/24} \times$$

$$\times \prod_{n=1}^{\infty} \left(1+q^{n+\theta-1/2} e^{2\pi i\phi}\right)\left(1+q^{n-\theta-1/2} e^{-2\pi i\phi}\right) \tag{2.11}$$

where

$$\eta = q^{1/24} \prod_{n=1}^{\infty} (1-q^n) \tag{2.12}$$

is Dedekind's η-function. The transition from product to sum-representation is a statement about bosonization in two-dimensional field theory. Parametrizing the two-dimensional world-sheet by $\sigma = \sigma_1 + \tau\sigma_2$ with $\tau = \tau_1 + i\tau_2$ (2.11) can be seen as the partition function of a complex fermion with boundary conditions

$$\begin{aligned} \psi(\sigma_1+2\pi, \sigma_2) &= -e^{-2\pi i\theta}\, \psi(\sigma_1, \sigma_2) \\ \psi(\sigma_1, \sigma_2+2\pi) &= -e^{-2\pi i\phi}\, \psi(\sigma_1, \sigma_2) \end{aligned} \tag{2.13}$$

For details see Ref. [5].

It will later be useful to know the behaviour of θ-functions under the change of the argument τ. We have

$$\begin{aligned} \vartheta\left[{\theta \atop \phi-\theta-1/2}\right](\tau+1) &= e^{-i\pi(\theta^2-\theta)}\, \vartheta\left[{\theta \atop \phi}\right](\tau) \\ \vartheta\left[{-\phi \atop \theta}\right](-1/\tau) &= (-i\tau)^{1/2} e^{-2\pi i\theta\phi}\, \vartheta\left[{\theta \atop \phi}\right](\tau) \end{aligned} \tag{2.14}$$

The corresponding transformations for η are

$$\eta(\tau+1) = e^{i\pi/12}\,\eta(\tau)$$
$$\eta(-1/\tau) = (-i\tau)^{1/2}\,\eta(\tau) \tag{2.15}$$

The two transformations $\tau \to \tau+1$ and $\tau \to -1/\tau$ generate the modular transformations

$$\tau \to \frac{a\tau+b}{c\tau+d} \qquad ad-bc=1 \tag{2.16}$$

and a,b,c,d integers.

Coming back to our lattice discussion, one might observe that the transformation $\tau \to -1/\tau$ corresponds to a transition from a certain lattice to its dual. It is just a Fourier transformation and we have the Poisson resummation formula

$$\sum_{p\in\Gamma} e^{i\pi\tau p^2} = \frac{1}{Vol(\Gamma)} \sum_{k\in\Gamma^*} e^{-i\pi k^2/\tau} \tag{2.17}$$

The reader can check this explicitly for (2.9) and (2.10) by using (2.14) and computing the volume of O(2N) lattices. The reciprocity of the number of points per unit volume for a lattice and its dual is therefore just expressed through the $\tau \to -1/\tau$ transformation. A self-dual lattice should therefore have a partition function that is invariant under this transformation (apart from the fact that the volume of such a lattice is unity). For a further check reconsider again SU(3)

$$Z_{Root} = \sum_{n,m} e^{i\pi\tau\frac{1}{2}(n^2+3m^2)}\,\frac{1}{2}\left(1+(-1)^{n+m}\right) \tag{2.18}$$

and

$$Z_{Weight} = \sum_{n,m} e^{i\pi\tau\left(\frac{1}{2}n^2+\frac{1}{6}m^2\right)} \frac{1}{2}\left(1+(-1)^{n+m}\right) \tag{2.19}$$

and show the validity of (2.17).

For group lattices, there is always this duality of root and weight lattices. What about self-dual lattices? The rectangular lattices consisting of scalar and vector conjugacy classes of O(2N) are self-dual

$$Z = \vartheta^N\begin{bmatrix}0\\0\end{bmatrix} \tag{2.20}$$

Even rarer are the even self-dual lattices. They only exist in 8n dimensions. In d = 8 there is only one such lattice: the E_8 root lattice. That such a thing can happen with E_8 can already be guessed from the fact that fundamental and adjoint representations coincide and that therefore root and weight lattices have to be the same. But let us be more explicit. Consider first the O(16) root lattice

$$Z = \frac{1}{2}\left[\vartheta^8\begin{bmatrix}0\\0\end{bmatrix} + \vartheta^8\begin{bmatrix}0\\1/2\end{bmatrix}\right] \tag{2.21}$$

Then observe that the shortest vectors in the spinor conjugacy class $(n_1+\frac{1}{2},\ldots,n_8+\frac{1}{2})$ have $(\text{length})^2$ two. The lattice

$$Z = \frac{1}{2}\left[\vartheta^8\begin{bmatrix}0\\0\end{bmatrix} + \vartheta^8\begin{bmatrix}0\\1/2\end{bmatrix} + \vartheta^8\begin{bmatrix}1/2\\0\end{bmatrix} + \vartheta^8\begin{bmatrix}1/2\\1/2\end{bmatrix}\right] \tag{2.22}$$

is even integral and self-dual. Vectors with $p^2 = 2$ come with multiplicity 240; the number of E_8 roots. Equation (2.22) defines the root-lattice of E_8. The self-duality of the lattice can best be seen by showing invariance of (2.22) under the transformation $\tau \to -1/\tau$. In d = 16 there are two even integral self-dual lattices: the $E_8 \times E_8$ root lattice and a spin(32)/Z_2 lattice given by

$$\frac{1}{2}\left[\vartheta^{16}\begin{bmatrix}0\\0\end{bmatrix} + \vartheta^{16}\begin{bmatrix}0\\1/2\end{bmatrix} + \vartheta^{16}\begin{bmatrix}1/2\\0\end{bmatrix} + \vartheta^{16}\begin{bmatrix}1/2\\1/2\end{bmatrix}\right] \tag{2.23}$$

States with $p^2 = 2$ correspond to the roots of O(32). In higher dimensions these lattices appear more frequently. Already in d = 24 one has 24 different choices - the Niemeyer lattices.

3. Modular Invariance and the Heterotic String

The heterotic string is left-right asymmetric. The left-handers $X(\sigma+t)$ correspond to the bosonic string in 26 dimensions, 16 of which are compactified on a lattice defined by vectors p. The mass spectrum is given by

$$\frac{1}{4}m_L^2 = N_L - 1 + P^2/2 \tag{3.1}$$

where N_L is integer (the oscillator number) and p is a vector of the even self-dual lattice $E_8\times E_8$ or spin(32)/Z_2. The right-handers correspond to the type I superstring, i.e., the GSO-projected NSR model. For the Neveu-Schwarz sector, we have

$$\frac{1}{4}m_R^2 = N_R - 1/2 \quad ; \quad N_R \in \mathbb{Z} + 1/2 \tag{3.2}$$

and in the Ramond sector

$$\frac{1}{4}m_R^2 = N_R \quad ; \quad N_R \in \mathbb{Z} \tag{3.3}$$

Consistency furthermore requires level matching, i.e.

$$m_L^2 = m_R^2 \tag{3.4}$$

These are a lot of rules and in the remainder of the chapter we want to analyze the special properties of the model, e.g., why does the gauge group have to be $E_8\times E_8$ or O(32) and what is the reason for the GSO projection?

A basic ingredient of string theory is the reparametrization invariance on the worldsheet swept by the string $X(\sigma,t)$. At the tree level of string scattering amplitudes, this worldsheet has the topology of the sphere and any σ-t reparametrization can be continuously connected to the identity. At higher level this is not true any longer. For example, at one loop the worldsheet has the topology of a torus

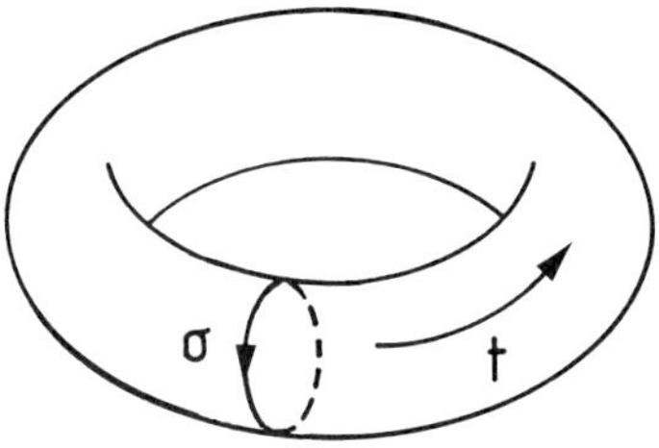

Figure 3

and certain reparametrizations, like the interchange of σ and t, are disconnected from the identity. Reparametrization invariance of string theory, of course, requires invariance under these so-called modular transformations. The discussion is usually done in Euclidean space. We pass from (σ,t) to the complex co-ordinate $\sigma = \sigma_1+\tau\sigma_2$ and τ is called the modular parameter. The modular transformations as defined in (2.16) correspond exactly to the reparametrizations on the torus, that cannot be continuously connected with the identity. The torus can be defined by the lattice as shown in Fig. 4

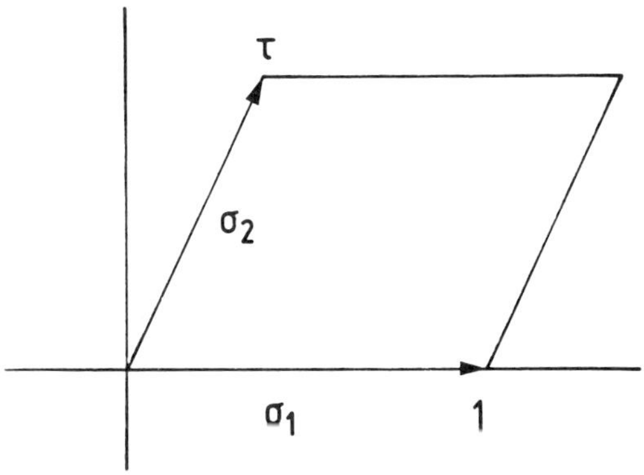

Figure 4

The modular transformations are generated by $\tau \to \tau+1$ and $\tau \to -1/\tau$. The first one leaves σ_1 unchanged and transforms σ_2 to $\sigma_1+\sigma_2$ while the second one essentially interchanges σ_1 and σ_2. In calculating loop amplitudes, we have to integrate over τ, the modular invariant measure being $d^2\tau/(\mathrm{Im}\tau)^2$. Because of (2.16) we cannot consider the full range of τ-

integration but "divide" by modular transformations and consider only a so-called fundamental domain [6]. One choice for such a domain is shown in Fig. 5

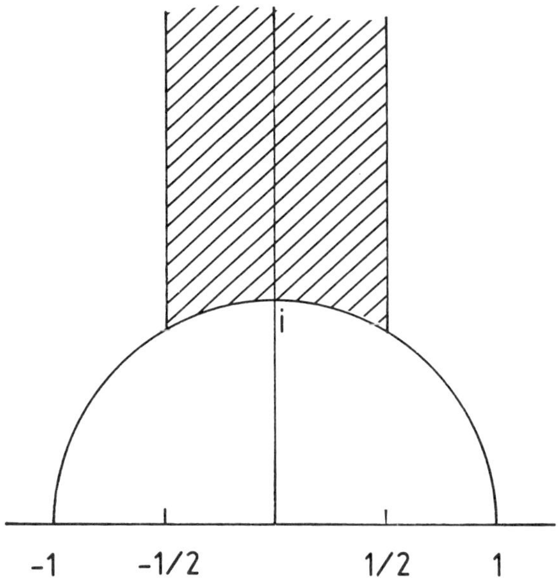

Figure 5

$-\frac{1}{2} \leqslant \mathrm{Re}\,\tau \leqslant +\frac{1}{2}$, $\mathrm{Im}\,\tau > 0$, $\tau^2 \geqslant 1$. The reader should check that all points in the upper half of the τ-plane can be reached by a modular transformation. This restriction of the τ-integration is very important for the discussion of finiteness of loop amplitudes in string theory.

The simplest loop diagram is the contribution to the vacuum energy. In field theory this gives usually a quartically divergent contribution. In string theory it corresponds to the τ-integration of the partition function

$$\int \frac{d^2\tau}{(\mathrm{Im}\,\tau)^2} \frac{1}{(\mathrm{Im}\,\tau)^4} Z(\tau)\,\bar{Z}(\bar{\tau}) \qquad (3.5)$$

where Z is the contribution of the right-movers (a function of τ) and $\bar{Z}$ the contribution of left-movers (function of $\bar{\tau}$). Given (3.1) for the left-movers, we have

$$\bar{Z}(\bar{\tau}) = \bar{\eta}(\bar{\tau})^{-24} \sum_{p \in \Lambda_{16}} \bar{q}^{\frac{1}{2}p^2} \tag{3.6}$$

We are now in a position to discuss the modular invariance of such an expression. Recall (2.15) to see that η^{24} is invariant under modular transformations [actually the $\tau \to -1/\tau$ transformation gives a prefactor but this cancels a factor arising in the transformation of the integration measure. Therefore, it would be better to say that η^{24} transforms covariantly. In the following we will just drop the $(-i\tau)^n$ factors]. The modular invariance of (3.6) requires therefore the invariance of the lattice sum, and we have discussed this question in the previous chapter. We have seen there that such transformations imply, in particular, a transition from a lattice to its dual lattice; invariance thus requires self-dual lattices. The restriction of the gauge group to $E_8 \times E_8$ or $O(32)$ comes from the requirement of modular invariance and is therefore a direct consequence of reparametrization invariance. For the lattice sum in (3.6), we have thus to insert either the square of (2.22) or (2.23). Observe also that in η^{24} it is crucial that the number of transverse oscillators is 24 by recalling the transformation of η under $\tau \to \tau+1$.

For the right-movers we have to discuss worldsheet fermions. In contrast to bosons, they can be either periodic or antiperiodic around the σ_1-cycle ($0 \leqslant \sigma_1 \leqslant \pi$) of the torus. As a result one obtains two sectors: the Neveu-Schwarz (NS) sector (antiperiodic in σ_1) and the Ramond (R) sector (periodic in σ_1). In addition we can consider periodicity and antiperiodicity in the σ_2-cycle. The partition function for a complex fermion with twisted boundary conditions is given by the θ-function in (2.11). From (2.13) we have $\theta,\phi = 0(\frac{1}{2})$ for antiperiodic (periodic) boundary conditions. Since the theory is defined in 8 transverse dimensions we have to consider as partition function

$$Z = \eta^{-12} \sum_{\theta_i,\phi_j=0}^{1/2} a_{ij}\, \vartheta^4\begin{bmatrix}\theta_i\\ \phi_j\end{bmatrix} \tag{3.7}$$

where 8 of the 12 η's correspond to the transverse bosonic excitations. For NS, $\theta_i = 0$ while for R, $\theta_i = \frac{1}{2}$. In the construction of the superstring the coefficients a_{ij} were chosen to have special values to give the desired result. We shall see now that modular invariance fixes the coefficients.

Let us start with the NS sector and consider antiperiodicity in σ_1 and σ_2, i.e., $\theta^4[{}^0_0]$. Using (2.14) for $\tau \to \tau+1$, we obtain $\theta^4[{}^{\,0}_{\frac{1}{2}}]$. Geometrically this is obvious by looking at Fig. 6:

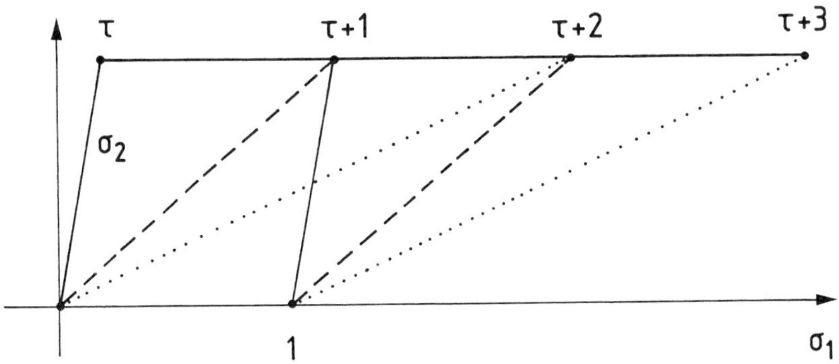

Figure 6

Under $\tau \to \tau+1$, η^{12} changes sign. A modular invariant partition function for the NS sector is therefore given by

$$Z_{NS} = \frac{1}{2}\eta^{-12}\left[\vartheta^4[{}^0_0] - \vartheta^4[{}^{\,0}_{1/2}]\right] \tag{3.8}$$

The second term in (3.8) with the - sign gives the GSO projection in the NS sector. Following our discussion in the last chapter we see that (3.8) represents the vector conjugacy class of an O(8) lattice. The allowed lattice vectors are $p = (n_1,\ldots,n_4)$ with n_i integer and sum of n_i odd. This implies odd p^2 and this is exactly the GSO projection.

The last term in (3.8) corresponds to periodic boundary conditions in the σ_2 direction. A $\tau \to -1/\tau$ transformation exchanges σ_1 and σ_2. Modular invariance teaches us that we cannot discuss the NS sector alone, it requires the existence of the R sector. Indeed, under $\tau \to -1/\tau$, $\theta[{}^{\,0}_{\frac{1}{2}}]$ transforms to $\theta[{}^{\frac{1}{2}}_{0}]$ following (2.14). Since η^{12} does not change sign in this transformation, the term inherits the - sign from the second term in (3.8). We have obtained

$$Z_R = -\frac{1}{2}\eta^{-12}\vartheta^4[{}^{1/2}_{0}] \tag{3.9}$$

via this construction. Z_R still has to fulfil a non-trivial constraint from modular invariance: the $\tau \to \tau+1$ constraint $\theta^4[{}^{\frac{1}{2}}_{0}]$ changes sign and cancels the sign obtained from the η^{-12} term, thus (3.9) is modular invariant. There are still some remarks to be made about (3.9):

(i) the factor $\frac{1}{2}$ signifies the GSO projection in the R sector. This factor comes from modular invariance, once we have normalized the NS sector correctly. Using the fact that $\theta[{}^{\frac{1}{2}}_{\frac{1}{2}}]$ vanishes, we can write (3.9) in a more symmetric form

$$Z_R = -\frac{1}{2}\eta^{-12}\left[\vartheta^4[{}^{1/2}_{0}] + \vartheta^4[{}^{1/2}_{1/2}]\right] \tag{3.10}$$

(ii) Observe that $\tau \to \tau+1$ invariance works just for θ^4 (or θ^{8n+4}) corresponding to eight transverse dimensions.

(iii) Equation (3.9) has an overall - sign. In view of (3.5) this signifies that the states in the R sector behave as space-time fermions. In the given model, modular invariance gives the correct relation between spin and statistics.

(iv) Observe that the sum $Z_{NS}+Z_R$ vanishes. The model is supersymmetric and there is no one-loop contribution to the vacuum energy (3.5).

To close this chapter let us stress again the importance of modular invariance for the construction of consistent string models. In heterotic string theory, it implies $E_8 \times E_8$ or $O(32)$ from the left-handers and gives the superstring on the right-hand side. Actually so far we have required modular invariance for left- and right-handers separately. In principle, it would be sufficient to require it only for the full theory, in which case several non-supersymmetric modular invariant $d = 10$ theories can be found. We will not discuss them here separately, since our discussion of strings on orbifolds will include these cases.

4. The Geometry of Orbifolds

Ten space-time dimensions seem to be required for the superstring. What about strings in four extended space-time dimensions? The easiest way to obtain such models is through torus compactification from d = 10 to d = 4. This construction is by now fully understood [7,8]. It can be characterized by an even integral self-dual Lorentzian lattice of signature (22,6). 22 = 16+6 for the left-movers and 6 for the right-movers. Such lattices [3] $\Lambda_{p,q}$ with signature (p,q) only exist if p-q = 0 mod 8. In a given dimension such a lattice is unique up to Lorentz transformations SO(p,q). Different models can be obtained through $SO(p,q)/SO(p)\times SO(q)$ transformations. For d = 4 this gives a 132 parameter family of consistent string models. The Lorentz transformations of the lattice are in one-to-one correspondence with background fields for the metric g_{ij} (i,j = 1,...,6), the antisymmetric tensor b_{ij} and gauge potentials a_i^I (I = 1,...,16) (corresponding to 132 degrees of freedom). The mass spectrum is given by

$$\begin{aligned} \frac{1}{4} m_L^2 &= N_L - 1 + L^2/2 \\ \frac{1}{4} m_R^2 &= N_R + \tilde{L}^2/2 \end{aligned} \tag{4.1}$$

where the six-dimensional vector $\tilde{L}$ is given by

$$\begin{aligned} \tilde{L} = -n^i R_i e_i + \frac{m_i}{2R_i} e^{*i} - \frac{b_{ij}}{R_i} e^{*i} n^j + \\ + \frac{a_i^I}{2R_i} e^{*i} [p^I - \tfrac{1}{2} a_j^I n^j] \end{aligned} \tag{4.2}$$

where n,m are integers and e_i defines the six-dimensional torus $e_i \cdot e_j = g_{ij}$. As normalization we have chosen $e_i^2 = 2$ and R_i denote the radii. The 16-dimensional vectors p^I represent the lattice of either the $E_8 \times E_8$ or the $spin(32)/Z_2$ lattice. For the left-movers we have to consider a 22-dimensional vector L. Its first six components are given by $\tilde{L}+2n^i R_i e_i$ and the last 16 by

$$L^I_{extra} = p^I - a_i^I n^i \tag{4.3}$$

The lattice $(L,\tilde{L})$ is even integral and self-dual. While in d = 10 we have gauge group $E_8 \times E_8$ or $O(32)$, we now obtain a gauge group of rank 22 from the left-movers. With suitable values of R_i, g_{ij}, b_{ij} any simply laced rank-six gauge group can be obtained. The Wilson lines a_i^I can further enhance the gauge group to, e.g., $O(44)$. For generic choices of the background fields, one obtains $U(1)^{22}$. Enhancement of the gauge group requires a careful choice of parameters. The potential has 132 flat directions. Given an enhanced gauge group, it can be broken smoothly by a variation of these parameters and a Higgs mechanism occurs.

Torus compactification of the heterotic string thus leads to a variety of d = 4 string models. Such a compactification, however, always leads to N = 4 supersymmetry (as in field theory) and thus to a non-chiral particle spectrum, not allowing a description of parity violation of the weak interactions.

To obtain chiral models we have to look for a more general construction principle, preferably one that is still simple enough to allow an exact treatment. Such models can be obtained via compactification on tori with twisted boundary conditions, so-called orbifolds. To construct these spaces one starts with a torus T^6 and then "divides" by a discrete group of symmetries P of the torus (the point group) to obtain $O^6 = T^6/P$. The geometrical object obtained in such a way is flat almost everywhere and therefore allows an exact treatment of string theory [1]. The curvature is concentrated on isolated points corresponding to the fixed points of P. Since T^6 is defined by a lattice Λ, we can equivalently define $O^6 = R^6/S$ where R^6 is flat space and the space group S contains P and the lattice shifts. To illustrate the concept let us discuss some two-dimensional examples. Start with an orthonormal lattice as shown in Fig. 7

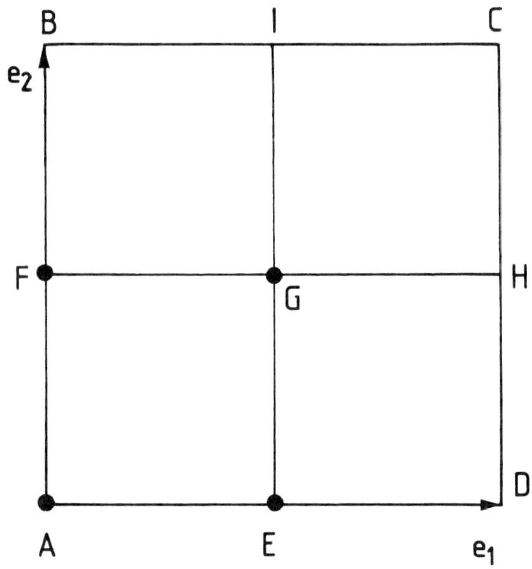

Figure 7

with T^2 defined by e_1 and e_2. We now divide by the group P of 180° rotations, i.e., we identify points connected by such a rotation. While the torus T^2 is given by the fundamental region ABCD (with AB ↔ CD and AD ↔ BC identified), the orbifold $O^2 = T^2/P$ needs only a subset of this domain. Let us first determine the fixed points of transformation P. A, of course, is trivially found to be a fixed point. More fixed points can be found by consideration of half lattice vectors. Reflecting E at the origin gives a point at distance e_1 from E and therefore E remains fixed under P. The same holds for F and G leading to a total of four fixed points. Take now the square AEGF and its reflection at the origin. By a shift e_1+e_2 it is identified with GHCI. The same is true for FGIB and EDHG. Thus the rectangle AEIB is sufficient to describe the orbifold. Furthermore, AE has to be identified with BI, AF with FB and EG with GI: i.e., take AEIB, fold along FG and glue the edges. The resulting "ravioli-like" object is an orbifold with conical singularities at the four fixed points. One could repair these singularities by cutting off the neighbourhoods of A, E, G and F and gluing in smooth disks. Topologically, this space would then be equivalent to a sphere.

Let us give a second example of a Z_2 orbifold (Z_2 because P was the two-element group). Start with the SU(3) root lattice as given in Fig. 8 by basis vectors e_1 and e_2

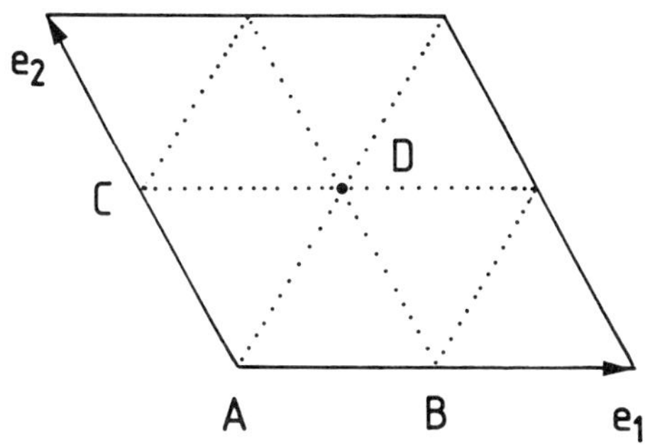

Figure 8

and identify points rotated by 180°. We have again four fixed points A, B, C and D. The reader should check that the orbifold looks as given in Fig. 9

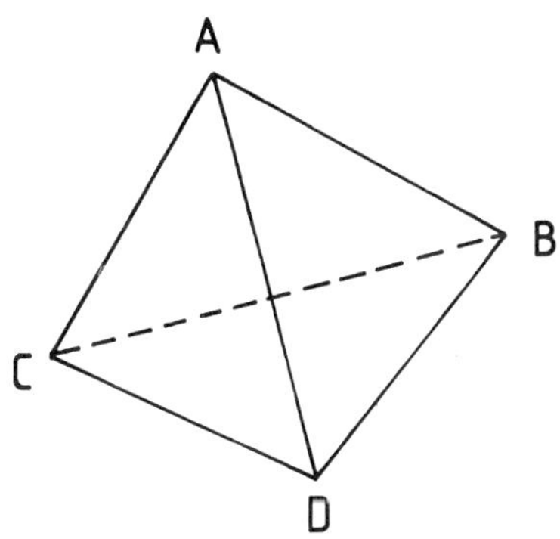

Figure 9

Observe that the lattice in Fig. 8 has also a Z_3 symmetry (rotations by 120°). $O^2 = T^2/Z_3$ will be discussed later in detail.

For Z_N orbifolds with N a prime number, we only need to know the number of fixed points to know the number of singular points. If N is non-prime this identification is not so simple (a fact that also leads to a subtlety discussing the massless spectrum of strings on such an orbifold). Let us see this in our final example. Start with the lattice of Fig. 7 and divide by $P = Z_4$, i.e., 90° rotations. Under those rotations only A and G are fixed points. But the group Z_4 also contains an element corresponding to 180° rotations with fixed points A, E, G and F. E and F have to be identified because of the 90° rotations. The square AEFG is enough to describe the Z_4 orbifold given in Fig. 10

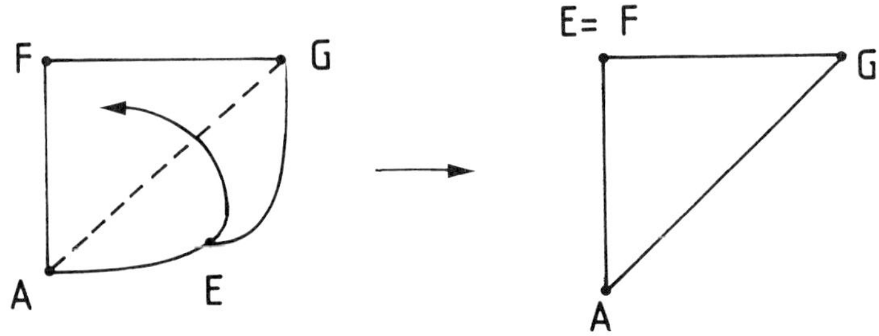

Figure 10

i.e., fold along AG and glue the edges.

5. Strings on Orbifolds

Before we discuss the motion of closed strings on orbifolds, let us recall the situation in the case of torus compactification. One starts with closed strings on flat space R^n and then identifies points that differ by lattice vectors: as a result we can distinguish two types of closed strings:

- those already closed on R^n;
- those not closed on R^n, but only on T^n; i.e., the two ends differ by a lattice vector.

This second type of states are winding states on the torus. It is not possible to consider a theory without those winding states. An unwound state can split into winding states of opposite winding number. A theory excluding the winding states will in general not be modular invariant. The Hilbert space of string states of the torus thus consists of an unwound sector and the various winding sectors. In all these sectors one has to project on states invariant under the translation by the lattice vectors defining the torus; the system has been put in a box, thus momenta in the compactified dimensions are quantized.

To discuss strings on orbifolds we start with a torus $T^n = R^n/\Lambda$ where Λ is a lattice generated by the basis vectors e_i^μ. Closed strings on the torus satisfy

$$X^\mu(\sigma=\pi) = X^\mu(\sigma=0) + 2\pi \ell^\mu \tag{5.1}$$

where $\ell^\mu = \Sigma\, m_i e_i^\mu$ ($m_i \in Z$) is a vector in Λ. Configurations with $\ell^\mu \neq 0$ are the winding states. To obtain an orbifold $O^n = T^n/P$, one further divides the torus by a group P of discrete transformations. The action of P should be consistent with the symmetry of the lattice. Alternatively we can, of course, describe the orbifold as $O^n = R^n/S$ where the space group S contains P and the translations e_i^μ. Closed strings on O^n satisfy

$$X^\mu(\sigma=\pi) = q X^\mu(\sigma=0) + 2\pi \ell^\mu \tag{5.2}$$

for $q \in P$ an element of the point group. As a result we can distinguish two types of closed strings:

- untwisted. These are the states where q in (5.2) corresponds to the identity element of the point group, i.e., those states that correspond to strings already closed on the torus T^n. Not all of the states are kept, however, one has to project on the states invariant under P.

- twisted. Here q in (5.2) represents a non-trivial element of the point group P, and these states are only closed on O^n but not on T^n (it would be an open string on T^n whose endpoints will be identified via the action of P). Again we have to project on P-invariant states.

The states in the untwisted sector correspond to a subset of states on the torus. What do the twisted states look like? Since the space is flat almost everywhere we can start with the oscillator expansion

$$X(\sigma,t) = x_0 + pt + 2\ell\sigma + \frac{i}{2}\sum\left[\frac{\alpha_n}{n} e^{-2in(t+\sigma)} + \frac{\tilde{\alpha}_n}{n} e^{-2in(t-\sigma)}\right] \tag{5.3}$$

Let us first consider the constraint through (5.2) for the centre-of-mass motion. Since (5.2) has to be true for all times t we obtain

$$X_0 + 2\pi\ell = q X_0 \tag{5.4}$$

This implies that X_o has to coincide with qX_o up to a lattice vector, i.e., X_o must be a fixed point of P. The centre-of-mass of twisted sector string states is located at the orbifold singularities and the twisted sectors are characterized by the fixed points of the transformations in P.

Relation (5.2) also gives constraints for the oscillator modes in (5.3) leading to relations like

$$\alpha_n e^{-2in\pi} = q\,\alpha_n \tag{5.5}$$

For Z_N transformations, we can represent q by $\exp(2\pi ik/N)$ $0 \leqslant k \leqslant N-1$ and (5.5) then leads to fractional oscillator numbers for the different twisted sector states. To illustrate the string configurations in the various sectors, consider again a simple two-dimensional example, the Z_2-orbifold from the SU(3) root lattice (Fig. 11)

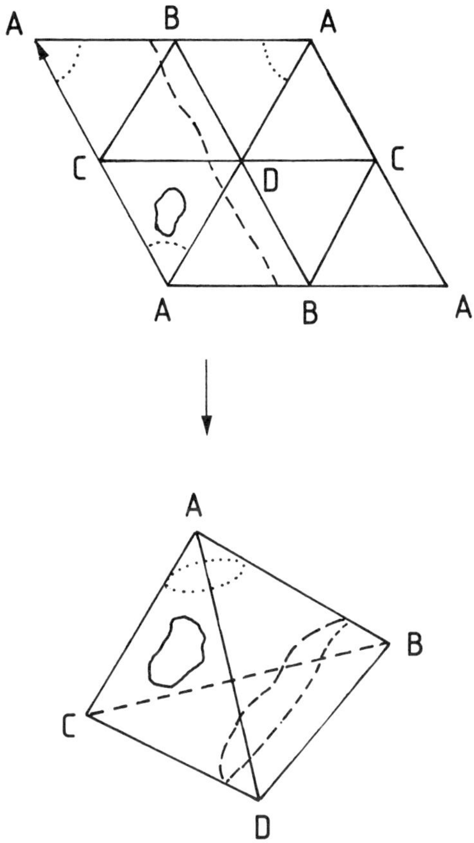

Figure 11

The solid line corresponds to a state in the untwisted sector, the dashed one to a winding state in the untwisted sector while the dotted line is a state in the twisted sector corresponding to fixed point A.

So much for the qualitative picture. Let us now discuss the spectrum of strings on orbifolds more explicitly and also formulate the constraints for consistent models. We have to satisfy the constraints of modular invariance and, actually, modular transformations can be used to construct a consistent string theory on an orbifold.

Let us again use complex co-ordinates as in Section 3, $\sigma = \sigma_1 + \tau\sigma_2$ where τ is the modular parameter. Consider the twisted string

$$X(\sigma_1+\pi,\sigma_2) = h\, X(\sigma_1,\sigma_2)$$

$$X(\sigma_1,\sigma_2+\pi) = g\, X(\sigma_1,\sigma_2) \qquad (5.6)$$

where g and h denote elements of the point group P. The transition from $\sigma = 0$ to $\sigma = \pi(1+\tau)$ can be done in two ways and consistency requires therefore that g and h commute. We now have to consider a variety of sectors denoted by (g,h) and h = 1 denotes the untwisted sector. Observe the analogy with our discussion of the NSR string in Chapter 3 where we had a twist of order two on the fermions. The untwisted sector splits up into several subsectors (g,1) for $g \in P$. We have here to consider

$$\sum_{g\in P} (g,1) \qquad (5.7)$$

and the sum projects on the invariant states. In the NS model we had seen this to give the GSO projection. For a Z_N twist we have

$$\sum_{k=0}^{N-1} \exp(2\pi i k/N) = 0 \qquad (5.8)$$

annihilating all the non-invariant states in the sum of (5.7). This concludes our discussion of the untwisted sector.

The twisted sectors can be obtained via modular transformations from the untwisted sector. Consider the sector (g,1) and recall that the transformation $\tau \to -1/\tau$ exchanges σ_1 and σ_2, thus leading to a sector (1,g). [Compare the discussion leading to formula (3.9) in Section 3.] This shows that a consistent string model on an orbifold has to include the twisted sectors. Other subsectors in the g-twisted sector can be obtained via a $\tau \to \tau+1$ transformation [see the transformation properties of θ-functions in (2.14) as well as in Fig. 6]. Starting from (1,g) the transformation $\tau \to \tau+1$ will lead to (g,g) and repeated application of $\tau \to \tau+n$ to (g^n,g) in the twisted sector. For a twist of order N we would require the sector (g^N,g) to coincide with (1,g) exactly and this is a restriction for the partition function of modular invariant models [9]. If this condition is satisfied, the model is one-loop modular invariant by construction. For the different twisted sectors we have to sum

$$\sum_h (h,g) \quad ; \quad hg = gh \tag{5.9}$$

only over elements h commuting with g. If P is non-Abelian, there might be new constraints from higher loop modular invariance [10]. Equation (5.9) again gives the correct projection on P-invariant states in the twisted sectors, as inherited from the correct projection in the untwisted sector.

This then gives the following construction scheme for string models on orbifolds. First choose a torus and a twist. Take the Hilbert space for strings on the torus and project on states invariant under the twist: this defines the untwisted sector. Then proceed with a modular transformation $\tau \to -1/\tau$ to obtain the twisted sectors. Check with $\tau \to \tau + n$ transformations whether each twisted sector closes under these transformations. We are now ready to discuss a special example in detail.

6. The Z-orbifold

Start with the $E_8 \times E_8$ heterotic string in d = 10. The mass formula is given by

$$\begin{aligned} m^2 &= 2m_R^2 = m_R^2 + m_L^2 \\ \tfrac{1}{4} m_R^2 &= N_R \\ \tfrac{1}{4} m_L^2 &= N_L - 1 + p^2/2 \end{aligned} \tag{6.1}$$

where N_L, N_R are integers and p is a vector of the $E_8 \times E_8$ lattice. (For the definition of this lattice see sections 2 and 3.) Compactify to d = 4 on a six torus given by the product of three SU(3) root lattices and the twist P is chosen to be a simultaneous 120° rotation of these three lattices. On an individual SU(3) root lattice this leaves three fixed points defining the weight lattice of SU(3) (see Fig. 12); thus in total we have $3^3 = 27$ fixed points.

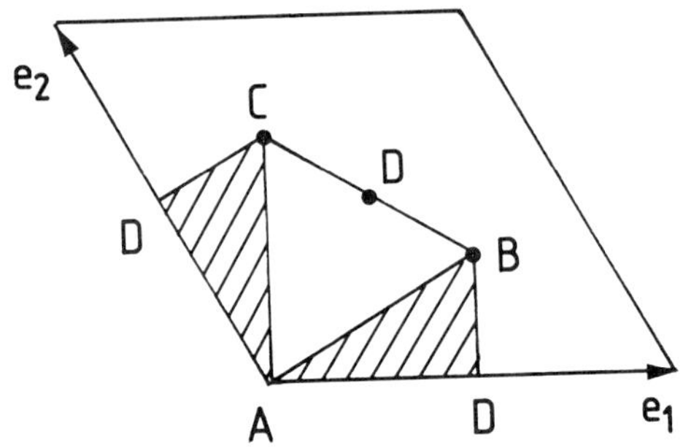

Figure 12

This completely specifies the twist on space-time bosons. We choose space-time fermions to be twisted in the same way, leading to a space-time supersymmetry.

In order to break the gauge group, one accompanies the space-time twist by a gauge transformation in $E_8 \times E_8$, i.e., we define a homomorphism of $P = Z_3$ in $G \subset E_8^2$. There are many ways to do this embedding of the twist in the gauge group. Let us here choose to represent the gauge transformation by a shift in the E_8 lattice (which in this case can be shown to be equivalent to a rotation [11]) $p \to p+v$. There are, of course, constraints on v. First of all, P is a twist of order three which implies that 3v should be an E_8 lattice vector. Moreover, the restriction from modular invariance discussed in the last section, (namely those from the $\tau \to \tau+1$ transformations in the twisted sectors) leave only a few possible choices. We choose here

$$v = \frac{1}{3}(1, 1, 2, 0, \cdots, 0) \tag{6.2}$$

which is called the standard embedding (why it is called so will become clear later). The projection for strings on the Z-orbifold will now be on states invariant under the combined action of P and G. Going to a complex notation the eigenvalues are third roots of unity represented by $\alpha = e^{2\pi i/3}$. For massless states in the untwisted sector this gives

Right		1	α	α^2
8_v		2_v	3	3
8_s		2_s	3	3
Left	$\rightarrow$			
8_v		2_v	3	3
16		16		
240		78	81	81
240'		240'		

(6.3)

$8_{v,s}$ denote the vector (spinor) with respect to the "transverse" Lorentz group O(8) in d = 10. Under the Z_3 twist 8_v splits up in an invariant O(2) vector (the d = 4 transverse Lorentz group) and six scalars transforming as α or α^{-1}. 8_s gives rise to an invariant Weyl spinor and six non-invariant states. Here the P eigenvalue is linked with helicity.

Upon multiplying left and right states, we see that only one of the four possible gravitini remain in the spectrum, thus we have N = 1 supersymmetry. We also see here explicitly that the embedding of the twist in the gauge group is necessary for the breakdown of gauge symmetry. Here the eigenvalues are obtained by $\exp(2\pi ipv)$ giving invariant states in the adjoint of $SU(3)\times E_6$ and the 81 states with α correspond to a (3,27) representation of this group. The ones with α^{-1} correspond to $(\overline{3},\overline{27})$ but to form invariant states they are combined with α states from the right-hand side having opposite helicity. We therefore obtain three chiral superfields transforming as (3,27). In addition we have nine singlet chiral superfields and of course the supergravity multiplet, as well as the gauge multiplet of $E_6\times SU(3)\times E_8$. The unbroken gauge group is of course the one that commutes with the twist. The twist chosen here was the centre of the SU(3). This concludes our discussion of massless states in the untwisted sector.

The twisted sector comes in 27 copies with identical spectrum associated with the 27 fixed points of Z_3. We have to distinguish the sectors twisted by 120° and by 240°. On the right-hand side these two sectors combine to give a chiral massless superfield at the lowest level (a fact which will become clear when we later derive the partition functions).

For the left-hand side we have six twisted oscillators: three with $N_L = Z+1/3$ and three with $Z+2/3$ and the mass formula

$$\frac{1}{4} m_L^2 = N_L - 2/3 + (p+v)^2/2 \tag{6.4}$$

where we have included the vacuum energy of twisted oscillators according to the formula [1]

$$-\frac{1}{24} + \frac{1}{4}\eta(1-\eta) \tag{6.5}$$

In the gauge group, we had represented the twist by a shift vector v. The shift in the vacuum energy is obtained here by the fact that $(p+v)^2 \geqslant 2/3$ has a minimal length. To obtain massless states one has two possibilities

- $N_L = 1/3$ with $(p+v)^2 = 2/3$

- $N_L = 0$ with $(p+v)^2 = 4/3$

Let us count the states using the representation of the E_8 lattice as discussed in Chapter 2. In the first case only the scalar conjugacy class of O(16) contributes $(n_1,\ldots,n_8) = p$ with Σn_i even. $(p+v)^2 = 2/3$ leads to three choices, corresponding to a $\bar{3}$ of SU(3). At each fixed point we have $3(1,\bar{3})$ because of the three oscillators with $N_L = 1/3$.

The second case $(p+v)^2 = 4/3$ is more complicated. In the O(16) scalar conjugacy class we have 11 solutions: $n_i = \pm 1$ for one i with $4 \leqslant i \leqslant 8$ and $n_3 = -1$, $n_1 = n_2 = 0$ gives ten possibilities corresponding to a 10 of O(10) as a subgroup of O(16) (compare Section 2) and the last one given by $n_1 = n_2 = -1$, $n_i = 0$ $i \geqslant 3$ a singlet of O(10). In the spinor conjugacy class we find 16 states $(n_1+5/6,\ n_2+5/6,\ n_3+7/6,\ n_4+\frac{1}{2},\ldots,\ n_8+\frac{1}{2})$ with Σn_i even and $(\text{length})^2 = 4/3$ corresponding to a spinor of O(10). These combine to a 27-representation of E_6. In the twisted sectors we therefore find in total chiral supermultiplets transforming as $(1,\bar{3})$ with multiplicity 81 and (27,1)) with multiplicity 27. Observe that gauge anomalies in untwisted and twisted sectors do not separately vanish, but only vanish in the complete theory. Let us also stress here that the obtained spectrum is chiral leading to 36 generations of E_6. We have found this spectrum by a succession of educated guesses. The result can of course be rigorously obtained using the modular invariant construction presented in the last section. In the remainder of this section I shall sketch this construction explicitly for the case of the Z-orbifold. This necessarily will become quite technical and the reader should be acquainted with the material presented in Sections 2 and 3.

Let us discuss left- and right-handed partition functions separately. In the untwisted sector we have to consider the subsectors (1,1), $(\theta,1)$ and $(\theta^2,1)$ where θ represents the 120° rotation. Let us concentrate on $P_{(\theta,1)}$ where we have the contribution from the E_8' lattice

$$A = \frac{1}{2\eta^8}\left[\vartheta^8\begin{bmatrix}0\\0\end{bmatrix} + \vartheta^8\begin{bmatrix}0\\1/2\end{bmatrix} + \vartheta^8\begin{bmatrix}1/2\\0\end{bmatrix} + \vartheta^8\begin{bmatrix}1/2\\1/2\end{bmatrix}\right] \tag{6.6}$$

The E_8 lattice with phases

$$B = \frac{1}{2\eta^8}\left[\vartheta^2\begin{bmatrix}0\\1/3\end{bmatrix}\vartheta\begin{bmatrix}0\\2/3\end{bmatrix}\vartheta^5\begin{bmatrix}0\\0\end{bmatrix} + \cdots\cdots\right] \tag{6.7}$$

and the space-time oscillators

$$C^{-1} = q^{8/24}\prod_{n=1}^{\infty}(1-q^n)^2(1-q^n e^{2\pi i/3})^3(1-q^n e^{-2\pi i/3})^3 \tag{6.8}$$

which, using θ-function identities, can be rewritten as

$$C = 3\sqrt{3}\,\eta\,\vartheta^{-3}\begin{bmatrix}1/2\\-1/6\end{bmatrix} \tag{6.9}$$

The partition function $P_{(\theta,1)}$ in the $(\theta,1)$ sector from the left-handers is then given by the product of A, B and C.

To obtain the twisted sectors, we have to perform a $\tau \to -1/\tau$ modular transformation. Except for the η's A will remain invariant as a self-dual lattice. For B we shall obtain

$$B(-1/\tau) \cong \sum q^{\frac{1}{2}(p+v)^2}\left(1+e^{2\pi i p\cdot w}\right) +$$

$$+ \sum q^{\frac{1}{2}(p+w+v)^2}\left(1+e^{2\pi i (p+w)w}\right) \tag{6.10}$$

where $w = \frac{1}{2}(1,1,\ldots,1)$. Next we have to check the phases obtained after a $\tau \to \tau+3$ transformation to check for modular invariance. Here $\tau \to \tau+3$ gives $\exp(3i\pi(p+v)^2)$. One checks easily that in the case under consideration $3(p+v)^2 = 0$ modulo 2 as well as $3(p+w+v) = 0$ mod 2. So for B there is no phase under the $\tau \to \tau+3$ transformation. In general this need not be the case. Only when all parts of the partition functions are multiplied does such a phase have to vanish. Next we have to discuss C including the Dedekind functions $C = \eta^{16}C'$. Dropping $(-i\tau)^{\frac{1}{2}}$ factors we obtain

$$C'(-1/\tau) = \frac{3\sqrt{3}}{4} \cdot \frac{\exp(-i\pi/2)}{\eta^{15}\vartheta^3\left[\begin{smallmatrix}-1/6\\-1/2\end{smallmatrix}\right]} \tag{6.11}$$

Again $\tau \to \tau+3$ does not give rise to a phase. Observe the factor $3\sqrt{3}$ in (6.11). From the right-handed oscillators we will get a similar factor leading to 27 the number of fixed points. In a left-right asymmetric context one side gives an overall multiplicity corresponding to the square root of the number of fixed points [12]. Assuming that there is no additional phase from the right-handers (which we will show to be true later) we can now look at the projections in the twisted sectors. B comes with a phase $\exp(i\pi(p+v)^2)$ under $\tau \to -1/\tau$. For C' we obtain in the $(1,\theta)$ sector (dropping prefactors)

$$\eta^{-18}\prod\left(1-q^{n-2/3}\right)^3\left(1-q^{n-1/3}\right)^3 \tag{6.12}$$

while in the (θ,θ) sector we have

$$e^{4\pi i/3}\eta^{-18}\prod\left(1-q^{n-2/3}e^{-4\pi i/3}\right)\left(1-q^{n-1/3}e^{4\pi i/3}\right) \tag{6.13}$$

Combining these results for C' and B we see that for the states with no oscillator excitation, one projects on $(p+v)^2-4/3 = 0 \bmod 2$, while for the states with one $q^{1/3}$ excited oscillator one obtains $(p+v)^2-8/3 = 0 \bmod 2$, corresponding to the massless spectrum previously given.

There remains the discussion of the right-handed sector. The bosonic oscillators behave as C discussed in the left-handed sector and need no further elaboration. We can thus focus on the NSR model. In d = 10 this partition function can be written in terms of θ-functions as given in (3.8) and (3.10), and again we would like to represent the twist as a shift on the "lattice". The twists we consider are, in general, given in a complex basis as $\exp 2\pi i(aJ_{12}+bJ_{34}+cJ_{56})$ (here a=b=1/3, c=2/3). The eigenvalues of the space-time fermions are then given by $\exp i\pi(\pm a\pm b\pm c)$. This formula is useful for the discussion of remaining supersymmetries [1]. In the untwisted sector one now shifts the second argument of the θ-functions by (a,b,c), e.g., in the NS sector, the (θ,1) sector is given by

$$\vartheta\begin{bmatrix}0\\0\end{bmatrix}\vartheta^2\begin{bmatrix}0\\1/3\end{bmatrix}\vartheta\begin{bmatrix}0\\2/3\end{bmatrix} - \vartheta^2\begin{bmatrix}0\\5/6\end{bmatrix}\vartheta\begin{bmatrix}0\\7/6\end{bmatrix}\vartheta\begin{bmatrix}0\\1/2\end{bmatrix} \qquad (6.14)$$

The sum of the various subsectors then gives the projection on invariant states. In (6.14) one can explicitly see that of the eight ground states, two remain invariant while three have phase α and three phase α^{-1}. Twisted sector contributions are again obtained by a $\tau \to -1/\tau$ transformation. Typically this transformation exchanges the two components of the θ-functions and we have to consider contributions like

$$\vartheta\begin{bmatrix}0\\0\end{bmatrix}\vartheta^2\begin{bmatrix}1/3\\0\end{bmatrix}\vartheta\begin{bmatrix}2/3\\0\end{bmatrix} + \cdots \qquad (6.15)$$

Working out the ground state shows that only one helicity component is selected. The CPT conjugate can be found in the θ^2 twisted sector. This concludes our sketch of the partition functions and the modular invariance of the Z-orbifold. Formulae valid in more general cases can be found in Ref. [13].

7. Further Developments

We have discussed in detail the basic facts and techniques of orbifold compactification and the reader should now be able to follow the original papers concerning further developments. These include the discussion of background fields [14], left-right asymmetric twists [12], methods to lower the rank of the gauge group [11] and many more technical details. Finally one would, of course, like to know whether such an approach can lead to a model that could realistically describe the spectrum and dynamics of quarks and leptons. Such attempts are reviewed in Ref. [15], where also a complete list of references can be found.

References

[1]. L.Dixon, J.Harvey, C.Vafa and E.Witten, Nucl.Phys. B261 (1985) 678;and B274 (1986) 285.

[2]. D.Gross, J.Harvey, E.Martinex and R.Rohm, Nucl.Phys. B260 (1985) 569; B267 (1986) 75.

[3]. P.Goddard and D.Olive in *Vertex Operators in Mathematics and Physics*, Springer (1984).

[4]. D.Mumford, Tata Lectures on Theta Functions, Birkhäuser (1983).

[5]. L.Alvarez-Gaumé, P.Ginsparg, G.Moore and C.Vafa, Comm. Math.Phys. 106 (1986) 1.

[6]. See, e.g., J.P.Serre, *A Course in Arithmetic*, Springer (1973).

[7]. K.S.Narain, Phys.Lett. 169B (1986) 41.

[8]. K.S.Narain, M.H.Sarmadi and E.Witten, Nucl.Phys. B279 (1987) 369.

[9]. C.Vafa, Nucl.Phys. B273 (1986) 592.

[10]. D.Freed and C.Vafa, Comm.Math.Phys. 110 (1987) 349.

[11]. L.E.Ibanez, H.P.Nilles and F.Quevedo, Phys.Lett. 192B (1987) 332.

[12]. K.S.Narain, M.H.Sarmadi and C.Vafa, preprint HUTP-86/A089 (1986).

[13]. L.E.Ibanez, J.Mas, H.P.Nilles and F.Quevedo, CERN preprint TH.4859 (1987) to appear in Nucl.Phys.

[14]. L.E.Ibanez, H.P.Nilles and F.Quevedo, Phys.Lett. 187B (1987) 25.

[15]. L.E.Ibanez, CERN preprint Th.4769 (1987).

Quantization of bosonic closed strings and the Liouville model

by

Sylvie Paycha
Mathematics Department, Ruhr-University,
Bochum, FRG

1. Introduction

This lecture is based on joint work with S. Albeverio, R. Høegh-Krohn, S. Scarlatti.

In the closed string picture, point particles moving in space time are replaced by loops moving in space-time. Let d be a natural number, $d \geq 2$. A closed string in $\mathbb{R}^d$ (the euclidean version of space-time) is a C^∞ map from the one-dimensional unit sphere S^1 into $\mathbb{R}^d$.

This string picture introduces a new symmetry which did not appear in the point particle picture; the group of C^∞ diffeomorphisms of S^1 denoted by $\mathcal{D}(S^1)$ acts on the set $\mathcal{L}(\mathbb{R}^d)$ of closed strings in $\mathbb{R}^d$ as follows:

$$\begin{aligned} \mathcal{D}(S^1) \times \mathcal{L}(\mathbb{R}^d) &\longrightarrow \mathcal{L}(\mathbb{R}^d) \\ (\varphi, \gamma) &\longmapsto \varphi \cdot \gamma \qquad \text{with} \quad \varphi \cdot \gamma(s) = \gamma(\varphi(s)) . \end{aligned}$$

In the quantization procedure through functional integration methods, this symmetry will yield a natural invariance property of the measures involved, namely a parametrization invariance of the surface described by the string moving in space time.

In analogy with point particles, a closed string moving in euclidean space-time $\mathbb{R}^d$ follows a variational principle. In the time $[0,T], T \in \mathbb{R}^+$, the string $\gamma \in \mathcal{L}(\mathbb{R}^d)$ describes a C^∞ 2-dimensional surface embedded in $\mathbb{R}^d$:

$$\begin{aligned} X : S^1 \times [0,T] &\longrightarrow \mathbb{R}^d \\ (s,t) &\longmapsto (X^\mu(s,t))_{\mu=1}^d \end{aligned} \tag{1.1}$$

where X denotes a map from $S^1 \times [0,T]$ in $\mathbb{R}^d$ which is C^∞, injective and has a non identically zero differential.

From the definitions (1.1), we see that a free string evolving in $\mathbb{R}^d$ can be described by an infinite cylinder embedded in $\mathbb{R}^d$ and hence can be seen as a sphere with two punctures embedded in $\mathbb{R}^d$. As described in [0], interacting strings can also be seen as compact surfaces with a finite number of punctures embedded in $\mathbb{R}^d$.

ISBN 0-12-218100-X

We shall denote by Λ° such surfaces and Λ the corresponding compactified surface obtained by adding a finite number of points to Λ°. In the same way this leads us to the following general description of closed bosonic strings:

$$\begin{aligned} X : \Lambda^\circ &\longrightarrow \mathbb{R}^d \\ (\xi_1, \xi_2) &\longmapsto (X^\mu(\xi_1, \xi_2))_{\mu=1}^d \end{aligned} \tag{1.2}$$

where X is a C^∞ map from Λ° in $\mathbb{R}^d$, which is injective and has a non indentically zero differential; Λ° is a C^∞ compact 2-dimensional real surface with a finite number of punctures, parametrized by $(\xi_1, \xi_2) \equiv \xi$.

A natural generalization of the classical action given by the length of the curve in the case of point particles moving in $\mathbb{R}^d$ is the area of the surface described by the closed strings in space time [1].

$$A_{N,G}(X) = \int_{\Lambda^\circ} \sqrt{-\left(\frac{\partial X^\mu}{\partial \xi_1} \cdot \frac{\partial X^\mu}{\partial \xi_2}\right)^2 + \left\|\frac{\partial X}{\partial \xi_1}\right\|^2 \left\|\frac{\partial X}{\partial \xi_2}\right\|^2} \, d\xi \tag{1.3}$$

A minimization of this action yields the following equations of motion

$$\partial_a \left(\sqrt{h}\, h^{ab}\, \partial_b\, X^\mu\right) = 0 \tag{1.4}$$

where ∂_a stands for $\frac{\partial}{\partial \xi_a}$, a varies between 1 and 2, and $h_{ab} = \partial_a\, X^\mu\, \partial_b\, X^\mu$ is the metric induced in the embedded surface in $\mathbb{R}^d$ by the natural euclidean metric on $\mathbb{R}^d$. h^{ab} is the metric given by the inverse matrix of h_{ab} and $\sqrt{h}$ is the root of the determinant of the matrix h_{ab}.

In 1977, Brink, Di Vecchia and Howe [2] showed that these equations of motion could also be obtained by an action whose expression is quadratic in the derivatives of X. Indeed, for a C^∞ Riemannian metric g on Λ°, the classical equations of motion (1.4) are obtained by minimising the following action in X and g:

$$A(X, g) = \frac{1}{2} \int_{\Lambda^\circ} \sqrt{g}\, g^{ab}\, \partial_a\, X^\mu\, \partial_b\, X^\mu\, d\xi \tag{1.5}$$

Here $\sqrt{g(\xi)} \equiv \sqrt{det\ g(\xi)}$ where $det\ g(\xi)$ is the determinant of the matrix $(g_{ab}(\xi))_{\substack{a=1,2\\ b=1,2}}$, $g^{ab}(\xi)$ is the inverse matrix of $g_{ab}(\xi)$.

This is an easy consequence of the fact that

$$\min_{X,g} A(X, g) = \min_{X} A(X, h)$$
$$\text{and } A_{NG}(X) = A(X, h).$$

The action $A(X, g)$ is invariant under a conformal rescaling of the metric $g \to e^\varphi g$, $\varphi \in C^\infty(\mathbb{R})$, where $C^\infty(\mathbb{R})$ are the C^∞ $\mathbb{R}$-valued functions on Λ and is also invariant under a smooth change of parametrization $(\xi_1, \xi_2) \to (\eta_1, \eta_2)$ of the surface Λ°. These two invariance properties permit us to transform the integration domain Λ° in (1.5) into a compact integration domain namely the surface Λ obtained from Λ° by adding a finite number of points.

For example, in the case of the cylinder $\Lambda^\circ = \mathbb{R} \times S_1$ parametrized by (ξ_1, ξ_2) equipped with the natural metric $g_{ab}(\xi_1, \xi_2)$ with $g_{ab}(\xi) = \delta_{ab}(\xi)$, we can set $(\eta_1, \eta_2) = (e^{\xi_1}, \xi_2)$ whereby the integration domain becomes $]0, +\infty[\times S_1 \equiv \Lambda_1$ and the metric $g_{ab}(\xi_1, \xi_2)$ is transformed into a metric $\tilde{g}_{ab}(\eta_1, \eta_2)$ represented by the matrix $\begin{bmatrix} \eta_1^{-2} & 0 \\ 0 & 1 \end{bmatrix}$. Then, by a conformal rescaling

$$\tilde{g}(\eta_1, \eta_2) \longrightarrow \frac{\eta_1^2}{\left(1 + \frac{\eta_1^2}{c^2}\right)^2} \tilde{g}(\eta_1, \eta_2) \qquad , \quad c \in \mathbb{R}^* .$$

we obtain a metric $\bar{g}(\eta)$ represented by the matrix

$$\begin{bmatrix} \frac{1}{\left(1 + \frac{\eta_1^2}{c^2}\right)^2} & 0 \\ 0 & \frac{\eta_1^2}{\left(1 + \frac{\eta_1^2}{c^2}\right)^2} \end{bmatrix}$$

Remark that $\bar{g}(\eta)$ is well defined for $\eta_1 \to 0$ and $\eta_1 \to +\infty$, so that we can extend $\bar{g}$ by continuity to $\bar{\mathbb{R}}^+ \times S_1$. Assuming that X has a C^∞ extension on Λ, we can look at the integral (1.5) as an integral on Λ [0]. As a consequence of the above remarks, in the following we shall assume that X is a C^∞ embedding from a 2-dimensional real orientated C^∞ compact connected surface Λ in $\mathbb{R}^d$ of genus $p > 0$ without boundary and we consider the action $A(X, g)$ as an integral over Λ instead of Λ°.

We set the following notations:

$$E \equiv \left\{ \begin{array}{ll} \text{one to one } C^\infty \text{ mappings} & \begin{array}{c} X : \Lambda \hookrightarrow \mathbb{R}^d \\ (\xi_1, \xi_2) \mapsto X^\mu(\xi_1, \xi_2) \end{array} \\ \text{with} & \left(\frac{\partial X}{\partial \xi_1}(\xi), \frac{\partial X}{\partial \xi_2}(\xi)\right) \neq (0,0) \quad \forall \xi = (\xi_1, \xi_2) \in \Lambda \end{array} \right\} \tag{1.6}$$

$T\ \Lambda$ is the tangent vector bundle to Λ

$T^*\ \Lambda$ is the cotangent vector bundle to Λ (1.7)

$S^2\, T^*$ is the bundle of 2 covariant symmetric tensors on Λ

$$C^\infty(S^2\, T^*) \equiv \bigcap_{k=0}^{\infty} C^k(S^2\, T^*) \tag{1.8}$$

where $C^k(S^2\, T^*)$ is the space of C^k sections of $S^2\, T^*$

$$M \equiv \left\{ \begin{array}{c} g \in C^\infty \left(S^2\, T^*\right),\; g \text{ induces a positive definite scalar product} \\ \text{on } T_m\, \Lambda, \;\text{ for every m in } \Lambda \end{array} \right\} \tag{1.9}$$

We then proceed to the quantization of closed bosonic strings following the scheme established by Polyakov in [3] which corresponds to a functional integration quantization method.

We consider the following heuristic partition function:

$$Z = \int_M D[g] \int_E D_g[X]\, e^{-\frac{1}{2} A(X,g)}\, e^{-\mu^2 \int_\Lambda \sqrt{g(\xi)} d\xi} \tag{1.10}$$

$D[g]$ and $D_g[x]$ are formal Lebesgue measures on M and E, the index g stresses the g dependance of the measure on E, μ is a formal infinite constant. Up to a cosmological term $e^{-\mu^2 \int_\Lambda \sqrt{g(\xi)} d\xi}$ which was introduced for renormalization purposes, the partition function is a formal mean value of $e^{-\ A(X,g)}$ on the space of embeddings and metrics.

In this article [3], Polyakov established in the case of genus $p = 0$ and $d < 26$, a heuristic connection between this quantization procedure for closed bosonic strings and the quantization of the classical Liouville model in two dimensions with zero mass by showing the following result:

$$Z \sim \int_{C^\infty(\mathbb{R})} e^{-\mathcal{L}(\varphi)} D[\varphi]$$

where the symbol $\sim$ stand for an identification up to infinite constant terms, $C^\infty(\mathbb{R})$ is the space of C^∞ real valued functions on Λ, $D[\varphi]$ a formal measure on $C^\infty(\mathbb{R})$ and $\mathcal{L}(\varphi) = \frac{26-d}{48\pi}\left(\int_\Lambda \left(\frac{1}{2}\left(\frac{\partial\varphi}{\partial\xi_1}\right)^2 + \frac{1}{2}\left(\frac{\partial\varphi}{\partial\xi_2}\right)^2\right) d\xi + \int_\Lambda \mu^2 e^{\varphi(\xi)} d\xi\right)$ is proportional to the classical Liouville action. $\mathcal{L}(\varphi)$ can be interpreted as the conformal anomaly term; it arises at the quantum level from a "breaking" of the invariance under conformal transformations that held at the classical level :

$$A(X,g) = A\left(X, e^\varphi g\right) \qquad \forall \varphi \in C^\infty(\mathbb{R})$$

through the quantization procedure.

This relationship between the Liouville model and closed strings has been generalized to the case of a surface of genus $p \geq 1$ [4,5,15]. However, the formal measure $D[\varphi]$ [13] involved in this description is φ dependent, as it can formally be written $\prod_{\xi\in\Lambda} e^{\varphi(\xi)} d\varphi(\xi)$ whereas the quantization of the Liouville model involves integration $e^{-\mathcal{L}(\varphi)}$ against a formal measure of the form $\prod_{\xi\in\Lambda} d\varphi(\xi)$.

In this article, we shall show that a reasonable interpretation of the term $\prod_{\xi\in\Lambda} e^{\varphi(\xi)}$ will enable us to put into correspondance the quantization of closed bosonic strings and the quantization of the Liouville model. We show that the Liouville measure formally given by $e^{-\mathcal{L}(\varphi)} \prod_{\xi\in\Lambda} d\varphi(\xi)$ is well defined for $d \leq 12$. This will give us for a surface Λ of genus $p > 1$ a mathematical realization in dimension $d \leq 12$ of the Polyakov measure given formally by:

$$\mu_p[g] = \left(\int_E e^{-A(X,g)} D_g[X]\right) e^{-\mu^2 \int_\Lambda \sqrt{g(\xi)}d\xi} D[g] \tag{1.11}$$

To establish these results we shall proceed in various steps, always assuming the surface Λ is of genus $p > 1$; in section 1, we define a parametrization of the space of metrics M. In section 2, we give a description of the heuristic measures involved in (1.10). In section 3, we describe how the formal measure $D_g[g]$ transforms through the reparametrization of the space of metrics. In section 4, we shall give the general lines of the computation of the various determinants arising in the heuristic measure induced by $\mu_p[g]$ through this reparametrization of the space of metrics. In section 5, for $d \leq 12$, we shall interprete this heuristic measure in terms of a Liouville measure which we show is well defined.

2. A parametrization of the space of metrics

As before, Λ is a C^∞ compact connected oriented 2-dimensional real manifold without boundary and genus $p > 1$. We first introduce some notations:

For a bundle E on Λ, $C^k(E)$ denotes the space of C^k-sections of E with the topology of uniform convergence in k derivatives. It has the topology of a Banach space. $C^\infty(E) \equiv \bigcap_{k=0}^\infty C^k(E)$ and it is a Fréchet manifold whose topology is given by the C^k-norms. M defined by (1.9) is an open cone in $C^\infty(S^2T^*)$ and is hence a Fréchet manifold equipped with the C^∞ Whitney topology; its tangent space at any point g in M is given by $T_gM = C^\infty(S^2T^*)$. It can in fact be equipped with a structure stronger than that of a Fréchet manifold, namely a C^∞ inverse limit Hilbert (I.L.H.) manifold structure [7]. For $E = \Lambda \times \Lambda$, we see that the sections of E are exactly the maps of Λ into Λ. We set:

$$D = \{ f \in C^\infty(\Lambda \times \Lambda),\ f \text{ is invertible, preserves orientation and } f^{-1} \in C^\infty(\Lambda \times \Lambda)\}$$

and

$$D_0 = \{ f \in D(\Lambda),\ f \text{ is homotopic to identity}\}$$

D and D_0 are Fréchet manifolds when equipped with the C^∞ Whitney topology and the tangent spaces at a point f in D respectively D_0 are given by $T_f\, D = f^*C^\infty(T\Lambda)$, $T_f\, D_0 = f^*C^\infty(T\Lambda)$ respectively. Again, they can further be equipped with a strong C^∞ I.L.H. structure [10]. D acts on $C^\infty(S^2T^*)$ in the following way; if $f \in D$ and $h \in C^\infty(S^2T^*)$, $m \in \Lambda$ and $X, Y \in T_m$, $f^*(h)_m(X,Y) = h_{f(m)}(L_fX, L_fY)$, where f^* is the pull back of f, $f^*(h)_m$ is considered as a bilinear form on T_m and L_f is the map from T_m to T_m induced by the diffeomorphism f. M is invariant by this right action that we denote by A: $D \times M \to M$,

We now describe the conformal action on M. $C^\infty(\mathbb{R})$ denotes the C^∞ $\mathbb{R}$-valued functions on Λ and $\mathcal{P} \equiv \{e^\varphi, \varphi \in C^\infty(\mathbb{R})\}$. $\mathcal{P}$ acts on M as follows:

$$\begin{aligned} &\mathcal{P} \times M \longrightarrow M \\ &(p, g) \longmapsto p \cdot g \qquad \text{where } (p \cdot g)(m) = p(m)\, g(m), \ \forall m \in \Lambda \end{aligned}$$

$\mathcal{P}$ is a C^∞ I.L.H. manifold and the quotient space $M/\mathcal{P}$ is also a C^∞ I.L.H. manifold [7].
The action A induces an action

$$\begin{aligned} \bar{A} : M/\mathcal{P} \times D &\longrightarrow M/\mathcal{P} \\ ([g], f) &\longmapsto f^*.[g] \equiv [f^* g]. \end{aligned}$$

where $[g] = \{p \cdot g, \ p \in \mathcal{P}\}$ for a metric g in M.
Since $f^*(p \cdot g) = (p\, f) \cdot f^* g$ it follows that $f^*.[g]$ is well defined. The resulting orbit space $M/\mathcal{P}/D_0$ is the space of C^∞ non equivalent conformal structures on Λ. Two metrics g_1 and g_2 in M are conformally equivalent if there exists an $f \in D_0$ and a $p \in \mathcal{P}$ such that $f^* g_1 = p g_2$. $M/\mathcal{P}/D_0$ can therefore be interpreted as the space of classes of conformally equivalent metrics.

We now introduce the space of C^∞ complex structures on Λ. Let T_1^1 be the vector bundle over Λ of tensors of type $(1,1)$ and $C^\infty(T_1^1)$ the corresponding space of C^∞ sections.

The space of C^∞ almost complex structures is defined as follows:
$\mathcal{A} = \{J \epsilon C^\infty(T_1^1) / J^2 = -I$ and the natural orientation induced by J is that of $\Lambda\}$.
The space $\mathcal{A}$ is a C^∞ Fréchet submanifold of the Fréchet space $C^\infty(T_1^1)$ and even a C^∞ I.L.H. manifold [7]. The space of complex structures C is the space of integrable almost complex structures and can be desribed as follows:

$$C = \{J \epsilon \mathcal{A}, \ \forall X, Y \epsilon C^\infty(T\Lambda), [JX, JY] - [X, Y] - J[X, JY] - J[JX, Y] = 0\}$$

As Λ is of dimension 2, $C = \mathcal{A}$. Furthermore, it is shown in [7] that:

Prop.2.1: There is a C^∞ I.L.H. diffeomorphism $\Phi : C \longrightarrow M/\mathcal{P}$ such that

$$\Phi(f^* J) = f^* \Phi(J) \quad \forall f \epsilon D \ \forall J \epsilon C$$

D acts on C by pullback: for $J \epsilon C$ and $f \epsilon D$, define the pullback $f^* J \epsilon C$ by:

$$(f^* J)(m) \cdot X = d\, f(m)^{-1} J\,(f\,(m))\, d\, f(m) \cdot X$$

when $X \epsilon T_m \Lambda$ and d denotes differentiation.

We introduce the Teichmüller space, i.e. the space of nonequivalent complex structures:

$$\mathcal{T} \equiv C/D_0 \tag{2.1}$$

It has been pointed out in [6] that this definition of Teichmüller space coincides with the usual definition given by Ahlfors and Bers [8][9]. It is by now a classical result that $\mathcal{T}$ is a real analytic submanifold of $SL(2, \mathbb{R})^{2p}$ [6] and it has been proved that it is diffeomorphic to $\mathbb{R}^{6p-6}$ [7].

We now state a result proved in [7] on the structure of C, using the fact that D_0 is a strong I.L.H. Lie group :

Prop.2.2: D_0 acts freely on C. If π denotes the quotient map $\pi : C \longrightarrow \mathcal{T}$, where $\mathcal{T}$ is given the quotient topology, π defines a strong C^∞ I.L.H. trivial principal bundle.

For a metric g we can define $\pi\left(\Phi^{-1}([g])\right)$ which is called the Teichmüller class of g and is denoted by $t(g)$. Thus for any $h \epsilon [g] : t(h) = t(g)$.

Remark: In the case of genus 0 or 1, the action is not free as there exist conformal killing diffeomorphisms, i.e. diffeomorphisms f such that there is a metric $g \epsilon M$ and a function $p \epsilon \mathcal{P}$ with the property $f^* \cdot g = p \cdot g$.

Let

$$\begin{aligned} s : \mathcal{T} &\longrightarrow C \\ t &\longmapsto s(t) \end{aligned} \tag{2.2}$$

denote a fixed C^∞ global section of the fibre bundle $\pi : C \longrightarrow \mathcal{T}$. This induces a C^∞ map:

$$\begin{aligned} \Phi \circ s : \mathcal{T} &\longrightarrow M/\mathcal{P} \\ t &\longmapsto \Phi \circ s(t) \end{aligned}$$

where $M/\mathcal{P}$ is looked upon as a C^∞ I.L.H. manifold.

For g in M, $R(g) \epsilon C^\infty(\mathbb{R})$ denotes the scalar curvature of g and $M_{-1} \equiv \{g \epsilon M, R(g) = -1\}$.

Following [7], we now set up a one to one correspondance between $M/\mathcal{P}$ and M_{-1}; we shall here only state the results refering the reader to [7] for the proofs.

M_{-1} is a C^∞ .I.L.H. manifold and we have :

Prop.2.3: For a metric g in M, there exists a unique $\psi(g) \epsilon \mathcal{P}$ such that $R(\psi(g) \cdot g) = -1$ and the map

$$\begin{aligned} M &\longrightarrow \mathcal{P} \\ g &\longmapsto \psi(g) \end{aligned}$$

is a C^∞ I.L.H. map.

It induces a C^∞ I.L.H. map $\Psi : M \longrightarrow M_{-1}$

$$g \longmapsto \psi(g) \cdot g$$

From the fact that $\Psi(g_1) = \Psi(g) \quad \forall g_1 \in [g]$, it follows that Ψ induces a C^∞ I.L.H. map :

$$\begin{aligned} \tilde{\Psi} : M/\mathcal{P} &\longrightarrow M_{-1} \\ [g] &\longmapsto \Psi(g) \end{aligned}$$

From the above discussion, we obtain a C^∞ map:

$$\begin{aligned} \Xi : \mathcal{T} &\longrightarrow M_{-1} \\ t &\longmapsto \tilde{\psi}\left(\Phi(s(t))\right) \end{aligned} \tag{2.3}$$

where M_{-1} is looked upon as a C^∞ I.L.H. manifold. From the fact that the I.L.H. structure is stronger than the Fréchet structure follows that Ξ can be seen as a C^∞ map from $\mathcal{T}$ to M_{-1} when M_{-1} is equipped with the C^∞ Whitney topology.

Before stating the theorem that gives a parametrisation of M, we shall show a preliminary easy lemma:

Lemma 2.4: For t in $\mathcal{T}$ the Teichmüller class $t(\Xi(t))$ of $\Xi(t)$ is equal to t.

Proof: Let g be a metric in M, for any $h \epsilon [g]$, we have $t(h) = \pi(\Phi^{-1}([g]))$ hence as $\tilde{\psi}(\Phi(s(t))) \epsilon \Phi(s(t))$ by construction we conclude that:

$$\begin{aligned} t\left(\tilde{\psi}(\Phi(s(t)))\right) &= \pi\left(\Phi^{-1}(\Phi(s(t)))\right) \\ &= \pi(s(t)) = t \end{aligned}$$

Theorem 2.5 Given a C^∞ section s of the fibre bundle $\pi : C \longrightarrow \mathcal{T}$, for a metric g in M, there exists a unique f in D_0 and a unique $\varphi \epsilon P$ such that

$$g = f^* e^\varphi g_t \quad \text{where} \quad g_t \equiv \Xi(t(g))$$

The map: $T : P \times D_0 \times \mathcal{T} \longrightarrow M$ is C^∞.

$$(\varphi, f, t) \longmapsto g \equiv f^* e^\varphi g_t$$

Proof: The existence property is an easy consequence of the lemma 2.4 :

For g in M, $g_t \equiv \Xi(t(g))$ is well defined and by the lemma 2.4 we have $t(g) = t(g_t)$. Thus $\pi(\Phi^{-1}(g_t)) = \pi(\Phi^{-1}(g))$ by definition of the Teichmüller class. As $\pi : C \longrightarrow \mathcal{T}$ is a principal fibre bundle with structure group D_0 , there exists $f \epsilon D_0$ such that

$$f^*\left(\Phi^{-1}[g_t]\right) = \Phi^{-1}([g])$$

but $f^* \Phi^{-1}[g_t] = \Phi^{-1}[f^* g_t]$ (this is easily verified from the equality in Prop.2.1). These two equalities imply that $[g] = [f^* g_t]$ and hence yield the existence of $\tilde{\varphi} \epsilon C^\infty(\mathbb{R})$ such that $g = e^{\tilde{\varphi}} f^* g_t$. Setting $\varphi = \tilde{\varphi} \circ f^{-1}$, the last equality reads $g = f^* e^\varphi g_t$ which yields the existence of the triple (f, φ, t) satisfying the theorem.

As for the uniqueness, let us assume that for a given C^∞ section s of $\pi : C \longrightarrow \mathcal{T}$ and a given g in M, there are two triples (f_1, φ_1, t_1) and (f_2, φ_2, t_2) in $D_0 \times P \times \mathcal{T}$ satisfying the conditions of the theorem. Then by definition

$$g_{t1} = g_{t2} = \Xi(t(g)) \equiv g_t$$

and hence $g = f_1^* e^{\varphi_1} g_t = f_2^* e^{\varphi_2} g_t$. (*)

From these equalities follows easily that:

$$\Phi^{-1}[g] = f_1^* \Phi^{-1}[g_t] = f_2^* \Phi^{-1}[g_t]$$

Let us denote by c_t the complex structure $\Phi^{-1}[g_t]$, then $\left(f_1^{*-1} \circ f_2^*\right) c_t = c_t$ which implies that $f_1 = f_2$ from the fact that D_0 acts freely on C.

Returning to $(*)$ we now write $e^{\varphi_1 - \varphi_2} g_t = g_t$ which again by the freedom of the action of P on M yields $\varphi_1 = \varphi_2$ which ends the proof of uniqueness.

Let us now investigate the regularity properties of the map T:
The action:

$$\begin{aligned} D_0 \times M &\longrightarrow M \\ (f, g) &\longmapsto f^* g \end{aligned} \tag{2.4}$$

as well as the action

$$\begin{aligned} P \times M &\longrightarrow M \\ (p, g) &\longmapsto p \cdot g \end{aligned} \tag{2.5}$$

are smooth. On the other hand the map $t \longrightarrow g_t \equiv \Xi(t(g))$ given by (2.3) is smooth. This accounts for the smoothness of the map T.

As we shall see in the next paragraph, the transformation of the heuristic measure $\mu_p[g]$ at a point given in M involves not the operator T but its differential which, by the above theorem is well defined. We shall denote by dT^g the differential at point g acting from $T_\varphi C^\infty(I\!R) \times T_f D_0 \times T_t \mathcal{T}$ onto $T_g M$, where $g = T(\varphi, f, t)$ We introduce the notation

$$E_1^g \equiv T_\varphi C^\infty(I\!R) \times T_f D_0 \times T_t \mathcal{T} \qquad \text{for} \quad g = T(\varphi, f, t) \tag{2.6}$$

and correspondingly

$$E_1^{\bar{g}_t} \equiv T_\varphi C^\infty(I\!R) \times T_{1\!\!1} D_0 \times T_t \mathcal{T} \qquad \text{for} \quad \bar{g}_t = T(\varphi, 1\!\!1, t) \tag{2.7}$$

We define, for a fixed metric g in M

$$\begin{aligned} T_1^g : C^\infty(I\!R) &\longrightarrow M \qquad & T_2^g : D_0 &\longrightarrow M \qquad & T_3^g : \mathcal{T} &\longrightarrow M \\ \varphi &\longmapsto e^\varphi g & f &\longmapsto f^* g & t &\longmapsto f^* e^\varphi \Xi(t). \end{aligned}$$

and the corresponding differentials (which are well defined by the smoothness of the actions given by (2.4) and (2.5)), $dT_1^g(\varphi), dT_2^g(f), dT_3^g(t)$ acting respectively from $T_\varphi C^\infty(I\!R)$ to $T_g M$, $T_f D_0$ to $T_g M$ and $T_t \mathcal{T}$ to $T_g M$. It is clear from the definition of T that $dT^g(\varphi, f, t)$ splits as follows:

$$dT^g(\varphi, f, t) = dT_1^{f^* g_t}(\varphi \circ f) + dT_2^{\bar{g}_t}(f) + dT_3^g(t)$$

where $\bar{g}_t \equiv e^\varphi g_t$ and $g_t \equiv \Xi(t)$.

We can compute the first two terms in this expression using Gâteaux derivatives; it is easily seen that

$$dT_1^{f^* g_t}(\varphi \circ f) \cdot \lambda = (\lambda \circ f)\left(e^{\varphi \circ f}\right) f^* g_t = f^*\left(\lambda e^\varphi g_t\right) = f^*\left(\lambda \bar{g}_t\right) \qquad \forall \lambda \epsilon T_\varphi C^\infty(I\!R)$$

Now, for v in $T_f D_0 = f^* C^\infty(T\Lambda)$, we can write $v = f^* u,\ u \epsilon C^\infty(T\Lambda)$. We define a local 1-parameter group of diffeomorphisms generated by the family f^u_λ of an open subset U of D such that, for $\lambda, \mu \epsilon I_\epsilon$ with $f^u_\lambda(\xi) \epsilon U$, $f^u_\mu(\xi) \epsilon U$, $f^u_{\lambda+\mu}(\xi) = f^u_\lambda \circ f^u_\mu(\xi)$ and $u(\xi)$ is the tangent vector to the curve $f^u_\lambda(\xi)$ at $f^u_0(\xi) = \xi$. This one parameter group generates a one parameter group $h^u_\lambda = f^u_\lambda \circ f$ with tangent vector v at $h^u_0(\xi) = f(\xi)$. The Fréchet derivative at point f in direction v of T^g_2 is given by

$$\lim_{t \to 0} \frac{(h^u_\lambda)^* \bar{g}_t(\xi) - f^* \bar{g}_t(\xi)}{t} = f^* \lim_{t \to 0} \frac{(f^u_\lambda)^* \bar{g}_t(\xi) - \bar{g}_t(\xi)}{t} = f^*(\nabla_{\bar{g}_t} u).$$

Finally, as $\mathcal{T}$ is a finite dimensional manifold, $d\Xi(t)$ at point t of $\mathcal{T}$ is given by the ordinary differential through a basis $(e^t_i)^n_{i=1}$ with $n \equiv \dim \mathcal{T} = 6p - 6$ of $T_t \mathcal{T}$ and by the smoothness property of the maps explicited in (2.4) and (2.5), we can write

$$dT^g_3(t)\, s = f^* e^\varphi \frac{\partial \Xi(t)}{\partial t_i} \cdot s_i = f^* e^\varphi \frac{\partial g_t}{\partial t_i} \cdot s_i$$

where $\quad s = s_i e^t_i \;\epsilon\; T_t \mathcal{T}$.

Summing up the results we have for $(\lambda, u, s) \epsilon E^{\bar{g}_t}_1$:

$$dT^g(\lambda, f^* u, s) = f^* \left[\lambda\, \bar{g}_t + \nabla_{\bar{g}_t} u + e^\varphi \frac{\partial g_t}{\partial t_i} \cdot s_i \right] \tag{2.8}$$

Let us remark that

$$dT^{\bar{g}_t}(\lambda, u, s) = \lambda\, \bar{g}_t + \nabla_{\bar{g}_t} u + e^\varphi \frac{\partial g_t}{\partial t_i} \cdot s_i$$

so that we have the relation

$$dT^g(\lambda, f^* u, s) = f^* \left(dT^{\bar{g}_t}(\lambda, u, s) \right) \qquad \forall (\lambda, u, s) \epsilon E^{\bar{g}_t}_1 \tag{2.9}$$

We therefore work with $dT^{\bar{g}_t}$ from which we shall get the corresponding results on dT^g through (2.9).

We shall need the following classical property of 2-tensor fields on Λ.

Lemma 2.6: Given a metric g in M, every covariant real 2-tensor field T_{ab} on Λ can be written in a unique way as the sum of a traceless covariant 2-tensor field S_{ab} (the trace being taken with respect to the metric g) and a covariant 2-tensor field R_{ab} on Λ of the form $R_{ab} = \alpha g_{ab}$ with α a real valued function on Λ.

Proof: The existence comes from the splitting of T_{ab}:

$$T_{ab} = \frac{1}{2} g_{ab} g^{cd} T_{cd} + T_{ab} - \frac{1}{2} g^{cd} T_{cd} g_{ab}$$

we then set $R_{ab} = \frac{1}{2} g_{ab} g^{cd} T_{cd}$ and $S_{ab} = T_{ab} - \frac{1}{2} g^{cd} T_{cd} g_{ab}$
The uniqueness of this decomposition is clear, for if $T_{ab} = \alpha g_{ab} + S_{ab}$ we immediately have $\alpha = \frac{1}{2} g_{ab} T_{ab}$ and $S_{ab} = T_{ab} - \alpha g_{ab}$.
This leads to the following definition; for a metric g in M, we set $S_{0,g}$ to be the space of tensors in $C^\infty (S^2 T^*)$ which have trace zero with respect to the metric g. The lemma 2.6 yields the decomposition: $C^\infty (S^2 T^*) = C^\infty (\mathbb{R}).g \oplus S_{0,g}$ for a metric g in M.
Using the results of lemma 2.6, we can decompose the third term in the bracket on the right hand side of (2.8) :

$$\left(e^\varphi \frac{\partial gt}{\partial t_i} s_i \right)_{ab} = \frac{1}{2} \bar{g}_t^{cd} \left(e^\varphi \frac{\partial g_t}{\partial t_i} s_i \right)_{cd} \bar{g}_{t\,ab} + \left(e^\varphi \frac{\partial g_t}{\partial t_i} s_i \right) - \frac{1}{2} \bar{g}_t^{cd} \left(e^\varphi \frac{\partial g_t}{\partial t_i} s_i \right)_{cd} \bar{g}_{t\,ab} \,.$$

and set for convenience :

$$\chi_i^{\bar{g}_t} \equiv e^\varphi \left(\frac{\partial gt}{\partial t_i} - \frac{1}{2} g_t^{cd} \left(\frac{\partial g_t}{\partial t_i} \right)_{cd} g_t \right) \equiv e^\varphi \chi_i^{gt} \tag{2.10}$$

Of course we have that $\chi_i^{\bar{g}_t} \in S_{0,\bar{g}_t}$ and $\chi_i^{gt} \in S_{0,gt}$. We apply the lemma again to the second term on the right hand side of (2.8) :

$$(\nabla_{\bar{g}_t} u)_{ab} = \frac{1}{2} \bar{g}_{t\,ab} \bar{g}_t^{cd} (\nabla_{\bar{g}_t} u)_{cd} + (P_{\bar{g}_t} u)_{ab}$$

where, for a metric g in M, P_g is an operator acting from $C^\infty (T\Lambda)$ into $S_{0,g}$ given by :

$$(P_g u)_{ab} \equiv (\nabla_g u)_{ab} - \frac{1}{2} g^{cd} (\nabla_g u)_{cd} g_{ab} \tag{2.11}$$

At this stage, for $(\lambda, u, s) \epsilon E_1^{\bar{g}_t}$, $dT^{\bar{g}_t}(\lambda, u, s)$ reads :

$$dT^{\bar{g}_t}(\lambda, u, s) = \bar{g}_t \left(\lambda + \frac{1}{2} \bar{g}_t^{cd} (\nabla_{\bar{g}_t} u)_{cd} + \frac{1}{2} \bar{g}_t^{cd} \left(e^\varphi \frac{\partial g_t}{\partial t_i} s_i \right)_{cd} \right) + P_{\bar{g}_t} u + \chi_i^{\bar{g}_t} s_i \tag{2.12}$$

We shall further split this sum using orthogonality properties :
For $h, k \in C^\infty (S^2 T^*)$ and for a metric g in M, we set :

$$\langle h, k \rangle^g \equiv \int_\Lambda d\xi \sqrt{g(\xi)} \{ G^{abcd}(\xi) + K g^{ab}(\xi) g^{cd}(\xi) \} h_{ab}(\xi) k_{cd}(\xi) \tag{2.13}$$

where K is a fixed constant such that $\langle h, h \rangle^g > 0 \quad \forall h \neq 0$,

$$G_{ab}^{cd} \equiv \frac{1}{2} \left(\delta_a^c \delta_b^d + \delta_b^c \delta_a^d - g_{ab} g^{cd} \right),$$

and

$$G^{abcd} \equiv g^{ae} g^{bf} G^{cd}_{ef}$$

We shall also need the following scalar product in $T\Lambda$: for u, v in $T\Lambda$, and for a metric g in M , we set :

$$\langle u, v\rangle_g \equiv \int_\Lambda d\xi \sqrt{g(\xi)} g^{ab}(\xi) u_a(\xi) u_b(\xi) \tag{2.14}$$

with $u_a = g_{ac} u^c$

We denote by $\overline{C^\infty(T\Lambda)}^g$ the closure of $C^\infty(T\Lambda)$ for the scalar product $\langle\cdot, \cdot\rangle_g$. P_g is densly defined in $\overline{C^\infty(T\Lambda)}^g$ and its adjoint operator P_g^* with respect to $\langle\cdot, \cdot\rangle_g$ and $\langle\cdot, \cdot\rangle^g$ is therefore well defined .
S_{0,g_t} splits into an orthogonal sum :

Lemma 2.7 : For $\bar{g}_t = e^\varphi \Xi(t)$ with $t \in \mathcal{T}$ and $\varphi \in C^\infty(I\!R)$, the following decomposition holds :

$$C^\infty(S^2T^*) = C^\infty(I\!R)_{g_t} \oplus Im P_{g_t} \oplus \text{Ker } P_{g_t}^*$$

and the sums are orthogonal sums with respect to $\langle\cdot, \cdot\rangle^{g_t}$

We shall only give a sketch of this proof :
It is clear that $\overline{S_{0,g_t}} = \overline{Im P_{g_t}} \oplus \overline{\text{Ker } P_{g_t}{}^*}$ where $\overline{S_{0,g_t}}, \overline{Im P_{g_t}}$ and $\overline{\text{Ker } P_{g_t}^*}$ are the closures of $S_{0,g_t}, Im P_{g_t}$ and Ker $P_{g_t}^*$ for the scalar product $\langle\cdot, \cdot\rangle^{g_t}$ defined in (2.13). On the other hand $P_{g_t}^*$ can be interpreted as an elliptic operator on the compact surface Λ ; let $\xi_t \equiv (\xi_t^1, \xi_t^2)$ be the isothermal atlas corresponding to the conformal structure $[\bar{g}_t]$ on Λ , then

$$\left(P_{g_t}^* h\right)_b = e^{\theta t} \frac{\partial}{\partial \xi_t^a} (h_{ab}) \tag{2.15}$$

for $h \in S_{0,g_t}$ where $\bar{g}_t(\xi_t) = e^{\theta_t} \delta_{ab}(\xi_t)$. Thus Ker $P_{g_t}^*$ is finite dimensional and $\overline{\text{Ker } P_{g_t}^*} = \text{Ker } P_{g_t}^*$. An element h in S_{0,g_t} can be written as $h = h_1 + h_2$ where $h_1 \in \text{Ker } P_{g_t}^*$ and $h_2 \in \overline{Im P_{g_t}}$. Thus h_2 is a C^∞ tensor and hence belongs to $Im\, P_{g_t}$. This yields the result :

$$S_{0,g_t} = Im P_{g_t} \oplus \text{Ker } P_{g_t}^*$$

and ends the proof of the lemma for the other direct sum arising in the decomposition of $C^\infty(S^2T^*)$ is a consequence of lemma 2.6 and is easily proven to be orthogonal.□

Now we can rewrite (2.12) using the orthogonal decomposition given by lemma 2.7:

$$\begin{aligned} dT^{g_t}(\lambda, u, s) =& \bar{g}_t \left(\lambda + \frac{1}{2}\bar{g}_t^{cd} (\nabla_{g_t} u)_{cd} + \frac{1}{2}\bar{g}_t^{cd} \left(e^\varphi \frac{\partial g_t}{\partial t_i}\right)_{cd} s_i\right) \\ &+ P_{g_t}\left(u + s_i V_i^{g_t}\right) + s_i \left\langle \chi_i^{g_t}, u_k^{g_t} \right\rangle^{g_t} u_k^{g_t} \end{aligned} \tag{2.16}$$

where $(u_k^{g_t}, k = 1, 2, .., n)$ is an orthogonal basis of the finite dimensional space Ker $P_{g_t}^*, V_i^{g_t} \in C^\infty(T\Lambda)$ for $i \in \{1, 2, .., n\}$, and $\chi_i^{g_t} = e^\varphi \chi_i^{gt}$ is as in (2.10).

Let us, at this stage, point out that Ker $P^*_{g_t}$ can be interpreted as the space of holomorphic quadratic differentials on Λ as can be seen from (2.15), see e.g. [7]. From (2.15) it is clear that Ker $P^*_{g_t}$ is independant of φ for it only depends on ξ_t and not on θ_t .
However we remark that $\left(u_k^{g_t}\right)_{k=1}^n$ is φ - dependent for $\bar{g}_t = e^\varphi \Xi(t)$ as a consequence of the φ - dependence of the scalar product $\langle \cdot , \cdot \rangle^{\bar{g}_t}$. We choose an arbitrary basis $\left(\varphi_j^{gt}\right)_{j=1}^n$ of Ker$P^*_{g_t}$ which will be independent of φ and set :

$$u_k^{g_t} = b_{kj}\varphi_j^{gt}$$

where $(b_{kj})_{\substack{k=1,\dots,n\\ j=1,\dots,n}}$ is a $n \times n$ matrix in $I\!R$ Then (2.15) reads :

$$\begin{aligned} dT^{g_t}(\lambda, u, s) =& \bar{g}_t\left(\lambda + \frac{1}{2}\bar{g}_t^{cd}(\nabla_{g_t}u)_{cd} + \frac{1}{2}\bar{g}_t^{cd}(e^\varphi \frac{\partial gt}{\partial t_i})_{cd}\, s_i\right) \\ &+ P_{g_t}(u + s_i V_i^{g_t}) + s_i b_{kl} b_{kj}\left\langle \chi_i^{g_t}, \varphi_j^{g_t}\right\rangle^{g_t} \varphi_l^{gt} \end{aligned} \tag{2.18}$$

This gives us an expression of the derivative of the operator T^{g_t} which we shall use in the following section.
We now proceed to the description of the heuristic measures involved in (1.10).

3. The Polyakov "measure" on the space of metrics

In this section, we give a formal description of μ_p described by (1.11) and study its behavior under the transformation T of th. 2.5. It is not in the scope of this article to give a precise description of this formal measure; this would involve introducing local cylinder set measures on the manifold induced by cylinder set measures on the tangent spaces through an exponential mapping, and the existence of an exponential mapping requires a stronger structure than just the Fréchet structure M is equipped with namely a Hilbert manifold structure. In what follows we only give a device as how to use these heuristic measures extending the integration rules used in the finite dimensional case to separable Hilbert spaces.

The parametrisation of the manifold of metrics given in th.2.5 induces a transformation of the measure $D[g]$ and hence $\mu_p[g]$, which will grossely become a tensor product of measures defined in D_0, $C^\infty(I\!R)$ and $\mathcal{T}$. The local heuristic measure $D[g]$ induces a formal measure $D_g[h]$ on the tangent space T_gM at point g , which transforms under the map dT^g when $D[g]$ is transformed through the map T ; we shall later on in this paragraph give a description of the formal measure induced by $D_g[h]$ on $T_\varphi C^\infty(I\!R) \times T_f D_0 \times T_t\mathcal{T}$ through dT^g.
We now set the general framework to describe these measures.
Let E be a topological space and H be its closure with respect to a continuous scalar product $\langle \cdot , \cdot \rangle$ defined on E. We shall make the assumption that H is a separable

Hilbert space and denote by $(H_n)_{n\in\mathbb{N}}$ an orthonormal basis of H for the scalar product $\langle\cdot,\cdot\rangle$.
We shall also supose that $H_n \in E \quad \forall n \in \mathbb{N}$. For $h \in H$, we can write :

$$h = \sum_{n=1}^{\infty} h_n H_n \tag{3.1}$$

where we have set

$$h_n \equiv \langle h, H_n\rangle \tag{3.2}$$

For a generic element h in H, we define $D[h]$ by the formal product measure :

$$\prod_{n=1}^{\infty} dh_n \tag{3.3}$$

where dh_n is the ordinary Lebesgue measure in $\mathbb{R}$.
Let H_1 be another separable Hilbert space with scalar product $\langle\cdot,\cdot\rangle_1$ and A an operator from H_1 to H defined on a topological space E_1 dense in H_1 for the topology given by the scalar product, we make the further assumption that $A(E_1)$ is dense in H. A admits an adjoint operator A^* from H to H_1 and we shall assume that A^*A is a strictly positive operator and has discrete spectrum $(\alpha_n)_{n\in\mathbb{N}}$ with an orthonormal basis of eigenvectors denoted by $(H_n^1)_{n\in\mathbb{N}}$. Then $\left(H_n \equiv \frac{AH_n^1}{\sqrt{\alpha_n}}\right)_{n\in\mathbb{N}}$ is an orthonormal basis of H. The formal product measure $\prod_{n=1}^{\infty} dh_n$ defined above with respect to this orthonormal basis induces a formal product measure on H_1 as follows; let $h = Ah^1$ for some h^1 in H_1 :

$$\begin{aligned} h_n &= \left\langle h, \frac{AH_n^1}{\sqrt{\alpha_n}}\right\rangle = \left\langle Ah^1, \frac{AH_n^1}{\sqrt{\alpha_n}}\right\rangle = \left\langle h^1, \frac{A^*AH_n^1}{\sqrt{\alpha_n}}\right\rangle_1 \\ &= \sqrt{\alpha_n}\langle h^1, H_n^1\rangle_1 = \sqrt{\alpha_n}\, h_n^1 \end{aligned}$$

where $h_n^1 \equiv \langle h^1, H_n^1\rangle_1$. We thus obtain the formal identifications :

$$\begin{aligned} D[h] &= \prod_{n=1}^{\infty} d\langle Ah_1, H_n\rangle \\ &= \prod_{n=1}^{\infty}\sqrt{\alpha_n}\prod_{n=1}^{\infty} dh_n^1 \\ &= \prod_{n=1}^{\infty}\sqrt{\alpha_n}\, D[h^1] \end{aligned} \tag{3.4}$$

where $D[h^1]$ is the formal product measure $\prod_{n=1}^{\infty} dh_n^1$.

We now apply this general framework to $E^g = T_g M$ and $H^g = \overline{T_g M}^g$ where g is a generic metric in M. We set for $g = T(\varphi, f, t), E_1^g \equiv T_\varphi C^\infty(\mathbb{R}) \times T_f D_0 \times T_t \mathcal{T}$ and $H_1^g = \overline{E_1^g}^g$ the closure of E_1^g for the scalar product $\{\cdot,\cdot\}_g \equiv (\cdot,\cdot)_g \otimes \langle \cdot,\cdot \rangle_g \otimes (\cdot,\cdot)$ where $(\cdot,\cdot)$ is the scalar product in $T_t \mathcal{T}$ given by $(s,\tilde{s}) \equiv s_i \tilde{s}_i$ for $s = s_i e_i^t$, $\tilde{s} = \tilde{s}_i e_i^t$. We want to transform a formal measure on $T_g M$ through the operator dT^g.
As dT^g is densely defined on H_1^g, we can consider its adjoint operator w.r.t. the scalar product $\{\cdot,\cdot\}_g$ which we shall denote by $(dT^g)^*$. From (2.9) and the invariance property of $\{\cdot,\cdot\}_g$ i.e.

$$\{(\lambda, f^* u, s); (\mu, f^* v, r)\}_g = \{(\lambda, u, s); (\mu, v, r)\}_{\bar{g}_t} \quad \forall (\lambda, u, s) \in E_1^{\bar{g}_t} \quad \forall f \in D,$$

and for $g = f^* \bar{g}_t$ we can write :

$$(dT^g)^* = f^* (dT^{\bar{g}_t})^* \qquad \text{and} \qquad (dT^g)^* (dT^g) = f^* \left[(dT^{\bar{g}_t})^* (dT^{\bar{g}_t}) \right] \tag{3.5}$$

Remark that this expression is well defined as $dT^g(E_1^g) \subset E^g$.
We shall now study the spectral properties of $(dT^{\bar{g}_t})^* (dT^{\bar{g}_t})$ in order to apply the transformation rules of the formal measures described by (3.4) to this specific transformation. We split dT^g into two operators :

$$dT^g = f^* (\Gamma_2^{\bar{g}_t} \circ \Gamma_1^{\bar{g}_t})$$

with

$$\Gamma_1^{\bar{g}_t}(\lambda, u, s) = \left(\lambda + \frac{1}{2} \bar{g}_t^{cd} (\nabla_{\bar{g}_t} u)_{cd} + s_i \frac{1}{2} \bar{g}_t^{cd} (e^\varphi \frac{\partial g t}{\partial t_i})_{cd} \, , \, u + s_i V_i^{\bar{g}_t} \, , \, s_i e_i^t \right)$$

and

$$\Gamma_2^{\bar{g}_t}(\lambda, u, s) = \lambda \bar{g}_t + P_{\bar{g}_t} u + W_{\bar{g}_t}(s)$$

and where $\quad W_{\bar{g}_t}(s) = s_i \langle \chi_i^{\bar{g}_t}, u_k^{\bar{g}_t} \rangle^{\bar{g}_t} u_k^{\bar{g}_t} \quad$ and $\quad s = s_i e_i^t$.
We have the formula :

$$(dT^g)^* (dT^g) = f^* \left((dT^{\bar{g}_t})^* (dT^{\bar{g}_t}) \right) \tag{3.6}$$

with

$$(dT^{\bar{g}_t})^* dT^{\bar{g}_t} = (\Gamma_1^{\bar{g}_t})^* \left(4K \mathbb{1}_{T_\varphi C^\infty(\mathbb{R})} \otimes P_{\bar{g}_t}^* P_{\bar{g}_t} \otimes Q_{\bar{g}_t} \right) \Gamma_1^{\bar{g}_t} \tag{3.7}$$

where K is the positive constant given in (2.13), and $Q_{\bar{g}_t}$ is given by :

$$Q_{\bar{g}_t}(s, \tilde{s}) = s_i \tilde{s}_j b_{ks} b_{kl} \langle \chi_i^{\bar{g}_t}, \varphi_s^{gt} \rangle^{\bar{g}_t} \left\langle \chi_j^{\bar{g}_t}, \varphi_l^{gt} \right\rangle^{\bar{g}_t}$$

The operator $(dT^{\bar{g}_t})^* (dT^{\bar{g}_t})$ as it stands does not have discrete spectrum; the operator $\mathbb{1}_{T_\varphi C^\infty(\mathbb{R})}$ seems to spell disaster ! We slightly modify the operator $dT^{\bar{g}_t}$ so as to get an operator with discrete spectrum as follows :

We set for $\eta > 0$

$$\Gamma^{g_t}_{2,\eta}(\lambda, u, s) \equiv (\mathbb{1} + \eta\Delta_{g_t})\lambda\bar{g}_t + P_{g_t}u + s_i\langle\chi_i^{g_t}, u_k^{g_t}\rangle^{g_t} u_k^{g_t}$$

Here Δ_{g_t} is the Laplace Beltrami operator w.r.t the metric $\bar{g}_t$. We denote by $A_\eta^{g_t}$ the operator $(\Gamma^{g_t}_{2,\eta})\Gamma_1^{g_t}$ and by $\Theta_\eta^{g_t}$ the operator $\left(A_\eta^{g_t}\right)^* A_\eta^{g_t}$ where $(A_\eta^{g_t})^*$ is the adjoint operator to $A_\eta^{g_t}$ which is well defined as $A_\eta^{g_t}$ is densely defined in $H_1^{g_t}$. $\Theta_\eta^{g_t}$ can be expressed as follows:

$$\Theta_\eta^{g_t} = \left(\Gamma_1^{g_t}\right)^* \left(4K\left(\mathbb{1} + \eta\Delta_{g_t}\right)^2 \otimes P^*_{g_t}P_{g_t} \otimes Q_{g_t}\right)\Gamma_1^{g_t} \tag{3.8}$$

The operator $\Theta_\eta^{g_t}$ is well defined on $E_1^{g_t}$, being constructed with differential operators and we can state the

Lemma 3.1 : For $\eta > 0$ and $\bar{g}_t = e^\varphi g_t$, with $g_t = \Xi(t)$, the operator $\theta_\eta^{g_t}$ is strictly positive, self adjoint, has purely discrete spectrum and all but a finite number of its eigenvalues are strictly larger than 1.

General lines of the proof : The strict positivity of $\Theta_\eta^{g_t}$ is essentially a result of the fact that Q_{g_t} is positive definite which can be seen from the definition of Q_{g_t}. $\Theta_\eta^{g_t}$ is therefore injective and we can study its inverse $(\Theta_\eta^{g_t})^{-1}$ which turns out to be a compact operator as a result of the compactness of $(\mathbb{1} + \eta\Delta_{g_t})^{-2}, (P_{g_t}P_{g_t})^{-1}, Q_{g_t}^{-1}$. The first two operators are compact because $\mathbb{1} + \eta\Delta_{g_t}$ and $P^*_{g_t}P_{g_t}$ are elliptic, the latter operator acting on a finite dimensional space.
Thus, $(\Theta_\eta^{g_t})^{-1}$ has spectrum $\{\lambda^{g_t}_{\eta,n}, n \in \mathbb{N}\}$ with no accumulation point apart from 0 and we can associate to it an orthonormal basis of eigenvectors $\{H^1_{\eta,n}, n \in \mathbb{N}\}$ so that $\Theta_\eta^{g_t}$ has discrete spectrum $\{\alpha^{g_t}_{\eta,n} \equiv \frac{1}{\lambda_{\eta,n}}, n \in \mathbb{N}\}$ and an orthonormal set of eigenvectors $\{H^1_{\eta,n}, n \in \mathbb{N}\}$. From the description of $\Theta_\eta^{g_t}$ it is clear that all but a finite number of the eigenvalues of $\Theta_\eta^{g_t}$ are strictly larger than 1.

For a metric $g = f^* e^\varphi g_t$, we now have two Hilbert spaces H^g and H_1^g together with an operator $f^* A_\eta^{g_t}$ acting from $E_1^{g_t}$ onto E^g. By the above Lemma $(f^* A_\eta^{g_t})^* \cdot (f^* A_\eta^{g_t}) = f^*\left(A_\eta^{g_t *} \cdot A_\eta^{g_t}\right)$ (on the right hand side the adjoint is taken with respect to $\{.,.\}_{g_t}$, on the left hand side with respect to $\{.,.\}_g$) is a strictly positive operator with discrete spectrum so that we can relate the transformation of a heuristic measure on $T_g M$ through $f^* A_\eta^{g_t}$ to the general framework set at the beginning of the paragraph.

We first construct a heuristic measure on $T_g M$. The closure H^g of E^g is a separable Hilbert space to which we shall associate an orthonormal basis as follows: on $H_1^{g_t}$, we consider an orthonormal basis of eigenvectors $\left\{H^1_{\eta,n}, n \epsilon \mathbb{N}\right\}$ of $\Theta_\eta^{g_t}$ associated to the eigenvalues $\left\{\alpha^{g_t}_{\eta,n}, n \epsilon \mathbb{N}\right\}$

Then following the general scheme, we set :

$$H_{\eta,n} \equiv \frac{f^*\left(A_\eta^{g_t} \cdot H^1_{\eta,n}\right)}{\sqrt{\alpha^{g_t}_{\eta,n}}} \tag{3.9}$$

From the fact that

$$\langle f^* h, f^* k\rangle^g = \langle h, k\rangle^{\bar g_t}$$

with $g = f^*\bar g_t$ and $f \epsilon D_0$, and the surjectivity of $A_\eta^{\bar g_t}$ we deduce that $\{H_{\eta,n};\, n \epsilon I\!N\}$ forms an orthonormal basis of H^g for the scalar product $(\cdot,\cdot)_g$.

For a generic h in H^g, we can write $h = \sum_{n\in I\!N} h_n^g H_{\eta,n}$ where $h_n^g \equiv \langle h; H_{\eta,n}\rangle^g$ for the scalar product $< \cdot,\cdot >^g$ defined in (2.13). This yields a formal product measure on H^g:

$$D_g[h] = \prod_{n=1}^{\infty} d\, h_n^g \tag{3.10}$$

This heuristic measure induces a formal measure on $H_1^{\bar g_t}$ as follows :
let $h = f^* A_\eta^{\bar g_t} h^1$ with $h^1 \epsilon E_1^{\bar g_t}$ We have

$$\begin{aligned} h_n^g &= \left(f^* A_\eta^{\bar g_t} h^1, H_{\eta,n}\right)^g \\ &= \left\langle A_\eta^{\bar g_t} h^1, \frac{A_\eta^{\bar g_t} H_{\eta,n}^1}{\sqrt{\alpha_{\eta,n}^{\bar g_t}}}\right\rangle^{\bar g_t} \\ &= \sqrt{\alpha_{\eta,n}^{\bar g_t}}\, h_n^{\bar g_t,1} \end{aligned}$$

with $h_n^{\bar g_t,1} \equiv \left\{h^1, H_{\eta,n}^1\right\}_{\bar g_t}$ so that $D_g[h]$ induces the formal measure :

$$D_{\bar g_t}^\eta[h_1] = \prod_{n\epsilon I\!N} \sqrt{\alpha_{\eta,n}^{\bar g_t}}\, D_{\bar g_t}[h_1] \tag{3.11}$$

where $D_{\bar g_t}[h_1] = \prod_{n\epsilon I\!N} d\, h_n^{\bar g_t,1}$ is the heuristic product measure on $H_1^{\bar g_t}$ given in terms of the orthonormal basis $\left\{H_{\eta,n}^1, n\epsilon I\!N\right\}$
Let us stress here that $D_{\bar g_t}[h_1]$ can be seen as a tensor product of heuristic measures:

$$D_{\bar g_t}[h_1] = D_{\bar g_t}[\lambda] \otimes D_{\bar g_t}[u] \otimes ds \tag{3.12}$$

where $h_1 = (\lambda, u, s)\epsilon E_1^{\bar g_t}$ and $H_{\eta,k}^1 \equiv \left(\Lambda_k^{\bar g_t}, U_k^{\bar g_t}, S_k^{\bar g_t}\right)$. Here $D_{\bar g_t}[\lambda] = \prod_{n=1}^{\infty} d\lambda_n^{\bar g_t}$ with $\lambda_n^{\bar g_t} = \left(\lambda, \Lambda_n^{\bar g_t}\right) g_t$ $D_{\bar g_t}[u] = \prod_{n=1}^{\infty} du_n^{\bar g_t}$ with $u_n^{\bar g_t} = \left\langle u, U_n^{\bar g_t}\right\rangle_{\bar g_t}$ and ds is the ordinary Lebesgue measure on $T_t\mathcal{T}$.

With these notations, the formal measure induced on $H_1^{\bar g_t}$ through $f^* A_\eta^{\bar g_t}$ by the heuristic measure $D_g[h]$ defined on H^g is

$$D_{\bar g_t}^\eta\,[(\lambda, u, s)] = \prod_{n\epsilon I\!N} \sqrt{\alpha_{\eta,n}^{\bar g_t}} \prod_{n=1}^{\infty} d\lambda_n^{\bar g_t} \prod_{n=1}^{\infty} du_n^{\bar g_t}\, ds \tag{3.13}$$

We now return to the heuristic Polyakov measure (1.11) to study its transformed measure under the map T. Let us first interpret the integral $\int_E e^{-A(X,g)} D_g[X]$ Recall that $A(X,g) = \frac{1}{2} \int_\Lambda \sqrt{g} g^{ab} \delta_a X^\mu \delta_b X^\mu d\xi$ for g in M.

Remark that, for a diffeomorphism f in D_0, we have $A(X, f^* g) = A(X,g)$ We can express $A(X,g)$ in terms of the Laplace Beltrami operator so that $\int_E e^{-A(X,g)} D_g(X)$ will appear as a Gaussian integral. We now make this description more precise :
$A(X,g) = (X, \Delta_g^d X)_g^d$ where $(\cdot,\cdot)_g^d$ is a scalar product on $C^\infty(\mathbb{R}^d)$ given by :

$$(X,Y)_g^d = \sum_{\mu=1}^{d} (X^\mu, Y^\mu)_g \tag{3.14}$$

for a metric g in M and $\Delta_g^d \equiv \Delta_g \otimes \cdots \otimes \Delta_g$ (Δ_g is the Laplace Beltrami operator to the metric g, the tensor product is taken d times).

Let $(\Delta_g^d)'$ denote the projection of Δ_g^d to the orthogonal subspace of $C^\infty(\mathbb{R}^d)$ to Ker Δ_g^d with respect to the scalar product $(\cdot,\cdot)_g{}^d$. It can be shown that $(\Delta_g^d)' = f^* (\Delta_{g_t}^d)'$ is a strictly positive self adjoint operator with discrete spectrum. This is a result of the ellipticity of Δ_{g_t} which can be written $\Delta_{g_t}\lambda(\xi) = e^{-\varphi(\xi_t)} \partial_{\xi_t^1} \partial_{\xi_t^2} \lambda(\xi)$ in terms of the isothermal coordinates $\xi_t \equiv (\xi_t^1, \xi_t^2)$ for the metric g_t. Let e_n^g be an orthonormal system of eigenvectors of $(\Delta_g)'$ with respect to the scalar product $(\cdot,\cdot)_g$ and let β_n^g be the corresponding eigenvalues.

We set for X in $C^\infty(\mathbb{R}^d)$

$$X_n^g \equiv (X_n^{g,\mu})_{\mu=1}^d \qquad \text{and} \qquad X_n^{g,\mu} \equiv (X_n^\mu, e_n^g)_g$$

We define the heuristic measure $D_g[X]$ by

$$D_g[X] = \prod_{n=1}^{\infty} \prod_{\mu=1}^{d} dX_n^{g,\mu}$$

We interpret $\int_E e^{-A(X,g)} D_g[X]$ as the infinite product

$$\int e^{-\sum_{n=1}^{\infty} \sum_{\mu=1}^{d} \beta_n^g (X_n^{g,\mu})^2} \prod_{n=1}^{\infty} \prod_{\mu=1}^{d} dX_n^{g,\mu}$$

which in turn can be integrated and yields the formal identification :

$$\int_E e^{-A(X,g)} D_g[X] = \prod_{n=1}^{\infty} (\beta_n^g)^{-\frac{d}{2}} \tag{3.15}$$

The last equality is a result of the invariance property of the scalar product $(\cdot,\cdot)_g$ with respect to the action of D_0.

Using (3.15), we rewrite $\mu_p[g]$ given by (1.11) as follows :

$$\mu_p[g] = \prod_{n=1}^{\infty} (\beta_n^g)^{-\frac{d}{2}} e^{-\mu^2 \int_\Lambda \sqrt{g(\xi)}\, d\xi} D[g].$$

To study the transformed measure under the map T, as before, we investigate the formal measure $\mu_p^g[h]$ induced by $\mu_p[g]$ on the tangent space $T_g M$ at point g and the limit when η goes to zero of the formal measure $\nu_p^{g,\eta}[h_1]$ it induces on $T_\varphi C^\infty(I\!R) \times T_{\mathrm{II}} D_0 \times T_t \mathcal{T}$ through the map $f^* A_\eta^{g_t}$

Explicitely this reads :

$$\mu_p^g[h] \equiv \prod_{n=1}^{\infty} (\beta_n^g)^{-\frac{d}{2}} e^{-\mu^2 \int_\Lambda \sqrt{g(\xi)}\, d\xi} D_g[h] \tag{3.16}$$

$$\begin{aligned} \nu_p^{g,\eta}[(\lambda,u,s)] = &\prod_{n=1}^{\infty} (\beta_n^g)^{-\frac{d}{2}} \prod_{n=1}^{\infty} (\alpha_{\eta,n}^{g_t})^{\frac{1}{2}} e^{-\mu^2 \int_\Lambda \sqrt{g(\xi)}\, d\xi} \\ &\times \prod_{n=1}^{\infty} d\lambda_n^{g_t} \prod_{n=1}^{\infty} du_n^{g_t}\, ds \end{aligned} \tag{3.17}$$

As $\prod_{n=1}^{\infty} (\beta_n^g)^{-\frac{d}{2}} = \prod_{n=1}^{\infty} (\beta_n^{g_t})^{-\frac{d}{2}}$ and $e^{-\mu^2 \int_\Lambda \sqrt{g(\xi)}\, d\xi} = e^{-\mu^2 \int_\Lambda \sqrt{g_t(\xi)}\, d\xi}$, the formal measure on $T_\varphi C^\infty(I\!R) \times T_{\mathrm{II}} D_0 \times T_t \mathcal{T}$ induced by the Polyakov measure through $f^* A_\eta^{g_t}$ reads:

$$\nu_p^{g,\eta}[(\lambda,u,s)] = \prod_{n=1}^{\infty} (\beta_n^{g_t})^{-\frac{d}{2}} \prod_{n=1}^{\infty} (\alpha_{\eta,n}^{g_t})^{\frac{1}{2}} e^{-\mu^2 \int_\Lambda \sqrt{g_t(\xi)}\, d\xi} D_{g_t}[\lambda] D_{g_t}[u] ds \tag{3.18}$$

The heuristic measure on $E_1^{g_t}$ corresponding to the Polyakov construction would be obtained as a formal weak limit when η goes to zero of $\nu_p^{g,\eta}$ giving :

$$\begin{aligned} \nu_p^g[(\lambda,u,s)] = &\prod_{n=1}^{\infty} (\beta_n^{g_t})^{\frac{d}{2}} \lim_{\eta\to 0} \prod_{n=1}^{\infty} (\alpha_{\eta,n}^{g_t})^{\frac{1}{2}} \\ &\times e^{-\mu^2 \int_\Lambda \sqrt{g_t(\xi)}\, d\xi} D_{g_t}[\lambda] D_{g_t}[u]\, ds \end{aligned} \tag{3.19}$$

The next stage is to interpret the formal infinite products $\prod_{n=1}^{\infty} (\beta_n^{g_t})^{-\frac{d}{2}}$ and $\prod_{n=1}^{\infty} (\alpha_{\eta,n}^{g_t})^{\frac{1}{2}}$. This leads us to the next paragraph that deals with infinite dimensional determinants.

4. Infinite dimensional determinants arising in the Polyakov partition function

We have seen in section 3 that the heuristic Polyakov measure $\nu_p^{g,\eta}$ involves two infinite products of eigenvalues $\prod_{n=1}^{\infty} \beta_n^{g_t}$ and $\prod_{n=1}^{\infty} \alpha_{\eta,n}^{g_t}$. In this section, we shall interpret these formal expressions as infinite dimensional determinants of the operators $(\Delta_{g_t})'$ and $(A_\eta^{g_t})^* A_\eta^{g_t}$ respectively. We shall be concerned more specifically with the φ-dependance of these determinants in order to compute the conformal anomaly term arising in this quantization procedure. We first start with a general definition of determinants.

Let H be a separable Hilbert space equipped with the scalar product $(\cdot,\cdot)$. We shall denote by $C(H)$ the class of selfadjoint operators acting in H which are strictly positive with discrete spectrum and such that all but a finite number of eigenvalues are greater than 1.

Let A be a selfadjoint operator acting in H with purely discrete spectrum and such that all but a finite number of eigenvalues are positive. Let us moreover assume there exists an orthonormal basis $(H_n)_{n\in\mathbb{N}}$ of H such that $H_n \epsilon D(A)$, $\forall n \epsilon \mathbb{N}$. Then $\sum_{n=1}^{\infty} (AH_n, H_n)$ is a number in $\mathbb{R} \cup \{\infty\}$ which is independant of the choice of the orthonormal basis; this is a direct generalisation of the same result for operators for which $\sum_{n=1}^{\infty} (AH_n, H_n) < \infty$; the only difference being that $\sum_{n=1}^{\infty} (AH_n, H_n)$ can be infinite. We shall call $\sum_{n=1}^{\infty} (AH_n, H_n)$ the trace of A and denote it by $tr(A)$.

We now take A in $C(H)$ and define the operator $\log A$ by:

$$D(\log A) = \{h \epsilon H,\ h = \sum_{n=1}^{\infty} h_n\, H_n \qquad \text{with} \quad \sum_{n=1}^{\infty} ((\log \lambda_n)\, h_n)^2 < \infty\}.$$

where $\{\lambda_n,\ n = 1, ..., \infty\}$ and $\{H_n,\ n = 1, ..., \infty\}$ are respectively the eigenvalues and the corresponding eigenvectors of A and

$$(\log A)\,(h) = \sum_{n=1}^{\infty} (\log \lambda_n)\, h_n\, H_n \qquad \forall h \epsilon D(\log A)$$

For A in $C(H)$, the trace of $\log A$ is well defined by the above disussion and we can set for A in $C(H)$

$$\det A \equiv \exp\,(tr\,(\log A)) \tag{4.1}$$

Let us remark that for A in $C(H)$, $\det A = \prod_{n=1}^{\infty} \lambda_n$ with $\{\lambda_n,\ n = 1, ..., \infty\}$ the set of eigenvalues of A.

We now return to the infinite products $\prod_{n=1}^{\infty} \alpha_{\eta,n}^{g_t}$ and $\prod_{n=1}^{\infty} \beta_n^{g_t}$ arising in the expression (3.17) of $\nu_p^{g,\eta}$ to interpret them as infinite dimensional determinants.

Lemma 3.1 shows that $\left(A_\eta^{g_t}\right)^* A_\eta^{g_t}$ lies in $C(H_1^{g_t})$. Then, by the definition of the determinant we have

$$\det\left(\left(A_\eta^{g_t}\right)^* A_\eta^{g_t}\right) = \prod_{n=1}^{\infty} \alpha_{\eta,n}^{g_t} \qquad \text{which lies in } \overline{\mathbb{R}^+} \tag{4.2}$$

In the same way, $(\Delta_{g_t})'$ lies in $C\left(\overline{C^\infty(\mathbb{R})}^{g_t}\right)$; $\mathrm{Ker}^\perp \Delta_{g_t}$ is the orthogonal supplementary in $\overline{C^\infty(\mathbb{R})}^{g_t}$ of $\mathrm{Ker}\,\Delta_{g_t}$ for the scalar product $(\cdot,\cdot)_{g_t}$ and as before

$$\det(\Delta_{g_t}{}') = \prod_{n=1}^{\infty} \beta_n^{g_t} \tag{4.3}$$

We shall now study the determinant (4.2).

In the description of $\left(A_\eta^{g_t}\right)^* A_\eta^{g_t}$ given in (3.8), arise the operators $\mathbb{1} + \eta\Delta_{g_t}$ and $P_{g_t}^* P_{g_t}$ which we shall now study in more details.

For this purpose, we introduce some auxiliary operators. $C^\infty(\mathbb{C})$ denotes the space of C^∞ $\mathbb{C}$-valued functions on Λ and H^j the closure of $C^\infty(\mathbb{C})$ for the scalar product:

$$(f,h)_{g_t}^j = \int_\Lambda \left(\sqrt{\bar{g}_t(\xi)}\right)^{j+1} f(\xi)\bar{h}(\xi)\, d\xi \qquad \text{where} \quad j \in \mathbb{Z} \tag{4.4}$$

which generalizes the scalar product $(f,h)_{g_t}$ which we recover setting $j = 0$. We recall at this point that $\bar{g}_t = e^{\varphi}\, g_t$.

We consider the operator on $C^\infty(\mathbb{C})$ given by:

$$\left(L^j f\right)(\xi_t) \equiv -e^{-(j+1)\vartheta_t(\xi_t)} \frac{\partial}{\partial z_t} e^{j\vartheta_t(\xi_t)} \frac{\partial}{\partial \bar{z}_t} f(\xi_t) \tag{4.5}$$

where $\bar{g}_{t\,ab}(\xi_t) = e^{\vartheta_t(\xi_t)}\delta_{ab}(\xi_t)$ for $a = 1,2$ and $b = 1,2$. Here ξ_t is the isothermal atlas corresponding to the conformal structure $[g_t]$ and $z_t = \xi_t^1 + i\xi_t^2$.

$(L^j)'$, the projection of L^j onto the orthogonal subspace $\mathrm{Ker}^\perp L^j$ of H^j to its kernel $\mathrm{Ker}\, L^j$ is a selfadjoint strictly positive elliptic operator in H^j and lies in $C(H^j)$.

We have the following identifications :

$$\left(P_{g_t}^* P_{g_t} u\right)^{z_t} = L^1 u^{z_t} \quad \forall u \in C^\infty(T\Lambda) \qquad \text{where} \quad u^{z_t} = u^1 + i\,u^2 \tag{4.6}$$

with u^1, u^2 the components of the vector u with respect to the natural basis $\frac{\partial}{\partial \xi_t^1}, \frac{\partial}{\partial \xi_t^2}$ and

$$L^0\left(f_1 + i\, f_2\right) = \Delta_{g_t} f_1 + i\,\Delta_{g_t} f_2 \tag{4.7}$$

with $f = f_1 + i f_2 \epsilon C^\infty(\mathbb{C})$

From these identifications follows that, the operator $\mathbb{1} + \eta\Delta_{\bar{g}_t}$ for $\eta > 0$ and $P^*_{\bar{g}_t} P_{\bar{g}_t}$ lie in $C\left(\overline{C^\infty(\mathbb{R})}^{\bar{g}_t}\right)$ and $C\left(\overline{C^\infty(T\Lambda)}^{\bar{g}_t}\right)$ respectively, where $\overline{C^\infty(T\Lambda)}^{\bar{g}_t}$ is the closure of $C^\infty(T\Lambda)$ for the scalar product $(\cdot,\cdot)_{\bar{g}_t}$. Hence $\det(\mathbb{1}+\eta\Delta_{\bar{g}_t})$ and $\det(P^*_{\bar{g}_t} P_{\bar{g}_t})$ are well defined.

In the following, we shall give a heuristic justification of the following identification :

$$\det\left(\left(A^{\bar{g}_t}_\eta\right)^*\left(A^{\bar{g}_t}_\eta\right)\right) = \det\left(\mathbb{1}+\eta\Delta_{\bar{g}_t}\right)^2 \times \det\left(P^*_{\bar{g}_t} P_{\bar{g}_t}\right) \times \det\left(Q_{\bar{g}_t}\right) \tag{4.8}$$

where $Q_{\bar{g}_t}$ is a finite dimensional matrix which depends on $\bar{g}_t$.

To justify this equality let us consider as in section 3 a basis $\left(H^1_{\eta,n}\right)_{n\in\mathbb{N}}$ of eigenvalues of $\left(A^{\bar{g}_t}_\eta\right)^*\left(A^{\bar{g}_t}_\eta\right)$ which is orthonormal with respect to the scalar product $\{\cdot,\cdot\}_{\bar{g}_t}$. R_N denotes the orthogonal projection with respect to $\{\cdot,\cdot\}_{\bar{g}_t}$ onto the first N eigenvectors of this basis. Let us point out that the finite rank operator $\left(\left(A^{\bar{g}_t}_\eta\right)^*\left(A^{\bar{g}_t}_\eta\right)\right)^{-1} R_N$ converges in the operator norm to the compact operator $\left(\left(A^{\bar{g}_t}_\eta\right)^*\left(A^{\bar{g}_t}_\eta\right)\right)^{-1}$. With the notations of section 3 and using the methods involved in the proof of the lemma 3.1, we can show that $\Gamma^{\bar{g}_t}_1\left(\left(A^{\bar{g}_t}_\eta\right)^*\left(A^{\bar{g}_t}_\eta\right)\right)^{-1} R_N\left(\Gamma^{\bar{g}_t}_1\right)^*$ converges pointwise to $\Gamma^{\bar{g}_t}_1\left(\left(A^{\bar{g}_t}_\eta\right)^*\left(A^{\bar{g}_t}_\eta\right)\right)^{-1}\left(\Gamma^{\bar{g}_t}_1\right)^* = \left(\left(\Gamma^{\bar{g}_t}_{2,\eta}\right)^*\Gamma^{\bar{g}_t}_{2,\eta}\right)^{-1}$. Hence it is natural to assume that there is also convergence of determinants. $\Gamma^{\bar{g}_t}_1 R_N$ has finite rank, as can easily be seen from the description of $\Gamma^{\bar{g}_t}_1$ given in section 3. We describe its finite dimensional image by G_N. Then $(\Gamma^{\bar{g}_t}_1)^*{}_{|R_N H^{\bar{g}_t}_1}$ can be seeen as an operator acting from $R_N H^{\bar{g}_t}_1$ to G_N and thus be represented by a finite dimensional matrix. From the matrix representation of $\Gamma^{\bar{g}_t}_1$ by an upper triangular matrix with ones on the diagonal it is clear that $(\Gamma^{\bar{g}_t}_1)^*{}_{|R_N H^{\bar{g}_t}_1}$ also be represented as an upper triangular matrix with ones on the diagonal. Thus $\det\left((\Gamma^{\bar{g}_t}_1)^*{}_{|R_N H^{\bar{g}_t}_1}\right) = 1$. Analogous treatment can be applied to $R_N\left(\Gamma^{\bar{g}_t}_1\right)^*$ which in turn acts on a finite dimensional subspace F_N of $H^{\bar{g}_t}_1$ and $\det\left(\left(\Gamma^{\bar{g}_t}_1\right)^*{}_{|F_N}\right) = 1$. By the usual rules for finite dimensional determinants we can write:

$$\begin{aligned}
&\det\left(\Gamma^{\bar{g}_t}_1\left(\left(A^{\bar{g}_t}_\eta\right)^*\left(A^{\bar{g}_t}_\eta\right)\right)^{-1} R_N\left(\Gamma^{\bar{g}_t}_1\right)^*\right)\\
&\quad= \det\left((\Gamma^{\bar{g}_t}_1)^*{}_{|R_N H^{\bar{g}_t}_1}\right)\det\left(\left(\left(A^{\bar{g}_t}_\eta\right)^*\left(A^{\bar{g}_t}_\eta\right)\right)^{-1}_{|R_N H^{\bar{g}_t}_1}\right)\det\left(\left(\Gamma^{\bar{g}_t}_1\right)^*{}_{|F_N}\right)\\
&\quad= \det\left(\left(\left(A^{\bar{g}_t}_\eta\right)^*\left(A^{\bar{g}_t}_\eta\right)\right)^{-1}_{|R_N H^{\bar{g}_t}_1}\right)\\
&\quad= \left(\prod_{n=1}^{N}\alpha^{\bar{g}_t}_{\eta,n}\right)^{-1}
\end{aligned}$$

This yields, with the assumption on the identification of the limits of determinants made above :

$$\lim_{N \to +\infty} \left(\prod_{n=1}^{N} \alpha_{\eta,n}^{g_t} \right)^{-1} = \det \left(\left(\Gamma_{2,\eta}^{g_t} \right)^* \Gamma_{2,\eta}^{g_t} \right)^{-1}$$

or in other words

$$\det \left(\left(A_\eta^{g_t} \right)^* \left(A_\eta^{g_t} \right) \right) = \det \left(\left(\Gamma_{2,\eta}^{g_t} \right)^* \Gamma_{2,\eta}^{g_t} \right).$$

The last step to justify (4.8) is to show that

$$\det \left(\left(\Gamma_{2,\eta}^{g_t} \right)^* \Gamma_{2,\eta}^{g_t} \right) = \det \left(\mathbb{1} + \eta \Delta_{g_t} \right)^2 \det \left(P_{g_t}^* P_{g_t} \right) \det \left(Q_{g_t} \right).$$

Again we look at the inverse operators which are compact operators; let Q_N be the orthogonal projection onto the finite dimensional subspace of $\overline{C^\infty(\mathbb{R})}^{g_t}$ spanned by the first N eigenvalues of $\mathbb{1} + \eta \Delta_{g_t}$, the order here being given by an order of the corresponding eigenvalues. Let P_N be the orthogonal projection onto the finite dimensional subspace of $\overline{C^\infty(T\Lambda)}^{g_t}$ spanned by the first N eigenvectors of $P_{g_t}^* P_{g_t}$. Then $Q_N \otimes P_N \otimes \mathbb{1} \equiv \tilde{R}_N$ is an orthogonal projector of $H_1^{g_t}$ for the scalar product $\{\cdot,\cdot\}_{g_t}$ onto a finite dimensional subspace of $H_1^{g_t}$ spanned by a finite number of eigenvectors of $\left(\Gamma_{2,\eta}^{g_t} \right)^* \Gamma_{2,\eta}^{g_t}$. By the definition of the determinant we have

$$\begin{aligned}
&\det \left(\left(\Gamma_{2,\eta}^{g_t} \right)^* \Gamma_{2,\eta}^{g_t} \right) \\
&= \lim_{N \to \infty} \det \left(\left(\Gamma_{2,\eta}^{g_t} \right)^* \Gamma_{2,\eta}^{g_t} {}_{\mid \tilde{R}_N H_1^{g_t}} \right) \\
&= \lim_{N \to \infty} \left(\det \left(\mathbb{1} + \eta \Delta_{g_t} \right)^2_{\mid Q_N \overline{C^\infty(\mathbb{R})}^{g_t}} \det \left(P_{g_t}^* P_{g_t} \right)_{\mid P_N \overline{C^\infty(T\Lambda)}^{g_t}} \det Q_{g_t} \right) \\
&= \left(\det \left(\mathbb{1} + \eta \Delta_{g_t} \right) \right)^2 \det \left(P_{g_t}^* P_{g_t} \right) \det Q_{g_t}.
\end{aligned}$$

This concludes the justification of the identification (4.8).

From (4.8) follows an estimate on the limit when η goes to zero of $\det \left(\left(A_\eta^{g_t} \right)^* \left(A_\eta^{g_t} \right) \right)$ namely:

$$\lim_{\eta \to 0} \det \left(\left(A_\eta^{g_t} \right)^* \left(A_\eta^{g_t} \right) \right) = \det \left(P_{g_t}^* P_{g_t} \right) \det \left(Q_{g_t} \right) \tag{4.9}$$

This follows from the fact that the eigenvalues of $\mathbb{1} + \eta \Delta_{g_t}$ are decreasing and tend to 1 when η goes to zero.

Returning to the heuristic transformed Polyakov measure (3.17) and using (4.2), (4.3) and (4.9), we see that we are left with the estimation of the heuristic density :

$$D_{g_t}^d \equiv \lim_{\eta \to 0} \det \left(\Delta'_{g_t} \right)^{-\frac{d}{2}} \det \left(\left(A_\eta^{g_t} \right)^* \left(A_\eta^{g_t} \right) \right)^{\frac{1}{2}} = \det \left(\Delta'_{g_t} \right)^{-\frac{d}{2}} \det \left(P_{g_t}^* P_{g_t} \right)^{\frac{1}{2}} \det \left(Q_{g_t} \right)^{\frac{1}{2}} \tag{4.10}$$

We shall use the heat kernel approximation methods to compute these determinants [11,12]. For this purpose, we shall need the following definition:

Definition 4.1 : For $\varepsilon > 0$, we define a function $f_\epsilon : \mathbb{R}^+/\{0\} \to \mathbb{R}$ with $f_\epsilon(\lambda) = \exp\left(-\int_\epsilon^{\epsilon/\lambda} t^{-1}\exp(-t\lambda)\,dt\right)$ and for an operator A in $C(H)$ the corresponding ε-regularized operator A_ϵ by:

$$D(A_\epsilon) = \left\{ h\epsilon H \mid h = \sum_{n=1}^{\infty} h_n H_n \quad \text{with} \quad \sum_{n=1}^{\infty} (f_\epsilon(\lambda_n) h_n)^2 < \infty \right\}$$

with $\quad A_\epsilon h = \sum_{n=1}^{\infty} f_\epsilon(\lambda_n) h_n H_n \quad$ for all $h\epsilon D(A)$ where $(\lambda_n)_{n\in\mathbb{N}}$ and $(H_n)_{n\in\mathbb{N}}$ are as before respectively the eigenvalues and eigenvectors of A.

Proposition 4.2 : Let A be in $C(H)$ and A_ϵ the corresponding ε-regulized operator. Then A_ϵ lies in $C(H)$ and $\det A_\epsilon$ converges to $\det A$ for $\varepsilon \longrightarrow 0$.

Proof: From the fact that $\lambda \gtreqless 1$ implies $f_\epsilon(\lambda) \lesseqgtr 1$ for all λ in $\mathbb{R}/\{0\}$ and $\varepsilon > 0$, and that $f_\epsilon(\lambda)$ is a strictly increasing function we see that $A_\epsilon \epsilon C(H)$.
As for the convergence of the determinants, we first observe that $\lim f_\epsilon(\lambda_n) = \log \lambda_n$ as $\varepsilon \to 0$ for all the eigenvalues $(\lambda_n)_{n\in\mathbb{N}}$ of A. As A_ϵ lies in $C(H)$, we can split $\log\det A_\epsilon = \sum_{n=1}^{\infty} \log f_\epsilon(\lambda_n)$ into

$$\log\det A_\epsilon = \sum_{\lambda_n \le 1} \log f_\epsilon(\lambda_n) + \sum_{\lambda_n > 1} \log f_\epsilon(\lambda_n)$$

The first sum is taken over a finite number of eigenvalues, the second sum converges to $\sum_{\lambda_n > 1} \log \lambda_n$ when ε goes to zero by dominated convergence if $\det A < \infty$ and by the Fatou lemma if $\det A = \infty$. Hence $\log\det A_\epsilon$ tends to $\log\det A$ when ε goes to zero.

Proposition 4.3 :

$$D_{g_t}^d = \lim_{\epsilon\to 0}\lim_{N\to\infty} D_{g_t,d}^{\epsilon,N} \tag{4.11}$$

with

$$D_{g_t,d}^{\epsilon,N} = \prod_{n=1}^{N} \left(f_\epsilon(\beta_n^{g_t})\right)^{-\frac{d}{2}} \left(f_\epsilon(\gamma_n^{g_t})\right)^{\frac{1}{2}} \det(Q_{g_t})^{\frac{1}{2}} \tag{4.12}$$

where $(\gamma_n^{g_t})_{n\epsilon\mathbb{N}}$ is the set of eigenvalues of $P_{g_t}^* P_{g_t}$.

Proof: Returning to (4.10) and recalling that $\Delta_{g_t}' \epsilon C(\overline{C^\infty(\mathbb{R})}^{g_t})$, $P_{g_t}^* P_{g_t} \epsilon C\left(\overline{C^\infty(\mathrm{T}\Lambda)}^{g_t}\right)$ we can write:

$$D_\epsilon^{g_t} = \lim_{\epsilon\to 0} \left(\det(\Delta_{g_t}')_\epsilon\right)^{-\frac{d}{2}} \left(\det(P_{g_t}^* P_{g_t})_\epsilon\right)^{\frac{1}{2}} \det(Q_{g_t})^{\frac{1}{2}} \tag{4.13}$$

As a consequence of the fact for A in $C(H)$, $\det f_\epsilon(A) = \prod_{n=1}^{\infty} f_\epsilon(\lambda_n)$, and that $\Delta_{\bar{g}_t}{}' \in C(\mathrm{Ker}^\perp \Delta_{\bar{g}_t})$ and $P^*_{\bar{g}_t} P_{\bar{g}_t} \in C(\overline{C^\infty(T\Lambda)}^{\bar{g}_t})$, we have the following equalities :

$$\det\left(\Delta_{\bar{g}_t}{}'\right)_\epsilon = \lim_{N\to\infty} \prod_{n=1}^{N} f_\epsilon\left(\beta_n^{\bar{g}_t}\right)$$

$$\det\left(P^*_{\bar{g}_t} P_{\bar{g}_t}\right)_\epsilon = \lim_{N\to\infty} \prod_{n=1}^{N} f_\epsilon\left(\gamma_n^{\bar{g}_t}\right)$$

These equalities together with (4.13) yield (4.11) and (4.12).

The next part of this paragraph is dedicated to the computation of $D^{\epsilon,N}_{\bar{g}_t,d}$ By (4.6) and (4.7), we get :

$$\det\left(P^*_{\bar{g}_t} P_{\bar{g}_t}\right) = \det\left(L^1\right)' \tag{4.14}$$

$$\det\left(\Delta_{\bar{g}_t}{}'\right) = \left(\det\left(L^0\right)'\right)^{\frac{1}{2}} \tag{4.15}$$

If $\left(\lambda_n^j\right)_{n\in\mathbb{N}}$ are the eigenvalues of the operator $(L^j)'$, we can write in the same way:

$$\prod_{n=1}^{\infty}\left(f_\epsilon\left(\beta_n^{\bar{g}_t}\right)\right)^{-\frac{d}{2}} = \left(\prod_{n=1}^{\infty} f_\epsilon\left(\lambda_n^0\right)\right)^{-\frac{d}{4}} \tag{4.16}$$

$$\prod_{n=1}^{\infty}\left(f_\epsilon\left(\gamma_n^{\bar{g}_t}\right)\right)^{\frac{1}{2}} = \left(\prod_{n=1}^{\infty} f_\epsilon\left(\lambda_n^1\right)\right)^{\frac{1}{2}} \tag{4.17}$$

so that

$$D^{\epsilon,N}_{\bar{g}_t,d} - \left(\prod_{n=1}^{\infty} f_\epsilon\left(\lambda_n^0\right)\right)^{-\frac{d}{4}} \times \left(\prod_{n=1}^{\infty} f_\epsilon\left(\lambda_n^1\right)\right)^{\frac{1}{2}} \times \det\left(Q_{\bar{g}_t}\right)^{\frac{1}{2}}$$

We now set :

$$F_j^{\epsilon,N} \equiv \sum_{n=1}^{N} \log f_\epsilon\left(\lambda_n^j\right) \tag{4.18}$$

so that

$$D^{\epsilon,N}_{\bar{g}_t,d} = e^{\frac{-d}{4}F_0^{\epsilon,N} + \frac{1}{2}F_1^{\epsilon,N}} \times \det\left(Q_{\bar{g}_t}\right)^{\frac{1}{2}} \tag{4.19}$$

Keeping in mind that $F_j^{\epsilon,N}$ depends on φ for $j = 0,1$ through $\bar{g}_t = e^\varphi \bar{g}_t$ arising in (4.4) and (4.5), we shall split $F_j^{\epsilon,N}$ into a part independent of φ which tends to infinity when N goes to $+\infty$ and a part which remains finite when N goes to $+\infty$ and carries the φ-dependance of $F_j^{\epsilon,N}$:

$$F_j^{\epsilon,N} = \sum_{n=0}^{N} \log h_\epsilon\left(\lambda_n^j\right) + NC_\epsilon$$

where $C_\epsilon \equiv \int_\epsilon^{+\infty} e^{-t} \frac{dt}{t}$ is independant of φ and

$$h_\epsilon(\lambda) = \exp\left(-\int_\epsilon^{+\infty} e^{-t\lambda} \frac{dt}{t}\right) \qquad \lambda > 0$$

We set

$$H_j^{\epsilon,N} = F_j^{\epsilon,N} - NC_\epsilon = \sum_{n=1}^{N} \log h_\epsilon\left(\lambda_n^j\right)$$

It can easily be shown that $\lim H_j^{\epsilon,N}$ when N goes to $+\infty$ exists and is finite for all $\varphi \epsilon C^\infty(I\!R)$, $\varepsilon > 0$. We shall denote this limit by H_j^ϵ.

We now state a result about the φ-dependance of $H_j^{\epsilon,N}$. As the proof is rather involved, we shall not give the proof here. We only mention that it uses the results of [11,12] on the heat kernel expansion. For more details we refer the reader to [4,13,14].

Proposition 4.4: For any $\chi, \varphi \epsilon C^\infty(I\!R)$ and $\epsilon > 0$, the Gâteaux derivative of H_j^ϵ with respect to φ in the direction χ exists and is given up to an additional term $\int_\Lambda \sqrt{g_t(\xi)}\chi(\xi)\,\theta_\varphi^j(\epsilon,\xi)\,d\xi$ where $\sup_{\xi\epsilon\Lambda} |\theta_\varphi^j(\epsilon,\xi)| \leq \sqrt{\epsilon}M_j(\varphi)$ for a constant $M_j(\varphi)\epsilon I\!R^+$ by the Gâteaux derivative with respect to φ in the direction χ of $-\tilde{H}_j^\epsilon$ with:

$$\begin{aligned}\tilde{H}_1^\epsilon \equiv & \frac{1}{2\pi\varepsilon}\int_\Lambda \sqrt{g_t(\xi)}e^{\varphi(\xi)}\,d\xi + \frac{13}{24\pi}\int_\Lambda \sqrt{g_t(\xi)}(\Delta_{g_t}\varphi)(\xi)\varphi(\xi)\,d\xi \\ & + \frac{13}{12\pi}\int_\Lambda d\xi\,\sqrt{g_t(\xi)}R_{g_t}(\xi)\varphi(\xi) - \log\det\left(\langle\varphi_m^{g_t},\varphi_r^{g_t}\rangle^{g_t}\right)_{mr}\end{aligned}$$

if $j = 1$ and

$$\begin{aligned}\tilde{H}_0^\epsilon \equiv & \frac{1}{2\pi\varepsilon}\int_\Lambda \sqrt{g_t(\xi)}e^{\varphi(\xi)}\,d\xi + \frac{1}{24\pi}\int_\Lambda \sqrt{g_t(\xi)}(\Delta_{g_t}\varphi)(\xi)\varphi(\xi)\,d\xi \\ & + \frac{1}{12\pi}\int_\Lambda d\xi\,\sqrt{g_t(\xi)}R_{g_t}(\xi)\varphi(\xi) - 2\log\int_\Lambda \sqrt{g_t(\xi)}e^{\varphi(\xi)}\,d\xi\end{aligned}$$

if $j = 0$

Here R_{g_t} stands for the scalar curvature of $g_t = \Xi(t)$ (with the notations of section 2).

This identification of the Gâteaux derivative with respect to φ of H_j^ϵ and $-\tilde{H}_j^\epsilon$ for $j = 0, 1$ leads to a corresponding identification of H_j^ϵ and $-\tilde{H}_j^\epsilon$ themselves and finally to the subsequent proposition:

Proposition 4.5: For $M > 0$, for φ in $C^\infty(\mathbb{R})$ with supremum norm in all derivatives $\|\varphi\|_\infty$ bounded by M, for all $\varepsilon > 0$, and setting $\bar{g}_t = e^\varphi g_t$ we have that $D^{\varepsilon,N}_{\bar{g}_t,d}\left(\det{}^N(P^*_{g_t}P_{g_t})_\varepsilon\right)^{-\frac{1}{2}}\left(\det{}^N(\Delta'_{g_t})_\varepsilon\right)^{\frac{d-2}{2}} e^{\frac{d-2}{8\pi\varepsilon}\int_\Lambda\sqrt{g_t(\xi)}d\xi}\, Vol_{g_t}(\Lambda)^{-\frac{d}{2}}$ converges when N goes to infinity and its limit is:

$$(\det Q_{\bar{g}_t})^{\frac{1}{2}} e^{\frac{d-2}{8\pi\varepsilon}\int_\Lambda\sqrt{\bar{g}_t(\xi)}d\xi} \times e^{\frac{d-26}{48\pi}\left[\frac{1}{2}\int_\Lambda\sqrt{g_t(\xi)}(\Delta_{g_t}\varphi)(\xi)\varphi(\xi)d\xi+\int_\Lambda\sqrt{g_t(\xi)}R_{g_t}(\xi)\varphi(\xi)d\xi\right]}$$

$$\times\left(\frac{\det\langle\varphi^{\bar{g}_t}_m,\varphi^{\bar{g}_t}_r\rangle^{\bar{g}_t}}{\det\langle\varphi^{g_t}_m,\varphi^{g_t}_r\rangle^{g_t}}\right)^{\frac{1}{2}}\times Vol_{\bar{g}_t}(\Lambda)^{-\frac{d}{2}}\times e^{O_M(\varepsilon)}$$

with $\lim_{\varepsilon\to 0} O_M(\varepsilon) = 0$ and $\det^N(\Delta'_{g_t})_\varepsilon \equiv \prod_{n=1}^N f_\varepsilon(\beta^{g_t}_n)$ ($(\beta^g_n)_{n\in\mathbb{N}}$ are the eigenvalues of (Δ'_g) as in section 3).

Remark 4.6 : This proposition is a result on the φ-dependance of the limit when N goes to infinity of $D^{\varepsilon,N}_{\bar{g}_t,d}$ for, even though $D^{\varepsilon,N}_{\bar{g}_t,d}$ itself doesn't converge to a finite expression, the infinite part of its limit can be compensated by terms independant of φ - ie. $\left(\det^N(\Delta'_{g_t})_\varepsilon\right)^{\frac{d-2}{2}} e^{\frac{d-2}{8\pi\varepsilon}\int_\Lambda\sqrt{g(\xi)}d\xi}\left(\det{}^N(P^*_{g_t}P_{g_t})_\varepsilon\right)^{-\frac{1}{2}}$.

5. The Polyakov bosonic strings and the Liouville measure

Before establishing a relationship between the quantization of bosonic strings and the quantization of the classical 2- dimensional Liouville model, we shall work out a heuristic expression of the transformed Polyakov measure under the map T with the help of the results of the preceeding section. We shall see that the density of this transformed measure consists essentially of the Liouville action combined with a term that only depends on the Teichmüller parameters. The next step will be to interpret this heuristic measure in terms of the Liouville measure.

In section 3, we had worked out the formal expression of the transformed Polyakov measure $\mu^p_g[h]$ on T_gM through the map $A^{\bar{g}_t}_\eta = \Gamma^{\bar{g}_t}_{2,\eta}\circ\Gamma^{\bar{g}_t}_1$. This gave us a heuristic measure $\nu^{g,\eta}_p$ on $E^{\bar{g}_t}_1 = T_\varphi C^\infty(\mathbb{R})\times T_{\mathrm{II}}D_0\times T_tT$ for each $\eta > 0$.

With the notations of section 4 and for $h^1\in E^{\bar{g}_t}_1$ this formal measure $\nu^{g,\eta}_p$ is expressed as follows:

$$\nu^{g,\eta}_p = e^{-\mu^2\int_\Lambda\sqrt{\bar{g}_t(\xi)}d\xi}\det(\Delta_{\bar{g}_t}{}')^{-\frac{d}{2}}\det\left((A^{\bar{g}_t}_\eta)^*(A^{\bar{g}_t}_\eta)\right)^{\frac{1}{2}} D_{\bar{g}_t}[\lambda]D_{\bar{g}_t}[u]\,ds \qquad (5.1)$$

As we are interested in the transformation of the formal measure μ_p through the map T, as pointed out at the end of section 3, we shall let η go to zero in the density of the measure $\nu_p^{g,n}$, subintending here a formal weak convergence of measures. We recall that the formal limit measure was denoted by ν_p^g and we can write:

$$\nu_p^g = e^{-\mu^2 \int_\Lambda \sqrt{g_t(\xi)}\, d\xi} D_{g_t}^d D_{g_t}[\lambda] D_{g_t}[u]\, ds \tag{5.2}$$

where $D_{g_t}^d$ is given by (4.11).

According to the results of section 4, the density standing on the right hand side of (5.2) is not a well defined expression for it is infinite. We saw in proposition 4.4, that this infinite term could be compensated by a purely t-dependant expression $H_{N,\epsilon}(t) \equiv \left(\det^N (\Delta_{g_t}{}')_\epsilon\right)^{\frac{d-2}{2}} e^{\frac{d-2}{8\pi\epsilon}\int_\Lambda \sqrt{g(\xi)}\, d\xi} \left(det^N (P_{g_t}^* P_{g_t})_\varepsilon\right)^{-\frac{1}{2}}$ yielding in the limit when ε goes to zero a finite φ-dependent expression (Here t runs over the Teichmuller space τ and φ in $C^\infty(\mathbb{R})$).

As we are concerned with the conformal anomaly, - ie. the φ-dependant part of the Polyakov measure - we extract the φ-dependant part of the measure ν_p^g. In this respect, introducing in ν_p^g the purely t-dependant term $H_{N,\epsilon}(t)$ explicited above will not modify the φ-dependance of the Polyakov measure. This leads us to the study of the heuristic measure on $C^\infty(\mathbb{R})$:

$$\begin{aligned}\nu_1^{g_t}[\lambda] \equiv \lim_{\epsilon\to 0} \Bigg[e^{-\mu_\epsilon^2 \int_\Lambda \sqrt{\bar g_t(\xi)}\, d\xi} \lim_{N\to\infty} \Big[D_{g_t,d}^{\epsilon,N} \left(\det^N (\Delta_{g_t}{}')_\epsilon\right)^{\frac{d-2}{2}} \\ \times \left(det^N (P_{g_t}^* P_{g_t})_\varepsilon\right)^{-\frac{1}{2}} \Big] \times \mathrm{Vol}_{g_t}(\Lambda)^{-\frac{d}{2}} \times e^{\frac{d-2}{8\pi\varepsilon}\int_\Lambda \sqrt{g_t(\xi)}\, d\xi} \Bigg] D_{g_t}[\lambda]\end{aligned} \tag{5.3}$$

(with $\mu_\epsilon \in \mathbb{R}^*$) which takes into account all the φ-dependance of $D_{g_t}[h_1]$ for both $D_{g_t}[u]$ and ds are independant of φ - the first one because of it being defined with a φ-independant scalar product $< \cdot,\cdot >$ and the second one because it is an ordinary Lebesgue measure.

If we take $\mu_\epsilon^2 \equiv \frac{d-2}{8\pi\epsilon} - \frac{d-26}{48\pi}\tilde K$ for $\tilde K \in \mathbb{R}^+$, this, by proposition 4.5 reads:

$$\begin{aligned}\nu_1^{g_t}[\lambda] = H(t) e^{\frac{d-26}{48\pi}\left[\tilde K \int_\Lambda \sqrt{\bar g_t(\xi)}\, d\xi + \frac{1}{2}\int_\Lambda \sqrt{g_t(\xi)}\, (\Delta_{g_t}\varphi)(\xi)\, \varphi(\xi)\, d\xi \right.} \\ \left. {}^{+ \int_\Lambda d\xi \sqrt{g_t(\xi)} R_{g_t}(\xi)\varphi(\xi)\, d\xi}\right] (\mathrm{Vol}_{g_t}(\Lambda))^{-\frac{d}{2}} D_{g_t}[\lambda]\end{aligned} \tag{5.4}$$

Remark 5.5 : The term $(\mathrm{Vol}_{g_t}(\Lambda))^{-\frac{d}{2}}$ can be compensated by introducing an appropriate normalization term in μ_p^g [4,13,14]. For simplicity of notations we just leave it out in what follows.

The next step is to evaluate the φ-dependance of the formal measure $D_{\bar{g}_t}[\lambda]$. The formula that gives $D_{\bar{g}_t}[\lambda]$ stated after (3.12) is only a prescription for the formal measure $D_{\bar{g}_t}[\lambda]$ but it does not define a measure in a mathematical sense (at most we could define cylinder set measures that are not σ-additive on $\overline{C^\infty(\mathbb{R})}^{\bar{g}_t}$). This formal measure can therefore be interpreted in various ways and this ambiguity in the interpretation of the measure leads to an ambiguity in the interpretation of the φ-dependance. One proposal [15] was to assume that $D_{\bar{g}_t}[\lambda]$ is absolutely continous with respect to $D_{\bar{g}_t}[\lambda]$ and to interpret the formal density by:

$$e^{\int_\Lambda \sqrt{g(\xi)} e^{\varphi}(\xi)\delta(0)\, d\xi} \tag{5.5}$$

where $\delta(0)$ stands for $\delta(\xi,\xi)$ with $f(\xi') = \int_\Lambda \sqrt{g(\xi)}\delta(\xi',\xi)f(\xi)\, d\xi$. for $f \in C^\infty(\mathbb{R})$.

This φ-contribution of the measure $D_{\bar{g}_t}[\lambda]$ can in this case be compensated by the cosmological term of the form $e^{-\mu^2 \int_\Lambda \sqrt{\bar{g}_t(\xi)}\, d\xi}$ which appears in Polyakov's scheme so that choosing μ accordingly we see that the φ-dependant part of the Polyakov measure is given by (5.4) with $D_{\bar{g}_t}[\lambda]$ substituted by $D_{g_t}[\lambda]$.

Another approach to the problem of the φ-dependance of $D_{\bar{g}_t}[\lambda]$ is to use formal gaussian measures to compare $D_{\bar{g}_t}[\lambda]$ and $D_{g_t}[\lambda]$. Let us briefly draw the lines of this point of view. We multiply both formal measures with the same density: $e^{-\frac{1}{2}((\Delta_{g_t})'\lambda,\lambda)_{g_t}} = e^{-\frac{1}{2}((\Delta_{g_t})'\lambda,\lambda)_{g_t}}$. Recall that Δ_g is the Laplace-Beltrami operator for the metric g in M and $(\lambda,\tilde{\lambda})_g \equiv \int_\Lambda \sqrt{g(\xi)}\,\lambda(\xi)\tilde{\lambda}(\xi)\, d\xi$ for $(\lambda,\tilde{\lambda})\epsilon C^\infty(\mathbb{R})\times C^\infty(\mathbb{R})$.

This yields: $\mu^1(\lambda) \equiv e^{-((\Delta_{g_t})'\lambda,\lambda)_{g_t}} D_{\bar{g}_t}[\lambda]$ and $\mu^2(\lambda) = e^{-\frac{1}{2}((\Delta_{g_t})'\lambda,\lambda)_{g_t}} D_{g_t}[\lambda]$. To compare these two heuristic measures on $\overline{C^\infty(\mathbb{R})}^{\bar{g}_t}$ we take their Fourier transforms with respect to the scalar product $(\cdot,\cdot)_{\bar{g}_t}$. It is easy to see that $\hat{\mu}^2(\tilde{\lambda}) = \left(\frac{\det(\Delta_{g_t})'}{\det(\Delta_{g_t})'}\right)^{-\frac{1}{2}} \hat{\mu}^1(\tilde{\lambda})$ where $\hat{\mu}^i(\lambda)$ $i = 1,2$ is the Fourier transform of μ^i taken at the point $\tilde{\lambda}$. Here we omit the zero modes whose corresponding φ-dependance can however be absorbed by renormalization terms of the form $\int_\Lambda \sqrt{g(\xi)}\, d\xi$ [16,4]. The injectivity of the map that to a measure associates its Fourier transform then gives:

$$\mu^1(\lambda) = \left(\frac{\det(\Delta_{\bar{g}_t})'}{\det(\Delta_{g_t})'}\right)^{-\frac{1}{2}} \mu^2(\lambda) \qquad \forall \lambda \epsilon C^\infty(\mathbb{R})$$

and hence

$$D_{\bar{g}_t}[\lambda] = \left(\frac{\det(\Delta_{\bar{g}_t})'}{\det(\Delta_{g_t})'}\right)^{\frac{1}{2}} D_{g_t}[\lambda]$$

Inserting this in (5.4), we obtain:

$$\nu_1^{\bar{g}_t}[\lambda] = H(t)e^{\frac{(d+1)-26}{48\pi}\left[\tilde{K}\int_\Lambda \sqrt{\bar{g}_t(\xi)}\, d\xi + \frac{1}{2}\int_\Lambda \sqrt{g_t(\xi)}\,(\Delta_{g_t}\varphi)(\xi)\varphi(\xi)\, d\xi + \int_\Lambda d\xi \sqrt{g_t(\xi)} R_{g_t}(\xi)\varphi(\xi)\right]} \mathrm{Vol}_{\bar{g}_t}(\Lambda)^{-\frac{d}{2}} D_{g_t}[\lambda] \tag{5.6}$$

because in (5.1) the density of $D_{g_t}[\lambda]$ with respect to $D_{g_t}[\lambda]$ will contribute by changing d into $d+1$ in the exponential factor.

In order to take into account both these interpretations we shall investigate, for fixed t in $\mathcal{T}$:

$$\bar{\nu}_1^{g_t}[\lambda] \equiv e^{\frac{\delta_d - 26}{48\pi}\left[\widetilde{K}\int_\Lambda \sqrt{\bar{g}_t(\xi)}\,d\xi + \frac{1}{2}\int_\Lambda \sqrt{g_t(\xi)}\,(\Delta_{g_t}\varphi(\xi)\,\varphi(\xi)\,d\xi + \int_\Lambda d\xi\sqrt{g_t(\xi)}R_{g_t}(\xi)\varphi(\xi)\right]} (\mathrm{Vol}_{g_t}(\Lambda))^{-\frac{d}{2}}\, D_{g_t}[\lambda] \tag{5.7}$$

setting $\delta_d \equiv d$ or $d+1$ according to the interpretation refered to.

We now propose a mathematical formulation of this "conformal anomaly" measure $\bar{\nu}_1^{g_t}[\lambda]$ through a well defined Liouville measure. (Referring to remark 5.5, we drop the term $(\mathrm{Vol}_{g_t}(\Lambda))^{-\frac{d}{2}}$ obtaining a measure $\tilde{\nu}_1^{g_t}$.)

For this, we introduce, for each t in $\mathcal{T}$, the Hilbert Schmidt operator $G_t \equiv \left(\Delta_{g_t}{}'\right)^{-1}$ and denote by χ_ρ for $\rho > 0$ a smooth function on Λ such that its Fourier transform vanishes outside a ball of radius $\frac{1}{\rho}$ and for any $f \epsilon C^\infty(\mathbb{R})$, $(f * \chi_\rho)(y) \to f(y)$, $y \epsilon \Lambda$ as $\rho \to 0$ ($*$ meaning convolution). We set $G_\rho^t \equiv \chi_\rho * \Gamma_t * \chi_\rho$ and take $\widetilde{K}$ to be the limit when ρ goes to zero of the constant $K_{\epsilon,\rho}^d \equiv \lambda e^{-\frac{\alpha^2}{2}G_\rho^t(0)} \times \frac{48\pi}{26-\delta_d}$ with $\lambda \geq 0$ and α in $\mathbb{R}$.

Regarding $\tilde{\nu}_1^{g_t}$ as a measure on $C^\infty(\mathbb{R})$ through the identification of the tangent space $T_\varphi C^\infty(\mathbb{R})$ with $C^\infty(\mathbb{R})$, we have a local measure $\nu_1^{g_t}[\varphi]$ around the point φ in $C^\infty(\mathbb{R})$. We set $\varphi \equiv \alpha\psi$ and obtain,

$$\begin{aligned}\tilde{\nu}_1^{g_t}[\psi] &= \lim_{\rho\to 0} e^{-\lambda\int_\Lambda e^{-\frac{\alpha^2}{2}G_\rho^t(0)} e^{\alpha\psi(\xi)}\sqrt{g_t(\xi)}\,d\xi} \\ &\times e^{-\frac{1}{2}\int_\Lambda \sqrt{g_t(\xi)}\,(\Delta_{g_t}\psi)(\xi)\psi(\xi)\,d\xi} \times e^{-\frac{1}{\alpha}\int_\Lambda \sqrt{g_t(\xi)}R_{g_t}(\xi)\psi(\xi)\,d\xi} \times D_{g_t}[\psi]\end{aligned} \tag{5.8}$$

Up to a cut off on ψ $\quad(\psi \longrightarrow \psi_\rho \equiv \psi * \chi_\rho)$, $\tilde{\nu}_1^{g_t}[\psi]$ can be interpreted as

$$e^{-\frac{1}{2}\left(R_{g_t}(\xi),\psi\right)_{g_t}} \lim_{\rho\to 0} e^{-\lambda\left(\chi_\Lambda,:e^{\alpha\psi_\rho}:\right)_{g_t}} \mu_{0,t}(\psi)$$

where: χ_Λ is the characteristic function of Λ, $:e^{\alpha\psi_\rho}: \equiv e^{-\frac{\alpha^2}{2}G_\rho^t(0)}\,e^{-\alpha\psi_\rho}$ and $\mu_{0,t}$ is the Gaussian measure with mean zero and covariance G_t.

We shall now state a result about the existence of such measures. For the sake of not making this section too long, we shall not give a proof of this result, which can be shown by using the methods of [17] (see also [18]).

$L^2(\mu_{0,t})$ denotes the space of real functions on Λ which are square invariant with respect to $\mu_{0,t}$:

Proposition 5.1:

i) For α in $\mathbb{R}$ and $\alpha^2 < 4\pi$, $\left(\chi_\Lambda,:e^{\psi_\rho}:\right)_{g_t}$ converges in $L^2(\mu_{0,t})$ when ρ goes to 0 to a positive random variable $U_t(\psi)$.

ii) For all ψ in the support of $\mu_{0,t}$, for all $\lambda \geq 0$, we set

$$\mu_t(\psi) \equiv \exp\left(-\lambda U_t(\psi)\right) \mu_{0,t}(\psi) \Big/ \int \exp\left(-\lambda U_t(\psi)\right) d\mu_{0,t}(\psi)$$

For $\alpha^2 < 4\pi$, μ_t is a well defined probability measure absolutely continuous with respect to $\mu_{0,t}$.

iii) For $\alpha^2 < 4\pi$, the function $\psi \longrightarrow \exp -\alpha \left(R_{g_t}, \psi\right)_{g_t}$ is μ_t measurable and in $L^p(\mu_t)$ for all $1 \leq p < \infty$. Moreover:

$$\tilde{\mu}_t(\psi) \equiv \exp\left(-\alpha \left(R_{g_t}, \psi\right)_{g_t}\right) \mu_t(\psi) \Big/ \int e^{-\alpha \left(R_{g_t}(\xi), \psi\right)_{g_t}} \mu_t(\psi)$$

is a well defined measure absolutely continuous with respect to $\mu_{0,t}$.

The proposition 5.1 yields the existence of $\tilde{\nu}_1^{g_t}[\psi]$ for $\alpha^2 < 4\pi$ i.e. $\delta_d \leq 13$ or $d \leq 12$. This gives a mathematical formulation when the dimension d is smaller than 13 of the part of the Polyakov measure corresponding to the conformal anomaly in the quantization of the bosonic strings. It can further be shown that the Liouville measure μ_t for $\alpha^2 < 4\pi$ satisfies the Osterwalder Schrader axioms of euclidian quantum field theory so that it gives a quantization of the Liouville model. Hence we have put the quantization of bosonic strings in correspondence with the quantization of the Liouville model, for strings moving in space-time dimensions not larger than 12.

Acknowledgements

It is a great pleasure to thank Professor Dita & Professor Georgescu for their kind invitation. I also would like to thank Professor Anne Marie Boutet de Monvel for interesting discussions and wish to express my gratitude to Professor Sergio Albeverio, Dr. Sergio Scarlatti and Professor Raphael Høegh Krohn for their cooperation and constant encouragements. A month after I had finished writing this paper I heard of the sad news about the sudden death of Raphael Høegh-Krohn. I wish to express here my deep sorrow and my admiration for the great mathematician he was.

References

[0] : M.G. Green, J.H. Schwartz & E. Witten : Superstring theory, Volume 1, Cambridge Monographs on mathematical physics (1987)

[1] : Y. Nambu in Symmetries and Quark Models, ed. R. Chand ,Gordon and Breach, New York 1970
T. Goto, Progr.-Theor. Phys. 46 (1971) p. 1560

[2] : Brink - Di Vecchia - Howe: Phys. Lett. *B65*, (1976), p. 471

[3] : Polyakov: Quantum geometry of bosonic strings. Phys. Lett. 103*B* (1981), p. 207

[4] : O. Alvarez, Theory of strings with boundaries: fluctuations topology and quantum geometry. Nucl. Phys. *B216* (1983), p. 125

[5] : G. Moore, P. Nelson: Measure for moduli. Nucl. Phys. *B266* (1986), p. 58

[6] : C.J. Earle, J. Eells: A fibre bundle description of Teichmüller theory. Jour. Diff. Geom., 3 (1969), p. 19 - 43

[7] : A.E. Fischer, A.J. Tromba: On a purely "Riemannian" proof of the structure and dimension of the unramified moduli space of a compact Riemann surface. Math. Ann. 267 (1984), p. 311 - 345

[8] : L. V. Ahlfors: Teichmüller spaces. Proc. Internat. Congress. Math. (1962), Institute Mittag Leffler, Djursholm, Sweden, (1963), p. 3 - 9

[9] : L. Bers: Spaces of Riemann surfaces. Proc. Internat. Congress Math., Edinburgh (1958), p. 349- 361

[10] : H. Omori: On the group of diffeomorphisms on a compact manifold. Proc. Symp. Pure Math. *AMS*15 (1970), p.167 - 183

[11] : R.T. Seeley: Amer. J. of Math 91 (1969) p. 889 - 963

[12] : P.B. Gilkey: The index theorem and the heat equation. Mathematics lecture series 4, Publish or Perish (1974)

[13] : B. Durhuus, P. Oleson, S.L. Peterson: Polyakov's quantized strings with boundary terms. Nucl. Phys. *B198* (1982), p. 157 - 188
B. Durhuus: Quantum theory of strings. Nordita preprint lectures given at NB I/NORDITA, autumn 1982 (non published)

[14] : S. Albeverio, R. Høegh-Krohn, S. Paycha, S. Scarlatti, paper in preparation

[15] : J. Polchinsky : Evaluation of the one loop string path integral, C.M.P. 104 (1986)

[16] : E. D'Hoker, D.H. Phong: Nucl. Physics *B269* , (1986), p. 205

[17] : S. Albeverio, R. Høegh-Krohn: The Wightman axioms and the mass gap for strong interactions of exponential type in two-dimensional space-time, I. Funct. Anal. 16 (1974), p. 39 - 82

[18] : S. Albeverio, R. Høegh-krohn, S. Paycha, S. Scarlatti : Path space measure for the Liouville quantum field theory and the construction of relativistic strings, Physics letters *B174* (1986)

The Covariant Lattice Construction of Four-Dimensional Strings

W. Lerche and A.N. Schellekens

CERN, 1211 Geneva 23, Switzerland

1. Introduction

The main goal of string theory is an ambitious one: to obtain a consistent theory of all interactions, including gravity. Just a few years ago it seemed that gravity was all it could describe, but that has changed drastically with the discovery of the heterotic string [1]. Although at first it seemed that the theory would fail its first confrontation with experiment by predicting the wrong space-time dimension, that too changed very quickly after its birth. Indeed, what once may have seemed an insurmountable obstacle is now a gaping hole: not only *can* we construct string theories in four dimensions, but it is also so easy that we can make them in embarrassing abundance. Meanwhile the miracles that started the excitement have survived, although we are now so used to them that we tend to forget them: chiral string theories in four dimensions exist and their spectra have the same general features (gauge groups, representations and family replication) as that of the standard model. If all this is realized in a class of theories that holds the promise of offering a finite and consistent theory of quantum gravity, then there is certainly reason for excitement. The dream of having a *unique* theory of everything must be shelved for the moment, but fortunately uniqueness is the aspect that we can most easily live without. On the other hand, giving up consistency would be fatal.

The floodgates of four-dimensional string construction were first opened in [2]. The initial trickle of Calabi-Yau compactifications quickly grew into a raging torrent, which broadened into orbifold compactifications [3]. The theories that are now usually called four-dimensional strings are mathematically not very different from these early ideas, and it doesn't make much sense to try to make a distinction. If there is a difference, then it is one of philosophy rather than physics. Whereas initially one tried to compactify ten-dimensional strings, nowadays the emphasis is more on constructing string theories directly in four dimensions. We believe that the idea of compactification is too limited, and too much linked to notions of space-time that may be unsuitable for string theory, to be of lasting value. But all this is just a matter of interpretation, and does not influence the facts, such as the spectrum and the S-matrix.

The change of philosophy from compactification to four-dimensional strings started

ISBN 0-12-218100-X

with the seminal paper of Narain [4], and gained momentum due to the work of Kawai. et. al. Soon thereafter many ways of constructing heterotic strings in four dimensions were discovered [5-8], the oldest being fermionic constructions, the covariant lattice construction, and asymmetric orbifolds, which are now understood to be close relatives. Several other ideas have surfaced more recently [9-12]. For recent reviews of aspects of string vacua see *e.g.* [13], [14].

These lectures are devoted to a detailed description of the covariant lattice construction and its underlying principles. The construction is based on the bosonic formulation of the fermionic string and its ghost system [15]. The possibility of using this formalism for the construction of new modular invariant string theories was first discussed in [16], and it was used for four-dimensional strings in [6]. A related idea is the "charge lattice" of [5], which these authors obtained by bosonizing their fermionic construction. The result is a light-cone version of covariant lattices.

Although we tried to keep these notes as elementary as possible, some basic knowledge of string theory is assumed, and may be gotten from one of the many reviews of the subject, some of which are listed under [17]. In the next chapter we begin with an introduction to the principles of string construction. This section serves also a guide to the rest of the paper.

2. Basic principles of string construction

The most fundamental quantity in any string theory is a function $X^\mu(\sigma^0,\sigma^1)$ that specifies the space-time position of every point of the string (denoted by σ^1) for any value of the proper-time, σ^0. Here μ labels the space-time coordinates. The space-time dimension is d, and our interest will ultimately be in the value $d=4$. Throughout this paper we will only consider *closed* strings, which means that $X^\mu(\sigma^0,0) = X^\mu(\sigma^0,\pi)$ (the choice of the σ_1-interval is a matter of convention). When it propagates through space-time the string sweeps out a two-dimensional surface parametrized by σ^0 and σ^1. This surface is called the *world-sheet*, in analogy to world-lines for point particles. For non-interacting closed strings it has the shape of a tube. If strings interact they do so by splitting and joining, and as a result they sweep out world-sheets that look like complicated pieces of plumbing.

The modern way of describing strings is to regard them as two-dimensional field theories. One regards $X^\mu(\sigma_0,\sigma_1)$ as a set of d scalar fields, which are functions of a two-dimensional coordinate $\sigma^\alpha, \alpha = 0,1$. Once one realizes that one is dealing with two-dimensional field theories, one is more or less naturally led to the questions: Which conditions should be imposed on such a field theory for it to describe a consistent string theory? And which two-dimensional field theories satisfy these conditions?

One has to be careful in formulating these conditions, because many claims about uniqueness of string theory where based on implicitly or explicitly imposed conditions which were unnecessarily strong. So we should begin by carefully listing and examining the conditions. Our list of what we believe to be necessary conditions for d-dimensional strings is:

1. The presence of d unconstrained scalar fields X^μ
2. Reparametrization invariance

3. Conformal invariance

4. Modular invariance

5. World sheet supersymmetry and superconformal invariance (for fermionic strings).

These are all conditions imposed on the world-sheet field theory. Remarkably, if one satisfies them one is guaranteed not to be confronted with most of the disasters that can occur in field theory in d-dimensional space-time. Reparametrization invariance means that the description of string propagation does not depend on how one chooses the coordinates σ_0 and σ_1 on the world-sheet. This has a simple interpretation in two-dimensional field theory language: it's just two-dimensional general coordinate invariance. The first two points are so basic to string theory that one would not even want to consider what would go wrong if they were not satisfied. For the last three conditions (which we will discuss in more detail in a moment) one can identify some desirable space-time properties that follow from them.

Conformal invariance guarantees Lorentz-invariance and absence of ghosts, general coordinate invariance in space-time and gauge invariance (if the theory has massless vector bosons, as is usually the case). Although it has been the subject of debate whether one can make sense of string theory without it, it seems clear that one gets into serious technical complications if one abandons it. The attempts to abandon it were motivated by the fact that conformal invariance restricts, naively at least, the space-time dimension to 26 for bosonic strings, and to 10 for fermionic ones. Since there are now other ways to get to lower dimensions, it seems highly preferable to keep it.

One of the nice consequences of modular invariance is that it ensures that chiral fermion representations coming from string theory are anomaly free. This is highly non-trivial, especially in higher dimensions, and the discovery of a rather spectacular cancellation in 10 dimensions [18] was the starting point of the current wave of interest in the subject. The precise relation with modular invariance was only understood much later [19]. Anomalies are certainly not the only problems one can expect in the absence of modular invariance, but it is the easiest one to identify.

World-sheet supersymmetry is essential for fermionic strings, and one of the problems one has in its absence is a violation of Lorentz invariance. We will discuss this in chapter 7. Here we will continue with a more detailed look at conformal and modular invariance.

2.1 Conformal invariance

A conformal transformation is a coordinate dependent rescaling of the metric on the world-sheet. It is this invariance which makes string theory so simple in comparison with, for example, membranes. The standard two-dimensional action for free bosons,

$$S = \int d^2\sigma \sqrt{(-g)} g^{\alpha\beta} \partial_\alpha X^\mu \partial_\beta X_\mu,$$

where $g_{\alpha\beta}$ is the world-sheet metric and g its determinant, is not only general coordinate invariant, but has an additional invariance under $g_{\alpha\beta} \rightarrow \Lambda(\sigma^\alpha) g_{\alpha\beta}$. This invariance makes it possible to choose a gauge (light-cone gauge) in which one gets an action which depends only on the $d-2$ "transverse" (*i.e.* transforming in the vector representation of the little group $SO(d-2)$ of a massless particle) degrees of freedom X^i, and which is non-interacting and manifestly ghost-free. In the absence of conformal invariance there is an extra mode of the metric, the Liouville mode, which enters into the action, and which makes life considerably more complicated. Thus we would like to maintain conformal invariance even if other terms are added to the action.

Although conformal invariance is a symmetry of the classical action, it is afflicted by an anomaly in the quantized theory. There are several contributions to this anomaly. In units in which a scalar contributes 1, the relevant ones are $\frac{1}{2}$ for a Majorana-Weyl fermion, -26 for the ghosts of general coordinate invariance, and +11 for the ghosts of local supersymmetry. Notice that though the graviton has no physical components in two dimensions, it still contributes via its ghosts. There are two obvious ways of cancelling these anomalies :

$N=0$: 26 scalars + gravity ghosts

$N=1$: 10 scalars and Majorana Weyl fermions + gravity ghosts + superghosts

Since there is always general coordinate invariance on the world sheet, one always has the corresponding ghosts. In these solutions one can replace a scalar by two fermions or vice versa, but that amounts just to two-dimensional bose-fermi equivalence and does not really bring anything new. We have made our choice in such a way that supersymmetry is manifest, but that is not necessary and we will in fact choose a purely bosonic realization in our construction. One can also try to have more supersymmetries, but the resulting theories turn out to be relevant only in two space-time dimensions or less.

At this point we should make a trivial but crucial remark. The traditional interpretation of these solutions is to associate all the scalars with space-time coordinates. In that case the $N=0$ system lives in 26 dimensions, and the $N=1$ system in $d=10$. This is not necessary however. One can equally well associate just a subset of the scalars with space-time coordinates, and regard the others as some sort of "internal" degrees of freedom. Then the numbers 26 and 10 are "maximal" rather than "critical" dimensions. In some cases, and in particular for bosonic strings, this means nothing more than that some dimensions are "compactified", but for fermionic strings there are more interesting possibilities, as we will see.

A special feature of conformally invariant closed strings is that the modes of vibration of the string can be unambiguously separated into *left-moving* ones and *right-moving* ones. Since the world-sheet theory is non-interacting, and because there are no end-points of the string that modes can bounce back from, a left-moving mode will move to the left forever. This left/right-mover separation goes so far that one can actually choose *different* conformal systems for them. This leads in an obvious way to three kinds of closed string theory:

Bosonic :	$N=0$ left-movers;	$N=0$ right-movers
Heterotic :	$N=0$ left-movers;	$N=1$ right-movers
Type II :	$N=1$ left-movers;	$N=1$ right-movers

This also specifies our convention for left- and right-movers in the heterotic string.

We will mainly consider heterotic strings in this paper, although everything we say is valid for type II strings as well. The latter seem however phenomenologically less promising at the moment [20], [21]. The heterotic string forces us to make a previous condition more precise. Since it has 26 bosons on the left and only 10 on the right, it is not *a priori* obvious in which dimensions it can exist. The precise requirement on the presence of a set of scalars X^{μ} is however that both its left and right-moving components are present in the theory. One cannot associate an unpaired left-moving mode with a flat space-time dimension (this would be in conflict with modular invariance). Thus the maximal dimension for the heterotic string is 10. Psychologically it was perhaps this heterotic string asymmetry that lead to a re-evaluation of the role of world-sheet scalars, making four-dimensional strings possible.

2.2 MODULAR INVARIANCE

So far we have imposed conditions to ensure that string theory is well-defined locally on the world-sheet. In the perturbation expansion of string theory one does however encounter world-sheets with non-trivial topology, where the *global* properties of the theory become important. The simplest of these diagrams is the torus, roughly corresponding to a one-loop diagram in field theory. Topologically the description of the multiloop diagrams in the expansion of closed (oriented) strings is very simple: they are all equivalent to multiple tori, and there is just one topology for each order in the expansion. These loops may have external lines (tubes) attached to them. Such surfaces are known as *Riemann surfaces.*

By "modular invariance" we mean absence of global anomalies on any of these surfaces. We also mean *factorization*, *i.e.* the amplitude for a multiloop diagram should equal the product of its subdiagrams in a limit where it is pulled apart into these subdiagrams. Although one should impose modular invariance as a condition on diagrams with and without external lines, it is sufficient to examine the ones without external lines, because all other diagrams can always be obtained from them as a subdiagram of some higher loop surface.

The exact conditions for modular invariance beyond one loop are not known for fermionic strings, since one does not know enough about the dependence on the "supermoduli", but reasonable *ansätze* exist, and it is under these assumptions that one usually checks modular invariance. For pedagogical reasons we will not say much about higher loop modular invariance except that it is a property of the class of theories we will consider (under the aforementioned assumptions, which have to be made for any fermionic string theory).

One-loop modular invariance is however rather easy to understand. A one-loop diagram is a torus, and it can be represented by points in the complex plane modulo a two-dimensional lattice. This corresponds to a choice of coordinates and a flat, constant metric on the torus. To regard the complex plane modulo a lattice simply means that points that differ by a lattice vector are identified with each other. A two dimensional lattice has two basis vectors, chosen so that all other vectors are integral linear combinations of them. By rotations and scale transformations one can always align one of the basis vectors with the real axis, and scale it to have length 1. The second one can always be put into the positive upper half plane, where it can be specified by a complex number τ, with $Im(\tau) > 0$. We have now used all the transformations at our disposal to bring the metric in the simplest possible form. The remaining free parameter of the torus, τ, parametrizes metrics that cannot be mapped into each other by general coordinate transformations or Weyl transformations.

Not all these metrics are inequivalent however. Although they cannot be related by *local* transformations, some can be related by *global* ones. For example, if one defines a torus by $\tau' = \tau + 1$ one makes the same identification in the complex plane, but the metric would look rather different, *i.e.* have a different but equivalent parametrization. On a real-life, donut-shaped torus this transformation corresponds to cutting the torus along a cycle which one cannot be continuously deformed to a point, rotating one edge by 2π, and glueing the edges back together. A torus has two independent cycles of this type, and hence there is a second transformation of τ which leads to equivalent metrics, namely $\tau \to -\frac{1}{\tau}$. Together these transformations generate the group of *modular transformations*. Although one can write down an explicit expression for all its elements, one needs in practice usually only its two generators,

$$\begin{aligned} \tau &\to \tau + 1 \\ \tau &\to -\frac{1}{\tau} \end{aligned} \tag{2.1}$$

To calculate loop-diagrams à la Polyakov [22] one performs a path-integral over all fields in the two-dimensional action (for a given metric on the world-sheet), and furthermore one integrates over all inequivalent metrics. The expression for a one loop diagram is (after some gauge fixing) of the form

$$\int \frac{d^2\tau}{(Im\tau)^2}\mathcal{F}(\tau),$$

where $\mathcal{F}$ is some function which depends on the theory, and the other factor is the invariant integration volume. The function $\mathcal{F}$ (or rather a slightly modified function, defined more precisely in section (4.3)) is called the *partition function.* The issue of modular invariance centers on the boundaries of the τ-integral. If one were to integrate over the entire positive half plane, one would encounter the same metric an infinite number of times. One would like to restrict the integration to a *fundamental domain*, which is defined as a region in the upper half plane which contains every metric just once (the canonical choice is $-1 \leq Re\tau < 1, |\tau| \geq 1$), but this can only be done in a meaningful way if $\mathcal{F}(\tau)$ does not depend on the region one chooses, since otherwise the result would be ambiguous. Thus $\mathcal{F}$ should be invariant under (2.1) . This is what is meant by modular invariance.

For an n-loop surface the problem is very similar. Instead of a single complex parameter τ one gets now a complex $n \times n$ matrix Ω, the *period matrix* . For discussions of modular transformations for higher genus Riemann surfaces we refer to [23].

2.3 FOUR DIMENSIONAL STRINGS: THE BASIC IDEA

There is a fairly logical path towards four-dimensional string constructions of the type we are going to consider, and it might be useful to have some idea about this before starting with the more technical aspects. Terms not explained below will be discussed in detail in the remaining chapters.

Consider first the ten-dimensional heterotic string. Its construction has been explained in much detail and with great clarity in [1], which is obligatory literature for anyone interested in string constructions. Here we concentrate on just one aspect, namely the role of the 16 left-moving bosons that are not paired with right-movers into space-time coordinates. Instead one regards them as compactified space-time coordinates (at least in the mathematical sense of the word, because as a physical concept this is somewhat inadequate (and unnecessary)). The center-of-mass coordinate of the string in these 16 dimensions is put in a periodic space (a torus) with periodicities specified by a 16 dimensional lattice. Modular invariance restricts the choice of this lattice (and in particular does not allow an infinite radius torus, which would correspond to flat space). One finds that it must be *even* and *self-dual*, and such 16-dimensional lattices have been classified, the only possibilities being $E_8 \times E_8$ or a certain subset of the $SO(32)$ weight lattice.

Narain took this construction one logical step further. Instead of insisting on 10-dimensional Lorentz-invariance he required only Lorentz-invariance in $d < 10$ dimensions (we will choose $d = 4$ here, for definiteness and other reasons), and consequently paired only 4 left- and right-moving bosons into space-time coordinates. The remaining ones, 22 left-movers and 6 right-movers are "compactified" (the physical meaning of this terminology has become less and less clear throughout the historical development, but the mathematical description is unambiguous) on a torus of total dimension 22+6. The condition for modular invariance is that the lattice defining this torus must be even and self-dual with respect to a *lorentzian metric*. There is a continuous infinity of such lattices!

A problem with this approach is that it does not lead to chiral theories. This is related to the fact that the 10 right-moving Majorana-Weyl fermions of the heterotic string have remained untouched. Since they are responsible for the existence of space-time fermions, what happens is that all ten-dimensional space-time fermions which are chiral are trivially compactified on a torus, and it is a well-known fact that this leads to non-chiral theories in lower dimensions.

The next logical step in the development avoids this problem by making the Majorana-Weyl fermions participate in the process. To do so, they have to be on equal footing with the other fields, which can be achieved by bosonizing them. Actually we are not going to bosonize anything, but we will just start with 15 bosons (rather than with 10 bosons and 10 fermions) to cancel the conformal anomaly. The five bosons replacing the 10 fermions are included in the same torus, which now has dimension 22+11. There is a small technical complication however: to get a lattice description of the theory one has to bosonize also the ghosts of local supersymmetry, and enlarge the lattice to include the charge of the resulting boson.

Now one can check the conditions for modular invariance, and the result is again that the torus has to correspond to a self-dual lattice, only this time it is not even but odd. This is all that needs to be said about modular invariance in these theories. It can be shown that (within the usual assumptions) this self-duality guarantees not just one-loop, but also higher-loop modular invariance.

Before we can make all this more precise we have to explain the corner stones of constructions of this type, and we will do so in the next three chapters on lattices, torus compactification and conformal field theory. These chapters are more or less independent, and readers sufficiently familiar with any of these subjects may want to skip some of them. Chapter 6 discusses the Neveu-Schwarz-Ramond model and its bosonic form, which is the heart of our construction. In particular we have to explain in some detail the properties of its bosonized ghost system. It is unfortunately not possible to explain this without the sophisticated techniques of conformal field theory, and although we have tried to supply the necessary information in chapter 5, we are aware of the fact that it may not be adequate to readers unfamiliar with it. Fortunately one finally arrives at a formulation of the theory where most of this can be forgotten again, at least as long as one is only interested in model building and spectrum calculations. For this reason we have formulated chapters 8 and 9 in such a way that they are self-contained, so that even readers who didn't fully grasp the ghost system should be able to apply the rules of model building. Meanwhile chapter 7 contains the discussion of world-sheet supersymmetry promised above.

3. Lattices

In this section we collect most of the known facts about self-dual lattices. In the mathematics literature, most work on this subject appears in journals on Number Theory, which are unfortunately not always easily available in physics libraries. Some of the relevant papers are listed under Ref. [24-26], and readers interested in proofs not provided below should consult these papers. We also recommend the reviews by Goddard and Olive [27].

3.1 DEFINITIONS

A lattice Λ is defined as a set of points in a vector space V of dimension N, of the form

$$\Lambda = \{\sum_{i=1}^{N} n_i \vec{e}_i | n_i \in Z\}, \tag{3.1}$$

where $\vec{e}_i$ $(i = 1, \ldots, N)$ is a set of basis vectors of V. The vector spaces we will consider are R^N, with a euclidean inner product, and $R^{p,q}, (p+q=N)$, with a "lorentzian" inner product. In terms of the components v^a of vectors $\vec{v} \in V$ we define these inner products as $\vec{v} \cdot \vec{w} = v^a v^b d_{ab}$, with $d_{ab} = \delta_{ab}$ for R^N, and $d_{ab} = \epsilon(a)\delta_{ab}, \epsilon(a) = 1$ for $1 \le a \le p, \epsilon(a) = -1$ otherwise, for $R^{p,q}$. In the following we will denote these inner products as $(+)^N$ and $(+)^p(-)^q$ respectively, and we will refer to d_{ab} as the *signature*.

The *unit cell* of a lattice is the set of points $\{\vec{x} = \sum_i x_i e_i | 0 \leq x_i < 1\}$. It contains precisely one lattice point (i.e $\vec{x} = 0$), and every vector of V has precisely one "image" in the unit cell, obtained by adding a lattice vector to it. The *volume* $vol(\Lambda)$ of the unit cell of Λ is obtained by defining a matrix

$$g_{ij} = \vec{e}_i \cdot \vec{e}_j, \tag{3.2}$$

in terms of which

$$vol(\Lambda) = \sqrt{|\det g|}. \tag{3.3}$$

The matrix g_{ij} is called the *metric* of the lattice. It contains all information regarding the angles of the basis vectors and their lengths. Notice that if we change the signature from lorentzian to euclidean (by changing $(+)^p(-)^q$ to $(+)^{p+q}$) then $\det g$ changes at most by a sign, so that the volume of the unit cell remains the same.

The *dual* Λ^* of Λ is defined as

$$\Lambda^* = \{\vec{w} \in V | \vec{w} \cdot \vec{v} \in Z, \forall \vec{v} \in \Lambda\}. \tag{3.4}$$

A basis for Λ^* is the set of vectors $\vec{e}_i^*$, with $\vec{e}_i^* \cdot \vec{e}_j = \delta_{ij}$. It is easy to see that this is indeed a basis for Λ^*, i.e.

$$\Lambda^* = \{\sum_{i=1}^{N} m_i \vec{e}_i^* | m_i \in Z\}. \tag{3.5}$$

To calculate the volume of Λ^* we define

$$g_{ij}^* = \vec{e}_i^* \cdot \vec{e}_j^*, \tag{3.6}$$

which is the metric of the dual lattice. Using the definition of the basis vectors one can easily show that $\sum_i e_i^{*a} e_i^b = d^{ab}$, and that therefore g^* is the inverse of g. Thus we have

$$vol(\Lambda) = (vol(\Lambda^*))^{-1}. \tag{3.7}$$

A lattice Λ is called

— *unimodular* if $vol(\Lambda) = 1$

— *integral* if $\vec{v} \cdot \vec{w} \in Z, \forall \vec{v}, \vec{w} \in \Lambda$

— *even* if Λ is integral, and $\vec{v}^2 = 0 \bmod 2, \forall \vec{v} \in \Lambda$

— *odd* if Λ is integral, but not even

— *self-dual* if $\Lambda = \Lambda^*$

Obviously Λ is integral if and only if $\Lambda \subset \Lambda^*$.

3.2 COSETS

If Λ_s is a sublattice of Λ, then we can decompose Λ in *cosets* with respect to Λ_s. To do so, choose a set of *coset representatives* $\vec{\beta}_\alpha$, so that

$$\begin{aligned} &\vec{\beta}_\alpha \in \Lambda, \vec{\beta}_\alpha \notin \Lambda_s \\ &\vec{\beta}_\alpha - \vec{\beta}_{\alpha'} \notin \Lambda_s \text{ if } \alpha \neq \alpha'. \end{aligned} \tag{3.8}$$

If we extend this set with the vector $\vec{\beta} = 0$ to represent Λ_s itself, we get

$$\Lambda = \sum_\alpha (\vec{\beta}_\alpha + \Lambda_s), \tag{3.9}$$

i.e. every vector in Λ can be written as $\vec{\beta}_\alpha + \vec{v}_s, \vec{v}_s \in \Lambda_s$. Every coset $\vec{\beta}_\alpha + \Lambda_s$ has precisely one point in the unit cell of Λ_s. Therefore, if there are N_s cosets (including Λ_s), we find

$$vol(\Lambda) = \frac{1}{N_s} vol(\Lambda_s). \tag{3.10}$$

This is a fairly trivial, but very useful observation. A direct consequence is that a lattice is self-dual if and only if it is both integral and unimodular.

3.3 LIE ALGEBRA LATTICES

It is often convenient to express a self-dual lattice in terms of Lie algebra lattices. In the following when we speak of a Lie algebra lattice we mean any sublattice of the weight lattice of a semi-simple Lie algebra (for a discussion of Lie algebras, roots and weights we refer to [28]). In particular we need the *simply laced* lattices, belonging to semi-simple Lie algebras which are products of D_n, A_n and E_6, E_7, E_8. They have the property that all their roots have the same length. If one normalizes them so that they have all norm 2, the weight lattice Λ_W is the dual of the root lattice Λ_R, which is itself a sublattice of the weight lattice (the *norm* of a vector is its squared length)

$$\Lambda_W = \Lambda_R^* \ ; \ \Lambda_W \supset \Lambda_R \tag{3.11}$$

Therefore we can decompose Λ_W into cosets with respect to Λ_R. In the case of Lie algebra lattices these cosets are called *conjugacy classes*. In general cosets form a group under addition, with the sublattice Λ_s as the unit element. In this case this group is called the *center* of the Lie algebra.

We will make extensive use of D_n lattices, for which we denote the conjugacy classes as follows

$$\begin{aligned} (0) &= \Lambda_R = \{(k_1, \ldots, k_n),\ k_i \in Z,\ \sum_{i=1}^n k_i = 0 \bmod 2\} \\ (v) &= \Lambda_R + (1, 0, 0, \ldots, 0) \\ (s) &= \Lambda_R + (\tfrac{1}{2}, \tfrac{1}{2}, \ldots, \tfrac{1}{2}) \\ (c) &= \Lambda_R + (-\tfrac{1}{2}, \tfrac{1}{2}, \ldots, \tfrac{1}{2}) \end{aligned} \tag{3.12}$$

One may check that the center is isomorphic to Z_4 if n is odd, and to $Z_2 \times Z_2$ if n is even.

The irreducible representations of the semi-simple Lie algebras are characterized by their weight vectors, which for a given representation all belong to the same conjugacy class of the lattice. Hence we can specify conjugacy classes also by giving representations that belong to them. In the case of D_n, (0) contains all even rank tensor products of the vector representation (which includes the singlet and the adjoint), (v) contains all odd-rank tensors, and (s) and (c) contain spinors of opposite chirality.

A useful property of conjugacy classes is that their mutual inner products modulo integers depend only on the class, and not on the individual vectors. If $\vec{\beta}_1, \vec{\beta}_2 \in \Lambda_W$, and $\vec{\lambda}_1, \vec{\lambda}_2 \in \Lambda_R$, we have

$$(\vec{\beta}_1 + \vec{\lambda}_1) \cdot (\vec{\beta}_2 + \vec{\lambda}_2) = \vec{\beta}_1 \cdot \vec{\beta}_2 \bmod 1,$$

because $\vec{\beta}_i \cdot \vec{\lambda}_j \in Z$, and because the root lattice of simply laced algebras is integral (this follows from the fact that it is contained in its own dual, see (3.11)). In fact Λ_R is an even lattice, because its basis vectors are roots, which have norm 2. Using this property one can show by a similar argument that the norms of two vectors in the same conjugacy class are always equal modulo 2. These properties are clearly of great help in constructing even integral lattices, because one only needs to check the lengths and inner products of conjugacy class representatives. For this reason we have collected some information concerning the conjugacy classes of the (simple) simply laced algebras in the following table.

Lie algebra	conjugacy class	smallest representation	weight norm (mod 2)
D_n	(0)	singlet	0
	(v)	vector	1
	(s)	spinor	$\frac{n}{4}$
	(c)	conjugate spinor	$\frac{n}{4}$
A_n	(p)	rank p tensor $(0 \le p < n)$	$\frac{p(n+1-p)}{n+1}$
E_6	(0)	singlet	0
	(1)	(27)	$\frac{4}{3}$
	(2)	$(\overline{27})$	$\frac{4}{3}$
E_7	(0)	singlet	0
	(1)	(56)	$\frac{3}{2}$
E_8	(0)	singlet	0

The volume of the unit cells of Λ_R and Λ_W are easy to obtain. Since they are each other's dual

$$vol(\Lambda_R) = vol(\Lambda_W)^{-1}$$

If we denote the order of the center (i.e. the number of conjugacy classes) by N_c, we learn from (3.10) that

$$vol(\Lambda_W) = \frac{1}{N_c} vol(\Lambda_R),$$

so that

$$vol(\Lambda_R) = \sqrt{N_c}$$
$$vol(\Lambda_W) = \frac{1}{\sqrt{N_c}}$$

All these results are valid for simple as well as semi-simple Lie algebras. The order of the center of a semi-simple Lie algebra is simply the product of the order of the centers of its simple factors. The same is true for the volume of the unit cell.

To specify a Lie algebra lattice completely, we have to supply two pieces of information: the semi-simple Lie algebra, and a list of the conjugacy classes that appear. For a given semi-simple Lie algebra there are in general many possible choices of conjugacy classes. This choice is restricted only by the requirement that the list must close under addition, which in particular means that the root lattice must always be present. The possible sets of conjugacy classes are thus in one-to-one correspondence with the subgroups of the center. Instead of giving a full list of conjugacy classes, one can give a shorter list of *generators*, which is a subset of conjugacy classes from which all others can be obtained by addition.

Since the volume of the unit cell of the root lattice is $\sqrt{N_c}$, it follows immediately that a Lie algebra lattice is unimodular if and only if it contains precisely $\sqrt{N_c}$ conjugacy classes. This is only possible if we choose a Lie algebra for which N_c is a perfect square. The lattice will be integral if and only if all conjugacy class generators have integral norm and integral mutual inner product. It will be even if and only if they have even norm and integral inner products.

Using the table one can now easily construct many examples of euclidean self-dual lattices. The reader is invited to check that

— D_n with conjugacy classes (0) and (v) is odd and self-dual (this lattice is also known as Z^n).

— D_n with (0) and (s) is odd self-dual for $n = 4 \bmod 8$. and even self-dual for $n = 0 \bmod 8$

— A_{n^2-1} with (0) and (n) is odd self-dual for n even and even self-dual for n odd.

In general we call vectors of norm 2 on euclidean lattices *roots*. The roots include in any case those of the Lie algebras explicitly used in the construction of the lattice (i.e. those in the conjugacy class (0)), but in some cases there may be additional norm 2 vectors in the other conjugacy classes. In that case we can always write the lattice in a different way, using larger Lie algebras and fewer conjugacy classes, although this is not always convenient. For example, the lattices D_8 with $(0)+(s)$ and A_8 with $(0)+(3)$

defined above are both equal (up to rotations) to E_8, which has only one conjugacy class, namely (0) (and which is therefore automatically self-dual!).

A large class of even lorentzian self-dual lattices in $R^{p,p}$ can be constructed as follows. Take a semi-simple Lie algebra of rank p. Consider the set of all vectors in $R^{p,p}$ of the form $(\vec{v}_1, \vec{v}_2)$, so that $\vec{v}_1$ and $\vec{v}_2$ belong to the *same* conjugacy class of the Lie algebra. With respect to the lorentzian metric any such vector has integral inner product with any other such vector, any furthermore they all have even length. It is also easy to see that the resulting lorentzian lattice is unimodular, so that it is a self-dual lattice. Lattices of this type were first described by Englert and Neveu [29].

To conclude this discussion of Lie algebra lattices, we want to introduce a minor, but useful extension of this set, namely the lattice D_1. Although it is not part of Cartan's classification of simple Lie algebra's, it is easy enough to define by simply extending the definition of D_n to $n = 1$. The resulting one-dimensional root lattice consists simply of the even integers. The shortest non-trivial vector has norm four, so there are no roots on this root lattice. The vector conjugacy class consists of the odd integers, and the two spinor conjugacy classes are obtained by shifting the root lattice by $+\frac{1}{2}$ and $-\frac{1}{2}$.

3.4 Lattice Engineering

There are several tricks for making new self-dual lattices out of already known ones. These tricks are all variations on the same basic theme: one adds new conjugacy classes and removes old ones, in such a way that their total number stays the same (i.e. the lattice remains unimodular), and that all inner products remain integral.

A very general method is the following. Let Λ be a self-dual lattice. Choose a vector $\vec{s}$ so that $k\vec{s} \notin \Lambda, 1 \leq k \leq p-1$, but $p\vec{s} \in \Lambda$. For convenience we will assume that p is prime for the moment. Now we can define a lattice Λ_0 as the sublattice of Λ which has integer inner product with $\vec{s}$. Since $\vec{s} \notin \Lambda$ there must be a vector $\vec{t} \in \Lambda$ such that $\vec{s} \cdot \vec{t}$ is not integer. But then $p\vec{t}$ belongs to Λ_0, and is the smallest multiple of $\vec{t}$ in Λ_0. Hence we can decompose Λ in cosets

$$\Lambda = \sum_{k=0}^{p-1} (k\vec{t} + \Lambda_0)$$

Consider now the new lattice

$$\Lambda' = \sum_{k=0}^{p-1} (k\vec{s} + \Lambda_0)$$

clearly Λ' is unimodular. If furthermore $\vec{s}$ has integral norm, it is self-dual. If Λ is even and $\vec{s}$ has even length, Λ' is even self-dual.

Another useful trick is the following. Consider a Lie algebra lattice containing, among others, a D_n factor. Now replace this D_n factor by D_m, where $m = n + 8l, l \in Z$. Since the unit cells of D_n and D_m have the same volume for *any* n and m, the new lattice is unimodular. Furthermore one can easily check that if $n = m \bmod 8$ all inner products and norms of the four conjugacy classes remain unchanged modulo two. Hence the new lattice will be (even) self-dual if the original lattice was (even) self-dual.

This transformation is of course meaningless if it results in $m = 0$. But one can give a meaning to it if m becomes negative. Consider for example a euclidean even self-dual lattice $\Lambda_p \times D_n$ with inner product with signature $(+)^{p+n}$. For $m < 0$ (with m defined as above) we can consider the lorentzian lattice $\Lambda_p \times D_m$ with signature $(+)^p(-)^{-m}$. This lattice is unimodular for the reasons explained above. Furthermore the lorentzian inner products of the new lattice are equal to the euclidean ones of the original one modulo two. Thus we see that shifting n by multiples of eight to negative values has the effect of changing a euclidean self-dual lattice to a lorentzian one.

3.5 Lattice classification theorems

A lot is known about the possibilities for constructing self-dual lattices. One should keep in mind here that euclidean lattices can only be classified up to rotations, because any rotation respects the self-duality conditions. Likewise, lorentzian lattices can only be classified modulo Lorentz transformations. The known results are as follows:

1. *Lorentzian, odd.* All such lattices are Lorentz transformations of the lattice $Z^{p,q}$, which is the set of points in $R^{p,q}$ defined as $\{(n_1, \ldots, n_p, m_1, \ldots, m_q), n_i, m_j \in Z\}$.

2. *Lorentzian, even.* Lattices of this type exist only for $|p - q| = 0 \bmod 8$. For $p = q + 8n$ they are Lorentz transformations of the lattice $nE_8 \times qP^{1,1}$. The E_8 lattice is most easily described as D_8 with conjugacy classes (0) and (s), while $P^{1,1}$ is a lorentzian lattice of signature $(+)(-)$, which may be described as $D_1 \times D_1$ with conjugacy classes $(0,0)+(v,v)+(s,s)+(c,c)$ (this is a lattice of the Englert-Neveu type, discussed above).

3. *Euclidean, odd.* A class of such lattices which exists in any dimension is Z^n. In general, if any odd self-dual lattice has vectors of length one, they can only have mutual inner products $\pm 1, 0$. Thus they form an orthonormal basis for a subspace of R^n, which must be part of the basis of the lattice. Any other basis vector of the lattice must be orthogonal to this subspace. Thus if there are $2p$ length one vectors, the lattice is of the form $Z^p \otimes \Lambda^{n-p}$, where the second factor does not contain any vectors of length one.

An odd euclidean self-dual lattice Λ can be related to an even self-dual one of higher dimension in the following way [24]. Consider the sublattice Δ_0 of even norm vectors. Obviously Δ_0 has only one non-trivial coset in Λ, because the sum of two odd norm vectors is even. Thus

$$\Lambda = \Delta_0 + \Delta_v,$$

where Δ_v denotes the odd-norm vectors. Since $vol(\Delta_0)/vol(\Delta_0^*) = 4$, the dual of Δ_0 can be decomposed into four cosets,

$$\Delta_0^* = \Delta_0 + \Delta_v + \Delta_s + \Delta_c$$

These four cosets generate an abelian group of order four, which can only be Z_4 or $Z_2 \times Z_2$. This allows us to determine the mutual inner products of the four cosets and their norms modulo 2. If the group is Z_4, one finds $\Delta_c = -\Delta_s$ and $2\Delta_s \subset \Delta_v$, so that the norm of all vectors in Δ_s and Δ_c is equal to $\frac{k}{4} \bmod 2, k$ odd. In the latter case $2\Delta_s = 2\Delta_c \subset \Delta_0$, so that the norm of all vectors in Δ_s and Δ_c is half-integer. One can then use $\Delta_s + \Delta_v = \Delta_c$, plus the fact that vectors in Δ_s cannot have integer inner products with those in Δ_v, to show that the norm of vectors in Δ_s and Δ_c must be equal modulo 2. Thus for both possible abelian groups the vectors in Δ_s and Δ_c have, modulo two, norms equal to each other and to $\frac{k}{4}$, $0 \leq k < 7$. Consider then the lattice obtained by adding a D_{8-k} factor, so that the entire lattice consists of the vectors

$$(\Delta_0, 0) + (\Delta_v, v) + (\Delta_s, s) + (\Delta_c, c)$$

This lattice is unimodular, integral and even, so that it is even self-dual. So all odd self-dual lattice can be obtained from the even self-dual ones by inverting this procedure to remove D_n factors. This leaves us with the classification problem of

4. *Euclidean, even* lattices. These exist only in dimensions which are a multiple of 8, and are known explicitly for 8,16 and 24 dimensions. In 8 dimensions, the only possibility is E_8; in 16 dimensions one can have $E_8 \otimes E_8$ and D_{16} with conjugacy classes (0) and (s). These two lattices cannot be rotated into each other. In 24 dimensions there 24 even self-dual lattices, which are usually called the *Niemeier lattices* [25], after the German mathematician who first classified them. A list of their conjugacy classes can be found in the first paper of Ref. [24].

One of the 24 lattices is the Leech-lattice [30], which differs from the other 23 because it contains no vectors of norm 2. Nevertheless, we can in a certain sense regard it as a Lie algebra lattice, which is a product of 24 D_1 factors, on which one has 2^{24} conjugacy classes. A list of these classes (or rather 12 classes that generate them) will appear in [31].

There is a rather amusing way of estimating the number of lattices in $8n$ dimensions. Every lattice has a discrete automorphism group. This is a discrete subgroup of $O(8n)$ which takes the lattice into itself. It consists of at least two elements, the identity and the reflection through the origin. Denote the total number of elements of this group for a lattice Λ as $g(\Lambda)$. Then the following identity can be proved [26], [32].

$$\sum \frac{1}{g(\Lambda)} = \frac{B_{4n}}{8n} \prod_{j=1}^{4n-1} \frac{B_{2j}}{4j},$$

where B_{2j} are the Bernoulli numbers. This formula is known as the Minkowski-Siegel mass formula. The sum is over all lattices in $8n$ dimensions, and adds up to a number that is less than half the total number of lattices in this dimension. Thus the right hand side, multiplied by 2, is a lower limit of the number of lattices. Calculating it for the lowest dimensions, one finds

dimension	lower limit	actual number
8	2.8×10^{-9}	1
16	4.9×10^{-18}	2
24	15.8×10^{-15}	24
32	8.0×10^{7}	?

This explains why the lattices of dimension 32 have not yet been classified. The explosive growth of the number of lattices that one observes here is related to the reason why the number of chiral four dimensional string theories is so much larger then the number of ten-dimensional ones, as we will see.

4. Torus Compactification of the Bosonic String

Armed with all relevant facts about lattices, we can now attack the problem of the bosonic string, propagating through a 26 dimensional space-time of which some dimensions have been "compactified". Of course our ultimate interest is in fermionic strings, but since we will use a bosonic approach to fermionic strings, a thorough understanding of the compactified bosonic string is essential. The presentation given here is essentially the one of [4], [33], although these authors compactified the heterotic string rather than the bosonic string. To warm up, and to introduce some notations, we will begin with a very brief reminder of the basics, the quantization of the uncompactified string.

4.1 The uncompactified closed bosonic string

The propagation of a string through d dimensional space-time is governed by the action

$$S = -\frac{1}{4\pi\alpha'}\int d\sigma_0 \int_0^\pi d\sigma_1 \partial_\alpha X^\mu \partial^\alpha X^\nu \eta_{\mu\nu}, \tag{4.1}$$

where σ_0 and σ_1 are the world-sheet coordinates, indicating respectively the time direction and the direction along the string (one often denotes these variables as τ and σ, but we won't do this to prevent confusion with the usual notation for the one-loop modular parameter). In (4.1) $\eta_{\mu\nu}$ is a flat metric and $X^\mu(\sigma_0, \sigma_1)$ is a map from the world-sheet to space-time, which specifies the position of every point of the string at any time. The parameter α' (the "Regge slope parameter") is usually set to the value $\frac{1}{2}$, but we will refrain from that for a while. We shall only consider closed strings, for which one has the boundary condition

$$X^\mu(\sigma_0, 0) = X^\mu(\sigma_0, \pi). \tag{4.2}$$

From the action one can in the usual way derive the operators H and P which generate translations in the σ_0 and σ_1 directions on the world-sheet:

$$H = \frac{1}{4\pi\alpha'}\int_0^\pi d\sigma_1 \left[(\partial_0 X^\mu)(\partial_0 X_\mu) + (\partial_1 X^\mu)(\partial_1 X_\mu)\right] \tag{4.3}$$

$$P = \frac{1}{2\pi\alpha'}\int_0^\pi d\sigma_1 (\partial_0 X^\mu \partial_1 X_\mu) \tag{4.4}$$

The quantization of this system is simplest in light-cone gauge,in which the dynamical variables are $X^i, i = 1, \ldots, d-2$. Their canonical momenta Π_i are easily obtained from the action

$$\Pi_i = \frac{1}{2\pi\alpha'}\partial_0 X_i.$$

The solutions of the equations of motion have the form

$$X^i(\sigma_0, \sigma_1) = q^i + 2\alpha' p^i \sigma_0 + \frac{i}{2}\sqrt{2\alpha'}\sum_{n\neq 0}\frac{1}{n}(\alpha_n^i e^{-2in(\sigma_0+\sigma_1)} + \tilde{\alpha}_n^i e^{-2in(\sigma_0-\sigma_1)}). \tag{4.5}$$

Using the canonical commutator

$$[X^i(\sigma_0, \sigma_1), \Pi^j(\sigma_0, \sigma_1')] = i\delta^{ij}\delta(\sigma_1 - \sigma_1') \tag{4.6}$$

one finds the commutation relations for the quantized modes

$$\begin{aligned} [q^i, p^j] &= i\delta^{ij} \\ [\alpha^i_m, \alpha^i_n] &= m\delta^{ij}\delta_{n+m,0} \\ [\tilde{\alpha}^i_m, \tilde{\alpha}^i_n] &= m\delta^{ij}\delta_{n+m,0} \\ [\alpha^i_m, \tilde{\alpha}^i_n] &= 0 \end{aligned} \tag{4.7}$$

The spectrum is obtained by requiring that both H and P vanish. This leads to the conditions

$$2p_+p_- - p_i^2 \quad (= p^2 = m^2) = \frac{2}{\alpha'}(\mathcal{N} + \tilde{\mathcal{N}} - 2), \tag{4.8}$$

where the -2 is due to the zero mode fluctuations of the oscillators, and we have defined the number operators

$$\mathcal{N} = \sum_{n=1}^{\infty} \alpha^i_{-n}\alpha^i_n$$

and similarly for $\tilde{\mathcal{N}}$. From the vanishing of P we get

$$\mathcal{N} = \tilde{\mathcal{N}}.$$

In the quantized theory this is imposed as a condition on the physical spectrum.

It is convenient to define an operator H', which is equal to H, but without the zero-mode terms (*i.e.* the operators p^i and q^i). It is also useful to define the linear combinations

$$\begin{aligned} H_L &= \tfrac{1}{2}(H' - P) = 2(\tilde{\mathcal{N}} - 1) \\ H_R &= \tfrac{1}{2}(H' + P) = 2(\mathcal{N} - 1) \end{aligned},$$

which are the generators of world-sheet translations in the $\sigma_0 - \sigma_1$ and $\sigma_0 + \sigma_1$ directions respectively. Their eigenstates are called left- and right-moving modes respectively. Denoting the eigenvalues of H_L' and H_R' on a state $|\psi\rangle$ as $\frac{1}{2}\alpha' m_L^2$ and $\frac{1}{2}\alpha' m_R^2$ we can write the result obtained above as*

$$\begin{aligned} H_L\,|\psi\rangle &= \tfrac{1}{2}\alpha' m_L^2\,|\psi\rangle \\ H_R\,|\psi\rangle &= \tfrac{1}{2}\alpha' m_R^2\,|\psi\rangle \\ m^2 &= \tfrac{1}{2}(m_L^2 + m_R^2) \end{aligned}$$

with the condition

$$m_L^2 = m_R^2.$$

* For comparison with [6] note that we have reversed the assignment of $\tilde{\alpha}$ and $\tilde{r}$ here from right to left-movers here. This is more convenient in heterotic strings, where we will choose the NSR system to be right-moving, just as in [6].

4.2 The compactified closed bosonic string

Now we turn to the case of a torus compactified space-time. A torus in N dimensions is simply R^N modulo a lattice Λ. This means that points in R^N which differ by a vector on Λ are identified. If a string propagates in a space-time with this topology it is not necessary anymore that it closes into the same space-time point, as in (4.2) . We should allow for the possibility that the points $\sigma_1 = 0$ and $\sigma_1 = \pi$ of the string are identified points in R^N:

$$X^i(\pi) = X^i(0) + 2\pi L^i,$$

where $L^i \in \Lambda$ and the factor 2π is put in for convenience. Physically, this corresponds to a string winding around the torus a certain number of times. To accommodate this possibility one modifies the string field X^i in the following way

$$X^i(\sigma_0, \sigma_1) = q^i + 2\alpha' p^i \sigma_0 + 2L^i \sigma_1 + \frac{i}{2}\sqrt{2\alpha'} \sum_{n \neq 0} \frac{1}{n}(\alpha_n^i e^{-2in(\sigma_0+\sigma_1)} + \tilde{\alpha}_n^i e^{-2in(\sigma_0-\sigma_1)}), \quad (4.9)$$

where L^i is an operator whose eigenvalues are the lattice vectors.

Having done this, we go now through the same exercise as for the uncompactified string. It is easy to verify that nothing changes in the canonical commutators (4.7) except for the trivial addition that L^i should commute with all other operators. The mass formula replacing (4.8) acquires on the right hand side an extra term $\frac{1}{\alpha'^2}L^2$. Furthermore, to interpret the left hand side as a mass in $d - N$ dimensions we have to move the internal components of p^i to the right, to get

$$p^2 = m^2 = \sum_{I=1}^{N} p_I^2 + \frac{1}{\alpha'^2}L^2 + \frac{2}{\alpha'}(\mathcal{N} + \tilde{\mathcal{N}} - 2).$$

Here we have chosen the first N components as the compactified ones. The momenta $p_I, I = 1, \ldots, N$ are canonical momenta of a periodic coordinate, and therefore (for reasons very familiar from solid state physics) they are quantized. Indeed, if we write their eigenfunctions of momentum k_I as $e^{ik_I x_I}$, then we see that these functions are periodic on the boundaries of the unit cell of Λ if and only if $\vec{k} \cdot \vec{l} \in Z$ for every $\vec{l} \in \Lambda$. This means that the eigenvalues of k_I must lie on the dual lattice Λ^* of Λ.

The condition obtained from the vanishing of P reads now

$$\mathcal{N} - \tilde{\mathcal{N}} + p_I L_I = 0$$

It is again useful to define left and right hamiltonians H' exactly as before, but with the understanding that only the space-time momenta are removed from H to get H', not the internal ones. Then we get

$$\begin{aligned} \tfrac{1}{4}\alpha' m_R^2 &= \tfrac{1}{2}p_R^2 + \mathcal{N} - 1 \\ \tfrac{1}{4}\alpha' m_L^2 &= \tfrac{1}{2}p_L^2 + \tilde{\mathcal{N}} - 1, \end{aligned} \quad (4.10)$$

where as before left and right masses must be equal to each other and to m^2. The left and right momenta are defined as

$$\begin{aligned} p_R^I &= \frac{1}{\sqrt{2}}(\sqrt{\alpha'}p^I + \frac{1}{\sqrt{\alpha'}}L^I) \\ p_L^I &= \frac{1}{\sqrt{2}}(\sqrt{\alpha'}p^I - \frac{1}{\sqrt{\alpha'}}L^I) \end{aligned}$$

Before discussing the meaning of these left and right momenta, we want to mention a simple generalization, obtained by adding a background antisymmetric tensor field. This background field is coupled to X^μ via an additional term in the action

$$S_B = \frac{1}{2\pi\alpha'}\int d\sigma_0 \int_0^\pi d\sigma_1 \epsilon^{\alpha\beta} B_{\mu\nu}\partial_\alpha X^\mu \partial_\beta X^\nu$$

We take $B_{\mu\nu}$ to be a constant background field, which has non-vanishing components only in the compactified dimensions. The extra term is then a total derivative, which does not affect H or P. So how can it make any difference? It can, because there is a change in the canonical momentum operator, which now becomes

$$\begin{aligned} \Pi_\mu &= \frac{1}{2\pi\alpha'}(\partial_0 + B_{\mu\nu}\partial_1)X^\nu \\ &= \frac{1}{\pi}(p_\mu + \frac{1}{\alpha'}B_{\mu\nu}L^\nu + \text{oscillators}) \end{aligned}$$

For the oscillator contributions, everything remains the same. Although $\partial_1 X^\nu$ contains oscillator terms, there is a cancellation between left– and right-mover contributions to $[\Pi, X]$, so that the canonical commutators of the oscillators remain unchanged. The difference occurs for the zero modes, where p_μ is replaced as the generator of translations by $\tilde{p}_\mu = p_\mu + \frac{1}{\alpha'}B_{\mu\nu}L^\nu$. So it is $\tilde{p}$ rather than p which has its eigenvalues quantized on the dual of Λ. Although the mass formula remains the same as a function of p, it makes now more sense to express it in terms of $\tilde{p}$. The result is the same as (4.10) above, but with

$$\begin{aligned} p_R^I &= \frac{1}{\sqrt{2}}(\sqrt{\alpha'}\tilde{p}^I + \frac{1}{\sqrt{\alpha'}}(L^I - B^{IJ}L_J)) \\ p_L^I &= \frac{1}{\sqrt{2}}(\sqrt{\alpha'}\tilde{p}^I - \frac{1}{\sqrt{\alpha'}}(L^I + B^{IJ}L_J)) \end{aligned} \tag{4.11}$$

where $\tilde{p} \in \Lambda^*$ and $L \in \Lambda$. In the following we will omit the tilde on p. It is clear that the dependence on α' is not very significant. We can absorb it by changing the length of all vectors on Λ by $\sqrt{\alpha'}$ (and hence those on Λ^* by $\frac{1}{\sqrt{\alpha'}}$). This amounts to a rescaling of the size of the torus. From now on we will use this freedom to set $\alpha' = \frac{1}{2}$.

To interpret the result it is convenient to express the left and right momenta in terms of the basis vectors e_a^I of the lattice:

$$\begin{aligned} L^I &= e_a^I n_a \\ p^I &= e_a^{*I} m_a = e_a^I g_{ab}^* m_b \end{aligned}$$

Then we get

$$p_R^I = e_a^I(\frac{1}{2}g_{ab}^* m_b - g_{ab}^* B_{bc} n_c + n_a)$$

$$p_L^I = e_a^I(\frac{1}{2}g_{ab}^* m_b + g_{ab}^* B_{bc} n_c - n_a)$$

(upper and lower indices have no special significance here.) One can regard g (or its inverse g^*) as a constant background field, just as B. This becomes evident if one performs this change of basis directly in the action (4.1) : the effect is to change the flat metric δ_{IJ} (in the compactified dimensions) to g_{IJ}.

It is now instructive to calculate the inner product of the $2N$ dimensional vectors* $P = (p_L; p_R)$ and $P' = (p'_L; p'_R)$ (the latter is expressed in terms of n' and m') with respect to the *lorentzian* signature $[(+)^N; (-)^N]$. The result is

$$\vec{P} \cdot \vec{P}' = nm' + mn'$$

From this expression we can learn many things. First of all, the lattice $\Gamma_{N;N}$ consisting of the vectors P defined above is an integral and even lorentzian lattice. Secondly, the dot-products do not depend on the background fields g and B. Therefore they must be parameters of a Lorentz transformation. Finally, using this background independence, we can make the choice $e_a^I = \delta_a^I$, $g_{IJ} = \delta_{IJ}$, $B_{IJ} = 0$. In this special case the lattice is $(D_1 \times D_1)^N$ with conjugacy classes $(0,0) + (v,v) + (s,c) + (c,s)$ in each factor, which is self-dual. Because Lorentz transformations preserve self-duality, we conclude that $\Gamma_{N;N}$ is self-dual for any value of the background fields.

Thus compactification of the bosonic string on a torus R^N/Λ, with an additional background field B_{IJ}, can be described in terms of a new lattice $\Gamma_{N;N}$ which for *any* Λ and B is lorentzian self-dual. It should be clear that the lorentzian signature is a mathematical device for characterizing the lattice; it has nothing to do with the metric of space-time and does not lead to ghosts.

The number of parameters of g and B is $\frac{1}{2}N(N+1) + \frac{1}{2}N(N-1) = N^2$. One can verify that they are precisely the N^2 parameters of the coset $\frac{SO(N,N)}{SO(N)\times SO(N)}$, which is the set of Lorentz transformations of the lattice modulo the rotations of the left and right sector. These rotations are a symmetry of the theory, because all masses depend on p_L and p_R only via their norms. The genuine lorentz transformations mixing p_L and p_R on the other hand are not symmetries. Although the self-duality condition is invariant under such Lorentz transformations, the spectrum is not. So we have an N^2 parameter family of different string theories.

As seems usual in string theory, these results were originally discovered in the opposite order: First Narain [4] constructed the lorentzian lattices, and afterwards Narain, Sarmadi and Witten [33] gave the background field interpretation from which we started. We still have not discussed the relevance of the self-dual lattice which emerges from this construction. To do so, we have to consider the partition function of the compactified theory.

---

* We separate left- and right-moving sectors by a semicolon in the following.

Consider first the uncompactified bosonic string at one loop. The amplitude is an expression of the form

$$A_{1-loop} = \int \frac{\mathcal{D}X\mathcal{D}g}{V_{Diff}V_C} \exp(-S_E(g,X)), \tag{4.12}$$

where V_{Diff} and V_C denote the volume of the set of local diffeomorphism and of the conformal transformations respectively, and S_E is the bosonic string action (4.1) , continued to euclidean space. What is meant by this expression is that one should integrate over all values of X and the world-sheet metric on the torus that are not equivalent by reparametrizations and conformal transformations. This is the analog of the process of gauge-fixing in gauge theories, and this problem can be tackled with very similar techniques. We will not (need to) explain this procedure here, but simply state the result (the reader interested in the details is referred to [34]):

$$A_{1-loop} = V \int \frac{d\tau}{(Im\tau)^2}(Im\tau)^{\frac{2-d}{2}} \int \mathcal{D}'X \exp[-S_E(g(\tau),X)] \tag{4.13}$$

The factor V denotes the volume of space-time, which of course is infinite. This infinity is easy to understand: since the action vanishes for constant X, one of the infinitely many integrals in $\mathcal{D}X$ is simply an integral of a constant over all of space-time. This infinite factor is nothing to worry about, and is simply to be regarded as a momentum conservation delta function with argument zero, which appears because we are considering a diagram without external lines. However, we have to make sure that we factor it out correctly, which can be achieved by putting the system in a box in order to regulate the infinity. After taking out the zero mode integral, we denote the remaining integral by $\mathcal{D}'X$. After gauge fixing, the components of the world-sheet metric are constants (*i.e.* they are independent of the world-sheet coordinates), which depend on the modular parameter τ. The remaining X integral is therefore a simple gaussian, and the result may be written as

$$\int \mathcal{D}'Xe^{-S_E} = \frac{1}{\det'(-\Delta)}, \tag{4.14}$$

where Δ is the laplacian on the torus, $\Delta = g_{\alpha\beta}\partial_\alpha\partial_\beta$, and the prime on the determinant indicates that the zero-mode should be omitted. For every component of X^μ that is factored out one is left with a factor $\frac{1}{\sqrt{Im\tau}}$, which is nothing but the normalization of the zero-mode of the laplacian on the skewed torus (which has volume $Im\tau$). This does not quite account for all of the factors of $Im\tau$ in (4.13) , but the extra one is a remnant of the gauge fixing procedure. The thing to remember is that every uncompactified X^i contributes a factor $(Im\tau)^{-\frac{1}{2}}$ to the integrand.

The factor (4.14) of (4.13) is called the (one-loop) *partition function*, and will be denoted as $\mathcal{F}(\tau)$. It is possible to evaluate it directly by regularization of the infinite product of eigenvalues of Δ (see the appendix of [34]), but at one loop there is a more intuitive way. Let us first recall a result for a one-dimensional bosonic system, specified by a coordinate $X(t)$ and an action $S(X(t))$ One can prove (if the action satisfies certain conditions) the following relation between the path integral for a euclidean action S_E and the trace of the exponential of the hamiltonian H of the system

$$\int_{PBC} \mathcal{D}X e^{-S_E} = \mathrm{Tr}\, e^{-\beta H}.$$

Here "PBC" means that $X(t)$ must satisfy periodic boundary conditions on the interval $[0, \beta]$. For a derivation of this relation see *e.g.* [35]. The result is easily generalized to X's having more components, or even an infinite number of components. The latter case is relevant for us, because the scalar field $X(\sigma_0, \sigma_1)$ appearing in string theory can be regarded as having an infinite number of components labeled by σ_1 (times a finite number of components because of the space-time index, which we omitted here). To use this relation, one should think of $Im\tau$ as β. There is a small complication if $Re\tau \neq 0$, because then the periodicity of X in $Im\tau$ depends on $Re\tau$. This can be taken into account by including a σ_1 shift operator in the trace, so that states at $\sigma_0 = 0$ are not identified with the same states at $\sigma_0 = Im\tau$, but with states of a string which is rotated in the σ_1 direction by an amount $Re\tau$. The generator of infinitesimal σ_1 translations is of course P, so that one gets

$$\begin{aligned} \mathcal{F}(\tau) = \int_{T(\tau)} \mathcal{D}' X e^{-S_E} &= \mathrm{Tr}\, e^{\pi Im\tau H'} e^{i\pi Re\tau P} \\ &= \mathrm{Tr}\, e^{i\pi\tau H_R} e^{-i\pi\bar{\tau} H_L} \end{aligned} \tag{4.15}$$

Here we have rescaled the variables σ_0 and σ_1 by factors of π. The integral is over all two-dimensional fields X that are periodic on a torus T specified by a modular parameter τ.

This way of writing the partition function admits a simple interpretation. Expanding it in terms of $q = e^{i\pi\tau}$ one gets a power series of the form $\mathcal{F}(\tau) = \sum d_{nm} q^n \bar{q}^m$. Obviously, d_{nm} is simply the number of states of left mass m and right mass n. (At this point we should add a word of warning. The states appearing in the spectrum are only those for which $n = m$. Notice however that these are *not* the only states contributing to the one-loop path integral. The reader who thinks that states with $n \neq m$ are projected out by the $Re\tau$ integral should realize that the lower boundary of the standard modular domain is not a straight line.)

It is usually a simple combinatorial exercise to read off the partition function from the expressions for H_L and H_R. Every *term* in the hamiltonian yields a *factor* in the partition function, so that one can derive the partition function term-by-term. Take for example a single bosonic oscillator α_n. It contributes to the right Hamiltonian via a term $2\alpha_{-n}\alpha_n$, and when it acts it increases the mass of a state by $2n$. Thus it contributes to the partition function a factor

$$(1 + q^{2n} + q^{4n} + q^{6n} + \ldots) = \frac{1}{(1 - q^{2n})},$$

where on the right the term q^{2kn} corresponds to the contribution of $(\alpha_{-n})^k$. The partition function of the bosonic string contains such a factor for every oscillator.

For the uncompactified string we leave out the zero modes in H_L and H_R as discussed before. The remaining trace, *i.e.* the contribution from the bosonic oscillators to the partition function, is equal to

$$\mathcal{F}_B(\tau) = Tr q^{H'_R} \bar{q}^{H'_L} = |\eta(\tau)|^{-48} \tag{4.16}$$

where η is the famous Dedekind η-function

$$\eta(\tau) = q^{\frac{1}{12}} \prod_{n=1}^{\infty} (1 - q^{2n})$$

(as demonstrated in these equations, we will freely mix τ's and q's in the following.)

Since the oscillator contribution to the mass formula is the same for compactified and uncompactified coordinates, the partition function for the compactified string contains the same factor. The only difference is that the zero modes of the compactified coordinates are now included in the determinant as well as the hamiltonian. So first of all we have to multiply the uncompactified string result by $\sqrt{Im\tau}$ for every compactified coordinate to remove the zero-mode normalizations in (4.12) . Secondly, we have to take into account the contribution to the trace due to the extra momentum terms in the mass formula (4.10). Since this is simply an extra term in the mass formulas, it produces an extra factor in the partition function, which together with the $\sqrt{Im\tau}$ factors is called the *correction factor*, because it gives the correction in the bosonic partition function for the fact that part of the zero-modes are compactified. From the mass formula (4.10) we get for a lattice $\Gamma_{N;N}$

$$C(\Gamma_{N;N}) = \mathcal{L}_{N;N}(\tau)(Im\tau)^{N/2}, \tag{4.17}$$

where $\mathcal{L}_{N;N}$ is the *lattice partition function*:

$$\mathcal{L}_{N;N} = \sum_{(p_L;p_R)} e^{i\pi\tau p_R^2} e^{-i\pi\bar{\tau} p_L^2}. \tag{4.18}$$

Here the sum is over all the vectors on the lattice.

4.4 Modular invariance

We can now discuss the modular transformations of these functions. As explained in chapter 2, we have two transformations to check, namely $\tau \to \tau + 1$ and $\tau \to -\frac{1}{\tau}$. The first one leaves the partition function invariant if and only if it is an even function of q. This is clearly the case for both compactified and uncompactified strings, as a consequence of the fact that there are 24 transverse dimensions, and because $\Gamma_{N;N}$ is even. The second modular transformation is a bit harder to check, and requires a trick, which goes by the name of "Poisson resummation formula".

Define instead of (4.18) the function

$$P(\tau, \vec{z}) = \sum_{(p_L;p_R)} e^{i\pi\tau(\vec{p}_R+\vec{z}_R)^2} e^{-i\pi\bar{\tau}(\vec{p}_L+\vec{z}_L)^2}, \tag{4.19}$$

where $\vec{z} = (\vec{z}_L, \vec{z}_R)$ is a $2N$ vector, which is not restricted to lie on the lattice. Since we sum over all vectors on the lattice, the function P is obviously invariant if we shift $\vec{z}$ by a lattice vector. In other words, it is periodic with the periodicity of the lattice, and hence we can Fourier transform it. Normally the Fourier analysis of lattice functions would involve the dual lattice, but because our lattice is self-dual, we don't have to bother with that. The expression for P in terms of its Fourier transform $\tilde{P}$ is

$$P(\tau, \vec{z}) = \sum_{\vec{q}} e^{-2\pi i \vec{z}\cdot\vec{q}} \tilde{P}(\tau, \vec{q}) \tag{4.20}$$

The inverse transformation is of course

$$\tilde{P}(\tau, \vec{q}) = \int_{unit\ cell} d^{2N}z \, e^{-2\pi i \vec{z}\cdot\vec{q}} P(\tau, \vec{z}),$$

i.e. we have a multiple integral over all components of $\vec{z}$ with boundaries given by the unit cell of the lattice. The function P contains a sum over all lattice vectors. We can take the exponential multiplying it inside the sum, and furthermore we can shift the variables $\vec{z}$ in this exponential factor by $\vec{p}$. This shift introduces only a factor $\exp(2\pi i\vec{q}\cdot\vec{p})$, which is equal to one because both $\vec{q}$ and $\vec{p}$ are lattice vectors. The result is thus

$$\tilde{P}(\tau, \vec{q}) = \sum_{\vec{p}} \int_{unit\ cell} d^{2N}z \, e^{-2\pi i\vec{q}\cdot(\vec{z}+\vec{p})} e^{i\pi\tau(p_R+z_R)^2} e^{-i\pi\bar{\tau}(p_L+z_L)^2}$$

The sum combined with the integral gives as an integral over *all* unit cells of the lattice, *i.e.* over the entire space. Thus we get

$$\tilde{P}(\tau, \vec{q}) = \int_{R^{N;N}} d^{2N}x \, e^{-2\pi i\vec{q}\cdot\vec{x}} e^{i\pi\tau(x_R)^2} e^{-i\pi\bar{\tau}(x_L)^2},$$

where we have introduced a new integration variable $\vec{x}$ representing, in each cell, $\vec{p} + \vec{z}$. This integral is a simple gaussian, which yields

$$\tilde{P}(\tau, \vec{q}) = |\tau|^{-N} e^{-\frac{i\pi}{\tau}(q_R)^2} e^{\frac{i\pi}{\bar{\tau}}(q_L)^2}$$

Notice that we have made use here of the fact that the lattice is unimodular, so that the usual determinant factors that one expects from gaussian integrals are equal to one. Substituting this result into (4.20) and putting $\vec{z}$ equal to zero again we get the identity

$$P(-\tfrac{1}{\tau}) = |\tau|^N P(\tau) \tag{4.21}$$

One can assemble all these ingredients into the one-loop partition function of the compactified bosonic string. To check the modular invariance of the complete answer one has to verify that the factors $|\tau|$ (which are called *modular weight* factors) cancel. They do, because there are also contribution from the transformation of $Im\tau$ and $\eta(\tau)$. The transformation of $Im\tau$ is straightfoward: $Im(-\frac{1}{\tau}) = |\tau|^{-2}Im\tau$, which can be used to show that the correction factor (4.17) is invariant. To check the invariance of the uncompactified bosonic string, one needs furthermore

$$\eta(-\frac{1}{\tau}) = \sqrt{-i\tau}\ \eta(\tau).$$

The reader who checks this will notice that again $d = 26$ in (4.13) is essential.

It is useful to know that all of this extends in several ways. First of all there is an extension to lattices with different left and right dimensions, which appear in compactifications of heterotic strings, and also in the covariant lattice construction to be discussed in the following. The derivations are completely analogous, and the only difference is that in (4.21) one gets a modular weight factor which has different powers of τ and $\bar{\tau}$. Since these lattice functions appear in the partition functions with different powers of $\eta(\tau)$ and $\eta(\bar{\tau})$, that is no problem.

The extra argument $\vec{z}$ which we added to P in the derivation given above is of interest in its own right. It defines a more general *character valued* partition function, which has been used to prove anomaly cancellation. We refer to [19] for an extensive discussion of these issues.

Finally we want to point out that there is a straightforward extension of the correction factor to higher loops. The η- functions are replaced by the determinant of the laplacian on higher genus Riemann surfaces. The modular properties of this object are such that the uncompactified bosonic string is modular invariant for any genus. For the correction factor we can write down an explicit expression (see *e.g.* [36])

$$C_g(\Gamma) = \sum_{p_i \in \Gamma} \exp(i\pi p_R^i \Omega_{ij} p_R^j - i\pi p_L^i \bar{\Omega}_{ij} p_L^j) \det(Im\Omega)^{\frac{N}{2}}.$$

Although we will not do very much with this object in these lectures, it is good to know that it can be shown, by very similar techniques, to be modular invariant [37]. Since the construction of fermionic strings which follows is heavily based on self-dual lattices, it is very easy to prove that they are also multi-loop modular invariant.

5. Conformal Field Theory and Bosonization

5.1 Generalities

In this section we recall elementary concepts of conformal field theory [38] [39] which will be relevant in our context. The list of more detailed reviews contains [40-45]. We emphasize that conformal field theory is *a priori* a more general notion than string theory. Rather, strings moving in a specific d-dimensional space-time represent particular examples of conformal field theories. We will of course restrict the discussion to that types of (free) conformal field theories we need for our particular construction of heterotic strings.

The basic objects are *conformal* or *primary fields* $\phi(z,\bar{z})$, which are defined to transform as follows under coordinate reparametrizations,

$$\phi(z,\bar{z}) = \left(\frac{\partial w}{\partial z}\right)^{h} \left(\frac{\partial \bar{w}}{\partial \bar{z}}\right)^{\bar{h}} \phi(w,\bar{w}), \tag{5.1}$$

such that $\phi(z,\bar{z})(dz)^h(d\bar{z})^{\bar{h}}$ is conformally invariant. The complex variable parametrizes either the world-sheet cylinder continued to euclidean space (in which case we use instead the variable $w = 2\sigma_0 - 2i\sigma_1$) or the conformally equivalent punctured complex plane (with $z = e^w$). Fields on the cylinder and on the plane are related by

$$\phi_{cyl}(z,\bar{z}) = z^h \bar{z}^{\bar{h}} \phi_{plane}(w,\bar{w}). \tag{5.2}$$

Any conformal field is thus characterized by a *conformal weight* $(h,\bar{h})$; the scaling weight is then given by $h+\bar{h}$ and the two-dimensional spin by $h-\bar{h}$. We will call fields with half-integer spin two-dimensional fermions, those with integer spin bosons (there exist however also objects which are neither bosons nor fermions). If $\bar{h} = 0$ ($h = 0$), a field is called *chiral* or *right-*(*left-*)moving; it depends then only (anti-)analytically on z. In the following, we will mainly be concerned with chiral fields. All relevant (local) properties of a given conformal field are characterized by operator product expansions with arbitrary other fields:

$$\phi_i(z)\cdot\phi_j(w) = \sum_k (z-w)^{-h_i-h_j+h_k}\ \alpha_{ijk}\ \phi_k(w). \tag{5.3}$$

The non-zero coefficients α_{ijk} define the *fusion rules*. Unless all powers in $(z-w)$ in all possible operator products are integral ($h_k - h_j - h_i \in \mathbf{Z}$), the theory is ill-defined (*non-local*), that is, the correlation functions cannot be unambiguously integrated because of branch cuts. As we will see, imposing locality is a quite strong constraint.

The conformal weight h of any field ϕ_h is determined by the operator product with the stress-energy tensor $T = T_{zz}$, as it generates (5.1):

$$T(z)\cdot\phi_h(w) = h\frac{\phi_h(w)}{(z-w)^2} + \frac{\partial_w \phi_h(w)}{(z-w)} + \text{ regular terms }. \tag{5.4}$$

Conformal fields have the special property that they obey (5.4) without creating higher poles in $(z-w)$. For example, the fields (basically derivatives of ϕ_h) appearing in the non-leading and regular terms in (5.4) are *not* conformal fields. Quite generally, taking the operator product (5.3) of any two conformal fields creates a series of other operators, of which however only a part are conformal fields.

$T(z)$ itself has weight $h = 2$,

$$T(z)\cdot T(w) = \frac{\frac{1}{2}c}{(z-w)^3} + 2\frac{T(w)}{(z-w)^2} + \frac{\partial_w T(w)}{(z-w)} + \text{ regular terms }, \tag{5.5}$$

but for $c \neq 0$ it is not a conformal field due to the 'anomalous' higher pole. To understand the meaning of the parameter c, let us expand $T(z)$ in modes

$$T(z) = \sum_n z^{-n-2} L_n \tag{5.6}$$

according to the general rule

$$\phi_h(z) = \sum_n z^{-n-h} \phi_{h,n} \ , \qquad \phi_{h,n} = \tfrac{1}{2\pi i} \oint dz \ z^{n+h-1} \phi(z) \tag{5.7}$$

for expansions on the plane. The values of n in must be such that the integral is well-defined. For example if $h \in Z$ and $\phi(z)$ is single valued around the origin, then n takes all integral values. Contour integrating (5.5) leads now to the following equivalent expression:

$$[L_n, L_m] = \frac{c}{12} n(n^2-1)\delta_{n,-m} + (n-m)L_{n+m}. \tag{5.8}$$

Thus, c represents a central charge (anomaly) in the Virasoro algebra, and the zero mode L_0 is precisely one-half the Hamiltonian of the right-moving (analytic) sector of the theory (of course, the full theory contains in addition an isomorphic Virasoro algebra in the left-moving sector).

Quite generally, as above, the information encoded in operator product expansions is equivalent to the collection of all (anti-)commutation relations among the modes; operator products are just an economic way for writing these.

One of the most important fields is the derivative of a free boson, ∂X, which has $h = 1$. An example of X is the string functional described in sect. 2.1, mapped to the complex plane (called the *Fubini-Veneziano field*). X is itself not a conformal field, a fact which is related to its highly non-local operator product with itself (basically the two-dimensional propagator on the complex plane):

$$X(z)\cdot X(w) = -\log(z-w). \tag{5.9}$$

In contrast, ∂X is a local field:

$$\partial X(z)\cdot \partial X(w) = -\frac{1}{(z-w)^2} + \text{ regular terms }. \tag{5.10}$$

For free bosons, the corresponding stress energy tensor is

$$T(z) = -\tfrac{1}{2}{:}\partial X \partial X{:} \ = \ -\tfrac{1}{2} \lim_{w\to z} \left(\partial X(z)\cdot \partial X(w) + \frac{1}{(z-w)^2} \right), \tag{5.11}$$

where the right hand side indicates what the normal ordering dots mean (similarly, in the following, products of operators taken at the same point are always implicitly understood to be normal ordered, ie., all singularities are subtracted). Inserting (5.11) into (5.5) one learns that in this case $c = 1$.

Another important type of conformal field is given by $\phi_\lambda =: e^{i\lambda X}:$. From $T(z)\cdot\phi_\lambda(w) = \frac{1}{2}\lambda^2 \frac{1}{(z-w)^2}\phi_\lambda(w) + \ldots$ follows $h = \frac{1}{2}\lambda^2$. Furthermore, we have from (5.9)

$$
\begin{aligned}
:e^{i\lambda\cdot X}:(z)\cdot :e^{i\lambda'\cdot X}:(w) &= (z-w)^{\lambda\cdot\lambda'} :e^{i\lambda\cdot X(z)+i\lambda'\cdot X(w)}: \\
&= (z-w)^{\lambda\cdot\lambda'} :e^{i(\lambda+\lambda')\cdot X(w)}(1+i(z-w)\lambda\cdot\partial X(w)+\ldots):
\end{aligned}
\tag{5.12}
$$

Nothing in the above formally changes if we allow $\lambda \rightarrow \lambda^\mu$ and $X \rightarrow X^\mu$ to become vectors (for instance, the space-time coordinates of a string). For $2n$ bosons, $\mu = 1, \ldots 2n$, $T(z) = -\frac{1}{2}\partial X^\mu \partial X_\mu$ and accordingly $c = 2n$.

In *fermionic* string theories, there is a supersymmetry between $2n$ bosons X^μ and $2n$ fermions ψ^μ. The latter obey

$$
\psi^\mu(z)\cdot\psi^\nu(w) = \frac{\delta^{\mu\nu}}{(z-w)} + \text{ regular terms } , \tag{5.13}
$$

which, defining*

$$
\psi^\mu(z) = \sum_{n\in\mathbf{Z}+\frac{1}{2}} z^{-n-\frac{1}{2}}\psi^\mu_n, \tag{5.14}
$$

is equivalent to $\{\psi^\mu_n, \psi^\nu_m\} = \delta^{\mu\nu}_{n+m}$. The free fermion stress energy tensor is $T_\psi = \frac{1}{2}(\partial\psi)\psi$, and from (5.13) we have

$$
T_\psi(z)\cdot\psi^\mu(w) = \frac{1}{2}\frac{\psi^\mu(w)}{(z-w)^2} + \frac{\partial\psi^\mu(w)}{(z-w)} + \text{ regular terms } , \tag{5.15}
$$

so that $h = \frac{1}{2}$. Furthermore, the operator product of T_ψ with itself tells that a collection of $2n$ fermions has $c = n$. From these basic building blocks one can construct other conformal fields by considering operator products. For instance, the first regular term in the expansion (5.13) is given by

$$
J^{\mu\nu}(z) =: \psi^\mu\psi^\nu: , \qquad \mu,\nu = 1,\ldots 2n. \tag{5.16}
$$

One finds that the modes $J^{\mu\nu}_m = \frac{1}{2\pi i}\oint dz\, z^m J^{\mu\nu}(z)$ realize a $\widehat{SO(2n)}$ Kac-Moody algebra; in particular, the zero modes $J^{\mu\nu}_0$ are the usual $SO(2n)$ generators, under which ψ^μ_m transform as vectors. As $J^{\mu\nu}(z)$ is given above in terms of fermions, one calls this a *fermionic* construction of the Kac-Moody algebra. Later we will discuss a *bosonic* version.

* We take ψ here to be single valued on the complex plane. A more detailed discussion of fermion boundary conditions will be given in the next section

A very important operator in superconformal theories, *i.e.*, in fermionic strings, is the superpartner of the stress energy tensor, which has in the simplest case the form

$$T_F(z) = i\psi^\mu \partial X_\mu(z) = \sum_m z^{-m-\frac{3}{2}}\, G_m \ . \tag{5.17}$$

It has weight $h = \frac{3}{2}$, and the corresponding equ. (5.4) together with (5.5) and

$$T_F(z)\cdot T_F(w) = \frac{\frac{1}{3}c}{(z-w)^3} + \frac{2T(w)}{(z-w)} + \text{regular terms} \tag{5.18}$$

constitute the $N = 1$ superconformal algebra. Equivalently, the mode components satisfy the super Virasoro algebra. More precisely, since we are dealing here only with analytical fields, we describe a $(0,1)$ (*right-moving*) supersymmetry. The zero mode $G_0 = \frac{1}{2\pi i}\oint dz\ z^{\frac{1}{2}} T_F(z)$ is precisely the two-dimensional supercharge; from (5.18) follows

$$\tfrac{1}{2}\{G_0, G_0\} = L_0 - \tfrac{c}{24}. \tag{5.19}$$

T_F indeed transforms fermions into bosons and vice versa:

$$\begin{aligned} T_F(z)\cdot\partial X^\mu(w) &= -\frac{i}{(z-w)^2}\psi^\mu \\ T_F(z)\cdot\psi^\mu(w) &= \frac{i}{(z-w)}\partial X^\mu \end{aligned} \tag{5.20}$$

5.2 States

The reason why we considered *conformal* fields is that there is a one to one correspondence between conformal fields ϕ_h of weight h and *highest weight states*† $|h\rangle = \phi_h(0)\,|0\rangle$ of the Virasoro algebra, defined by

$$\begin{aligned} L_0\,|h\rangle &= h\,|h\rangle\,, \\ L_{+n}\,|h\rangle &= 0. \end{aligned} \tag{5.21}$$

Thus, h basically is the *mass* of the state $|h\rangle$. Such highest weight states characterize representations of the Virasoro algebra. The other states in a given representation (the *descendants*, which complete the *Verma module*) are obtained by acting with products of L_{-n} to $|h\rangle$; the corresponding operators (*e.g.*, derivatives of ϕ_h) are however not conformal fields, as they create higher poles in (5.4).

Physical states in a string theory are generated by certain combinations of conformal fields called *vertex operators*. They describe the emission of states from the world-sheet,

† $\phi_h(0)$ on the plane corresponds to an incoming state at $t = -\infty$ on the world-sheet cylinder.

i.e., the in- or out-coming asymptotic states. For open strings, they have the typical form

$$V(z) = \phi_h(z)e^{ikX}(z) \ . \tag{5.22}$$

In order to produce conformally invariant correlation functions $\langle V(z_1)...V(z_k)\rangle$, these objects must have conformal weight one. Thus, only if the on-shell condition $\frac{1}{2}k^2 = 1-h$ is satisfied (plus, in general, additional on-shell polarization conditions to prevent higher poles in (5.4)) $V(z)$ is a good vertex operator and $|\phi, k\rangle = V(0)\,|0\rangle$ a highest weight state. For instance, the tachyon vertex is given by (5.22) with $\phi(z) \equiv 1$, $(k^2 = -m^2 = 2)$, and the vector emission vertex in the bosonic string has $\phi(z) \equiv \xi^\mu \partial X_\mu(z)$ with $k^2 = \xi^\mu k_\mu = 0$.

In superconformal theories, matters are a little bit more subtle. Let us consider for a while a fermionic string theory involving a collection of world-sheet fermions ψ^μ in addition to the bosons, where μ is considered a space-time index: $\mu = 1, \dots 2n$. In addition to states created by purely bosonic operators (*i.e.*, by ∂X and exponentials of X), there exist also highest weight states of the form (see (5.14))

$$(\psi^{\mu_1}\psi^{\mu_2}\dots\psi^{\mu_k})(0)\,|0\rangle \;=\; \psi^{\mu_1}_{-\frac{1}{2}}\psi^{\mu_2}_{-\frac{1}{2}}\cdots\psi^{\mu_k}_{-\frac{1}{2}}\,|0\rangle \tag{5.23}$$

corresponding to antisymmetric tensors of rank k ($h = \frac{1}{2}k$). Taking $\phi_h = \psi^{\mu_1}\psi^{\mu_2}...\psi^{\mu_3}$ in (5.22) yields (roughly) vertex operators describing the emission of antisymmetric tensor states; in particular,

$$V = \xi^\mu \psi_\mu e^{ikX} \tag{5.24}$$

is related to the vector boson emission vertex. A quite useful object is the G-parity operator

$$G_{NS} = (-1)^F, \tag{5.25}$$

(F is the two-dimensional fermion number) which is $+1$ (-1) on even (odd) antisymmetric tensors (5.23) .

The crucial difference to the bosonic case is that the Hilbert space (*i.e.*, the representation space of the super Virasoro algebra) splits into two sectors, called *Neveu-Schwarz* (NS) and *Ramond* (R). This is due to the fact that there are two *spin structures* on the cylinder: in the NS-sector the fermions ψ have antiperiodic, while in the R-sector they have periodic boundary conditions. Performing a conformal transformation $z = e^w$ we see that (because of the factor $e^{\frac{1}{2}w} = z^{\frac{1}{2}}$ in (5.2)) fermions in the NS-sector are periodic (single-valued), and those in the R-sector antiperiodic (double-valued) along cycles enclosing the origin of the complex plane. Thus in the NS-sector the mode numbers n of ψ_n are half-integral, and in the R-sector integral. In particular, the zero modes in the R-sector obey

$$\{\psi_0^\mu, \psi_0^\nu\} = \delta^{\mu\nu} \ , \tag{5.26}$$

so they represent space-time ($SO(2n)$-) γ-matrices. Accordingly the representation space on which ψ^μ act must be spinorial. Thus, denoting by $|\alpha\rangle$ (α = spinor index) the ground state in the R-sector, a generic R-state (corresponding to (5.23)) looks like

$$(\psi^{\mu_1}\psi^{\mu_2}\dots\psi^{\mu_k})(0)\,|\alpha\rangle\,. \tag{5.27}$$

Clearly, while states in the NS-sector (5.23) correspond to space-time antisymmetric *tensors*, R-states (5.27) lead to space-time *spinors*. Like in the NS-sector, one can define a grading defined by the operator

$$G_R = \gamma_{2n+1}\,\otimes\,(-1)^F \tag{5.28}$$

with eigenvalues ± 1 (counting space-time helicity).

How can we create states in the R-sector via vertex operators ? If we view $\psi(z)$ as an analytic function on the plane, we can describe the double-valuedness around the origin (in the R-sector) as due to a square-root branch cut emanating from there: it flips the boundary conditions between periodic and antiperiodic. This cut creates stress energy (that is, it breaks translation invariance on the plane), and thus corresponds to an operator insertion at $z = 0$. This operator has to be an ordinary (super-)conformal field $S^\alpha(z)$, such as to generate a superconformal highest weight state:

$$|\alpha\rangle = S^\alpha(0)\,|0\rangle \tag{5.29}$$

S^α is called a *spin* field; it has a very complicated description in terms of the fundamental field ψ^μ. Basically it is the vertex operator for the emission of space-time spinors. In an unitary theory $G_0^2 \geq 0$, hence it follows from (5.19) that the ground state in the R-sector $|\alpha\rangle$ has weight $h = \frac{c}{24}$:

$$L_0\,|\alpha\rangle = \tfrac{c}{24}\,|\alpha\rangle\ . \tag{5.30}$$

We thus encounter a problem: for $|\alpha\rangle$ to be a highest weight state and S^α a conformal field, we need $h = 1$. Furthermore, the property that S^α produces a branch cut with ψ is imposed by requiring that the operator product expansion of $\psi(z)$ with $S^\alpha(w)$ contains half-integral powers in $(z - w)$:

$$\psi^\mu(z)\cdot S^\alpha(w) = \frac{1}{(z-w)^{\frac{1}{2}}}(\gamma^\mu)^\alpha_\beta S^\beta + \text{ regular terms }\ . \tag{5.31}$$

Thus naively the theory is *non-local* and as such ill-defined. We will discuss later how these difficulties can be resolved.

The above-explained construction is a special case of what one calls *twisting*: the R-sector is obtained from the NS-sector by a Z_2-twist. One can of course consider more general, *e.g.*, Z_N-twists, acting in particular also on the bosons ∂X; this is precisely what happens in the case of *orbifolds*.

5.3 Basics of Bosonization

Bosonization, a feature very special to two-dimensions, simply amounts to expressing fermionic field operators like ψ in terms of bosonic ones. The crucial point is that two conformal field theories are physically equivalent, *i.e.*, indistinguishable, if all correlation functions are identical. Now, (the singular parts of) all correlation functions are characterized by all possible operator product expansions (5.3) between the various fields. Thus bosonizing a theory essentially amounts to expressing any given two-dimensional fermionic operator by a certain bosonic expression in such a way that it reproduces all local operator product expansions. In particular, the realization of certain symmetries (*e.g.*, generated by $J^{\mu\nu}$ or T_F) depends only on operator products involving the symmetry currents and not on the way how these are represented.

To bosonize ψ^μ, $\mu = 1, \ldots, 2n$, we first switch to a complex basis by defining n pairs

$$\begin{aligned} \tfrac{1}{\sqrt{2}}(\psi^1 + i\psi^2) &\rightarrow \psi^1, \\ \tfrac{1}{\sqrt{2}}(\psi^1 - i\psi^2) &\rightarrow \psi^{-1}, \\ \tfrac{1}{\sqrt{2}}(\psi^3 + i\psi^4) &\rightarrow \psi^2, \\ \tfrac{1}{\sqrt{2}}(\psi^3 - i\psi^4) &\rightarrow \psi^{-2}, \end{aligned} \tag{5.32}$$

and so on. Then (5.13) takes the form ($j, k = 1, ..n$):

$$\begin{aligned} \psi^k(z)\cdot\psi^{-j}(w) &= \frac{\delta_{kj}}{(z-w)} + J^{k,-j}(w) \\ &\quad + \text{ other terms containing descendant fields .} \end{aligned} \tag{5.33}$$

Using (5.12) it is easy to see that we can express our $2n$ fermions by n bosons H_k in a way such that (5.33) is reproduced:

$$\psi^k = e^{iH_k} c_k \ , \quad \psi^{-k} = e^{-iH_k} c_{-k}. \tag{5.34}$$

(c_k are certain cocycle-phase generating operators which ensure that fermions with different indices really anticommute; in the following, we often will not write such factors explicitly.) Both conformal systems ($\{\psi^\mu\}$, $\{H_k\}$) have central charge $c = n$; in fact, it is only the total value of c that matters.

One can easily bosonize $J^{\mu\nu}$ by just reading it off from (5.33). Switching to the Cartan-Weyl-basis, we have for the off-diagonal ($k \neq j$) currents

$$\begin{aligned} J^{\pm k, \pm j}(z) &\equiv :\psi^{\pm k}\psi^{\pm j}:(z) \\ &= :e^{\pm iH_k \pm iH_j}:(z) c_{\pm k} c_{\pm j} \ , \end{aligned} \tag{5.35}$$

and for the diagonal currents, which span the Cartan subalgebra:

$$\begin{aligned} J^{k,-k}(z) &=: e^{iH_k} e^{-iH_k}:(z) \\ &= \lim_{w \to z} \left(e^{iH_k}(z) e^{-iH_k}(w) - \frac{1}{(z-w)} \right) \\ &= i\partial H_k(z) \ . \end{aligned} \tag{5.36}$$

It is easy to check that these currents represent a bosonic realization of the $\widehat{SO(2n)}$-

Kac-Moody algebra. Note that the price to pay for bosonization is non-covariance of the formalism: The off-diagonal currents are realized by solitons, *i.e.*, circle-valued bosons, while the diagonal, abelian currents take the canonical form. (5.35) and (5.36) are only a particular example of the more general *Frenkel-Kac* construction (For a detailed discussion we refer to [27].); the latter works for any (simply-laced) Lie algebra, if the root vectors $(\pm e_k \pm e_j)$ of $SO(2n)$ are replaced by the length-squared two root vectors of the algebra in question.

In the last section we have given further examples of operators in the fermionic sector. For instance, in bosonic language the operator corresponding to $\psi^\mu\psi^\nu...\psi^\rho$ (5.23) is (a linear combination of operators) of the form

$$\phi_\lambda(z) = e^{i\lambda\cdot H}(z) \;, \tag{5.37}$$

where λ is an n-dimensional vector of the form $(\pm 1, 0, \pm 1, \dots)$, depending on the precise tensor structure. The G-parity operator (5.25) takes the form

$$G = e^{i\pi\left(\sum_{k=1}^{n} \lambda_k\right)} \;, \tag{5.38}$$

which is ± 1 depending on if $\sum \lambda_i$ is even or odd. Operators describing (anti-)commuting objects are related to vectors λ of even (odd) length-squares; one can easily see this from equ. (5.12) upon exchanging z and w.

Is is clear that the collection of all operators of the form $e^{i\lambda H}$ is related to a *lattice*. This follows simply from the fact that in any operator product (5.12) the vectors add to give a new one: $\phi_\lambda(z)\cdot\phi_{\lambda'}(w) = \phi_{\lambda+\lambda'}(z-w)^{\lambda\lambda'} + \dots$ Consistency, that is, closure of the operator product algebra (5.3), then requires that the vertex operator $\phi_{\lambda+\lambda'}$ exists and generates a state $|\lambda+\lambda'\rangle$ of the theory. The lattice thus is formed by the set of vectors λ obtained by all possible additions of some given basic vectors. (Of course, there exist also vertex operators containing bosons ∂H; however, these play only a trivial role in the following).

What we are after is a classification scheme for vertex operators or states, in order to characterize the spectrum of a given theory. Given the above lattice, operator products preserve a natural Z_2-grading, that is, a split into vectors of even and odd length-squares; equivalently, a split into eigenvalues ± 1 of G_{NS}. More specifically, the lattice characterized above is precisely the weight lattice of D_n generated by the conjugacy classes (0) and (v). In particular, the non-zero roots of D_n $(\pm e_i \pm e_j)$ which appear in the bosonized currents (5.35) belong to the neutral class (0), while the fermions $\psi^{\pm i}$ belong to (v); the operator product expansion (5.13) of two fermions, which contains $J^{\mu\nu}$ in its regular terms, clearly reflects the rule $(v)+(v)=(0)$.

In this framework it is very easy to describe states in the R-sector. In the R-sector, the fields ψ^μ are twisted as explained in the previous section. In bosonic language such a twist corresponds to a *shift* of the lattice, a fact that is made plausible by the identity

$$\left(\frac{1}{\prod(1+q^{2n})}\right)\sum_{n\in\mathbf{Z}} q^{(n+\theta)^2} = q^{\theta^2}\prod_{n\geq 0}(1+q^{2n+2\theta+1})(1+q^{2n-2\theta+1}) \;, \tag{5.39}$$

which is a mathematical manifestation of bosonization: the RHS is the partition function of two fermions with twist θ, while the LHS represents the partition function of one compact boson taking values on a (one-dimensional) integer lattice shifted by θ. More precisely, the sum at the LHS gives the contribution of all solitonic states created by $\{\phi_\lambda\}$, $\lambda = n+\theta, n \in \mathbf{Z}$, and the product is the contribution of all possible combinations of derivatives of ∂H. The case at hand corresponds to n copies of (5.39), *i.e.*, to $2n$ fermions and to an integer lattice shifted by $\theta_1 = \ldots = \theta_n = \pm\frac{1}{2}$ mod 2. In other words, the R-sector is described by the spinor conjugacy classes (s) and (c) of D_n. To which particular spinor class a given state $|\lambda\rangle$ belongs is determined by the eigenvalue of G_R (5.28) which is, in bosonic language, up to a phase, identical to G_{NS} given in (5.38).

Every given conjugacy class can of course be represented by a particular vector belonging to that class; all other vectors in the same class are obtained by adding the root lattice, *i.e.*, all vectors in the neutral class (0). Thus all states (up to derivatives) in the R-sector are of the form

$$\left(e^{i\lambda\cdot H}S^\alpha\right)(0)\,|0\rangle\ , \quad \lambda \in (0)\ of\ D_n\ , \tag{5.40}$$

and we can think of the ground states of the R-sector $|\alpha\rangle = S^\alpha(0)\,|0\rangle$ as being generated by shift vectors of minimal lengths in the spinor conjugacy classes of D_n. Hence we find the following bosonic expression for the spin fields:

$$S^\alpha(z) = e^{i\lambda_s\cdot H}(z), \qquad \lambda_s = (\pm\tfrac{1}{2}, \pm\tfrac{1}{2}, \ldots, \pm\tfrac{1}{2}). \tag{5.41}$$

In fact, there are two spin fields, S^α and $S^{\dot\alpha}$, with opposite $SO(2n)$ helicity belonging to (s) and (c) of D_n; they are distinguished by an even or odd number of $-$ signs in λ_s, respectively. The number 2^{n-1} of possible sign choices for λ_s in each case is precisely the dimension of the two spinor representations of $SO(2n)$: $\alpha, \dot\alpha = 1, .., 2^{n-1}$. The conformal weights of S^α, $S^{\dot\alpha}$ are $h_s = \frac{1}{2}\lambda_s^2 = \frac{1}{8}n$; since in a supersymmetric theory $n_f = n_b = 2n$, and $c_b = 2c_f = 2n$, we find $h_s = \frac{c}{24}$ in accordance with (5.30).

Many properties of states in the various conjugacy classes can be determined solely in terms of addition and multiplication rules of $(0), (v), (s)$ and (c), in connection with the basic formula (5.12) for operator products of exponentials. For instance, the twisted boundary conditions for fermions in the R-sector follow from the branch cut arising from $(z-w)^{\lambda\cdot\lambda'}$ in (5.12) by the multiplication rule $(v)\cdot(s) \in \mathbf{Z}+\frac{1}{2}$.* (Anti-)commutation properties are also easily determined from (5.12) upon exchanging z and w; one finds that while all states in (0) commute and those in (v) anticommute, states in the spinor conjugacy classes (for generic n) have no definite (anti-)commutation properties.

The above discussion dealt only with periodic and antiperiodic fermionic boundary conditions, *i.e.*, with Z_2-twisted fermions. We have just seen that these can be described, in bosonic language, in terms of the weight lattice D_n of $SO(2n)$. It is obvious that one can easily generalize these concepts to other lattices (for instance, the weight lattices A_n of $SU(n+1)$). (One can, of course, consider also lattices which are products of weight lattices of several Lie groups, and/or lattices which are not at all related to Lie groups.) One then describes fermions with more complicated boundary conditions, and other objects, which need not be either fermionic or bosonic. Like the spin fields,

* This is nicely illustrated by (5.31), which also reflects the fusion rule $(v)+(s)=(c)$.

such operators are quite difficult to describe in terms of the fermionic variables ψ^μ. In bosonic language, however, all operators have the same (exponentiated) form, and ψ^μ are no more fundamental than S^α. Clearly, the bosonic lattice formulation provides the most simple and elegant description of such objects.

5.4 Compactification Revisited

We have seen that certain conformal fields (like ψ^μ, $J^{\mu\nu}$ and spin fields) can easily described in terms of free compact bosons. In fact, it is sometimes possible to interpret these as (torus-)compactified higher dimensional coordinates.

Consider, for example, torus compactification (as discussed in chapter 4) of the 26-dimensional closed bosonic string on an 8-dimensional even lorentzian self-dual lattice $(E_8)_L \times (E_8)_R$. This is the same as $(D_8)_L \times (D_8)_R$ with conjugacy classes $(0,0)$, $(0,s)$, $(s,0)$ and (s,s). Using the compactified coordinates $X^I(z), X^I(\bar{z})$ in the left- and right-moving sectors, it is easy to discuss the conformal field theory of this model in terms of operators like the currents of $(E_8)_L \times (E_8)_R$.

The currents belonging to (0) of the subgroup $D_8 \sim SO(16)$ are precisely given by (5.35) and (5.36); the currents (5.36) belonging to the Cartan subalgebra correspond to Kaluza-Klein states, the off-diagonal ones to winding states, *i.e.*, solitons. The remaining currents belonging to (s), *i.e.*, to $E_8/SO(16)$ are realized by spin fields (5.41) of $SO(16)$. This is possible since a spinor weight of D_8 has length-squared two, so that it can represent a root vector (we say that the spinor conjugacy class of D_8 *extends* D_8 to E_8). Now, what is relevant in the two dimensional sense are the properties of the conformal fields of the theory, and not if they can be interpreted in terms of compactified higher dimensions. According to the fermion-boson equivalence in two dimensions, in order to describe the same theory we could also start with a fermionic version of the above objects. The $SO(16)$ currents are then represented by bilinears as in (5.16), however, the currents represented by the spin fields have a very complicated, transcendental dependence on the fermions [46]). In this formulation, there is no trace of higher dimensions left! Apparently the concept of compactification is quite delicate in the string context.

The above example represents a compactified theory in the very sense. There exist however more general conformal field theories, which do not admit a compactification interpretation (we will encounter such theories in the next chapter). One should therefore not confuse compact bosons (equivalent to fermions) with compactified space-time coordinates (which are a special case).

For closed strings, which are described by a combination of a right-moving (z-dependent) and a left-moving ($\bar{z}$-dependent) conformal field theory, we require vertex operators to have conformal weight $(\bar{h}, h) = (1,1)$. Consequently, these consist of a left- and a right-moving piece together with the exponentiated momentum,

$$V(\bar{z}, z) = \phi_h(z)\phi'_{\bar{h}}(\bar{z})e^{ik\cdot X}(\bar{z}, z), \tag{5.42}$$

with the condition $\bar{h} = h$. ϕ_h and $\phi'_{\bar{h}}$, belonging to independent left- and right-moving sectors, are a priori not related to each other. For instance, for the compactified bosonic string above,

$$V_{\mu}^{ab}(\bar{z},z) = J^{ab}(\bar{z})\partial X_{\mu}(z)e^{ikX} \tag{5.43}$$

($\mu = 1, \ldots 18$) describes a gauge vector boson belonging to the adjoint representation of the left-moving E_8 with generators J^{ab}, whereas

$$V_{\mu}^{a'b'}(\bar{z},z) = J^{a'b'}(z)\bar{\partial} X_{\mu}(\bar{z})e^{ikX} \tag{5.44}$$

describes the emission of a vector boson of the right-moving E_8. On the other hand,

$$V_{\mu\nu}(\bar{z},z) = \bar{\partial} X_{\nu}(\bar{z})\partial X_{\mu}(z)e^{ikX} \tag{5.45}$$

describes the emission of gravitons, antisymmetric tensor fields and dilatons in $d = 18$.

In the next section we will apply the above ideas to a bosonic description of heterotic strings, which in general do not admit a compactification interpretation.

6. NSR Model, Superstring and Heterotic String

6.1 The Covariant NSR Model in Bosonic Lattice Formulation

We will first introduce the covariant ten-dimensional *NSR* model [47] (the *spinning string*) in bosonic language, which will allow a straightforward generalization to lower dimensions in the next chapter.

The covariant *NSR* model can be described by a specific conformal field theory whose main variables are ten bosons X^{μ} and ten fermions ψ^{μ}, which play the role of space-time coordinates. Thus, the basic conformal operators are $\partial X^{\mu}(z)$ and $\psi^{\mu}(z)$, whose properties have been discussed already in the foregoing chapter. Since these fields depend *analytically* on z, we deal only with right-moving fields. What we are after is however closed, in particular, heterotic string theory. Thus we will eventually have to pair the right-moving *NSR* model with an a priori independent left-moving ($\bar{z}$-dependent) conformal field theory. This will be explained further below.

On a flat background, and in superconformal gauge, the action takes the free form

$$S_{NSR} = -\tfrac{1}{2\pi} \int d^2 z \left(\partial X^{\mu} \bar{\partial} X_{\mu} - \psi^{\mu} \bar{\partial} \psi_{\mu} + \mathcal{L}_{ghost} \right) . \tag{6.1}$$

Let us denote by c and b the (anticommuting) ghost and antighost fields for fixing conformal invariance, and by γ and β the (commuting) ghost and antighost fields for fixing the fermionic part of superconformal invariance. Then the ghost lagrangian is given by

$$\tfrac{1}{2}\mathcal{L}_{ghost} = b\bar{\partial} c + \beta \bar{\partial} \gamma \ . \tag{6.2}$$

The conformal field theory of the X^{μ} and ψ^{μ} system was developed in the foregoing chapter; we indicated already, that this part of the theory by itself suffers from severe deficiencies. For example, ψ^{μ} as a two-dimensional spinor anticommutes with itself (cf.(5.13)). Since it carries however a space-time *vector* index (it appears in emission vertices for space-time tensors), one would rather like to have it generating space-time bosons. Moreover, in order to describe the R-sector, we had to introduce the spin field S^α. This field has conformal weight $h = \frac{n}{8} = \frac{5}{8}$ and not $h = 1$, what would be needed for a vertex operator describing massless space-time fermions. Inserted at some point on the complex plane, S^α creates a branch cut (5.31) in order to change the boundary conditions of ψ^μ (in bosonic formulation, this is related to the rule $(v)\cdot(s) \in \mathbf{Z}+\frac{1}{2}$ for the conjugacy classes of D_5, which is the Lie algebra of the Wick-rotated Lorentz group in ten dimensions). But the very presence of this cut makes the theory non-local. Another problem we encounter is that S^α does not anticommute with itself as it should as a space-time fermion. This fact is reflected by the rule $(s)\cdot(s) \in 2n + \frac{5}{4}, n \in \mathbf{Z}$ in D_5, which also reveals the presence of even worse cuts.

It is important to notice that many of these and other unpleasent features trace back to properties of the various conjugacy classes of the lattice D_5, if one uses the bosonic description of the ψ^μ. We thus expect that a resolution lies in the modification of this lattice.

The only new ingredient of the *NSR* model (6.1) compared to the conformal field theory of X^μ and ψ^μ as described in the last chapter is the ghost system. It is in fact a crucial ingredient, which can cure all above-mentioned diseases. To be more precise, let us first display some essential properties of the ghosts: $b(z)\cdot c(w) = c(w)\cdot b(z) = \frac{1}{(z-w)}$, $\beta(z)\cdot\gamma(w) = -\gamma(w)\cdot\beta(z) = -\frac{1}{(z-w)}$. Their conformal weights are

$$\begin{array}{ll} c: & h=-1, \\ b: & h=2, \\ \gamma: & h=-\frac{1}{2}, \\ \beta: & h=\frac{3}{2}, \end{array} \tag{6.3}$$

The contribution to the full energy momentum tensor is $T_{gh} = c\partial b + 2\partial c b - \frac{1}{2}\gamma\partial\beta - \frac{3}{2}\partial\gamma\beta$. Substituting this into (5.5), we learn that the contribution to the central charge of the b, c system is $c_{bc} = -26$, and similarly $c_{\beta\gamma} = 11$; thus altogether we have $c_{ghost} = -15$ and this cancels precisely the contribution of the ten X^μ plus ten ψ^μ. So the conformal anomaly disappears, which seems to be crucial for the consistency of the theory. This does of course *not* imply that the string theory can only be consistent in ten dimensions. Rather, what counts here is only the *total* value of $c_{matter} = 15$, and not any particular decomposition in terms of contributions of X^μ and ψ^μ. Allowing for internal degrees of freedom, which carry 'missing' central charge, will allow us later to construct string theories in lower dimensions than ten.

As we intend to discuss only the spectrum of the theory and not other aspects (like correlation functions), we will explain only aspects of the ghost system relevant here. The role of the b, c system is then quite simple: For our purposes, all what the b, c system does is to provide the correct vacuum energy. In sect. 5.2 we described how the states of the theory are created by acting with conformal fields on some vacuum. We did however not specify the vacuum itself. The correct vacuum (of the bosonic string) is simply $|0\rangle = c(0)\,|0\rangle_{SL(2,C)}$, where $|0\rangle_{SL(2,C)}$ is the true $SL(2,C)$ invariant vacuum of the theory, which is annihilated by $L_n, n \geq -1$. (Another way of describing the same situation is to redefine the vertex operators of the theory $V \to cV$; these objects then act on the $SL(2,C)$ invariant vacuum state.) Since c has weight $h = -1$, the level of $|0\rangle$ is -1; it represents the tachyon state of the bosonic string. Massless states, therefore, are build upon $|0\rangle$ using conformal fields of weight $h = 1$. This gives us back the the mass formula of the bosonic string (4.10) ($\mathcal{N}$ counts the number of derivatives in a vertex operator).

In fermionic strings, we have in addition the commuting ghosts β, γ. Their role is much more involved. As we cannot discuss all issues in great detail, we will just give recipes for handling these ghosts (details can be found *e.g.*in [39]). It is very convenient to bosonize also these fields:

$$\beta = e^{-i\phi}\partial\xi \ , \qquad \gamma = e^{i\phi}\eta \ . \tag{6.4}$$

Here, ϕ is a chiral boson with opposite norm, $\phi(z)\cdot\phi(w) = \log(z-w)$, and ξ and η are auxiliary fermions, which will play no role in the following. They are needed above since one cannot bosonize the commuting fields β, γ solely in terms of an anticommuting exponential of one boson.

One important feature of the ϕ-ghost is that the conformal field $e^{iq\phi}$ has a peculiar conformal weight:

$$\begin{aligned} T_{\beta\gamma}(z)\cdot e^{iq\phi}(w) &= -\tfrac{1}{2}q(q+2)\frac{e^{iq\phi}(w)}{(z-w)^2} \ + \ \dots \\ \to \quad h &= -\tfrac{1}{2}q(q+2) \ . \end{aligned} \tag{6.5}$$

The linear term $(-q)$ originates in an anomaly of the ghost number current $j = \beta\gamma$, and the negative sign is due to the negative metric of ϕ, which also implies

$$e^{iq\phi}(z)\cdot e^{ip\phi}(w) = (z-w)^{-qp}\, e^{i(q+p)\phi}(w) \ + \ \dots \tag{6.6}$$

Similar to the b, c system, also in the β, γ system the vacuum gets modified: it is characterized by a certain *bose ghost sea charge* q:

$$|0\rangle_{\beta\gamma} \equiv |q\rangle = e^{iq\phi}(0)\,|0\rangle \ . \tag{6.7}$$

To specify the possible values of q, note first that the $\beta\gamma$ ghosts are related to gauge degrees of the (non-dynamical) world-sheet gravitino, hence should have the same periodicity properties. Since that couples to the supercurrent $T_F = i\psi\partial X$, it follows that because of world-sheet supersymmetry the $\beta\gamma$ ghosts should have the same boundary conditions as the *NSR* fermions ψ^μ. This implies that we need *different* vacua in the *NS*- and *R*-sectors of ψ^μ: in the *NS*-sector $q \in \mathbf{Z}$ while in the *R*-sector $q \in \mathbf{Z}+\frac{1}{2}$. There exists now still an ∞-fold ambiguity in what values of q we actually choose. In fact, all values differing by integers are equivalent. Thus, for several reasons, let us choose as *'canonical'* ghost vacua

$$\begin{aligned} |0\rangle_{\beta\gamma}^{NS} &= |-1\rangle \ , \\ |0\rangle_{\beta\gamma}^{R} &= |-\tfrac{1}{2}\rangle \ . \end{aligned} \tag{6.8}$$

Later we will comment on other choices.

Similar to the b, c system, it is sometimes more convenient (in particular, for computing correlation functions) to redefine the vertex operators so that they act on a simple vacuum. Let us take here $|0\rangle = c(0)\,|0\rangle_{SL(2,C)}$ with level -1. Then

$$V_{NS}\, e^{-i\phi}(z) \quad \text{and} \quad V_R\, e^{-i\frac{1}{2}\phi}(z) \tag{6.9}$$

are the vertex operators we will consider from now on. By (6.5), the mass formula reads

$$\begin{aligned} \tfrac{1}{8}m^2 &= h - 1\ , \\ h &= \tfrac{1}{2}\lambda^2 + \mathcal{N} - 1 + \begin{cases} \tfrac{1}{2} & NS\text{-sector} \\ \tfrac{3}{8} & R\text{-sector} \end{cases} \end{aligned} \tag{6.10}$$

which displays two different normal ordering constants in the NS- and R-sectors. This formula already cures some of our problems: the fermion emission vertices for both chiralities now become

$$V_F^{spinor(s)} = u_\alpha S^\alpha e^{-i\frac{1}{2}\phi} e^{ikX}\ , \qquad V_F^{spinor(c)} = u_{\dot\alpha} S^{\dot\alpha} e^{-i\frac{1}{2}\phi} e^{ikX}\ , \tag{6.11}$$

and therefore (if $k^2 = 0$) have weight $h = 1$ (as opposed to $h = \frac{5}{8}$ without ghosts) what is needed for good vertex operators. Similarly, the vector boson emission vertex

$$V_B^{vector} = \xi_\mu \psi^\mu e^{-i\phi} e^{ikX}\ , \tag{6.12}$$

now also has $h = 1$ (as opposed to $h = \frac{1}{2}$) for vanishing momentum. V_F^{spinor} and V_B^{vector} therefore describe massless particles in ten dimensions. We can, of course, also build other operators like $V_B^{tachyon} = e^{-i\phi} e^{ikX}$ or $V_B^{tensor} = \xi_{\mu\nu}\psi^\mu\psi^\nu e^{-i\phi} e^{ikX}$, with mass $-\frac{1}{2}$ and $+\frac{1}{2}$, respectively. The ground states in the NS- and R-sectors are $\left|V_B^{tachyon}\right\rangle$ and $\left|V_F^{spinor}\right\rangle$, respectively.

We mentioned above that our other problems could possibly be solved by modifying the lattice D_5 related to the bosonized fermions. Contemplating on (6.9), the clever reader recognizes that part of these problems is solved already. To make this more clear, let us use a trick, invented by [15]: just consider the ghost charge q as part of an enlarged *covariant* lattice. That is, associate to any of the above (and all other possible) bosonized vertex operators (*i.e.*, states) a six-dimensional vector $w = (\lambda \mid q)$, *e.g.*,

$$\begin{aligned} w^{spinor} &= (\pm\tfrac{1}{2}, \pm\tfrac{1}{2}, \pm\tfrac{1}{2}, \pm\tfrac{1}{2}, \pm\tfrac{1}{2} \mid -\tfrac{1}{2}) \\ w^{vector} &= (\ 0, \ldots \pm 1, \ldots 0, \ldots \mid -1) \\ w^{tachyon} &= (\ 0,\ 0,\ 0,\ 0,\ 0 \mid -1) \end{aligned} \tag{6.13}$$

and $w^{tensor} = (\ 0, \ldots \pm 1, \ldots \pm 1, \ldots \mid -1)$, etc. In order to efficiently characterize the theory, let us investigate the nature of this six dimensional lattice Γ_6. It looks rather like an extension $D_5 \to D_6$, but we have to take the switched sign in the ghost metric into account. The natural inner product on this lattice is defined by the power of $(z - w)$ in operator product expansions, because that determines the various (anti-)commutation properties of the fields and measures locality. From (5.12) and (6.6) we have

$$V_\lambda e^{iq\phi}(z)\cdot V_{\lambda'} e^{iq'\phi}(w) = (z-w)^{\lambda\cdot\lambda' - qq'}\, V_{\lambda+\lambda'} e^{i(q+q')\phi}(w) \;+\; \dots \;. \tag{6.14}$$

Therefore the natural inner product has *lorentzian* metric $(+++++|-)$:

$$< w, w' >= \lambda\cdot\lambda' - qq' \;. \tag{6.15}$$

The same inner product appears also in the mass formula for arbitrary ghost charges:

$$\tfrac{1}{8}m^2 = \tfrac{1}{2} < w, w > -q + \mathcal{N} - 1 \;. \tag{6.16}$$

Accordingly, we denote the lorentzian lattice by $\Gamma_{5,1} = D_{5,1}$, with the comma indicating the sign flip in the metric.

As the ϕ-charge adds in (6.14), it is clear that for closure of the operator product algebra one needs also vertex operators with different ghost charges than the canonical ones (6.9). However, as mentioned above, vertex operators with ghost charges differing by integral units can represent (or generate) the same physical state. That is, there is a copy of every canonical vertex operator in every ghost sector; the different copies are also called *pictures*. There exists in fact an operation which maps a given vertex operator with ghost charge q into one with $q+1$:

$$V_{q+1}(z) =: \mathcal{P}_{+1} V_q :(z) \;. \tag{6.17}$$

The *picture changing operator* carries one unit of ghost charge, and is given by

$$\mathcal{P}_{+1}(z) = e^{i\phi} T_F(z) = ie^{i\phi}\, \psi\cdot\partial X(z) + \dots \;. \tag{6.18}$$

Thus, apart from the ghost part, picture changing is more or less the same as performing a two dimensional supersymmetry transformation*.

For instance, picture changing the vector emission vertex V_B^{vector} in the (-1)-picture (6.12) leads to its copy in the (0)-picture:

$$:\mathcal{P}_{+1} V_{B(-1)}^{vector}:(z) = V_{B(0)}^{vector}(z) = \xi\cdot[\partial X + i\psi(k\cdot\psi)]\, e^{ikX}(z) \;, \tag{6.19}$$

which (apart from the ψ-dependent piece) is the same as the vector emission vertex of the bosonic string (5.43). One easily checks that the mass formula (6.16) gives the same answer for both $V_{B(-1)}^{vector}$ and $V_{B(0)}^{vector}$. This already follows from the fact that the conformal weight h of $\mathcal{P}_{+1}(z)$ vanishes. In general, $\mathcal{P}_{+1}$ always acts in a way that compensates the effect of changing the q-charge. Thus the mass spectrum of the theory is bounded from below, a subtle feature which is not at all manifest in the mass formula (6.16).

---

* In fact, the ghost part is necessary in order to have a well-defined local action of T_F on states in the R-sector; remember the branch cut of ψ in T_F. Roughly speaking, one may regard $e^{i\phi}$ as a bosonized anticommuting coordinate, playing the role of ϵ in $[\epsilon Q^{SUSY}, \dots]$.

Upon bosonization, we can represent $\mathcal{P}_{+1}$ by the vector $w^{PC} = (0\ldots,\pm1,0\ldots \mid +1) \in D_{5,1}$. Picture changing then acts on the lattice(-part of the theory) simply as shifting by w^{PC}. In order for this operation to be well-defined and local, w^{PC} has to have integral inner product with all other lattice vectors. Although trivially satisfied in the ten dimensional *NSR* model, this condition will prove quite non-trivial for the lower dimensional theories we will consider below, as for these theories the supercurrent T_F (*i.e.*, w^{PC}) can take a different form.

The structure of (6.19) is more complicated than that of (6.12) (and even more complicated in higher pictures). In general, the canonical ghost sectors (6.9) have the virtue of being simplest in the sense that the vertex operators are related to lattice vectors in the most simple way. In the other ghost sectors, there is no manifest relation between lattice vectors and physical vertex operators; they are in general given by more complicated linear combinations. For our purposes, however, these complications play no role, as features like locality or (anti-)commutativity do not depend on them. That is, all relevant information about state classification is carried by the lattice vectors (oscillators, *i.e.*, derivatives, also play no role). For our purposes, we can therefore represent the spectrum of vertex operator in all ghost sectors by the lattice $D_{5,1}$.

This certainly does *not* mean that the partition function of the physical states is given by the partition function of the lattice! There is a priori no one-to-one relation between lattice vectors and physical states[†] . There exists however a brute force method to count the physical degrees of freedom correctly: let us split every lattice vector $w = (\lambda \mid q) \in D_{5,1}$ into a four dimensional 'light-cone' (describing the Wick-rotated transverse Lorentz group $SO(8)$) and a two dimensional 'ghost' part, according to the decomposition

$$\begin{aligned} D_{5,1} &= D_4^{light-cone} \times D_{1,1}^{ghost} \\ w &= (\,u,\,x\,)\ , \qquad u \in D_4^{light-cone}\ , \quad x \in D_{1,1}^{ghost}\ . \end{aligned} \tag{6.20}$$

Then, by definition, the physical states are those associated with

$$x - x_0 = \begin{cases} (\ 0 \mid -1) & NS\text{-sector} \\ (-\frac{1}{2} \mid -\frac{1}{2}) & R\text{-sector}\ . \end{cases} \tag{6.21}$$

One easily checks that this *physical state selection rule* is just a projection to light-cone states, in the canonical pictures (the astute reader also notices that for the above choice of x_0 above the mass formula (6.16) turns precisely into the light-cone mass formula). From the standpoint of conformal field theory, this separation of vectors into 'physical' and 'unphysical' ones is quite artificial (and ambiguous), as vertex operators are equally good in arbitrary pictures. The above selection rule is nothing but a help to identify and count the physical states of the theory.

† For example, w^{vector} (6.13) associated with $V^{vector}_{B(-1)}$ given in (6.12) describes ten degrees of freedom (which is the correct count), while $w^{vector} = (\ldots \pm 1, \ldots \pm 1, .. \mid 0)$ associated with the picture changed, equivalent vertex operator $V^{vector}_{B(0)}$ (6.19) has 40 components.

Having now everything reduced to the properties of the lattice $D_{5,1}$, let us split it (*i.e.*, the set of vertex operators) into cosets or conjugacy classes, much like we did for the case of D_n in section 5.3. In particular, as it is obvious from (6.13), V_B^{vector} belongs to (0) of $D_{5,1}$, $V_B^{tachyon}$ and V_B^{tensor} to (v) and V_F^{spinor} to (s) or (c), respectively. Any other vertex operator of the theory belongs to one of these four classes, and their lattice parts can be obtained by adding 'root' vectors belonging to the neutral class (0) to the vectors (6.13); these are simply ground state (lowest mass) representatives of the conjugacy classes (note in particular that the action of the picture changing operator $\mathcal{P}_{+1}$ belonging to (0) does not change the class of a vertex operator).

Then, like for D_n, we can represent all (anti-)commutation relations, fusion rules and possible nonlocalities of the *NSR* model in terms of addition and multiplication rules of the conjugacy classes of $D_{5,1}$ (the dot denotes the lorentzian inner product):

$$\begin{aligned}
(0)+(0)&=(0), &\quad (0)\cdot(0)&\in 2\mathbf{Z}\\
(0)+(v)&=(v), & (0)\cdot(v)&\in \mathbf{Z}\\
(0)+(s)&=(s), & (0)\cdot(s)&\in \mathbf{Z}\\
(0)+(c)&=(c), & (0)\cdot(c)&\in \mathbf{Z}\\
(v)+(v)&=(0), & (v)\cdot(v)&\in \mathbf{Z}\\
(v)+(s)&=(c), & (v)\cdot(s)&\in \mathbf{Z}+\tfrac{1}{2}\\
(v)+(c)&=(s), & (v)\cdot(c)&\in \mathbf{Z}+\tfrac{1}{2}\\
(s)+(s)&=(0), & (s)\cdot(s)&\in \mathbf{Z}\\
(c)+(c)&=(0), & (c)\cdot(c)&\in \mathbf{Z}\\
(s)+(c)&=(v), & (s)\cdot(c)&\in \mathbf{Z}+\tfrac{1}{2}
\end{aligned} \tag{6.22}$$

Note that all length(-squares) are integral, but not even; $D_{5,1}$ is therefore an *odd* lattice. (6.22) tells us that operators in (0) (even lengths) commute, those in $(s),(c)$ (odd lengths) anticommute with each other. This is precisly what we like to have, since (0) contains (odd-rank antisymmetric) tensors, and $(s),(c)$ spinors in space-time. Furthermore, $V_B^{vector}\in(0)$ and $V_F^{spinor}\in(s)$ are now local with respect to each other: the branch cut in $\psi(z)\cdot S(w)$ is cancelled by that of $e^{-i\phi}(z)\cdot e^{-i\phi/2}(w)$.

However, we are not yet completely done: the (even-rank antisymmetric) space-time tensors belonging to (v) anticommute and thus violate spin-statistics, and they are also non-local with respect to the spinors (as can be inferred from the $\frac{1}{2}$'s in (6.22)). Finally, the class (v) contains the tachyonic ground state of the *NS*-sector —an unpleasant feature. Now, what we have constructed so far is the spinning string, which is known to be inconsistent. One has to do little more to really obtain the superstring.

The remaining diseases of the spinning string are simply eliminated by restricting to a subset of the possible vertex operators, characterized by (0) and (s) (or (c)). One easily concludes from (6.22) that all operator products are local and close within this subset. The spin-statistic relation is satisfied as (v) is absent now: tensors and spinors are related to even and odd vectors on $D_{5,1}$. In particular, as massless states we keep $\left|V_B^{vector}\right\rangle$ and $\left|V_F^{spinor}\right\rangle$ which form a supersymmetry multiplet. Actually one can prove (we will do this explicitly below) that the whole spectrum is supersymmetric. Accordingly, one calls this theory the superstring.

Keeping only states in (0) and (s) can also be expressed as projecting onto states $|w\rangle$ with 'positive G-parity' $G = 1$, where

$$G\,|w\rangle \equiv (-1)^{F+F^{\phi}}\,|w\rangle = e^{2\pi i < w^{spinor(s)}, w>}\,|w\rangle \tag{6.23}$$

generalizes (5.38) by including the ϕ-ghost charge F^{ϕ}. This is called *GSO-projection* [48].

The *GSO*-projected lattice is smaller in the sense that it contains less conjugacy classes than the lattice we had before. Nevertheless we still will call it $D_{5,1}$, as according to usual notation a lattice is named after its root sublattice (0). (Remember that in order to characterize a lattice completely, one has to provide in addition a list of the conjugacy classes actually present.)

In fact, the projected lattice $D_{5,1}$ is *self-dual* as one can easily check. This self-duality is actually necessary for the consistency of the superstring: it ensures modular invariance of the partition function, which contains a lattice factor

$$\mathcal{L}_{5,1}(\tau) = \sum_{\substack{w=(u,x)\in D_{5,1} \\ x=x_0}} e^{i\pi\tau u^2} e^{2\pi i q}\ . \tag{6.24}$$

The sum is only over the states satisfying the selection rule (6.21), and the phase counts space-time fermions and bosons with opposite signs. The lattice function $\mathcal{L}_{5,1}$ differs in further ways from lattice partition functions (4.18) we encountered earlier. In particular, the lattice is odd rather than even, and in writing the exponent above we used the peculiar mass formula (6.16). It was shown in [16] that all these features conspire in precisely the right way to make the theory modular invariant. In the next section we will present an independent proof, which furthermore extends to higher loops.

Knowing this, we could have started already with a self-dual lattice from the very beginning and would never had encountered the problems of the spinning string; locality is guaranteed by the fact that on a self-dual lattice all inner products are integral. This will be our philosophy for constructing further string theories.

We have now everything prepared for generalizing the above to heterotic strings. We simply have to combine the above-explained, right-moving *NSR* model (not necessarily the superstring!) with a left-moving bosonic string. As the conformal anomaly $c_{bc} = -26$

for the bosonic string, we need in addition to the ten space-time coordinates X^μ 16 compact bosons $X^I(\bar{z})$ (or equivalently, 32 fermions):

$$S_{het} = -\frac{1}{2\pi}\int d^2z \left(\partial X^\mu \bar{\partial} X_\mu + \psi^\mu \bar{\partial}\psi_\mu + 2b\bar{\partial}c + 2\beta\bar{\partial}\gamma + \partial X^I \partial X_I + 2\bar{b}\partial\bar{c} \right) . \quad (6.25)$$

($I = 1\ldots16$). What is the momentum lattice of the compact bosons like ? It is convenient to combine this 16-dimensional lattice Γ_{16} with that related to the right-moving *NSR* model:

$$\Gamma_{16;5,1} \equiv (\Gamma_{16})_L \times (D_{5,1})_R . \quad (6.26)$$

We cannot strongly enough emphasize that this does *not* necessarily represent a direct product decomposition! As (6.26) denotes only the root sublattices, there may still exist non-trivial correlations between the conjugacy classes of Γ_{16} and $D_{5,1}$. In order to characterize (6.26), let us consider (in analogy to (6.14)) the operator product of two *closed string* vertex operators (5.42):

$$V_w(\bar{z},z)\cdot V_{w'}(\bar{w},w) = (\bar{z}-\bar{w})^{w_L\cdot w'_L}(z-w)^{\lambda\cdot\lambda'-qq'}\, V_{w+w'}(w) + \ \ldots \ ,$$

$$V_w(\bar{z},z) = e^{iw_L^I X^I}(\bar{z})\, e^{i\lambda\cdot H}\, e^{iq\phi}(z), \qquad w = (w_L; w_R), \qquad w_R = (\lambda \mid q) . \quad (6.27)$$

Upon rewriting the z- and w-dependent part using $(\bar{z}-\bar{w}) = \frac{|z-w|}{(z-w)}$, the condition for locality, *i.e.*, the absence of branch cuts becomes

$$< w, w' > \in \mathbf{Z} , \quad (6.28)$$

where the inner product has metric $\left[(-)^{16};(+)^5 \mid (-)\right]$. Thus, $\Gamma_{16;5,1}$ has to be an integral, lorentzian lattice with respect to this metric. Note however that the two sets of $(-)$-signs have completely different origins: the $(-)$-sign in the left-moving sector is the same as that arising in the lattice of a compactified closed bosonic string, as discussed in chapter 4: w_L and w_R play the role of p_L and p_R in sect.4.2, respectively.

The full partition function of such a theory has the form

$$\mathrm{Tr}\left[e^{i\pi\tau H'_R}\, e^{-i\pi\bar{\tau}H'_L}\, e^{2\pi i F^\phi}\right] = \frac{1}{\eta^{12}\bar{\eta}^{24}} \sum_{\substack{w=(w_L;w_R)\in\Gamma_{16;5,1}\\ w_R=(\lambda|q)=(u,x_0)}} e^{-i\pi\bar{\tau}w_L^2}\, e^{i\pi\tau u^2}\, e^{2\pi i q} \quad (6.29)$$

Modular invariance, as we will see, forces again this lattice to be (odd) self-dual. This does certainly not imply that Γ_{16} and $D_{5,1}$ should separately be self-dual lattices. That would correspond only to the simplest cases: taking in the right-moving sector the odd self-dual lattice $D_{5,1}$ corresponding to the superstring, requires Γ_{16} to be even self-dual. As explained in chapter 3, there exist only two such lattices, namely the root lattice of $E_8 \times E_8$ and the weight lattice of $\frac{Spin(32)}{Z_2}$. These choices lead to the well-known supersymmetric string theories with gauge groups $E_8 \times E_8$ and $SO(32)$ (the vertex operators for the emission of the corresponding gauge bosons are accordingly given by $V^{ab}(\bar{z},z) = J^{ab}(\bar{z})V^{vector}_{B(-1)}$).

Allowing for a non-trivial correlation between conjugacy classes, the two factors in (6.26) need not be separately self-dual; only the combined lattice needs to be. This leads to further possibilities of constructing theories like the non-supersymmetric $SO(16) \times SO(16)$-string [49] [50]: it is given by a lattice $\Gamma_{16;5,1} = (D_8 D_8)_L (D_{5,1})_R$ equipped with a certain list of correlated conjugacy classes. In this list, *all* conjugacy classes of $D_{5,1}$ of the spinning string do occur; however, the problems of the spinning string do not show up due to the particular, non-trivial interrelation with the left-moving sector. In particular, all branch cuts are cancelled by virtue of (6.28).

The above construction generalizes straightforwardly to the torus-compactified heterotic strings as introduced by [4]. In reducing the number of dimensions, the lattice gets larger as we include also the $2n$ torus-compactified left- and right-moving space-time coordinates. More specifically, in $d = 10 - 2n$ dimensions we need to have an odd self-dual lorentzian lattice

$$\Gamma_{16+2n;5+2n,1} = \Gamma_{16+2n;2n} \times D_{5,1} \ . \tag{6.30}$$

Here, $\Gamma_{16+2n;2n}$ is an even, lorentzian self-dual lattice as described by [4].There exists a continuous variety of such lattices: for example, four dimensional theories are characterized by 132 parameters. These are related to background values of the various fields, precisely as in the case of the torus-compactified bosonic string described in chapter 4. As $D_{5,1}$ in (6.30) is odd self-dual, it represents the torus-compactified fermionic degrees of the superstring. Therefore this kind of theories is non-chiral.

It is now clear how to construct more general, possibly chiral theories: we simply have to look out for odd self-dual lattices (6.30), where the two factors are correlated in a non-trivial way. These theories then will not, in general, be describable as compactifications in the very sense. What we have to make sure of, of course, is that we correctly assign even and odd vectors to space-time tensors and spinors; in lower dimensional theories, there are then also additional constraints related to world-sheet supersymmetry in the internal sector (this will be explained below in chapters 7 and 8). The structure of the covariant lattice is more general than that of (6.30):

$$\Gamma_{16+2n;5+2n,1} = (\Gamma_{16+2n})_L \times \left(\Gamma_{3n}^{int} D_{5-n,1}^{spacetime}\right)_R \tag{6.31}$$

Having found this simple structure, we can forget now all subtleties related to the superconformal ghost system.

What remains to be proven is the connection between (multiloop) modular invariance of the partition function and self-duality of the lattice (6.31). Before this will be attacked, let us first explain and switch to an alternative lattice formulation which is easier to deal with.

Although the *covariant lattices* introduced in the previous section specify the fermionic string theory completely, they are not very pleasant to deal with, because of their rather cumbersome metric, and because some of their vectors have odd norm. To recapitulate, what we will be interested in for string theories in $d = 10$ dimensions are odd self-dual lattices of the form $\Gamma_{16;5,1}$, where the six right-moving components belong to one of the four conjugacy classes of $D_{5,1}$. Furthermore the space-time spinors (*i.e.* spinor conjugacy classes of $D_{5;1}$) have to correspond to the vectors of odd length.

Now we will introduce a map of any such lattice to an *even* self-dual lattice. This map will have the useful property that it preserves modular invariance at all loop orders. Essentially, what we are going to do is to associate with any heterotic string theory (or type-II theory) a bosonic string theory, by mapping the right (or left) moving superconformal system to a bosonic one. This map works in general only in this direction.

This idea originates from [51], where it was used to argue that the fermionic strings are "vacua" of the bosonic string. Although this remains an attractive idea, this is not our interest here. For us the lattice map will merely be a useful tool for proving certain properties of fermionic strings. The form of the lattice map that we will use appeared first in [52], and in [44] it was shown that it preserves one-loop modular invariance. This was extended to higher loops in [53].

Basically, the idea is very simple. If we decompose one of the lattices of interest with respect to the $D_{5,1}$ conjugacy classes, we see that it consists of four sets of vectors

$$(\Delta_0, 0) + (\Delta_v, v) + (\Delta_s, s) + (\Delta_c, c), \tag{6.32}$$

where the second entry denotes the $D_{5,1}$ conjugacy classes, and the Δ's are simply the 16-dimensional vectors associated with them on the right lattice. Now define a new lattice $\Gamma_{16;8}$ by writing down a decomposition identical to (6.32), but with the second entry interpreted as a conjugacy class of D_8 instead of $D_{5,1}$. That is, we map

$$D_{5,1} = D_4^{light-cone} D_{1,1}^{ghost} \longleftrightarrow D_4^{light-cone} D_4^{ghost} = D_8 \ . \tag{6.33}$$

Observe that the old and the new lattice have the same number of conjugacy classes, so that if the old lattice was unimodular, so is the new one. Now compute the list of norms and inner products of the conjugacy classes of D_8. In comparison with (6.22) one finds that the only change is in the norm of the spinors, which change from odd into even length. But the spinors of $D_{5,1}$ were precisely the odd vectors on $\Gamma_{16;5,1}$. Hence the map preserves unimodularity and all mutual inner products, but it maps all the odd vectors into even ones. The result is a lattice $\Gamma_{16;8}$ which is *even* self-dual with respect to the metric $\left[(-)^{16};(+)^{8}\right]$. Thus, by use of this one-to-one map, we are able to find an alternative *even lattice formulation* of heterotic string theories.

If we extend the lattice with an E_8 factor on the right, we get a $\Gamma_{16;16}$ lattice of that type, on which the bosonic string can be compactified (this can also be achieved by using D_{16} in the above construction, but that is less useful for discussing the spectrum). The bosonic string compactified on the corresponding torus is all-loop modular invariant. Is this related to the modular invariance of the fermionic string from which it came?

To answer this question, one has to write down the multiloop partition function for the fermionic string, which has not been done yet. However, one usually assumes that as far as its modular transformation properties are concerned, it has the form (including only right-movers)

$$(DetD_{1/2})^5_\alpha (DetD_{3/2})^{-1}_\alpha,$$

where $D_{1/2}$ denotes the Dirac determinant, and represents the contribution of *two* Majorana Weyl fermions and $D_{3/2}$ is the Rarita-Schwinger determinant (which only has ghost contributions). The subscript α denotes the spin-structure. In general, this combination of determinants is multiplied with some α dependent factors from the right-movers.

The point is now, that the modular transformation properties of this object are known explicitly [54]. Using them, we can show that the following two spin-structure dependent quantities have identical modular transformations at any loop order

$$\begin{aligned} Y_\alpha &= \left(\frac{Det_{1/2}}{Det_{3/2}} \right)_\alpha \\ X_\alpha &= (Det_{1/2})^4_\alpha \sum_\beta (Det_{1/2})^8_\beta \end{aligned} \tag{6.34}$$

Thus if one replaces in a fermionic string partition function Y_α by X_α, the new one will be modular invariant if and only if the old one was.

This replacement is precisely what the lattice map achieves. To see that one has to make a simple basis transformation from spin-structure basis to conjugacy class basis. We will not explain this in detail here, but refer to [53]. Details aside, comparing Y_α and X_α the reader wil undoubtedly recognize that the first represents the $D_{1,1}$ ghost and longitudinal part of the covariant lattice (cf. (6.20)), while the second represents D_4 conjugacy classes, multiplied with the E_8 partition function.

Let us apply the above to the the superstring. As discussed in sect.6.2, the superstring is characterized by a $D_{5,1}$-lattice with conjugacy classes (0) and (s). By use of (6.33), this can be mapped onto D_8 with the same conjugacy classes, which is the same as E_8! Thus, (odd) self-duality of $D_{5,1}^{superstring}$ can be inferred from the (even) self-duality of E_8. This proves not only multiloop modular invariance of the superstring, but it also shows that (6.23) is the only projection leading to a modular invariant theory, as E_8 is the unique even self-dual lattice in 8 dimensions.

The beauty of the above reasoning lies in the fact that the ghost system is determined by the $N = 1$ conformal system, *i.e.* by a *world-sheet* property, that is the same for any fermionic string. Hence the replacement of Y by X, or the lattice map, is equally useful for any fermionic string theory, and in any space-time dimension.

So we might as well directly go to four-dimensional strings. As indicated at the end of the previous section, four-dimensional strings are specified by lattices of the form given in the left-hand side of (6.30), but without the restriction that it contains a $D_{5,1}$ root lattice. Since we are in four dimensions, it would be foolish to insist on ten-dimensional Lorentz-invariance. Instead, we only need to require that $D_{2,1}$ roots be present (the Wick-rotated four-dimensional Lorentz-group is $SO(4)$, with Lie algebra D_2). Thus in $10-2n$ dimensions we would be looking for lattices $\Gamma_{16+2n;5+2n,1}$ of the form (6.31), with the requirement that the last $6-n$ components belong to conjugacy classes of $D_{5-n,1}^{spacetime}$. We will again require that the lattice is odd self-dual, with odd lengths for spinors.

Using precisely the same lattice map, we replace now the $D_{5-n,1}^{spacetime}$ conjugacy classes by $D_{8-n}^{spacetime}$ conjugacy classes, that is

$$\begin{aligned}\Gamma_{16+2n;5+2n,1} &= (\Gamma_{16+2n})_L\left(\Gamma_{3n}^{int}D_{5-n,1}^{spacetime}\right)_R \\ &\longleftrightarrow (\Gamma_{16+2n})_L\left(\Gamma_{3n}^{int}D_{8-n}^{spacetime}\right)_R = \Gamma_{16+2n;8+2n}\ .\end{aligned} \tag{6.35}$$

Using the above we conclude then that the following statements are equivalent:

— The covariant lattice $\Gamma_{16+2n;5+2n,1}$ is odd self-dual (with odd norms for the spinors, as discussed above).

— The lattice $\Gamma_{16+2n;8+2n}\times E_8$ obtained by means of the lattice map is even self-dual.

— The bosonic string, compactified on the latter lattice is multi-loop modular invariant.

— The fermionic string defined by the odd self-dual covariant lattice is multi-loop modular invariant.

This is all that needs to be said about modular invariance. What corresponds to complicated spin-structure summations in fermionic constructions, is encoded in the simple statement that the lattice must be self-dual.

To some extent, we can drop the covariant lattice altogether, and work with even self-dual lattices. This is unfortunately not possible if one wants to study scattering amplitudes. For this purpose one needs the complete machinery of conformal field theory, and we have not succeeded to cast that into even lattice form. However, the even lattice formulation is extremely useful for discussing the spectrum.

Notice the map (6.35) above is a map between *lattices* (even with different dimensions) and *not* a map between individual vectors. The physical light-cone spectrum is obtained by applying a set of truncation rules to the bosonic string spectrum, which are designed to mimic exactly the corresponding procedure for the covariant, odd lattice. To get these rules, consider the physical state selection rule (6.21). It tell us to keep only states with special components in the $D_{1,1}$ sublattice, belonging to conjugacy classes (v) and (s) of that lattice. Since $D_{1,1}$ is mapped to D_4 by the lattice map, we take the identical conjugacy classes of D_4 to identify the physical states. Thus one physical state is represented by an entire conjugacy class of D_4. One can choose a representative of the D_4 conjugacy class that makes all this even simpler. Namely, if one chooses some weight vector of 8_s and 8_v to represent the spinor and vector conjugacy class, then the mass of the light-cone states is correctly given by the compactified bosonic string mass formula (4.10), in which one should include the contribution from the D_4^{ghost} components. In other words, if we make an (arbitrary) choice for the 8_v and 8_s weights, the physical state selection rule (6.21) for covariant lattices is replaced by

$$x = x_0 = \begin{cases} (\ 0,0,0,\ 1) & NS\text{-sector} \\ (\frac{1}{2},\frac{1}{2},\frac{1}{2},\frac{1}{2}) & R\text{-sector} \end{cases} \tag{6.36}$$

in the even lattice formulation ($w = (u,x) \in D_{8-n}^{spacetime}$, $u \in D_{4-n}^{light-cone}$, $x \in D_4^{ghost}$). The physical light-cone spectra are identical.

The above rules were first discovered in [51] and [52], and were a completely *ad hoc* prescription for getting the fermionic string from the bosonic one. Now we see that they do in fact follow from the physical state projection rule explained in sect. 6.1.

7. World-sheet supersymmetry

Up to now we have freely replaced world-sheet fermion by bosons. One has to be a little bit more careful about this, because one may endanger world-sheet supersymmetry. In the *NSR* sector of heterotic strings world-sheet supersymmetry is a manifest (local) $N = 1$ supersymmetry, which transforms the fermions ψ^μ into the bosons X^μ and vice-versa. The transformation is

$$\begin{aligned} \delta_\epsilon \psi^\mu &= i\epsilon\partial X^\mu \\ \delta_\epsilon X^\mu &= -i\epsilon\psi^\mu \end{aligned} \tag{7.1}$$

where ϵ is an infinitesimal grassmannian parameter. This transformation is generated by the operator

$$\mathcal{O} = \oint \frac{dz}{2\pi i} i\epsilon\psi^\mu \partial X_\nu = \oint \frac{dz}{2\pi i} \epsilon T_F \ , \tag{7.2}$$

in other words, the supersymmetry transformation of any conformal field $\Psi(z)$ is obtained by acting with $\mathcal{O}$

$$\delta_\epsilon \Psi = \mathcal{O}\Psi.$$

One may check that on ψ^μ and X^μ, $\mathcal{O}$ does indeed produce (7.1) .

Replacing ψ^μ by bosonic fields does not in itself destroy world-sheet supersymmetry. In two dimensions one can realize supersymmetry entirely in terms of bosons. Indeed, by simply replacing ψ^μ in (7.2) by its bosonized form (5.32) one obtains a generator of world-sheet supersymmetry which is expressed entirely in terms of d bosons X^μ (where d is the number of space-time dimensions) and $\frac{d}{2}$ bosons H_i replacing the fermions.

In constructing four-dimensional strings we do however something more drastic than just bosonization. We allow the bosons related to the internal world-sheet fermions to mix freely with the internal bosons (we will use the word "internal" to indicate those components of ψ and X that do not have space-time indices). Since our philosophy is to construct new string theories directly in four dimensions, we start always with a completely bosonic formulation, *i.e.* in practice we never bosonize anything. Therefore it is not automatically guaranteed that the result respects world-sheet supersymmetry.

Why do we need world-sheet supersymmetry? One could answer this question by stating that world-sheet supersymmetry is an essential part of superconformal field theory, which is the foundation of fermionic string theory, but there is a more explicit way of seeing what goes wrong if one does not have it.

One thing that goes wrong is Lorentz-invariance. World-sheet supersymmetry is used to remove unphysical degrees of freedom from the covariant action of fermionic strings, so as to write them entirely in terms of the physical, light-cone degrees of freedom. Although one can always write down such a light-cone action, it is important that it came from a covariant action. This makes it possible to write down a Lorentz-algebra which does not just generate the transverse rotations of the light-cone states, but extends to the full Lorentz-algebra. This extension involves the familiar generators L^{i-}, whose commutators have to be checked to see if the algebra closes. In supersymmetric strings one of the terms appearing in the expression for L^{i-} is $\oint \psi^i T_F$.

One of the things that can go wrong with Lorentz invariance is that $[L^{i-}, L^{j-}]$ simply does not yield the answer required by the algebra of the Lorentz group. This situation occurs in theories with a conformal anomaly, and can be avoided by choosing the field content so that the conformal anomaly cancels. This is one of the oldest ways of obtaining the critical dimension.

It might appear that nothing prevents us from simply using the Lorentz algebra of ten-dimensional strings, regarding 6 of the eight 8 transverse Lorentz indices as internal. This algebra closes, and it is indeed a valid choice in torus compactifications. However, closure of the algebra is not enough. In addition, one has to require that L^{i-} takes physical states into physical states. Otherwise one faces a problem which is even worse than the conformal anomaly, namely the states of the theory won't even fit into representations of the Lorentz group. (In the presence of a conformal anomaly, the Lorentz algebra does not close, but for the bosonic string and the *NSR* formulation of the superstring it is still possible to assemble the massive states into representations of $SO(d-1)$, the little group of massive particles)

This is indeed precisely what happens, and one sees this most easily in the chiral sector of chiral string theories. At first sight, chirality and string theory are almost contradictory, because one would think that chiral states can always be excited into massive states. Massive chiral states would however violate Lorentz invariance, so a really clever mechanism is needed to ensure that all massive states combine into non-chiral pairs, while the massless ones do not. This is achieved by world-sheet supersymmetry. Indeed, this restriction on massive chiral states turned out to be powerful enough to determine all the conditions of world-sheet supersymmetry for the class of theories considered in [6].

A second way in which world-sheet supersymmetry enters is through the picture changing operator. This operator has the form $\mathcal{P}_{+1} = e^{i\phi} T_F$, where ϕ is the bosonized ghost of section (6.1). It is essential for the calculation of scattering amplitudes, and it too should take physical states into physical states, and have a well-defined action.

The problem of finding a bosonic realization of world-sheet supersymmetry consists thus of two parts

1. Find a purely bosonic expression for T_F that satisfies the supersymmetry algebra.
2. Make sure that it acts properly on all states.

The algebra that the supercurrent T_F must satisfy can be encoded in operator products, and reads (with finite terms omitted)

$$\begin{aligned} T(z)T_F(w) &= \frac{3/2}{(z-w)^2}T_F(w) + \frac{1}{(z-w)^2}\partial_w T_F(w) \\ T_F(z)T_F(w) &= \frac{10}{(z-w)^3} + \frac{2}{(z-w)}T(w) \end{aligned} \tag{7.3}$$

The first of these equations states that it must have conformal weight $\frac{3}{2}$. The second one is just (5.18) , with $c = 15$, the conformal anomaly of 10 bosons and 10 Majorana-Weyl fermions. The form of the supercurrent is further restricted because we want it to be Lorentz invariant in d dimensions. This means that it must have the form

$$T_F = \psi^\mu \partial X_\mu + T_F^{int},$$

where the first term is just (5.17) , and the second is some expression of conformal weight $\frac{3}{2}$, constructed out of $3n$ bosons. Here n is defined by $d = 10 - 2n$. The $3n$ bosons consist of $2n$ internal X's and n bosons H_i representing the ψ's as in (5.34). From now on, let us denote *all* compact bosons generically by X. Subtracting the space-time part of the supercurrent in (7.3) , we see that T_F^{int} has to satisfy it with "10" replaced by $10 - d$, and $T(w)$ by $T(w)^{int}$, the energy momentum tensor of $3n$ bosons.

The first equation is quite easy to satisfy. The general solution has the form

$$T_F^{int} = \sum_{\vec{l}} \vec{B}(\vec{l})\cdot\partial\vec{X}\, e^{i\vec{l}\cdot\vec{X}} + \sum_{\vec{t}} A(\vec{t})\, e^{i\vec{t}\cdot\vec{X}}, \tag{7.4}$$

where $\vec{l}^2 = 1, \vec{t}^2 = 3$, and $\vec{l}\cdot\vec{B} = 0$. One can verify this by substituting (7.4) into the first equation of (7.3) , using the standard energy momentum tensor for $3n$ bosons (see (5.11)).

The second equation in (7.3) imposes more complicated conditions on A and $\vec{B}$, and we do not know its general solution. However, as we will show in a moment, we do know many special solutions. First however we want to address the second point mentioned above, the action on T_F on the states in the theory.

If an operator of the form $\exp(i\vec{w}'\cdot\vec{X})$ acts on a state $|\vec{w}>$ with lattice momentum $\vec{w}$, then it produces a state with momentum $\vec{w} + \vec{w}'$. Earlier in this section we have identified two operators whose action should take physical states into physical states, namely L^{i-} and $\mathcal{P}_{+1}$. If we write the right-moving part of the vectors of the covariant lattice in the form

$$w_R = (\ \textit{internal components},\ \ \textit{weight of } D_{d/2} \mid \textit{ghost charge})\ ,$$

according to the decomposition

$$\Gamma_{16+2n;5+2n,1} = (\Gamma_{16+2n})_L \times \left(\Gamma_{3n}^{int} D_{5\ \ n,1}^{spacetime}\right)_R, \qquad D_{5\ \ n,1}^{spacetime} = D_{d/2}^{lorentz}\Gamma_1^{ghost},$$

then we see that the internal part of the supercurrent contains exponentials of the above form, with $\vec{w}'$ equal to

$$w^{L^{i-}} = (\vec{l}, \vec{v} \mid 0) \text{ or } (\vec{t}, \vec{v} \mid 0) \qquad \text{for } L^{i-}$$
$$w^{PC} = (\vec{l}, 0 \mid -1) \text{ or } (\vec{t}, 0 \mid -1) \qquad \text{for } \mathcal{P}_{+1}$$

($\vec{v}$ denotes a vector weight of $D_{d/2}$.) If adding a vector $\vec{w}'$ to a lattice vector should give another lattice vector, then $\vec{w}'$ should lie on the lattice itself. For the picture changing operator one should furthermore require that its action on all states is *local*, *i.e.* does not introduce any branch cuts. In other words, if $e^{i\vec{w}\cdot\vec{X}}$ is the vertex operator for some physical state with lattice momentum $\vec{w}$, then we should require that the operator product

$$e^{i\vec{w}'\cdot\vec{X}}e^{i\vec{w}\cdot\vec{X}} = (z-w)^{\vec{w}'\cdot\vec{w}}e^{i(\vec{w}'+\vec{w})\cdot\vec{X}} + \ldots$$

does not contain any fractional powers of $(z-w)$. This implies that $\vec{w}'$ must have integral dot product with all lattice vectors, so that it must lie on the *dual* of the lattice. Fortunately our lattices are self-dual, so that these requirements on $\vec{w}'$ are compatible.

The requirements one gets from L^{i-} and $\mathcal{P}_{+1}$ are also compatible with each other, because in covariant lattices the $D_{d/2}$ components and the ghost charge belong to the same conjugacy class of $D_{d/2,1}$. What we conclude from this is that for every vector $\vec{l}$ and $\vec{t}$ appearing in T_F^{int} the lattice must contain vectors of the form $w_R = (\vec{l}, (v))$ and $(\vec{t}, (v))$, where (v) denotes generically the vector conjugacy class of $D_{d/2,1}$. This describes of course only the right-moving part of the lattice. The left components w_L of these vectors, which we will call *constraint vectors* in the following, must vanish.

There is yet another way to arrive at this conclusion, which we mention here because it uses more directly the requirement of supersymmetry, and can also be used to identify other world-sheet supersymmetries on the lattice. What these conditions are imposing on the theory is that the supersymmetry transformation not only respect the *local* properties of the fields, but also, on Riemann surfaces with non-trivial topology, their *global* properties. In (7.1) that would amount to requiring that both sides of the equation have the same periodicity. In the superstring, for which one only has (7.1) , this is simply achieved by choosing the boundary conditions for ϵ equal to those of the *NSR* fermions on any Riemann surface with given spin structure. In bosonic language choosing a spin-structure is replaced by choosing a *soliton sector*, which simply means to consider vacuum with a certain lattice momentum vector $\vec{w}$. States in this sector which are created by vertex operators of the form $\exp(i\vec{w}'\cdot\vec{X})$ pick up a phase $\exp(2\pi i\vec{w}'\cdot\vec{w})$ if one takes σ_1 to $\sigma_1+\pi$. When one considers one term of the supercurrent acting on it, then such a state is mapped into a state which picks up a different phase when moved around a closed loop. This phase change is $\exp(2\pi i\vec{p}\cdot\vec{w} + 2\pi i\phi(\vec{w}))$, where we have denoted the phase of ϵ in the sector $\vec{w}$ as $\phi(\vec{w})$. The vector $\vec{p}$ appearing here represents just one term in the supercurrent (*i.e.* it is one of the vectors $\vec{l}$ or $\vec{t}$ used above). If there were just one term, one can always adapt $\phi(\vec{w})$ to cancel the phase, so that supersymmetry respects all boundary conditions. If the supercurrent has a second term with a vector $\vec{p}'$, then this cancellation of the phase by $\phi(\vec{w})$ can work if and only if $\vec{p}-\vec{p}'$ has integer dot-product with any $\vec{w}$ on the lattice. But then it must lie on the lattice itself.

This argument tells us thus that the *difference* of any two vectors appearing in the supercurrent must lie on the lattice. This includes also the space-time part of the supercurrent. For that part the vector $\vec{p}$ belongs to the vector conjugacy class of $D_{d/2}$, so that we arrive at the same conclusions mentioned earlier.

There are two simple ways of obtaining supercurrents that satisfy the second equation of (7.3) . The first is to construct one out of free fermions, and bosonizing it. Such a free fermion supercurrent has the form

$$T_F(z) = \frac{1}{6} f_{abc} \psi^a \psi^b \psi^c, \tag{7.5}$$

where the $\psi's$ are $6n$ fermions related to the $3n$ internal bosons by bosonization. If f_{abc} are the structure constants of a semi-simple Lie algebra, then one can check that (7.3) is satisfied [55]. The generic choice for this Lie algebra (which has to be of dimension $6n$) is $SU(2)^{2n}$. In dimensions below five there are additional solutions, which are of some interest for type-II strings, but not for chiral heterotic strings. To obtain a supercurrent of the form (7.4) one simply bosonizes the $6n$ fermions into $3n$ bosons using (5.32) . There is some freedom in doing that, because on may associate the fermions in different ways with the $6n$ $SU(2)$ generators, but there is one especially simple way which yields only norm 3 vectors in the supercurrent. This is to take

$$T_F = N \sum_{i=1}^{n} [\psi_i^1 \psi_i^2 \psi_i^3 + \psi_i^{-1} \psi_i^{-2} \psi_i^{-3}]$$

where the notation is as in (5.32)). We leave it as an exercise to determine the normalization constant N. The vectors $\vec{t}$ one gets from this supercurrent are, for $n = 1$,

$$\vec{t} = (\pm 1, \pm 1, \pm 1), \tag{7.6}$$

For larger n one just adds sets of 8 such vectors in orthogonal three-dimensional subspaces, so that altogether one has $8n$ vectors. This supercurrent was first used by Kawai et. al., and led to their "triplet constraint".

Although one might think that all supercurrents that satisfy (7.3) can be written as a trilinear expression of free fermions, this is not the case. A counterexample, found in [56] and [57] has vectors $\vec{t}$ of the form (for $n = 1$)

$$\begin{aligned} \vec{t}_1 &= (\sqrt{3}, 0, 0) \\ \vec{t}_2 &= (0, \sqrt{3}, 0) \, . \\ \vec{t}_3 &= (0, 0, \sqrt{3}) \end{aligned} \tag{7.7}$$

This pattern can be repeated n times if $n > 1$. By bosonizing the free fermionic supercurrent, one can only get vectors of the form $(\pm 1, \pm 1, \pm 1, 0, \ldots, 0)$, *i.e.*, there must exist a basis in which three entries are ± 1, and the others zero. For one norm 3 vector one can always achieve that, but not necessarily simultaneously for several vectors. In particular it is not possible for (7.7) .

The mistaken impression that any supercurrent can be written in terms of free fermion trilinears, is based on the proof of [58] that the Fock space of any torus-compactified theory can always be written in terms of a fermionic representation. This may involve either fermions with Thirring interactions (to which the above does not apply, since one would not have a *free* fermion supercurrent), or even free fermions with complicated boundary conditions. Now of course a bosonic theory described in this fermionic way will still have world-sheet supersymmetry, since it is equivalent. But one cannot conclude from this argument that there must always exist a representation of the supercurrent of the trilinear form (7.5) .

An illustrative example is a string theory on an E_8 lattice. The Fock space can be described either by means of fermionic oscillators with certain GSO projections, or with bosonic oscillators and lattice momenta. The Frenkel-Kac generators of the D_8 subalgebra of E_8 can also be described both bosonically and fermionically, the latter being fermion bilinears. But whereas the bosonic description extends trivially to the remaining spinor roots of $E_8/SO(16)$, the fermionic one does not. To describe these generators one would need a *transcendental* expression in terms of the fermions rather than a bilinear one (see the discussion in section (5.4)). We believe that the same applies to the supercurrent, with "bilinear" replaced by "trilinear".

Finally we mention a second method for generating supercurrents, which owes its inspiration to asymmetric orbifolds. In this case one starts by taking the ten-dimensional form also in the internal space, *i.e.*

$$T_F^{int} = \sum_{I=1}^{2n} \psi^I \partial X_I .$$

The fermions can be bosonized trivialy. Although the rest of the supercurrent is already bosonic, we can try to re-express the factors ∂X as sums of exponentials of the form $\exp(i\vec{r}\cdot\vec{X})$, where $\vec{r}^2 = 2$. These expressions are constrained by the fact that they must satisfy the same operator products as the ∂X's. An equivalent way of saying this is that they must be the generators of a $U(1)^{2n}$ algebra. One can find such subalgebras by rotating the ∂X's into the root system of a Lie algebra. One way of finding such solutions is to use the Weyl rotations of a Lie algebra, and apply the conjugation that maps twists into shifts. We will not discuss this in more detail here; for details we refer to [59]. The solutions already given above can be obtained in this way (this has been shown in [57]) , as well as many others, listed in [59].

8. The Spectrum

To discuss the spectrum of covariant lattice theories we will use the even lattice formulation. It may be useful to summarize the requirements that we have derived in the foregoing chapters. To specify a covariant lattice theory in $d = 10 - 2n$ dimensions one must provide

— A lattice $\Gamma_{16+2n;8+2n}$ with left- and right dimensions as indicated

— A supercurrent that satisfies the $N = 1$ supersymmetry algebra, and is consistent with the lattice.

The supercurrent has the general form

$$T_F = \psi^\mu \partial X_\mu + T_F^{int}$$

where the internal part appears only below 10 dimensions. The fermions ψ^μ can be bosonized, and the momenta of the resulting bosons occupy $\frac{d}{2} = 5 - n$ of the $8 + 2n$ dimensions of the right part of the lattice. The internal part of the supercurrent is entirely bosonic, and has the form

$$T_F^{int} = \sum_{\vec{t}} A(\vec{t})\, e^{i\vec{t}\cdot\vec{X}} + \sum_{\vec{l}} \vec{B}(\vec{l})\cdot\partial\vec{X}\, e^{i\vec{l}\cdot\vec{X}},$$

where $\vec{t}^2 = 3$, $\vec{l}^2 = 1$ and $\vec{l}\cdot\vec{B} = 0$, and the coefficients satisfy the conditions discussed in the previous chapter. The momenta of the bosons $\vec{X}$ occupy $3n$ dimensions of the right part of the lattice. The remaining 3 components of the right lattice are related, in a somewhat indirect way, to the β, γ ghost system, as explained in chapter 6.

The lattice must satisfy the following requirements.

1. It must be even and self-dual with respect to a Lorentzian metric of the form $\mathrm{diag}((-)^{16+2n};(+)^{8+2n})$
2. The last $8 - n$ components of any vector must belong to one of the four conjugacy classes $(0), (v), (s)$ or (c) of the Lie algebra $D_{8-n}^{spacetime}$.
3. It must contain all vectors of the form $\vec{w}_L = 0, \vec{w}_R = (\vec{l}, (v))$ and $(\vec{t}, (v))$ for all $\vec{l}$ and $\vec{t}$ appearing in the supercurrent.

The last condition applies obviously only to dimensions less than 10.

As explained before, not all the states that one obtains from such a lattice are physical. The physical states are obtained by making a decomposition of D_{8-n} to its subalgebra $D_{4-n} \times D_4$. One keeps only states of which the last four components are fundamental vector or spinor weights of D_4, and counts them with multiplicity one.

This decomposition of $D_{8-n}^{spacetime} \rightarrow D_{4-n}^{light-cone} \times D_4^{ghost}$ is very simple. The branching rules are as follows:

$$(0) \rightarrow (0,0) + (v,v)$$
$$(v) \rightarrow (v,0) + (0,v)$$
$$(s) \rightarrow (c,c) + (s,s)$$
$$(c) \rightarrow (s,c) + (c,s)$$

We have written them in such a way that the second term is the one that survives the physical state projection (6.36), because it has as its second ($= D_4$) entry (v) or (s). Of course we could have chosen the other spinor weight of D_4 as the one that gives physical states, but this merely amounts to a chirality flip of the entire theory.

Considering now the first entry, we can read off the spin of the particles in each conjugacy class of D_{8-n}. The first entry corresponds to D_{4-n}, which is the transverse (Wick-rotated) Lorentz group in $10-2n$ dimensions. We see that the following general rule holds for D_{8-n} conjugacy classes

$$\begin{aligned} (0) &\rightarrow \text{ odd rank tensors} \\ (v) &\rightarrow \text{ even rank tensors} \\ (s) &\rightarrow \text{ spinors of chirality } + \\ (c) &\rightarrow \text{ spinors of chirality } - \end{aligned} \tag{8.1}$$

The assignment of positive or negative chirality to the spinors is a matter of convention.

This list gives us the Lorentz representations of the pure lattice states in light-cone gauge. These are states that have a certain lattice momentum, but no bosonic oscillator excitations. The complete Fock space of physical states is obtained as follows

$$(\alpha\text{-oscillators})(\tilde{\alpha}\text{-oscillators})|\vec{w}_L;\vec{w}_R> |_{m_L=m_R}.$$

Here α denotes all 24 left-moving oscillators, of which $d-2$ have a vector index of the transverse Lorentz group and will be denoted α^i. The remaining, "internal" ones will be denoted $\alpha^I, I = 1,\ldots,26-d$.

For the $\tilde{\alpha}$ oscillators things are a bit more complicated. One has $d-2 = 8-2n$ right-moving oscillators $\tilde{\alpha}^i$ in the vector representation of the transverse Lorentz group, which are simply the right-moving modes of the uncompactified coordinates. Then there are the oscillators belonging to the lattice degrees of freedom. Four of the $8+2n$ right-moving lattice dimensions are associated with ghosts and longitudinal modes, and the corresponding four bosonic oscillators should not be included; the remaining $2n+4$ oscillators ($3n$ internal ones and $4-n$ in the Cartan subalgebra of $D_{4-n}^{light-cone}$) should be allowed to act on the pure lattice states. They do not alter the Lorentz representation of the state on which they act. Obviously the $d-2$ left and right-moving bosonic oscillators α^i and $\tilde{\alpha}^i$ *do* change the Lorentz representation. Thus, as we already pointed out, (8.1) is valid only for pure lattice states.

The last fact one has to know about the spectrum is the mass formula. This is extremely simple: It is just the mass formula of the compactified bosonic string:

$$\begin{aligned} \frac{1}{8}m_L^2 &= \frac{1}{2}w_L^2 + \mathcal{N} - 1 \\ \frac{1}{8}m_R^2 &= \frac{1}{2}w_R^2 + \tilde{\mathcal{N}} - 1 \end{aligned}, \tag{8.2}$$

Here $(\vec{w}_L;\vec{w}_R)$ denotes a vector on $\Gamma_{16+2n;8+2n}$. Notice that the entire vector contributes to the mass formula, including the D_4 components. Since every physical state necessarily is a fundamental vector or spinor of this D_4, the lowest possible value of $\frac{1}{8}m_R^2$ is $-\frac{1}{2}$.

It follows that if we are only interested in massless states and tachyons, then we can ignore the right-moving oscillators, since they would increase the mass by too much. In the following we will use $M_R(M_L)$ to denote $\frac{1}{8}m_R^2$ ($\frac{1}{8}m_L^2$), and we will refer to this quantity simply as "mass". Thus we will quote masses in units in which the bosonic string tachyon has mass -1.

8.1 TEN-DIMENSIONAL STRINGS

Now we have all the ingredients necessary to determine the spectrum. Since the ten-dimensional case is somewhat simpler, we will consider it first. The lowest states one can have in the four conjugacy classes of D_8 are discussed below.

The conjugacy class (0)

The shortest vectors in this conjugacy class are $\vec{w}_R = 0$ and $\vec{w}_R$ equal to a root of D_8. The first cannot satisfy the physical state criterion, so that the physical state with the lowest value of m_R is obtained by taking $\vec{w}_R$ to be a root. Its mass is equal to zero, and according to (8.1) it yields a Lorentz vector. To satisfy the condition $M_L = M_R$ we have to pair it with a massless state from the left sector. There are two ways to achieve that: either by taking $\vec{w}_L^2 = 2$ or by taking $\vec{w}_L = 0$, and acting with one of the 24 bosonic oscillators. Eight of these oscillators are in the vector representation of the transverse Lorentz group $SO(8)$. Its tensor product with the right-moving sector yields the $SO(8)$ representations $(35) \oplus (28) \oplus (1)$, corresponding respectively to the graviton, the antisymmetric tensor B_{ij}, and a dilaton. The remaining 16 bosonic oscillators yield upon multiplication with the right sector 16 vector bosons. These turn out to be gauge bosons of a $U(1)^{16}$ gauge group. If there are any vectors $\vec{w}_L$ of norm 2 that combine on the right lattice with the (0) conjugacy class of D_8, then they will give rise to additional gauge bosons, extending $U(1)^{16}$ to a non-abelian gauge group of rank 16.

Note that if the lattice contains vectors of the form $(\vec{r};(0))$, with $\vec{r}^2 = 2$, then it contains in particular vectors of the form $(\vec{r};0)$. We will call such vectors, and also vectors of the form $(0;\vec{r})$ *roots* of the lorentzian lattice. Notice that all the physical states we have this discussed so far owe their existence to roots: the graviton and the 16 commuting gauge bosons to the roots of D_8, and the remaining gauge bosons to roots of the left lattice. The graviton (and also the 16 abelian gauge bosons) are inevitable, since the D_8 roots are always present.

It can be shown that the roots of the left (right) part of the lattice must always form a root system of some simply laced Lie algebra. In the subspace spanned by these root systems, all lattice vectors must have components that have integer dot products with the root lattice of this Lie algebra. Hence in this subspace they must be weights.

If the roots of the left lattice span an l-dimensional subspace of the 16 left-moving dimensions, then the gauge bosons belong to a non-abelian rank l group, times an abelian group $U(1)^{16-l}$. We will denote this entire group as G_L. All other states in the theory belong to certain representations of this group.

The conjugacy class (v)

The lowest state that can appear here is the vector representation of D_8. If it appears, there is a potential problem, because for this lattice vector $M_R = -\frac{1}{2}$. Because the lattice is even, we know that $\vec{w}_L^2 = 1 \bmod 2$. Should there exist a lattice vector with both $\vec{w}_R^2 = 1$ and $\vec{w}_L^2 = 1$ then the theory would have a tachyon. It might happen that the lowest value of $\vec{w}_L^2$ is in fact 3 rather than 1, and indeed there is one example of that, the $O(16) \times O(16)$ string (see the section 9.1 for a discussion). Furthermore there are lattices on which the conjugacy class (v) is simply not present. This is the case for the supersymmetric strings (see below).

The conjugacy classes (s) and (c)

The lowest possible states are the D_8 spinors, which have $M_R = 0$. As was the case for the conjugacy class (0), they can become physical either by pairing with a left-moving oscillator excitation or with a norm two vector. If the first happens, then in particular it can combine with an oscillator α_i. The left-right tensor product contains then a vector-spinor, usually called a gravitino.

If a gravitino is present, then the lattice state excited by the left-moving oscillators must have $\vec{w}_L = 0$, combined with a D_8 spinor. This lattice vector is a root of the *right* lattice, and by the arguments given above, it follows that the right lattice must have a larger root system than that of D_8. It does not require a lot of thought to realize what it must be, namely E_8. The presence of a gravitino implies that one must obtain supergravity in space-time. Again we see that a local symmetry of the theory is related to a root of the lattice.

Once one realizes that the spectrum has a gravitino, one would clearly expect it to be supersymmetric. For example, one would expect to find gaugino's, the superpartners of the gauge bosons. These are easy to identify: one can excite vectors with right-moving components in the (s) conjugacy class of D_8 in precisely the same way as those in the (0) conjugacy class discussed above. A little bit of thought will make it clear that these are the only massless spinors one can get; in ten-dimensional superstrings all massless fermions are either singlets or adjoints of the gauge group.

Of course one can also have non-supersymmetric strings. In that case, massless fermions can only arise from vectors of the form $(\vec{w}_L; (s))$, with $\vec{w}_L^2 = 2$. It should be emphasized that $\vec{w}_L$ cannot be equal to a root of the right lattice. If it were, one could subtract $(\vec{w}_L; 0)$ from the spinor, and obtain a gravitino by exciting the result with a bosonic oscillator. Thus in non-supersymmetric string theories in ten dimensions one cannot have spinors in the adjoint of the gauge group.

Up to now we have not distinguished the (s) and the (c) conjugacy class of D_8. They give spinors with opposite space-time chirality. Whether the theory is chiral depends on the representations (*i.e.* the weights) with which the two conjugacy classes combine on the left lattice. If these are always the same, the theory is not chiral. Otherwise it is.

After this rather lengthy discussion of ten-dimensional strings, let us now see what changes if one goes to lower dimensions. The main difference is that now there can be additional contributions to M_R from the components of $\vec{w}_R$ which extend beyond the space-time lattice D_{8-n}. For definiteness (and for phenomenological reasons) we will consider here four dimensions. The discussion of 6 and 8 dimensions is not very different, the main difficulty being the mod-8 properties of spinors in different dimensions.

In four dimensions, the lattice is $\Gamma_{22;14}$, and the right part contains a D_5 space-time lattice. Again we will consider each of the conjugacy classes of D_5.

The conjugacy class (0)

Here the discussion can be very brief, because there are almost no changes with respect to the ten-dimensional case. Roots of D_5 again give zero mass gravitons and gauge bosons, this time of course of a rank 22 gauge group. The 9 "internal" components of the right lattice play no role in the massless spectrum, as they would excite the mass above zero.

The conjugacy class (v)

As in ten dimensions, the lowest possible state here is a tachyon of mass $-\frac{1}{2}$. It originates from any lattice vectors of the form $(\vec{w}_L; 0, (v)), \vec{w}_L^2 = 1$, where the middle component refers to the nine internal dimensions of the lattice, according to the decomposition $\Gamma_{22;14} = (\Gamma_{22})_L \times \left(\Gamma_9^{int} D_5^{spacetime}\right)_R$. Clearly we have now the possibility to excite these nine components. In other words, the lattice may contain vectors of the form

$$(\vec{w}_L; \vec{u}_R, (v)), \quad \vec{w}_L^2 = 1 + \vec{u}_R^2.$$

This state would have mass $M_L = M_R = \frac{1}{2}\vec{u}_R^2 - \frac{1}{2}$. The possibilities that one has for $\vec{u}_R$ are severely restricted by the constraint vectors of the supercurrent. Since the lattice is self-dual, they must have integer dot-product with all the vectors $\vec{t}$ and $\vec{l}$ appearing in (7.4). For example, for the supercurrent (7.6) it is easy to see that the norm of $\vec{u}_R$ must be a multiple of $\frac{1}{4}$, so that in addition to the tachyon at mass $-\frac{1}{2}$, one may (depending on the lattice one chooses, of course) find tachyons of mass $-\frac{1}{4}$.

None of this is very important though, because one would normally not consider such tachyonic theories. More interesting is the possibility of having a vector $\vec{u}_R$ of norm 1. (Of course it is possible that the lattice has such vectors, without containing tachyonic vectors) In that case one obtains a state with $M_L = M_R = 0$, which is a massless scalar in a representation of G_L with weight vector $\vec{w}_L$, of norm 2.

This raises of course the possibility that $\vec{w}_L$ is actually a root of the left lattice. One can then go through an interesting chain of arguments. If $\vec{w}_L$ is a root (or, in other words, if one has a massless scalar in the adjoint representation of G_L), then one can subtract a root from $(\vec{w}_L; \vec{u}_R, (v))$, to obtain $(0; \vec{u}_R, (v))$. This is a root of the right part of the lattice, which has non-vanishing dot-product with the roots of D_5. Again it is not hard to see that the lattice must contain a larger rootsystem than that of D_5, of which the D_5 roots form a subset. This means that D_5 must be a *regular subalgebra* of some larger Lie algebra. There are only a few possibilities: This algebra must be E_6, E_7, E_8, or $D_{5+n}, n \geq 1$. Since the decomposition of the larger algebra must yield a vector representation of D_5, it can only be D_{5+n}. We will see in a moment that this destroys any possibility of getting chiral fermions out of such a theory, but let us first consider another consequence of this extension of D_5.

Of course we can excite the state corresponding to the extra roots by a right-moving oscillator α_i. This gives a massless space-time vector, and it should therefore be a gauge boson. That is indeed what it turns out to be. We will not go into the details of its interactions here, but we will just mention that its gauge symmetries originate from the right lattice, and involve the supercurrent in a non-trivial way. In fact, the gauge group turns out to be a non-regular subgroup of the Frenkel-Kac group of the right lattice. This group is of little relevance in heterotic strings, since it inhibits chirality, but it is more important in type-II strings; see [20], [21].

In chiral heterotic strings there is thus never a gauge group corresponding to the right-movers (this seems to have been proved more generally by Friedan and Shenker). As a consequence, the vectors $\vec{u}_R$ on the internal part of the right lattice only affect the *multiplicities* of the particles in the spectrum, not the representation they are in. There is no direct way of "measuring" $\vec{u}_R$, because there are no gauge symmetries coupling to it. Nevertheless these vectors do affect the theory in less direct ways, because the Yukawa couplings and Higgs self interactions depend on them.

The existence of this seemingly redundant nine-dimensional internal space is noteworthy for phenomenological reasons, because it makes it possible to have particles in the theory which have identical properties, but differ only in this part of the lattice. We need not point out that this might help explain the existence of fermion generations. Furthermore there can be interesting continuous and/or discrete *global* symmetries originating from this internal space.

The spinor conjugacy classes

Since a spinor of D_5 has norm $\frac{5}{4}$, it would appear that one could get tachyonic spinors of mass $-\frac{3}{8}$. The existence of tachyons in non-singlet representation of the Lorentz group signals an instability of the vacuum which would cause a breakdown of the Lorentz group. One might expect that world-sheet supersymmetry should prevent that. This does indeed appear to be the case. Observe that the D_5 spinor has half-integral dot-product with the constraint vectors of world sheet supersymmetry, because those vectors have (v) components in D_5. If a spinor were not to have any components in the internal dimensions, this would be in disagreement with the fact that the lattice is self-dual. Therefore, any D_5 spinor must appear in a vector of the form $(\vec{w}_L; \vec{u}_R, (s))$, and $\vec{u}_R$ must have half-integral dot-product with all the vectors $\vec{l}$ or $\vec{t}$ in the supercurrent. In all cases we have examined it turns out that to satisfy this condition, $\vec{u}_R$ must have norm greater than $\frac{3}{4}$. Furthermore there seems always to exist a vector $\vec{u}_R$ which saturates this lower limit. The reader may find it instructive to verify this for the two supercurrents (7.6) and (7.7) which we discussed in the previous section. Although it ought to be possible to prove this without explicitly writing down the supercurrent, for the moment we will accept as a fact that fermionic tachyons do not appear in the spectrum.

About the massless fermions one can say roughly the same as about the ten-dimensional case. The right-moving lattice vectors describing them must have norm 2. Because the lattice is lorentzian even, they appear in combination with even norm vectors $\vec{w}_L$ on the left. Two possibilities are of interest: $\vec{w}_L^2 = 0$ and $\vec{w}_L^2 = 2$.

In the first case one can excite the right-moving Fock space by means of bosonic oscillators α_i to obtain a gravitino. One can also excite it with an internal oscillator α_I or add a root of the left lattice (if it has any), to obtain a gaugino. The appearance of the gravitino tells us that the theory has a certain number of supersymmetries (more about this below). Again we see that the existence of local-space time symmetries is linked to the existence of certain roots of the right lattice.

In the second case, we may assume that $\vec{w}_L$ is not a root of the left lattice, for if were we could subtract $\vec{w}_L$ and obtain the case already discussed. If $\vec{w}_L$ is not a root, it is some vector weight of norm 2 of G_L, and one gets a massless spinor in the corresponding representation. If G_L has $U(1)$ factors part of $\vec{w}_L$ may correspond to $U(1)$-charges, and the remainder is a weight of the non-abelian part of G_L. In any case the only representations of the non-abelian part of G_L that one can get are those with weights of norm less than or equal to two. Fortunately the representations of the standard model satisfy this bound.

Nothing we have said so far addresses the question of chirality. The situation is a bit different here in comparison with ten dimensions, and this has simply to do with the difference in CPT transformations of spinors in these dimensions. One can easily understand that there is difference by changing the sign of a spinor in the conjugacy class (s) of D_8 or D_5: in the first case one gets a spinor in the same conjugacy class, whereas in the second case one obtains a spinor in (c). What does this have to with CPT? In four dimensions, a CPT transformation send a left-handed spinor in some representation (R) are of some gauge group into a right-handed anti-spinor in the complex conjugate representation $(\bar{R})$. Remarkably, there is an automorphism of every lattice that achieves precisely the same, namely multiplying all lattice vectors with -1. As discussed above, this flips the chirality of the D_5 spinor conjugacy classes, and hence the space-time chirality, while at the same time changing the signs of all vectors on the left lattice. As is well-known, changing the sign of a weight changes it to a weight of the complex conjugate representation.

Thus first of all CPT is a symmetry of any theory we construct. Secondly, to get the usual counting of four-dimensional fermions, we should consider only one D_5 chirality, (s), say, and regard states from the other conjugacy class as anti-particles. Thus if the lattice contains three vectors $(\vec{w}; \vec{u}_R^l, (s)), l = 1, \ldots, 3$, where $\vec{w}$ is a weight of a complex representation (R), we have three generations of fermions in the representation (R). The lattice will automatically also contain $(-\vec{w}; -\vec{u}_R^l, (c))$, but these are just the CPT conjugates of the three generations, and are not to be counted separately.

With this understanding, it is obvious how to count chiral fermions. If there are N generations of fermions in (R) and $\bar{N}$ in the complex conjugate $(\bar{R})$, then there are $N - \bar{N}$ chiral generations. This way of counting them assumes implicitly that pairs of opposite chirality will become massive, because there is no symmetry that would forbid such a mass term. We know however of no mechanism in string perturbation theory that would actually achieve this. Most likely, the massless spectrum is not altered to all orders in perturbation theory, and even the non-chiral fermions will then simply stay massless. It is however quite clear that non-perturbative effects have a role to play in obtaining low-energy phenomenology, and in particular they have to break supersymmetry and/or the weak interaction gauge symmetry. Since we know nothing about non-perturbative effects, we can take the path of maximal optimism, and assume that they will do whatever we need them to do.

There does not exist a simple criterion for determining *a priori* which lattices can give chiral fermions, and which cannot. There is however a simple necessary condition, namely that the D_5 root system should not be part of the root system of a larger D_n group. If that were the case, all other D_5 representations originate from D_n representations. For the spinors that means that they always come in pairs of opposite chirality, because a $D_n(n > 5)$ spinor (s) contains both (s) and (c) of D_5. These spinors belong to the same representation of any gauge group, and hence they are chiral conjugates (More precisely, if one applies a CPT conjugation to the spinors in the (c) conjugacy class of D_5, one can easily see that in the notation used above one would have $N = \bar{N}$). This loss of chirality is a well-known phenomenon in torus compactifications of higher dimensional chiral theories, and that is in fact precisely what is happening here: If the D_5 lattice extends to D_n one can regard the theory as a torus compactification of a higher dimensional one.

There are two situations in which such an extension of D_5 is forced on us. First of all, it occurs when we want to have gauge bosons belonging to G_R, as discussed above. Secondly, if the supercurrent contains any vectors $\vec{l}$ of length 1, the corresponding constraint vectors are roots, extending D_5. Thus to get chiral fermions, one needs $\vec{B}(\vec{l}) = 0$.

This concludes the discussion of the spectrum of covariant lattice theories. We emphasize once more a feature of this discussion which we think is rather elegant, and is unique to the *even* lattice formulation: All local symmetries (gravity, gauge symmetries of G_L and G_R and supergravity) are related to roots on the lattice. The only massless states not originating from roots are the scalars and the fermions (other than gravitino's and gaugino's).

8.3 SPACE-TIME SUPERSYMMETRY

Space-time supersymmetry is a subject which merits a separate discusion. As we have seen, gravitino's originate from roots on the lattice which are spinors of D_5. For reasons that should by now be familiar, this implies that D_5 is regular subalgebra of a larger Lie algebra. It is now quite clear that there is a restriction on the number of supersymmetries one can have, because there are not that many Lie algebras that extend D_5 by spinor roots; the only ones are E_6, E_7 and E_8. By simply counting the number of gravitino's one gets, we deduce that these three possibilities correspond respectively to $N = 1, N = 2$ and $N = 4$ supergravity. Similarly, for type-II theories one can get this from both the left and the right-moving sector, yielding $N = 1,2,3,4,5,6$ and 8 as the possible supersymmetries. We find it very intriguing that such relatively recent results as the classification of supersymmetries can be obtained from the classic work of Cartan!

It is not difficult to extend this to other dimensions, and for the dimensions above four one gets then*

$d=4$	D_5	$\subset$	E_6	$N=1$
			E_7	$N=2$
			E_8	$N=4$
$d=6$	D_6	$\subset$	E_7	$N=1$
			E_8	$N=2$
$d=8$	D_7	$\subset$	E_8	$N=1$
$d=10$	D_8	$\subset$	E_8	$N=1$

(Some people seem to use different conventions for counting the number of supersymmetries, in particular in six dimensions. To compare your favorite convention with ours use the torus compactifications of the $N=1$ theory in 10 dimensions; these correspond to E_8 in the table). For even more intriguing results in two dimensions we refer to [6].

It seems that everything that's known about supersymmetry and supermultiplets can be found back in the representation theory of the exceptional groups. For example, consider $E_6 \supset D_5$ in four dimensions. The smallest representations of E_6 are the singlet (1), the (27) and the (78), which is the adjoint. Their D_5 content is

$$\begin{aligned}(1) &\rightarrow (1)\\ (27) &\rightarrow (10) \oplus (16) \oplus (1)\\ (78) &\rightarrow (45) \oplus (16) \oplus (\overline{16}) \oplus (1)\end{aligned}$$

According to our rules, the D_5 singlets contain no physical states, while the (10) yields a scalar and the (16) a fermion. Thus the (27) contains the scalar multiplet. (On the lattice the representation (27), which has weights of norm $\frac{4}{3}$, must appear in combination with an internal space vector $\vec{u}_R$). In a similar way one can see that (78) contains the vector multiplet. Note that to count the spinors we should consider only the (16), because the $(\overline{16})$ is just a CPT conjugate. Note also that for every (27) the lattice necessarily contains a CPT conjugate $(\overline{27})$.

In fact, one can push this further, as it is possible to relate the "unphysical" left-over singlet components to the auxiliary fields of $N=1$ supergravity. Vertex operators for such auxiliary fields have recently been investigated [60]. It is conceivable that the above procedure works also for extended supergravities and/or higher dimensions, as well as for massive multiplets with arbitrary high spins, by using the representation theory of the appropriate exceptional group. We encourage the industrious student to work out the details.

---

* This relation between space-time supersymmetry and exceptional groups is in fact completely general, and applies for instance also to Calabi-Yau compactified string theories. The reason is that space-time supersymmetry constrains the vertex operator algebra so that it can always be mapped to an exceptional algebra.

We have not said much about massive states, but something needs to be said about them in supersymmetric theories. One would expect that they too belong to supermultiplets. This can in fact be proved by using the triality properties in a somewhat unexpected way. These properties are known to be related to supersymmetry in ten dimensions, and to clarify the relation between the Neveu-Schwarz-Ramond and the Green-Schwarz formalism. In those arguments, one uses triality rotations in the transverse Lorentz group. Clearly that is of no use below ten dimensions, since one has a different Lorentz group. Instead we will show that one can use the D_4 that in our formulation describes the ghosts and longitudinal modes for this purpose. This part of the lattice is dimension independent, so that what follows holds in arbitrary dimensions.

The D_4 factor plays a role in distinguishing fermions from bosons, because the former have as their D_4 components weights of the spinor representation (8_s), whereas the latter have weights of the vector representation (8_v). These vectors can be mapped into each other by an $SO(4)$ rotation which maps all the entire D_4 weight lattice onto itself. This is called an automorphism of the lattice. If such an automorphism maps weights of a given representation into weights of *the same* representation it is called an *inner* automorphism. Clearly, if one has a Lie algebra lattice (*i.e.* a sublattice of the complete weight lattice, containing at least the root lattice), then all inner automorphisms are of course symmetries of it. There are also automorphisms that map weights of *different* representations to each other. Such automorphism are called *outer*. In particular, D_4 has an outer automorphism which cyclically permutes $(8_s), (8_c)$ and (8_v), known as triality. Clearly an arbitrary D_4 lattice (take for example $(0) + (s)$) is not invariant under such an automorphism.

It is a special property of the embedding of D_4 in the exceptional groups that this outer automorphism becomes an inner automorphism of the exceptional group. This means in particular that for every (8_v) of D_4 contained in some representation of an exceptional group, there is also an (8_s) and (8_c). For the spectrum that statement translates into having a fermion for every boson, and hence into supersymmetry at all levels (at least as far as multiplicities are concerned).

9. Model building

There are many ways of finding solutions to the conditions listed in the beginning of the previous section, but unfortunately none is systematic enough to achieve a complete classification. Only in ten dimensions the complete list of solutions is known (by this we mean the complete list of solutions that can be obtained with the covariant lattice construction, and which therefore have a rank 16 gauge group. In ten dimensions one knows just one additional solution, which has an E_8 gauge group and is tachyonic [49], [61]). We will consider this case first.

9.1 TEN-DIMENSIONAL STRINGS

To obtain ten-dimensional strings we are looking for lattices $\Gamma_{16;8}$, where the right-moving components are D_8 weights. To find them, we can make use of a trick explained in section (3.4). Given a lorentzian even self-dual lattice of the required form, we can make an even self-dual *euclidean* lattice out of it by simply changing the sign of the last eight components of the metric. This seemingly absurd operation works only because the right lattice factor is known to be D_8; certainly changing the sign of the metric on an arbitrary lorentzian even self-dual does not preserve the self-duality. The result of this map is an even self-dual Euclidean lattice of dimension 24. But such lattices have been classified by Niemeier, so that classifying all ten-dimensional strings is reduced to merely finding the right mathematical literature (which in some libraries is still no small obstacle).

Not every Niemeier lattice yields a good string theory. To go back to a lorentzian lattice one has to be able to find a D_8 subalgebra. Obviously this must be a regular subalgebra, because we want to have a D_8 sub*lattice*, not just a sub*algebra*. This is not difficult, and yields the following solutions

$$
\begin{array}{rcl}
E_8 \times E_8 \times E_8 & \rightarrow & E_8 \times E_8 \\
D_{16} \times E_8 & \rightarrow & SO(32) \\
D_8 \times D_8 \times D_8 & \rightarrow & SO(16) \times SO(16) \\
D_{24} & \rightarrow & SO(32) \\
D_{12} \times D_{12} & \rightarrow & SO(8) \times SO(24) \\
D_{16} \times E_8 & \rightarrow & SO(16) \times E_8 \\
D_{10} \times E_7 \times E_7 & \rightarrow & SO(4) \times E_7 \times E_7 \\
D_9 \times A_{15} & \rightarrow & SU(16) \times U(1)
\end{array}
$$

Here we have indicated on the left the Niemeier lattice, and on the right the gauge group it produces. Notice that there is one Niemeier lattice, $D_{16} \times E_8$, that produces two different string theories, because there are two ways of embedding D_8.

The first two theories are supersymmetric, because $D_8 \subset E_8$. The last five are tachyonic, because D_8 is embedded in a D_n factor with $n > 8$. When the Niemeier lattice is mapped to $\Gamma_{16;8}$, this embedding leaves a trace in the form of lattice vectors $((v),(v))$, belonging to a $(D_{n-8})_L \times (D_8)_R$ sublattice originating from the D_n factor. Such vectors lead to tachyons. Finally, the third theory is the rather well-known $SO(16) \times SO(16)$ string, which does not have supersymmetry, and also no tachyons. When it was found it was the first of its kind, but in four dimensions such models are very common.

It is a useful exercise to determine the spectrum of some of these models, using the conjugacy classes of the Niemeier lattices. The answers can be found in [62].

In constructing lattices for four-dimensional strings we face a new technical difficulty, namely the constraint vectors for the supercurrent. Consider for example the choice (7.6). The first thing one should observe is that the form of this supercurrent forces all vectors in the nine internal dimensions to have as their components integers or half-integers. This is true because by addition of constraint vectors one gets, among others, vectors of the form

$$(0,\ldots,0,2,0,\ldots,0)$$

in the nine-dimensional internal space, and zero elsewhere. The structure of the lattice in this nine-dimensional space can then be summarized in a nice way by saying that all its vectors belong to conjugacy classes of $(D_1)^9$, where D_1 is the lattice introduced at the end of section (3.3).

Now the entire right lattice is of the form $D_5 \times (D_1)^9$, and it is not unreasonable to assume that also the left lattice can be written as a product of D_n factors. Since even under this assumption the lattices will turn out to be too numerous to be classified, whether one misses any solutions in this way is an academic question (our guess is that one does not). Any D_n factor on the left can always be decomposed into a product of D_1 factors, so on could try to enumerate all the even self-dual lattices based on

$$[(D_1)^{22}]_L[(D_1)^9 D_5]_R \tag{9.1}$$

conjugacy classes. One can imagine a *gedanken* computer program that does this. It would start with a lattice containing the conjugacy classes of the constraint vectors

$$\begin{gathered}[(0)^{22},(v),(v),(v),(0),(0),(0),(0),(0),(0),(v)]\\ [(0)^{22},(0),(0),(0),(v),(v),(v),(0),(0),(0),(v)]\\ [(0)^{22},(0),(0),(0),(0),(0),(0),(v),(v),(v),(v)],\end{gathered}$$

and then it would go through the list of conjugacy classes of (9.1) to find one of even lorentzian length, and with even inner product with all the vectors that are already on the the lattice. This procedure can be repeated to find a next conjugacy class, and when no new conjugacy class can be found, the lattice that one has is self-dual by definition.

Although writing this program is quite trivial, one is discouraged from running it because the total number of conjugacy classes of (9.1) is gigantic: 4^{32}. As explained in section (3.3), this means that a self-dual lattice based on (9.1) contains 2^{32} conjugacy classes, so that even checking them for tachyons, say, would cost an enormous amount of time.

Obviously this approach is quite useless in practice, but it does give us some insight in the extent of the problem. This can be made more precise by comparing it to a similar combinatorial problem, that of constructing all Euclidean even self-dual lattices of dimension 32. One could do that in an analogous way, namely by starting with $(D_1)^{32}$ (in 24 dimensions this approach would indeed give all 24 Niemeier lattices). As a combinatorial problem this is not too different from the one we were facing, because the number of conjugacy classes is the same*. But for this problem we know a (conservative) lower limit of the number of solutions, namely 8×10^7 (In [6] we gave an often misquoted rigorous upper limit on the number of theories, namely the number of even self-dual lattices of dimension 88. This number is finite, but larger than 10^{1500}. This was only intended to show that the number of solutions is finite, and does not in any way represent the actual number of solutions). This shows that we can expect to find a very large number of four-dimensional string theories for this choice of supercurrent.

A different, and more effective method for constructing lattices (for this supercurrent) was found by the authors of [63], who observed that they can all be obtained from *odd* self-dual lattices of dimension 22. By using techniques of the kind described in section (3.4), one can glue to this lattice a right-moving sector which contains the supercurrent constraint vectors. However from one odd self-dual lattice one gets many covariant lattices $\Gamma_{22;14}$, and a complete classification along these lines appears quite difficult. Of course the situation is improved considerably by imposing additional requirements, such as space-time supersymmetry.

Knowing the extent of the problem makes model building a somewhat arbitrary exercise. In fact "model building" is not even the correct terminology; "model finding" would be more appropriate. The objective is to *find* in the gigantic set of solutions one that corresponds to the standard model. It does not help to be just close to the standard model, since in string theory (ideally at least) one cannot adjust the model by adding a few higgses and tune a few parameters or by doing some of the other things model builders like to do.

This is a rather extreme view of the present situation, and in practice one can of course learn a lot from semi-realistic examples, and we will give some below. They are only intended as an illustration of the construction, and should not be confused with phenomenology. In [6] we produced many examples by modifying some of the Niemeier lattices. Here we will follow a perhaps more pedagogical route, the shift vector method. Any lattice of the form (9.1) can be transformed into any other one by adding a suitable set of shift vectors constructed out of D_1 "weights". In many cases these D_1 conjugacy classes contain roots, which form the root system of some simply laced Lie algebra. It is then convenient to write the the lattice in terms of the corresponding Lie algebra lattice to reduce the list of conjugacy classes. The simplest roots appear in conjugacy classes of the form $(0,\ldots,0,v,v,0,\ldots,0)$, which combine two D_1 factors into a D_2. However, it is certainly not true that from (9.1) one can only get D_n gauge groups. For example, the Niemeier lattices, which can have A_n or E_n factors, can be written as D_1^{24}.

A useful starting point is the lattice $(D_{22})_L \times (D_{14})_R$. To the root lattice one has to add four conjugacy classes to obtain a unimodular lattice, and if one furthermore requires it to be even, there is not much choice: The conjugacy classes must be $(0,0)+(v,v)+(s,s)+(c,c)$ (this is just an Englert-Neveu lattice with the left factor stretched by 8 dimensions). The phenomenology is extremely simple: The first term gives a graviton (plus $B_{\mu\nu}$ and a dilaton), and the gauge bosonss of SO(44) and the second

---

* The fact that the constraint vectors must be on the lattice reduces the number of solutions, but the fact that not all 32 factors are on equal footing in (9.1) leads to an enhancement.

one gives a tachyon in the vector representation of SO(44). The spinors are all massive and non-chiral, so that this example is certainly not of much interest. Note that the D_{14} lattice on the right contains the constraint vectors of world-sheet supersymmetry. In fact there are many possible sets of constraint vectors that one could choose. If one adds shift vectors, one has to make sure that they have integral dot product with one set of constraint vectors, so that these vectors are on the invariant sublattice Λ_0 defined in section (3.4).

Let us now make the model supersymmetric. This is easily achieved by adding a gravitino as a shift vector. To do so, split D_{14} into $D_6 \times D_8$ and add the conjugacy class $(0, 0, s)$ of $(D_{22})_L \times (D_6 \times D_8)_R$, where we embed the D_5 space-time lattice into D_8 (in the following we adopt the convention of embedding the space-time lattice in the last factor). Notice that two shift vectors in the same conjugacy class have the same effect, so that we can specify the shift vector by giving its conjugacy class with respect to the *sub*lattice.

Using the shift vector method explained in section (3.4) we can easily determine the conjugacy classes that appear. Since we have three D_n factors, there are in fact $2^3 = 8$ of them, and they are

$$(0,0,0) + (v,v,0) + (s,s,s) + (c,c,s) + (0,0,s) + (v,v,s) + (s,s,0) + (c,c,0)$$

What we have now obtained is a ten-dimensional heterotic string, compactified on a lattice $\Gamma_{22;6} = D_{22} \times D_6$. The D_8 factor of the covariant lattice has been embedded in E_8 because of the addition of the gravitino. This example has in fact more supersymmetry than we bargained for, namely $N = 4$. It is of course not chiral.

This illustrates several general principles. First of all, any theory can be made supersymmetric by simply adding as a shift vector a lattice vector which produces a gravitino in the spectrum. More generally, one can enforce the appearance of *any* state one likes by simply adding the corresponding lattice vector as a shift vector. One may however get unwanted states in addition. Another noteworthy aspect is the use of spinor weights to break D_n factors into smaller ones.

There are two obvious obstacles to chirality in our example. The first is that the D_5 root system is part of a larger D_n root system (that of D_8 in this case), which always inhibits chirality as explained in the previous section. The second is that that the $SO(44)$ gauge group simply doesn't have any complex representations, which is necessary for chirality in four dimensions. The only $SO(N)$ groups with complex representations are $SO(4k+2), k \in Z$. Their chiral representations are the spinors, and the fundamental spinor representations have weight length $\frac{2k+1}{4}$. Clearly only the fundamental spinors of $SO(2), SO(6), SO(10)$ and $SO(14)$ can occur for massless fermions, since all others would have to high a mass. Furthermore a fermion can not be chiral due to spinors of $SO(6)$, because the $SO(6)$ spinor representations are anomalous, and modular invariant string theory is anomaly free. (Note that $SO(2)$ anomalies can be cancelled by the four-dimensional analog of the Green-Schwarz mechanism [18], [64].)

These chiral representations can only appear in combinations with total weight length equal to two. This still leaves many possibilities, and one of the simplest is the

$SO(10) \times SO(6)$ representation $(16,4) + (16,\bar{4})$. The necessity to cancel the $SO(6)$ anomaly implies that the number of generations is a multiple of eight in this case.

Breaking $SO(44)$ into such small groups cannot be achieved by adding just one spinor shift vector, but two turn out to be sufficient. First decompose

$$(D_{22})_L(D_6D_8)_R \rightarrow (D_{14}D_8)_L(D_4D_2D_2D_6)_R$$

by adding a shift vector $(0,s;0,s,0,c)$. This shift vector projects out half of the gravitino's, so that we get an $N=2$ model with an $SO(16) \times SO(28)$ gauge group. It is still not chiral, but we can now break the lattice root system further down to

$$(D_7D_7D_5D_3)_L(D_2D_2D_2D_1D_1D_1D_5)_R$$

by adding the shift vector $(s,0,0,s;s,0,c,s,0,0,c)$. This again removes one gravitino, so that we have $N=1$ supersymmetry. The remaining lattice vector giving a gravitino is $(0,0,0,0;0,0,0,s,s,s,s)$. The constraint vectors appear in the conjugacy classes

$$\begin{gathered}(0,0,0,0;v,0,v,v,0,0,v)\\(0,0,0,0;0,v,v,0,v,0,v)\\(0,0,0,0;v,v,0,0,0,v,v)\end{gathered}$$

To check if these vectors are on the lattice, it suffices to verify that they have integer dot product with the shift vectors we used.

Working out the spectrum takes some effort, since the lattice has 11 factors, and hence 2^{11} conjugacy classes. Most of them contribute only to massive states, and with some ingenuity one can determine the massless states by hand. However, this kind of labor is easy to computerize, which makes it much easier to experiment with other lattice constructions. The model we just constructed turns out to be chiral, and its complete fermion spectrum consists of the $SO(14) \times SO(14) \times SO(10) \times SO(6)$ representations

$$\begin{gathered}2(1,1,16,4) + 2(1,1,16,\bar{4})\\+(14,14,1,1) + (1,1,10,6)\\+(14,1,10,1) + (1,14,10,1)\\+(14,1,1,6) + (1,14,1,6)\end{gathered}$$

Only the fermions in the first line are chiral, and give 16 generations of (16)'s of $SO(10)$. These fermions originate from the lattice vectors $(0,0,s,s,0,0,c,0,0,s,c)$ and $(0,0,s,c,0,0,s,0,0,s,c)$. Notice that the $(D_2)^3(D_1)^3$ factor of the right lattice only determines the multiplicities of the states in the theory, not their gauge quantum numbers. Since the model is supersymmetric, there are scalars in the same representations as the fermions.

This model is a well-known one, and appears in most of the papers on four-dimensional strings because it is one of the simplest to obtain. The construction given here closely resembles the one used in the fermionic constructions, and we have presented it in this way to allow the reader to make a comparison. Indeed, there is a one-to-one correspondence between the class of models based on (9.1) and fermionic models with "complex" boundary conditions, and (anti)-periodic boundary conditions.

One can continue from here, and add more shift vectors to slice the lattice into still smaller D_n chunks. It should be clear from the above that the reason for the rather large number of generations is the fact that their multiplicity is due to a spinor of $SO(6)$ and one of $SO(4)$. Thus the key to lowering the number of generations is to break these factors. This works indeed as expected, but we don't think it would be illuminating to show the gory details of such models. The model given above is just the tip of a gigantic iceberg. If one splits the lattice into more and more factors, there are more and more possibilities for doing this. Clearly the bulk of the models correspond to lattices which are split as much as possible. Unfortunately this "base of the iceberg" is the hardest part to explore, because of the huge number of conjugacy classes one has to deal with.

To conclude this chapter we will briefly discuss models for another choice of the supercurrent, namely (7.7). Just as our previous choice enforced a $(D_1)^9$ structure on the right lattice, the choice (7.7) restricts it as well. World-sheet supersymmetry requires that the lattice contain nine vectors of the form $(0, \ldots, 0, \sqrt{3}, 0, \ldots, 0, (v))$, where the last entry refers to D_5, and the others to the remaining nine dimensions of the right lattice. Any other lattice vector in the conjugacy classes (0) or (v) of D_5 must have entries that are a multiple of $\frac{1}{3}\sqrt{3}$ in these nine dimensions; if it is a spinor of D_5 its entries in these nine dimensions must be an odd multiple of $\frac{1}{6}\sqrt{3}$. Vectors of the form $(0, \ldots, 0, 2\sqrt{3}, 0, \ldots, 0, (0))$ must always be on the lattice, because they are equal to twice the constraint vectors.

This suggests to define a one-dimensional "root"-lattice

$$\mathcal{U} = \{2k\sqrt{3} \mid k \in Z\}$$

and its dual, or "weight"-lattice

$$\mathcal{U}^* = \{\tfrac{k}{6}\sqrt{3} \mid k \in Z\}.$$

This lattice $\mathcal{U}^*$ contains $\mathcal{U}$ as a sublattice, and has 12 cosets (which, by analogy with lie-algebra lattices we shall call "conjugacy classes") with respect to it. What we have just learned can then be rephrased as follows: the right lattice must be of the form $\mathcal{U}^9 \times D_5$. This notation must be interpreted in the usual sense, *i.e.* in addition to the root-lattice a certain set of conjugacy classes can appear.

The volume of the unit cell of the root-lattice $\mathcal{U}^9 \times D_5$ is $2(2\sqrt{3})^9$. To have a chance to get a self-dual lattice, we have to combine it a root-lattice on the left which completes this volume to an integer. This is easy: for example A_2, E_6 have a root lattice of volume $\sqrt{3}$, and $\mathcal{U}$ itself has volume $2\sqrt{3}$. Valid choices for the left lattice are for example $\mathcal{U} \times D_{21}$, or $A_2 \times D_{20}$, or $(A_2)^4 \times E_6 \times E_8$, or $\mathcal{U}^{21} \times D_1$. These are only a few possibilities from an obviously long list, and it should be clear that now one can once again start to construct models as we did above. This class is known to contain the Z_3 asymmetric orbifolds [8], [57]. We leave it as an exercise to find their conjugacy classes in this basis (the answer is essentially in [8], although in a different basis). They can then be used as a starting point for finding new theories by the shift vector method, yielding yet another, presumably very large set of theories. And this is still nowhere near the complete set, because there are many other supercurrents one can choose.

10. Final remarks

Despite the considerable length of this paper, there are still many things that we would have liked to discuss in some detail. Here we will merely list some of them with some brief comments and speculations.

— *Interactions* were not discussed in great detail, because this is not necessary with our construction. Since the entire method is based on conformal field theory, the formalism for calculating correlation functions on any Riemann surfaces is already built in. We don't expect these theories to be less consistent then the ten-dimensional ones. Explicit calculations of tree and one-loop diagrams have been done in [65], [64].

— *Overlaps* with other approaches are only partly understood. By construction, covariant lattice theories in four dimensions have gauge groups of rank 22. To the best of our knowledge other constructions yielding rank 22 gauge groups can be re-expressed in terms of covariant lattices. This has been shown for fermionic construction with complex fermion boundary conditions [5], and for asymmetric orbifolds with twists in the Weyl-group of a Lie-algebra [66],[59]. Furthermore the set of covariant lattice theories seems larger than that of fermionic constructions (of rank 22) with a trilinear supercurrent. The restriction to rank 22 can be overcome by considering

— *Twisted Covariant Lattices*, a new class of four-dimensional string theories, obtained by applying orbifold twists to covariant lattices. Since covariant lattices include Narain lattices, twisted covariant lattices include asymmetric orbifolds. The question is, does this class include anything more ? Since one is twisting a much larger class of lattices, there is certainly reason to speculate that it might. The most interesting twists would be those that map bosonized world-sheet fermions into world-sheet scalars, something that fits quite naturaly within our philosophy, and does not seem to be included in orbifold constructions. Obviously such twists will be severely restricted by modular invariance and world-sheet supersymmetry, but as far as we can tell there is no reason why these obstacles should be insurmountable.

— *The selection of the vacuum* has not been discussed, because we know of no principle that might distinguish one supersymmetric four-dimensional string from any other. Non-perturbative effects are usually invoked as a *deus ex machina* that will somehow achieve this, but this remains wishful thinking at present. A related question is whether four-dimensional strings are

— *Different vacua or different theories?* Probably regarding them as different vacua of the same, generic heterotic string theory is the best way of thinking about them, since after all different four-dimensional strings are nothing but the same conformal field theory with different boundary conditions. It is however not clear to us exactly in what sense one can regard all these theories as ground states of some space-time theory. Thinking about this question, one is automatically led to the question whether there are

— *Continuous parameters* connecting the theories considered here to other ones. One set of obvious candidates, namely Lorentz-rotations of the lattice, does not yield continuous parameters because in chiral theories they would ruin the constraints imposed on the right lattice by (among others) world-sheet supersymmetry. There could however very well be continuous parameters in the direction of lower rank theories.

Finally, there are many phenomenological questions that we wished we could answer. It would be very interesting to have an example that can approach the standards set in [67] for Calabi-Yau compactification, but the gigantic number of possibilities doesn't make the prospect of searching for one look very inviting. Probably the time is not yet ripe for genuine predictions from string theory, and it might be better to try to gain more insight into the vacuum structure first. In view of the extent of that problem, patience will be required.

ACKNOWLEDGEMENTS

We have benefited from discussions with many people, and are grateful for their help in improving our understanding of the subjects discussed here. We can't name them all without taking the risk of forgetting some, but we make an exception for Dieter Lüst, Bengt Nilsson and Nick Warner, whom we thank for the pleasant collaborations.

REFERENCES

[1]. D. Gross, J. Harvey, E. Martinec and R. Rohm, Nucl. Phys. B256 (1985) 253; Nucl. Phys. B267 (1986) 75.

[2]. P. Candelas, G. Horowitz, A. Strominger and E. Witten, Nucl. Phys. B258 (1985) 46.

[3]. L. Dixon, J. Harvey, C. Vafa and E. Witten, Nucl. Phys. B261 (1985) 620; Nucl. Phys. B274 (1986) 285; M. Mueller and E. Witten, Phys. Lett. 182B (1986) 28; L. Ibañez, H. Nilles and F. Quevedo, Phys. Lett. 187B (1987) 25.

[4]. K. Narain, Phys. Lett. 169B (1986) 41.

[5]. H. Kawai, D. Lewellen and S. Tye, Nucl. Phys. B288 (1987) 1.

[6]. W. Lerche, D. Lüst and A.N. Schellekens, Nucl. Phys. B287 (1987) 477.

[7]. I. Antoniadis, C. Bachas and C. Kounnas, Nucl. Phys. B289 (1987) 87.

[8]. K. Narain, M. Sarmadi and C. Vafa, Nucl. Phys. B288 (1987) 551.

[9]. D. Gepner, Nucl. Phys. B287 (1987) 111; *Exactly Solvable String Compactifications on Manifolds of $SU(N)$ Holonomy*, Princeton preprint PUPT-1066.

[10]. K. Bardacki, E. Rabinovici and B. Säring, *String models with $c < 1$ components*, CERN-TH.4760/87 (1987).

[11]. C.S. Lam, *GSO Projections of modular invariant aperiodic strings*, Princeton preprint IASSNS-HEP-87/12 (1987); A. Chamseddine and J.-P. Derendinger, *Twisted superstrings in four dimensions* Zürich preprint ETH PT 87-2.

[12]. J. Distler and B.R. Greene, *Aspects of (2,0) Compactifications*, Harvard preprint HUTP-87/A065

[13]. J.H. Schwarz, *Review of Recent Developments in Superstring Theory*, CALTECH preprint CALT-68-1432; L. Dixon, *Some World-Sheet Properties of Superstring Compactifications on Orbifolds and Otherwise*, Princeton preprint PUPT-1074.

[14]. C.A. Lütken, *A Large Class of Calabi-Yau Vacua*, CERN-TH.4900/87.

[15]. A. Cohn, D. Friedan, Z. Qiu and S. Shenker, Nucl. Phys. B278 (1986) 577.

[16]. W. Lerche and D. Lüst, Phys. Lett. 187B (1987) 45.

[17]. M. B. Green, J. H. Schwarz and E. Witten, *Superstring Theory*, Cambridge University Press, (1987); J. H. Schwarz, Phys. Rep. 89 (1982) 223.

[18]. M. Green and J. Schwarz, Phys. Lett. 149B (194) 117.

[19]. A.N. Schellekens and N.P. Warner, Phys. Lett. 177B (1986) 317; Phys. Lett. 181B (1986) 339; Nucl. Phys. B287 (1987) 87.

[20]. L. Castellani, R. d'Auria,F. Gliozzi and G. Sciuto, Phys. Lett. 168B (1986) 47; R. Bluhm, L. Dolan and P. Goddard, Nucl. Phys. B289 (1987) 364.

[21]. H. Kawai, D. Lewellen and S. Tye, Phys. Lett. 191B (1987) 63; W. Lerche, B.E.W. Nilsson and A.N. Schellekens, Nucl. Phys. B294 (1987) 136; L. Dixon, V. Kaplunowsky and C. Vafa, Nucl. Phys. B294 (1987) 43.

[22]. A. Polyakov,Phys. Lett. 103B (1981),207; 211.

[23]. L. Alvarez-Gaumé, G. Moore and C. Vafa, Comm. Math. Phys. 106 (1986) 1.

[24]. J. Conway and M. Sloane, Journal of Number Theory 15 (1982) 83; Europ. J. Combinatorics 3 (1982) 219.

[25]. H. Niemeier, Journal of Number Theory 5 (1973) 142.

[26]. J.-P. Serre, *A Course in Arithmetic*, Springer Verlag, New York, (1973).

[27]. P. Goddard and D. Olive, in *Vertex Operators in Mathematics and Physics*, Springer Verlag, (1984); Int. J. Mod. Phys. A (1986) 303.

[28]. R. Slansky, Phys. Rep. 79 (1981) 1.

[29]. F. Englert and A. Nevcu, Phys. Lett. 163B (1985) 349.

[30]. J. Leech, Canadian Journal of Mathematics 19 (1967) 251.

[31]. A.N. Schellekens, contribution to *Perspectives in string theory*, Copenhagen (1987), CERN-preprint, to appear.

[32]. C. Siegel, Ann. Math. 36 (1935) 527.

[33]. K. Narain, Sarmadi, E. Witten Nucl. Phys. B279 (1986) 93.

[34]. J. Polchinsky Comm. Math. Phys. 104 (1986) 37.

[35]. L. Alvarez-Gaumé, *Supersymmetry and Index theory*, Bonn lectures (1984).

[36]. P. Ginsparg and C. Vafa Nucl. Phys. B289 (1987) 414.

[37]. Th. Ohrndorf, Nucl. Phys. B281 (1987) 470.

[38]. A. Belavin, A. Polyakov and A. Zamolodchikov, Nucl. Phys. B241 (1984) 333.

[39]. D. Friedan, E. Martinec and S. Shenker, Nucl. Phys. B271 (1986) 93.

[40]. D. Friedan, in *Unified String Theories* (Proc. Santa Barbara workshop 1985), M. Green and D. Gross, eds. , World Scientific (1986).

[41]. S. Shenker in *Unified String Theories* (Proc. Santa Barbara workshop 1985), M. Green and D. Gross, eds. , World Scientific (1986).

[42]. E. Martinec, University of Chicago preprint EFI 87-45, Lectures given at the 1987 spring school on Superstrings, Trieste.

[43]. M. Peskin, SLAC-PUB 4251, lectures presented at the 3th Advanced Study Institute in elementary particle physics, Santa Cruz (1986).

[44]. H. Nicolai and A.N. Schellekens, in *Proceedings of the 5th Adriatic Meeting on Superstrings, Anomalies and Unification*, Dubrovnik (1986).

[45]. A. Kostelecky, O. Lechtenfeld, W. Lerche, S. Samuel and S. Watamura, Nucl. Phys. B238 (1987) 173.

[46]. P. Goddard, D. Olive and A. Schwimmer Phys. Lett. 157B (1985) 393.

[47]. P. Ramond, Phys. Rev D3 (1971) 2415; A. Neveu and J. H. Schwarz Nucl. Phys. B31 (1971).

[48]. F. Gliozzi, J. Scherk and D. Olive, Nucl. Phys. B122 (1977) 253

[49]. L. Dixon and J. Harvey, Nucl. Phys. B274 (1986) 93.

[50]. L. Alvarez-Gaumé, P. Ginsparg, G. Moore and C. Vafa, Phys. Lett. 171B (1986) 155.

[51]. A. Casher, F. Englert, H. Nicolai and A. Taormina, Phys. Lett. 162B (1985) 121.

[52]. F. Englert, H. Nicolai and A.N. Schellekens, Nucl. Phys. B274 (1986) 315.

[53]. A.N. Schellekens, Phys. Lett. 199B (1987) 427.

[54]. E. Verlinde and H. Verlinde,Nucl. Phys. B288 (1987) 357.

[55]. P. di Vecchia, V. G. Kniznhik, J. L. Petersen and P. Rossi, Nucl. Phys. B253 (1985) 701; P. Goddard and D. Olive, Nucl. Phys. B257 (1985) 245; I. Antoniadis, C. Bachas, C. Kounnas and P. Windey, Phys. Lett. 171B (1986) 51;

[56]. W. Lerche, B.E.W. Nilsson and A.N. Schellekens, Nucl. Phys. B294 (1987) 136.

[57]. H. Kawai, D. Lewellen and S. Tye, *Construction of four-dimensional fermionic strings with a modified supercurrent*, Cornell preprint CLNS 87/69.

[58]. J. Bagger, D. Nemeschansky, N. Seiberg and S. Yankielowicz, Nucl. Phys. B289 (1987) 349.

[59]. A.N. Schellekens and N.P. Warner, *Weyl groups, Supercurrents and Covariant Lattices*, CERN-TH.4916/87 (1987).

[60]. J. Atick, L. Dixon, and A. Sen Nucl. Phys. B292 (1987) 109; M. Dine, I. Ichinose and N. Seiberg, Nucl. Phys. B293 (1987) 253; S. Cecotti, S. Ferrara and L. Girardello, *Gravitational Supervertices in N=1, 4D-Superstrings*, CERN-TH.4802/87 (1987).

[61]. H. Kawai, D. Lewellen and S. Tye, Phys. Rev. D34 (1986) 3794; D. Benett, M. Brene L. Mizrachi and H. Nielsen, Phys. Lett. 187B (1986) 179.

[62]. W. Lerche, D. Lüst and A.N. Schellekens, Phys. Lett. 181B (1986) 71.

[63]. J. Balog, P. Forgacs, Z. Horvath and P. Vecsernyés, Phys. Lett. 197B (1987) 395.

[64]. W. Lerche, B.E.W. Nilsson, and A.N. Schellekens Nucl. Phys. B289 (1987) 609; W. Lerche, B.E.W. Nilsson, A.N. Schellekens and N.P. Warner, *Anomaly cancelling terms from the elliptic genus*, CERN-TH.4765/87 (1987).

[65]. D. Lüst, S. Theisen and G. Zoupanos, *Four-dimensional heterotic strings and conformal field theory* preprint MPI PAE PTh 83-87 (1987).

[66]. D. Lüst and S. Theisen, *Four dimensional strings - Orbifolds and Covariant Lattices* preprint MPI-PAE-PTh83/87 (1987).

[67]. B. Green, K.H. Kirklin, P.J. Miron and C.G. Ross Nucl. Phys. B278 (1986) 667.

A COURSE ON DETERMINANTS

P. Cartier

INTRODUCTION

Determinants enter the field of quantum physics via *Pauli's exclusion principle*. The mathematical expression of this principle is as follows: denoting by $\vec{r}_j$ the coordinates of the j-th electron (three space coordinates taking arbitrary real values, plus a spin coordinate taking the values $+\frac{1}{2}$ and $-\frac{1}{2}$), the wave function $\psi(\vec{r}_1,\ldots,\vec{r}_p)$ for a system of p electrons is *antisymmetric in its arguments* $\vec{r}_1,\ldots,\vec{r}_p$. If, for instance, $\psi_1(\vec{r}),\ldots,\psi_p(\vec{r})$ are normalized one-electron wave-functions, *mutually orthogonal*, corresponding to energy levels $E_1,\ldots,E_p$, then the normalized p-electron wave function $\psi(\vec{r}_1,\ldots,\vec{r}_p) = (p!)^{-1/2} \det \psi_i(\vec{r}_j)$ satisfies the exclusion principle and corresponds to the total energy $E_1+\ldots+E_p$. Similar constructions occur in Fock space or statistical mechanics. It is not the place to review the manifold physical implications of the exclusion principle; it suffices to say that the stability of matter as we know it depends strongly on this principle. At the mathematical level, the basic estimate was provided by Hadamard: if $A = (a_{ij})$ is a p x p matrix with complex elements, and if C is the maximum of the numbers $|a_{ij}|$, then the determinant D of A satisfies $|D| \leq p^{p/2}C^p$. This result shows a remarkable compensation occuring among the p! products of size C^p (approximately) which compose D, since $p^{p/2}$ is roughly of the order of $(p!)^{1/2}$ for large p. Nothing similar could occur in the case of *Bose-Einstein statistics*, where $\psi(\vec{r}_1,\ldots,\vec{r}_p)$ is symmetrical in its arguments $\vec{r}_1,\ldots,\vec{r}_p$ and the determinant should be replaced by a *permanent*

$$\sum_{\sigma\in S_p} \psi_1(\vec{r}_{\sigma(1)})\ldots\psi_p(\vec{r}_{\sigma(p)})$$

(symmetrization of the product $\psi_1(\vec{r}_1)\ \ldots\ \psi_p(\vec{r}_p)$).

ISBN 0-12-218100-X

A glance of the table of contents will reveal the organization of this paper. It is essentially a leisurely exposition of the basic properties of determinants, with special emphasis on the infinite-dimensional case. In a venerable subject like this, it is hard to innovate. In *part one*, we mostly review the properties of finite determinants in a form most suitable for generalizations. One of the novel features is our use of *volume forms* in subsections 1.4 and 1.5; we aim at giving characterizations not only of ordinary determinants, but also of powers of (absolute value of) determinants, and we provide a link with a non-commutative determinant introduced by Dieudonné around 1940. The connection between determinants and *antisymmetric tensors* is well-known. We made some efforts to present this (classical) theory in the spirit of *supersymmetry*. The analogy between the symmetric (Bose-Einstein statistics) and antisymmetric (Fermi-Dirac statistics) cases is especially transparent in the so-called *Mac Mahon's master theorem*, connecting various generating series of interest in statistical mechanics. We end part one by reviewing various formulas about Gaussian integrals; they are all classical and provide useful integral representations for various determinants. These formulas should be compared with the ones derived in part three using Berezin integration of functions of Grassmann variables.

Part two is devoted to the infinite-dimensional determinants which occur as variants of the *Fredholm determinants* for integral operators. We begin an exposition of the classical results of Fredholm. We endeavoured at motivating, as far as possible, the definitions by analogy with the finite-dimensional case. With the notable exception of *Fredholm's alternative*, we favoured the constructive proofs over the purely existential ones. *The basic formula*, which is used to define the determinant in various contexts is the following

$$\det(1 + A) = \sum_{n\geq 0} \operatorname{Tr}(\Lambda^n A) \quad ,$$

where $\Lambda^n A$ is the operator acting on the antisymmetric tensor space $\Lambda^n V$ by mapping $x_1 \wedge \ldots \wedge x_n$ into $Ax_1 \wedge \ldots \wedge Ax_n$. In physical slang, $\Lambda^n V$ is the n-particle Fock space for fermions, if V is the one-particle state space. The theory is especially smooth in the case of operators in Hilbert spaces - incidentally, this is the case of greatest relevance in quantum physics. But the Hilbert space theory does not contain the original case of integral operators with continuous kernels. The main difficulty to be overcome is that the series $\sum_n \lambda_n$ of the eigenvalues of an integral operator does not always converge, but instead $\sum_n |\lambda_n|^2$ is finite. Various authors (Grothendieck, Ruston) made efferts to extend the definition of Fredholm determinants to suitable classes of operators in Banach spaces. We present here a novel version, which depends strongly on the properties of Hilbert-Schmidt operators. This can be considered as the beginning of a theory of *renormalized determinants*.

Acknowledgements. I thank especially my dear friend Iouri Manin. Our many conversations about the mutual roles of mathematics and physics had a lasting influence on my thinking. His influence can be felt at many places in this paper. I also thank the organizers of the conference at Poiana Braşov, above all Radu Purice who provided me with a copy of his notes taken during the lectures, and Vladimir Georgescu. Without his endless patience, and the friendly pressure he exterted on me, this paper would have never been written up.

TABLE OF CONTENTS

PART ONE:

DETERMINANTS IN THE FINITE DIMENSIONAL CASE

1. A Review of the Elementary Theory

1.1. Let A be a *square matrix* of size n, with complex entries a_{ij} (for $1 \leq i \leq n$, $1 \leq j \leq n$). We denote by $a_1,\ldots,a_n$ its column vectors, so

$$\vec{a}_1 = \begin{pmatrix} a_{11} \\ a_{21} \\ \vdots \\ a_{n1} \end{pmatrix}, \quad \vec{a}_2 = \begin{pmatrix} a_{12} \\ a_{22} \\ \vdots \\ a_{n2} \end{pmatrix}, \quad \ldots, \quad \vec{a}_n = \begin{pmatrix} a_{1n} \\ a_{2n} \\ \vdots \\ a_{nn} \end{pmatrix}.$$

Therefore $\vec{a}_1,\ldots,\vec{a}_n$ are elements of the complex n-space $\mathbb{C}^n$. We also denote by I (or I_n) the *unit matrix*, by δ_{ij} its elements (Kronecker symbol) and $\vec{e}_1,\ldots,\vec{e}_n$ the column vectors of I_n. Therefore e_i is the vector whose only nonzero component is the i-th one, which is equal to 1. Any vector $\vec{x}$ in $\mathbb{C}^n$ with components $x_1,\ldots,x_n$ is therefore written as the linear combination $x_1\vec{e}_1 + \ldots + x_n\vec{e}_n$, and $\vec{e}_1,\ldots,\vec{e}_n$ is the so-called *canonical basis* of $\mathbb{C}_n$.

1.2. The *determinant* of A is a complex number, denoted usually by det A (or sometimes $|A|$). Viewed as a function of the columns $\vec{a}_1,\ldots,\vec{a}_n$ of A, we denote it as $\Delta(\vec{a}_1,\ldots,\vec{a}_n)$. We can express as follows the basic properties of the determinant:

a) $\Delta(\vec{e}_1,\ldots,\vec{e}_n) = 1$ — *normalisation*

b) for $\vec{a}_1,\ldots,\vec{a}_{i-1}, \vec{a}_{i+1},\ldots\vec{a}_n$ fixed, $\Delta(\vec{a}_1,\ldots,\vec{a}_i,\ldots,\vec{a}_n)$ is a linear function of the vector $\vec{a}_i$, hence a linear combination — *n-linearity* of $a_{1i},\ldots,a_{ni}$ with coefficients depending only on the other columns.

c) if we exchange $\vec{a}_i$ and $\vec{a}_{i+1}$ and keep the other vectors fixed, $\Delta(\vec{a}_1,\ldots,\vec{a}_n)$ gets multiplied by -1. *antisymmetry.*

These properties completely characterize the determinant.

It is easy to give an inductive construction of the determinant. Denote by $A^{(i)}$ (for $1 \leq i \leq n$) the square matrix of size n-1 obtained by erasing from A its first column and its i-th row. Then we have

$$\det A = \sum_{i=1}^{n} (-1)^{i-1} a_{i1} \cdot \det A^{(i)} \quad . \tag{1.1}$$

1.3. A slightly more invariant presentation is as follows. We consider a vector space V of finite dimension n over the field $\mathbb{C}$ of complex numbers. A *volume form* on V is a function $\omega(\vec{x}_1,\ldots,\vec{x}_n)$ depending on n vectors $\vec{x}_1,\ldots,\vec{x}_n$ in V, with complex values, which is multilinear and antisymmetric, namely

$$\omega(\ldots,\vec{x}_i,\ldots) = \lambda'\omega(\ldots,\vec{x}_i',\ldots) + \lambda''\omega(\ldots,\vec{x}_i'',\ldots) \tag{1.2}$$

if $\vec{x}_i = \lambda'\vec{x}_i' + \lambda''\vec{x}_i''$,

$$\omega(\ldots,\vec{x}_i,\vec{x}_{i+1},\ldots) = -\,\omega(\ldots,\vec{x}_{i+1},\vec{x}_i,\ldots) \quad . \tag{1.3}$$

Given any basis $\vec{e}_1,\ldots,\vec{e}_n$ of V, there exists a unique volume form ω_0 normalized by $\omega_0(\vec{e}_1,\ldots,\vec{e}_n) = 1$. Then, for any volume form, we get

$$\omega(\vec{x}_1,\ldots,\vec{x}_n) = t\omega_0(\vec{x}_1,\ldots,\vec{x}_n) \tag{1.4}$$

for $\vec{x}_1,\ldots,\vec{x}_n$ arbitrary in V, with a constant $t = \omega(e_1,\ldots,e_n)$. Hence, up to a scaling factor, volume forms are unique.

If there is a non-trivial linear relation $\lambda_1\vec{x}_1 + \ldots + \lambda_n\vec{x}_n = 0$ among the vectors $\vec{x}_1,\ldots,\vec{x}_n$, it follows immediately from (1.2) and (1.3) that $\omega(\vec{x}_1,\ldots,\vec{x}_n) = 0$ for any volume form ω on V. On the other hand, if a volume from ω assumes a nonzero value on some basis $\vec{e}_1,\ldots,\vec{e}_n$, then it is nonzero on any basis $\vec{x}_1,\ldots,\vec{x}_n$ whatsoever.

1.4. Let us denote by $B(V)$ the set of all (ordered) basis $\vec{x}_1,\ldots,\vec{x}_n$ of V, by $\Omega(V)$ the set of volume forms, and $\Omega^*(V)$ the set of nonzero volume forms. Any element ω of $\Omega^*(V)$ can be viewed as a function from $B(V)$ to the set $\mathbb{C}^\times$ of nonzero complex numbers. As such, it is characterized by the following variants of properties (1.2) and (1.3)

$$\omega(\lambda_1\vec{x}_1,\ldots,\lambda_n\vec{x}_n) = \lambda_1 \ldots \lambda_n\, \omega(\vec{x}_1,\ldots,\vec{x}_n) \tag{1.5}$$

if $\lambda_1,\ldots,\lambda_n$ are in $\mathbb{C}^\times$, and

$$\omega(\vec{x}_1,\ldots,\vec{x}_i + \vec{x}_k, \ldots, \vec{x}_n) = \omega(\vec{x}_1,\ldots,\vec{x}_i,\ldots,\vec{x}_n) \tag{1.6}$$

if k is different from i.

Let us check for instance the antisymmetry; it suffices to write the proof in the case $n = 2$. Hence

$$\begin{aligned}\omega(\vec{x}_2,\vec{x}_1) &= \omega(\vec{x}_2+\vec{x}_1,\vec{x}_1) = \omega((\vec{x}_2+\vec{x}_1), -\vec{x}_2 + (\vec{x}_2+\vec{x}_1))\\ &= \omega((\vec{x}_2+\vec{x}_1),-\vec{x}_2) = \omega(\vec{x}_2+\vec{x}_1 +(-\vec{x}_2), -\vec{x}_2)\\ &= \omega(\vec{x}_1, -\vec{x}_2)\end{aligned}$$

by repeated application of (1.6) and $\omega(\vec{x}_1,-\vec{x}_2) = -\omega(\vec{x}_1,\vec{x}_2)$ by (1.5).

To construct a volume form ω in $\Omega^*(V)$, we can proceed inductively as follows. Decompose V as the direct sum of a line D (of dimension 1) and a hyperplane H (of dimension $n-1$). Choose a vector $\vec{e} \neq 0$ in D and a volume form φ in $\Omega^*(H)$. Then there exists a unique ω in $\Omega^*(V)$, such that

$$\omega(\vec{e}, \vec{y}_1,\ldots,\vec{y}_{n-1}) = \varphi(\vec{y}_1,\ldots,\vec{y}_{n-1}) \tag{1.7}$$

for any basis $\vec{y}_1,\ldots,\vec{y}_{n-1}$ of H. Let $\vec{x}_1,\ldots,\vec{x}_n$ be a basis of V, and express $\vec{e}$ as the linear combination $\lambda_1\vec{x}_1,+\ldots+\lambda_n\vec{x}_n$.

From (1.5) and (1.6) it follows that one doesn't change the value $\omega(\vec{x}_1,\ldots,\vec{x}_n)$ if one adds to any $\vec{x}_i$ a linear combination of the other vectors. From (1.7) one derives

$$\omega(\vec{x}_1,\ldots,\vec{x}_n) = (-1)^{i-1}\lambda_i^{-1}\ \varphi(p(\vec{x}_1),\ldots,p(\vec{x}_{i-1}),p(\vec{x}_{i+1}),\ldots,p(\vec{x}_n)) \tag{1.8}$$

for any index i such that $1 \leq i \leq n$ and $\lambda_i \neq 0$; for every vector $\vec{x}$ in V, $p(\vec{x})$ is the unique vector in H such that $\vec{x}-p(\vec{x})$ lies in D (that is, is proportional to $\vec{e}$). It is easy to check that *the right-hand side of* (1.8) *is independent of the index* i *as long as* $\lambda_i \neq 0$, hence we have a construction of ω; checking properties (1.5) and (1.6) for ω is easy if a little tedious.

If $\vec{e}_1,\ldots,\vec{e}_n$ is a fixed basis of V, let $\vec{e} = \vec{e}_1$ and H be the subspace of V with basis $\vec{e}_2,\ldots,\vec{e}_n$. If we accept that there exists a unique φ_0 in $\Omega^*(H)$ normalized by $\varphi_0(\vec{e}_2,\ldots,\vec{e}_n) = 1$, it follows from (1.7) that there exists a unique ω_0 in $\Omega^*(V)$ normalized by $\omega_0(\vec{e}_1,\vec{e}_2,\ldots,\vec{e}_n) = 1$ and that $\omega_0(\vec{e}_1,\vec{y}_2,\ldots,\vec{y}_n) = \varphi_0(\vec{y}_2,\ldots,\vec{y}_n)$ for every basis $\vec{y}_2,\ldots,\vec{y}_n$ of H.

1.5. The previous construction of volume forms has some advantages on the more orthodox ones. For instance, we can modify the homogeneity property (1.5) into

$$\omega(\lambda_1\vec{x}_1,\ldots,\lambda_n\vec{x}_n) = |\lambda_1 \ldots\lambda_n|^s\ \omega(\vec{x}_1,\ldots,\vec{x}_n) \tag{1.9}$$

where s is a complex number. If s is an integer, we can also consider $(\lambda_1 \ldots \lambda_n)^s$ instead of $|\lambda_1 \ldots \lambda_n|^s$. Then it follows that any function $\omega\colon B(V) \to \mathbb{C}^\times$ satisfying (1.9) and (1.6) is of the form

$$\omega(\vec{x}_1,\ldots,\vec{x}_n) = t|\omega_0(\vec{x}_1,\ldots,\vec{x}_n)|^s \tag{1.10}$$

where ω_0 is as before and $t = \omega(\vec{e}_1,\ldots,\vec{e}_n)$ is a constant.

The previous considerations are valid *verbatim* in the real case. If we denote by $\Delta(\vec{x}_1,\vec{x}_2)$ the area of the parallelogram built in the plane $\mathbb{R}^2$ on the vectors $\vec{x}_1$, $\vec{x}_2$, it follows from geometric reasons that Δ satisfies the following rules (fig. 1):

$$\Delta(\lambda_1\vec{x}_1,\lambda_2\vec{x}_2) = |\lambda_1||\lambda_2| \; \Delta(\vec{x}_1,\vec{x}_2) \quad , \tag{1.11}$$

$$\Delta(\vec{x}_1,\vec{x}_2) = \Delta(\vec{x}_1 + \vec{x}_2, \vec{x}_2) = \Delta(\vec{x}_1, \vec{x}_2 + \vec{x}_1) \quad . \tag{1.12}$$

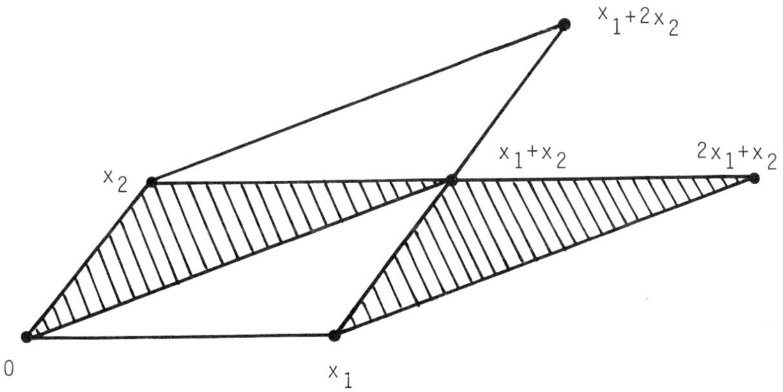

Fig. 1

Hence $\Delta(\vec{x}_1,\vec{x}_2) = |\det X|$ where X is the 2 x 2 matrix with columns $\vec{x}_1$ and $\vec{x}_2$ (a well-known result !). This extends immediately to the volume of the parallelotop in the space $\mathbb{R}^n$ spanned by n vectors $\vec{x}_1,\ldots,\vec{x}_n$; again this is equal to $|\det X|$ if X is the n x n matrix with columns $\vec{x}_1,\ldots,\vec{x}_n$.

We can also consider vectors in the *quaternionic* space $\mathbb{H}^n$, whose n components are quaternions. If $\vec{e}_1,\ldots,\vec{e}_n$ are the columns of the unit matrix I_n(as in section 1.1), we can associate to any invertible n x n matrix $A = (a_{ij})$ with quaternionic entries a determinant $\Delta(A)$ characterized by the following properties

$$\Delta(\vec{e}_1,\ldots,\vec{e}_n) = 1 \quad , \tag{1.13}$$

$$\Delta(\vec{a}_1\lambda_1,\ldots,\vec{a}_n\lambda_n) = |\lambda_1 \ldots \lambda_n|\Delta(\vec{a}_1,\ldots,\vec{a}_n) \tag{1.14}$$

$$\Delta(\vec{a}_1,\ldots,\vec{a}_i + \vec{a}_k,\ldots,\vec{a}_n) = \Delta(\vec{a}_1,\ldots,\vec{a}_i,\ldots,\vec{a}_n) \quad (i \neq k). \tag{1.15}$$

Here the determinant $\Delta(A)$ is expressed as a function of the columns $\vec{a}_1,\ldots,\vec{a}_n$ of A, a quaternionic vector $\vec{a} = {}^t(a_1,\ldots,a_n)$ is multiplied on the right by a quaternion λ, hence

$$\vec{a}\lambda = \begin{pmatrix} a_1\lambda \\ \vdots \\ a_n\lambda \end{pmatrix} \tag{1.16}$$

and the modulus of a quaternion $q = a + bi + cj + dk$ is as usual $|q| = (a^2 + b^2 + c^2 + d^2)^{1/2}$. This quaternionic determinant is a particular case of a non-commutative determinant defined by Dieudonné [5, tome II, p.67].

1.6. The basic property of determinants is *multiplicativity*. Let again V be a vector space of finite dimension n. The volume forms on V form a one-dimensional vector space $\Omega(V)$. If A is any linear operator on V, it acts on $\Omega(V)$ by

$$(\omega\cdot A)(\vec{x}_1,\ldots,\vec{x}_n) = \omega(A\vec{x}_1,\ldots,A\vec{x}_n) \tag{1.17}$$

for $\vec{x}_1,\ldots,\vec{x}_n$ in V; to verify this, it suffices to check that if ω satisfies the conditions (1.2) and (1.3), so does $\omega\cdot A$, and this is obvious. Since $\Omega(V)$ is one-dimensional, A acts on $\Omega(V)$ via multiplication by a scalar $\det A$, hence

$$\omega(A\vec{x}_1,\ldots,A\vec{x}_n) = (\det A)\ \omega(\vec{x}_1,\ldots,\vec{x}_n) \quad . \tag{1.18}$$

The multiplicative property

$$\det(AB) = \det A \cdot \det B \tag{1.19}$$

follows immediately.

A matrix $A = (a_{ij})$ of size n can be viewed as an operator acting on $\mathbb{C}^n$, transforming the vector $\vec{x}$ with components $x_1,\ldots,x_n$ into the vector $\vec{y} = A\vec{x}$ with components $y_1,\ldots,y_n$ given by

$$y_i = \sum_{j=1}^{n} a_{ij} x_j \quad . \tag{1.20}$$

In particular, the columns of A are $\vec{a}_1 = A\vec{e}_1, \ldots, \vec{a}_n = A\vec{e}_n$, and the determinant of A as an operator in $\mathbb{C}^n$ is the one defined in section 1.2. The multiplicative property (1.19) therefore applies to matrices.

1.7. It is not the place to review the *numerical methods* used to evaluate determinants. Needless to say, the inductive definition afforded by (1.1) is not practical unless n is very small, since it requires n! operations.

The multiplicative property can be used to give various characterizations of determinants. By the elementary matrix $M_{ij}(\lambda)$ we mean the matrix differing from the unit matrix I_n by the entry in row i and column j being put equal to λ (here $i \neq j$). If we denote the diagonal matrix with diagonal entries $c_1, \ldots, c_n$ as $\mathrm{diag}(c_1, \ldots, c_n)$, we have

$$\det M_{ij}(\lambda) = 1 \quad , \tag{1.21}$$

$$\det \mathrm{diag}(c_1, \ldots, c_n) = c_1 \ldots c_n \quad . \tag{1.22}$$

If a square matrix A has columns $\vec{a}_1, \ldots, \vec{a}_n$, the columns of $AM_{ij}(\lambda)$ are $\vec{a}_1, \ldots, \vec{a}_n$ except that $\vec{a}_j$ is replaced by $\vec{a}_j + \vec{a}_i\lambda$. Similarly, if $\vec{r}_1, \ldots, \vec{r}_n$ are the rows of A, the matrix $M_{ij}(\lambda)A$ differs from A by replacing the row $\vec{r}_i$ by $\vec{r}_i + \lambda\vec{r}_j$. By pre- or postmultiplying the matrix A by elementary matrices, we can therefore achieve to transform A into a diagonal matrix. Noticing that the matrices $AM_{ij}(\lambda)$, $M_{ij}(\lambda)A$ and A have all the same determinant, we conclude that the multiplicativity property together with formulas (1.21) and (1.22) characterizes the determinant.

1.8. A systematic procedure of this kind is known as *Gauss' pivoting method*. Put generally, assume that A is given in block form

$$A = \begin{pmatrix} M & N \\ P & Q \end{pmatrix}$$

where the sizes of the blocks are as follows

$$M \text{ is } p \times p, \quad N \text{ is } p \times q, \qquad (p + q = n)$$
$$P \text{ is } q \times p, \quad Q \text{ is } q \times q \quad .$$

If X if any q x p matrix and Y any p x q matrix, one gets

$$\begin{pmatrix} I_p & 0 \\ X & I_q \end{pmatrix}\begin{pmatrix} M & N \\ P & Q \end{pmatrix}\begin{pmatrix} I_p & Y \\ 0 & I_q \end{pmatrix} = \begin{pmatrix} M & MY+N \\ XM+P & XMY+XN+PY+Q \end{pmatrix} .$$

Moreover, the matrices $\begin{pmatrix} I_p & 0 \\ X & I_q \end{pmatrix}$ and $\begin{pmatrix} I_p & Y \\ 0 & I_q \end{pmatrix}$ are products of elementary matrices, hence of determinant 1. If M is invertible, choose X and Y such that XM+P = 0 and MY+N = 0. Hence

$$\begin{pmatrix} I_p & 0 \\ -PM^{-1} & I_q \end{pmatrix}\begin{pmatrix} M & N \\ P & Q \end{pmatrix}\begin{pmatrix} I_p & -M^{-1}N \\ 0 & I_q \end{pmatrix} = \begin{pmatrix} M & 0 \\ 0 & Q - PM^{-1}N \end{pmatrix}$$

and the final result reads as follows

$$\det\begin{pmatrix} M & N \\ P & Q \end{pmatrix} = \det M \cdot \det(Q - PM^{-1}N) \quad . \tag{1.23}$$

Hence an n-th order determinant is expressed as product of determinants of order p and q respectively.

The case p = 1 is worth recording. After permuting if necessary some rows and columns, one obtains the rule:

- choose a nonzero entry a_{ij} in A, known as *pivot* ;
- erase from A the row and column containing the pivot;
- replace any remaining element a_{kl} by

  $$b_{kl} = a_{kl} - a_{kj}\, a_{ij}^{-1}\, a_{il} \quad ;$$
- compute the determinant b of the (n-1)(n-1) matrix with entries b_{kl};
- multiply b by $(-1)^{i+j}\, a_{ij}$ to get the determinant of A.

1.9. Two final remarks are in order. The previous methods can be used to compute the *quaternionic determinants* with two small changes. Namely change $c_1 \ldots c_n$ into $|c_1 \ldots c_n|$ into formula (1.22) and use $\det A = b \cdot |a_{ij}|$ in the last step of the pivoting method.

Moreover, in formula (1.23) assume that $p = q$ and that the square matrices M, N, P, Q of size p commute pairwise. Then one gets

$$\det \begin{pmatrix} M & N \\ P & Q \end{pmatrix} = \det (MQ - NP) \quad . \tag{1.24}$$

Using a limiting argument, one gets rid of the restriction $\det M \neq 0$. Now the determinant of a 2 x 2 matrix $\begin{pmatrix} m & n \\ p & q \end{pmatrix}$ is $mq - np$.

Therefore the rule: use the formula to compute a 2 x 2 determinant replacing the scalar entries by the commuting $p \times p$ matrices, then take the determinant of the resulting $p \times p$ matrix. This rule can be generalized when the matrix A is in block form

$$A = \begin{pmatrix} A_{11} & \cdots & A_{1r} \\ \cdots & \cdots & \cdots \\ \cdots & \cdots & \cdots \\ A_{r1} & \cdots & A_{rr} \end{pmatrix}$$

whenever the $p \times p$ matrices A_{ij} commute pairwise (and $pr = n$). This is known as *Williamson's theorem.*

2. Symmetry Properties of Tensors

2.1. Let us recall the notion of *tensor product* of two vector spaces V and W. This is a new vector space denoted by $V \otimes W$; moreover, to any pair of vectors (x,y) in $V \times W$ is associated an element $x \otimes y$ of $V \otimes W$ and the following properties hold:

a) *for any* y *in* W*, the map* $x \mapsto x \otimes y$ *from* V *into* $V \otimes W$ *is linear;*

b) *as* a) *with* V *and* W *interchanged;*

c) *if* (e_α) *is a basis of* V *and* (f_β) *a basis of* W *, then the vectors* $e_\alpha \otimes f_\beta$ *form a basis of* $V \otimes W$.

With the previous notations, any element x of V has components x^α such that $x = \sum_\alpha x^\alpha e_\alpha$; similarly for any element $y = \sum_\beta y^\beta f_\beta$ of W and any element $t = \sum_{\alpha,\beta} t^{\alpha\beta} e_\alpha \otimes f_\beta$ of $V \otimes W$. Moreover if $t = x \otimes y$, one gets $t^{\alpha\beta} = x^\alpha y^\beta$.

Notice that we do not assume the dimensions of V and W to be finite. In a similar way, one defines the triple tensor product $U \otimes V \otimes W$ as $(U \otimes V) \otimes W$, etc.

Tensor products of spaces help transform multilinear functions into linear functions. For instance, if $\varphi(x,y)$ is an element of a space T depending linearly on x in V for y fixed in W, and symmetrically depending linearly on y in W for x fixed in V, there exists a unique linear map Φ from $V \otimes W$ into T such that $\varphi(x,y) = \Phi(x \otimes y)$. This follows from c) above, namely

$$\Phi(t) = \sum_{\alpha,\beta} t^{\alpha\beta}\varphi(e_\alpha, f_\beta) \tag{2.1}$$

for $t = \sum_{\alpha,\beta} t^{\alpha\beta} e_\alpha \otimes f_\beta$.

2.2. We fix now a vector space V of finite dimension n, and if necessary we use coordinates with respect to a fixed basis $e_1,\ldots,e_n$ of V. Fix an integer $k \geq 2$ and denote by $V^{\otimes k}$ the tensor product $V_1 \otimes \ldots \otimes V_k$ where all spaces $V_1,\ldots,V_k$ are put equal to V. By convention we set $V^{\otimes 0} = \mathbb{C}$ and $V^{\otimes 1} = V$. A general element in $V^{\otimes k}$ is written as a tensor

$$t = \sum_{\alpha_1\ldots\alpha_k} t^{\alpha_1\ldots\alpha_k} e_{\alpha_1} \otimes \ldots \otimes e_{\alpha_k} \tag{2.2}$$

where the indices $\alpha_1, \ldots, \alpha_k$ run over $1, \ldots, n$ independently.

Let us denote by S_k *the group of permutations* of the integers $1,2,\ldots,k$. *It operates on* $V^{\otimes k}$ in such a way that

$$\sigma\cdot(x_1\otimes \ldots\otimes x_k) = x_{\sigma^{-1}(1)} \otimes \ldots \otimes x_{\sigma^{-1}(k)} \tag{2.3}$$

for $x_1,\ldots,x_k$ in V. In components:

$$(\sigma\cdot t)^{\alpha_1\ldots\alpha_k} = t^{\alpha_{\sigma(1)}\ldots\alpha_{\sigma(k)}} \quad . \tag{2.4}$$

Moreover, the rule of operation is satisfied

$$(\sigma\cdot\tau)\cdot t = \sigma\cdot(\tau\cdot t) \tag{2.5}$$

for t in $V^{\otimes k}$ and σ, τ in S_k.

2.3. The action of the symmetry group on $V^{\otimes k}$ enables one to define the *symmetric part* S^kV of $V^{\otimes k}$ and the *antisymmetric part* Λ^kV of $V^{\otimes k}$.

By definition, S^kV consists of the tensors t in $V^{\otimes k}$ such that $\sigma\cdot t = t$ for all σ in S_k, or what is the same, the components $t^{\alpha_1 \cdot\cdot \alpha_k}$ are unchanged by any permutation of the indices $\alpha_1,\ldots,\alpha_k$. If $x_1,\ldots,x_k$ are vectors in V, we denote simply by $x_1 \ldots x_k$ their *symmetric product* $\sum_{\sigma\in S_k} \sigma\cdot(x_1\otimes \ldots\otimes x_k)$. Then the vector space S^kV has a basis consisting of the symmetric products $e_{\alpha_1} \ldots e_{\alpha_k}$ for $\alpha_1, \ldots,\alpha_k$ in ascending order $\alpha_1 \leqq \ldots \leqq \alpha_k$. These products can also be expressed as monomials

$$\underbrace{e_1 \ldots e_1}_{\beta_1} \underbrace{e_2 \ldots e_2}_{\beta_2} \cdots\cdots \underbrace{e_n \ldots e_n}_{\beta_n}$$

with integers $\beta_1,\ldots,\beta_n$ such that $\beta_1 \geqq 0, \ldots, \beta_n \geqq 0$, $\beta_1+ \ldots +\beta_n = k$.

Otherwise stated, S^kV can be interpreted as the set of polynomials in $e_1,\ldots,e_n$, homogeneous of degree k. Its dimension is equal to $\binom{n+k-1}{k}$.

2.4. For a permutation σ, the *number* $I(\sigma)$ *of inversions* is the number of pairs of integers i,j such that $1 \leqq i < j \leqq k$, $\sigma(i) > \sigma(j)$.

The *signature* $\operatorname{sgn}\sigma$ is defined as $(-1)^{I(\sigma)}$; its main property is expressed by

$$\operatorname{sgn}(\sigma\tau) = \operatorname{sgn}\sigma \cdot \operatorname{sgn}\tau \quad . \tag{2.6}$$

Then Λ^kV consists of the tensors t such that $\sigma\cdot t = (\operatorname{sgn}\sigma)\cdot t$ for all σ in S_k; alternatively, the component $t^{\alpha_1 \ldots\alpha_k}$ vanishes when two of the indices $\alpha_1,\ldots,\alpha_k$ are equal and is multiplied by - 1 if one interchanges two indices. Hence, up to a sign, the nonzero components of t are the $t^{\alpha_1\ldots\alpha_k}$ for $1 \leqq \alpha_1 < \ldots < \alpha_k \leqq n$. Put in another form introduce the *wedge product* $x_1\wedge \ldots \wedge x_k$ of the vectors $x_1,\ldots,x_k$ by

$$x_1\wedge \ldots \wedge x_k = \sum_{\sigma\in S_k} (\operatorname{sgn}\sigma)\, \sigma\cdot(x_1\otimes \ldots \otimes x_k) \quad . \tag{2.7}$$

Then the vector space $\Lambda^k V$ is spanned by such products and a basis of $\Lambda^k V$ consists of the products $e_{\alpha_1} \Lambda \ldots \Lambda e_{\alpha_k}$ for $1 \leq \alpha_1 < \ldots < \alpha_k \leq n$. Hence $\Lambda^k V$ is of dimension $\binom{n}{k}$.

2.5. *The case* k=n *is particularly interesting*. The vector space $\Lambda^n V$ is one-dimensional, spanned by the products $x_1 \Lambda \ldots \Lambda x_n$ for $x_1, \ldots, x_n$ running over V, and a basis is given by $e_1 \Lambda \ldots \Lambda e_n$. We shall follow Grothendieck's convention and denote $\Lambda^n V$ by det V (recall n is the dimension of V).

The volume forms on V can be interpreted as linear forms on det V. More precisely, for any volume form ω on V there exists a unique linear form $\tilde{\omega}$ on det V such that

$$\omega(x_1, \ldots, x_n) = \tilde{\omega}(x_1 \Lambda \ldots \Lambda x_n) \quad . \tag{2.8}$$

Conversely, for any linear form φ on det V, the formula

$$\omega(x_1, \ldots, x_n) = \varphi(x_1 \Lambda \ldots \Lambda x_n) \tag{2.9}$$

defines a volume form ω on V.

2.6. Let W be a subspace of V. Denote by m the dimension of W, and by V/W the factor space (of dimension n-m). Let p be the natural projection of V onto V/W. Then one can identify the spaces det V and det W $\otimes$ det(V/W) in such a way that the wedge product $x_1 \Lambda \ldots \Lambda x_n$ in det V is identified to $(x_1 \Lambda \ldots \Lambda x_m) \otimes (p(x_{m+1}) \Lambda \ldots \Lambda p(x_n))$ in case $x_1, \ldots, x_m$ belong to W.

Dually, if ω_W is a volume form on W and $\omega_{V/W}$ a volume form on V/W, there exists a unique volume form ω_V on V such that

$$\omega_V(x_1, \ldots, x_n) = \omega_W(x_1, \ldots, x_m)\omega_{V/W}(p(x_{m+1}), \ldots, p(x_n)) \tag{2.10}$$

for $x_1, \ldots, x_m$ in W and $x_{m+1}, \ldots, x_n$ in V.

A variant is obtained by choosing a subspace H of V such that V be a direct sum of W and H. We can identify det V to det W $\otimes$ det H in such a way that $x_1 \Lambda \ldots \Lambda x_m \Lambda y_1 \Lambda \ldots \Lambda y_{n-m}$ corresponds to $(x_1 \Lambda \ldots \Lambda x_m) \otimes (y_1 \Lambda \ldots \Lambda y_{n-m})$ for $x_1, \ldots, x_m$

in W and $y_1,\ldots,y_{n-m}$ in H. Similarly, a volume form ω_V on V is obtained from a volume form ω_W on W and a volume form ω_H on H. Namely

$$\omega_V(x_1,\ldots,x_m,y_1,\ldots,y_{n-m}) = \omega_W(x_1,\ldots,x_m)\omega_H(y_1,\ldots,y_{n-m}) \tag{2.11}$$

for $x_1,\ldots,x_m$ in W and $y_1,\ldots,y_{n-m}$ in H. The construction given in section 1.4 corresponds to the particular case $m = 1$.

2.7. *Let* A *be a linear operator acting on* V. Then A acts on $V^{\otimes k}$ in such a way that

$$A\cdot(x_1\otimes \ldots \otimes x_k) = Ax_1\otimes \ldots \otimes Ax_k \tag{2.12}$$

for $x_1,\ldots,x_k$ in V. In components, if A is represented by the matrix (a_α^β) in such a way that $Ae_\alpha = \sum_\beta a_\alpha^\beta e_\beta$, then a tensor t with components $t^{\alpha_1\ldots\alpha_k}$ is transformed by A into a tensor At with components

$$(A\cdot t)^{\alpha_1\ldots\alpha_k} = \sum_{\beta_1,\ldots,\beta_k} a_{\beta_1}^{\alpha_1} \ldots a_{\beta_k}^{\alpha_k}\, t^{\beta_1\ldots\beta_k} . \tag{2.13}$$

This is in accordance with the standard rules of tensor calculus.

It follows from the formulas (2.3) and (2.12) that A acting on $V^{\otimes k}$ commutes with the action of the symmetry operators. Hence it leaves invariant the subspaces S^kV and Λ^kV. As a matter of notation, we denote by $A^{\otimes k}$ the operator in $V^{\otimes k}$ given by the action of A, by S^kA its restriction to S^kV and by Λ^kA its restriction to Λ^kV. For the three kinds of products of vectors, we have therefore the rules

$$A^{\otimes k}(x_1\otimes \ldots \otimes x_k) = Ax_1\otimes \ldots \otimes Ax_k , \tag{2.14}$$

$$S^kA(x_1 \ldots x_k) = Ax_1 \ldots Ax_k , \tag{2.15}$$

$$\Lambda^kA(x_1\wedge \ldots \wedge x_k) = Ax_1\wedge \ldots \wedge Ax_k . \tag{2.16}$$

If we restrict A to the group GL(V) of invertible linear transformations in V, we get thus three linear representations of GL(V) in the spaces $V^{\otimes k}$, S^kV and Λ^kV respectively.

In particular, $\Lambda^n A$ is an operator in the one-dimensional space $\Lambda^n V$ = det V. It follows from the duality of det V and $\Omega(V)$ and from formula (1.20) that $\Lambda^n A$ acts by multiplication by det A on det V. Otherwise stated, one has

$$Ax_1 \wedge \ldots \wedge Ax_n = (\det A) \cdot (x_1 \wedge \ldots \wedge x_n) \tag{2.17}$$

for $x_1, \ldots, x_n$ in V. This is yet another characterization of the determinant.

Assume that W is a subspace of V, stable under A. Choose the basis $e_1, \ldots, e_n$ of V in such a way that $e_1, \ldots, e_m$ constitute a basis of W. The matrix of A in this basis is in block form

$$A = \begin{pmatrix} M & N \\ 0 & P \end{pmatrix} ,$$

where the sizes are as follows: M is m x m, N is m x (n-m), and P is (n-m) x (n-m).

Then the identification of det V and det$W \otimes$ det(V/W) amounts more or less to the classical determinant formula

$$\det A = \det M \cdot \det P \quad . \tag{2.18}$$

2.8. It is the aim of *supersymmetry* to unify bosons and fermions, or in algebraic terms to unify symmetric and antisymmetric tensors. We present here a general method inspired by recent work on *Yang-Baxter equation.*

Fix an integer $k \geq 2$, and define elements $s_1, \ldots, s_{k-1}$ in the symmetric group S_k by

$$s_i(j) = \begin{cases} j & \text{if } j \neq i, i+1 \\ i+1 & \text{if } j = i \\ i & \text{if } j = i+1 \end{cases} \tag{2.19}$$

(s_i is the interchange, or transposition, of i and i+1). It is a classical theorem in group theory (Moore, 1894) that these elements generate the group S_k and that a complete list of relations among these generators is as follows:

$$s_i^2 = 1 \qquad \text{for} \quad i = 1,\ldots,k-1 \tag{2.20}$$

$$s_i s_j = s_j s_i \qquad \text{if} \quad |i-j| \geq 2 \tag{2.21}$$

$$s_i s_{i+1} s_i = s_{i+1} s_i s_{i+1} \qquad \text{for} \quad i = 1,\ldots,k-2 \tag{2.22}$$

In graphical terms, the generators $s_1,\ldots,s_{k-1}$ correspond to the nodes of a chain

1 2 3 k-2 k-1

two generators commute if the corresponding nodes are not directly connected in the chain, and the relation (2.22) holds for pairs of generators corresponding to nodes adjacent to the same edge.

Let now T be any operator acting on $V^{\otimes 2}$. We denote by $T^{i,i+1}$ the operator acting on $V^{\otimes k}$ in such a way that

$$T^{i,i+1}(x_1 \otimes \ldots \otimes x_k) = x_1 \otimes \ldots \otimes x_{i-1} \otimes T(x_i \otimes x_{i+1}) \otimes \otimes x_{i+2} \otimes \ldots \otimes x_k \quad . \tag{2.23}$$

One could in a similar way define more generally operators $T^{i,j}$ acting on the factors of ranks i and j of $V^{\otimes k}$. This construction can be illustrated by a *simple quantum-mechanical model*. We consider a system of k particles labelled 1,2,...,k, subjected to *pair interaction*. If V is the space of one-particle states, $V^{\otimes k}$ describes the states of the system. The elementary interaction law is expressed by a *two-body potential*, an operator T in $V^{\otimes 2}$; then $T^{i,j}$ is the contribution to the potential energy stemming from the pair of particles labelled i and j.

When $|i-j|$ is at least 2, the sets $\{i,i+1\}$ and $j\{j,j+1\}$ are disjoint and the operators $T^{i,i+1}$ and $T^{j,j+1}$ commute. From the equations (2.20) to (2.22), one concludes that there exists a linear representation π_T of the group S_k in the space $V^{\otimes k}$ such that $\pi_T(s_i) = T^{i,i+1}$ (for $1 \leq i \leq k-1$) iff the following conditions are satisfied

We denote by $\Sigma^k V$ the subspace of $V^{\otimes k}$ consisting of the tensors t such that $\sigma \cdot t = t$ for every σ in S_k. We define a product $x_1 \ldots x_k$ for vectors by

$$x_1 \ldots x_k = \sum_{\sigma \in S_k} \sigma \cdot (x_1 \otimes \ldots \otimes x_k) \tag{2.30}$$

It reduces to the symmetric product of vectors if $x_1, \ldots, x_k$ are all even, to the wedge product $x_1 \wedge \ldots \wedge x_k$ if $x_1, \ldots, x_k$ are all odd. The space $\Sigma^k V$ decomposes as a direct sum of subspaces $\Sigma^{B,F} V$ where $B \geq 0$, $F \geq 0$, $B + F = k$. The number B is the *bosonic number* and F the *fermionic number*, and $\Sigma^{B,F} V$ is generated by the products $(x_1 \ldots x_B) \cdot (y_1 \wedge \ldots \wedge y_F)$ where $x_1, \ldots, x_B$ are even and $y_1, \ldots, y_F$ are odd; it can be identified in a natural way with $S^B V^+ \otimes \Lambda^F V^-$. In particular, if $V = V^+$ (that is $\varepsilon = 1$) then $\Sigma^k V$ reduces to $S^k V$, and if $V = V^-$ (that is $\varepsilon = -1$) then $\Sigma^k V$ reduces to $\Lambda^k V$. The elements in $\Sigma^{B,F} V$ are considered as even or odd according as F is even or odd (parity by fermionic number).

3. Mac Mahon's Master Theorem

3.1. We keep the previous notations. For instance, V is a complex vector space of finite dimension n. We defined previously the antisymmetric tensor spaces $\Lambda^0 V = C$, $\Lambda^1 V = V$, $\Lambda^2 V, \ldots, \Lambda^n V = \det V$. Moreover, $\Lambda^k V$ is reduced to 0 for $k > n$. Denote by ΛV the direct sum of the space $\Lambda^0 V, \ldots, \Lambda^n V$. We define parity on ΛV by the rules

$$\Lambda^+ V = \bigoplus_{k \geq 0} \Lambda^{2k} V \tag{3.1}$$

$$\Lambda^- V = \bigoplus_{k \geq 0} \Lambda^{2k+1} V \quad . \tag{3.2}$$

The wedge product of vectors can be extended to a multiplication in ΛV, which is bilinear, associative, complying with the signe rule

$$\xi \wedge \eta = (-1)^{pq} \eta \wedge \xi \tag{3.3}$$

$$TT = 1 \quad \text{in } V^{\otimes 2} \quad \textit{"Unitarity"} \tag{2.24}$$

$$T^{12}T^{23}T^{12} = T^{23}T^{12}T^{23} \quad \text{in } V^{\otimes 3} \quad \textit{"Yang-Baxter equation"} \tag{2.25}$$

2.9. Let us assume that ε is an operator on V such that $\varepsilon^2=1$. Then V is the direct sum of V^+ and V^-, where V^+ corresponds to the eigenvalue +1 of ε and V^- to the eigenvalue -1. A vector in V^+ is called *even*(or bosonic) and a vector in V^- is called *odd*(or fermionic). We consider the operator T in $V^{\otimes 2} \equiv V \otimes V$ characterized by

$$\begin{aligned} T(x_+ \otimes y_+) &= y_+ \otimes x_+ \\ T(x_+ \otimes y_-) &= y_- \otimes x_+ \\ T(x_- \otimes y_+) &= y_+ \otimes x_- \\ T(x_- \otimes y_-) &= -y_- \otimes x_- \end{aligned} \tag{2.26}$$

where x_+ and y_+ are even, and x_-, y_- are odd. This is the well-known *Koszul's sign rule*: "insert a factor -1 each time you permute two odd factors". The properties (2.24) and (2.25) are easily checked, hence we get an action of the symmetric group S_k on $V^{\otimes k}$. Explicitely, one gets

$$\sigma\cdot(x_1\otimes \ldots \otimes x_k) = (-1)^I \, x_{\sigma^{-1}(1)}\otimes \ldots \otimes x_{\sigma^{-1}(k)} \tag{2.27}$$

where I is the number of pairs (i,j) of integers such that $1 \leq i < j \leq k$, $\sigma(i) > \sigma(j)$ and x_i, x_j are both odd (we assume that $x_1,\ldots,x_k$ have well-defined parities, either even or odd). In particular

$$\sigma\cdot(x_1\otimes \ldots \otimes x_k) = x_{\sigma^{-1}(1)}\otimes \ldots \otimes x_{\sigma^{-1}(k)} \quad \text{if } x_1, \ldots, x_k \text{ are all even} \tag{2.28}$$

$$\sigma\cdot(x_1\otimes \ldots \otimes x_k) = \operatorname{sgn}\sigma\cdot(x_{\sigma^{-1}(1)}\otimes \ldots \otimes x_{\sigma^{-1}(k)}) \quad \text{if } x_1,\ldots,x_k \text{ are all odd.} \tag{2.29}$$

if ξ belongs to $\Lambda^p V$ and η to $\Lambda^q V$. Even elements commute to even or odd elements and $\xi \wedge \eta = -\eta \wedge \xi$ if both ξ and η are odd. If $\xi = x_1 \wedge \ldots \wedge x_p$ and $\eta = y_1 \wedge \ldots \wedge y_q$, then $\xi \wedge \eta$ is by definition the wedge product of vectors $x_1 \wedge \ldots \wedge x_p \wedge y_1 \wedge \ldots \wedge y_q$.

3.2. Let A be an operator acting linearly on V. We defined the operator $\Lambda^k A$ acting on $\Lambda^k V$. We denote by ΛA the operator on ΛV which coincides with $\Lambda^k A$ on $\Lambda^k V$ for $k = 0,1,\ldots,n$. Then ΛA extends the operator A on $V = \Lambda^1 V$ and respects multiplication

$$\Lambda A(\xi \wedge \eta) = (\Lambda A)\xi \wedge (\Lambda A)\cdot\eta \quad . \tag{3.4}$$

The *first statement* in Mac Mahon's theorem is the following formula

$$\det(1 + tA) = \sum_{k=0}^{n} t^k \operatorname{Tr}(\Lambda^k A) \quad , \tag{3.5}$$

where t is a complex parameter (or a formal variable).

The proof is especially simple if A can be *diagonalized*. Assume that there exists a basis $e_1,\ldots,e_n$ of V consisting of eigenvectors of A, namely

$$Ae_1 = \lambda_1 e_1, \ \ldots, \ Ae_n = \lambda_n e_n \quad . \tag{3.6}$$

Then the tensors $e_{i_1} \wedge \ldots \wedge e_{i_k}$ for $1 \leqq i_1 < \ldots < i_k \leqq n$ form a basis of $\Lambda^k V$ and $\Lambda^k A$ multiplies such a tensor by $\lambda_{i_1} \ldots \lambda_{i_k}$. Hence the trace of $\Lambda^k A$ is the elementary symmetric function $\sum_{i_1<\ldots<i_k} \lambda_{i_1} \ldots \lambda_{i_k} = c_k$ of $\lambda_1, \ldots, \lambda_n$. Moreover, the operator $1+tA$ multiplies every basis vector e_i by $1 + t\lambda_i$. Hence formula (3.5) reduces to the classical statement

$$\prod_{i=1}^{n} (1 + t\lambda_i) = \sum_{k=0}^{n} c_k t^k \quad . \tag{3.7}$$

3.3. In the general case, we could argue by continuity since it can be proved that any matrix (or operator) is a limit of matrices conjugate to diagonal ones. Instead, we shall rely on the possibility of finding for the operator A a basis $e_1,\ldots,e_n$

of V such that the matrix (a_i^j) expressing A in this basis be *upper triangular* ($a_i^j = 0$ if $i < j$).

Let us not assume at first that the matrix (a_i^j) be triangular. For every increasing sequence $I = (i_1 < \ldots < i_k)$ of indices between 1 and n, denote by e_I the wedge product $e_{i_1} \wedge \ldots \wedge e_{i_k}$. By definition, we have $Ae_i = \sum_j a_i^j e_j$. It follows

$$(\Lambda^k A) \cdot e_I = \sum_J a_I^J e_J \tag{3.8}$$

where a_I^J is the minor of A corresponding to the set I of columns and the set J of rows. In particular, we get

$$\operatorname{Tr}(\Lambda^k A) = \sum_I a_I^I \quad , \tag{3.9}$$

where the sum is extended over all sequences $I = (i_1 < \ldots < i_k)$ of length k.

Assuming now the matrix (a_i^j) to be triangular with diagonal elements $\lambda_1, \ldots, \lambda_n$, then a_I^I is the determinant of a triangular matrix with diagonal elements $\lambda_{i_1}, \ldots, \lambda_{i_k}$. By a well-known generalization of formula (2.18), we get $a_I^I = \lambda_{i_1} \ldots \lambda_{i_k}$, hence $\sum_I a_I^I = c_k$ and one concludes the proof as before.

We conclude with two remarks:

a) Putting together formulas (3.5) and (3.9) we get

$$\det(1 + tA) = \sum_{k=0}^{n} t^k \sum_{|I|=k} a_I^I \quad , \tag{3.10}$$

a well-known formula for the characteristic determinant. Here $|I|$ is the length of the increasing sequence I.

b) We can put t = 1 in formula (3.5). Hence we get

$$\det(1 + A) = \operatorname{Tr}(\Lambda A) = \operatorname{Tr}(\Lambda^+ A) + \operatorname{Tr}(\Lambda^- A) \quad . \tag{3.11}$$

Putting t = -1, we get

$$\det(1 - A) = \operatorname{Tr}(\Lambda^+ A) - \operatorname{Tr}(\Lambda^- A) \tag{3.12}$$

The right-hand side is the so-called *supertrace* of A acting on the space ΛV with even part $\Lambda^+ V$ and odd part $\Lambda^- V$.

3.4. We study now the action of A in the *symmetric algebra* $SV = \bigoplus_{k\geq 0} S^kV$. The *symmetric product of vectors* can be extended to a multiplication in SV which is bilinear, associative as well as commutative. Choosing a basis $e_1,\ldots,e_n$ of V enables one to consider the elements of SV as the polynomials with complex coefficients in $e_1,\ldots,e_n$.

The operator A on V defines operators S^kA acting on S^kV. We denote by SA the operator on SV which restricts to S^kA on the subspace S^kV of polynomials homogeneous of degree k. It respects the symmetric product.

With these notations, the *second statement* in Mac Mahon's theorem is the following formula

$$\det(1 - tA)^{-1} = \sum_{k\geq 0} t^k \operatorname{Tr}(S^kA) \quad . \tag{3.13}$$

The proof is quite similar to the previous one. Assume first that A can be put in diagonal form as in formula (3.6). Then the monomials $e_1^{\beta_1} \ldots e_n^{\beta_n}$ with $\beta_1 + \ldots + \beta_n = k$ form a basis of S^kV, and A multiplies such a monomial by $\lambda_1^{\beta_1} \ldots \lambda_n^{\beta_n}$. Hence we get for the trace

$$\operatorname{Tr}(S^kA) = \sum_{\beta_1+ \ldots +\beta_n=k} \lambda_1^{\beta_1} \ldots \lambda_n^{\beta_n} \quad , \tag{3.14}$$

and therefore

$$\begin{aligned}
\sum_{k\geq 0} t^k \operatorname{Tr}(S^kA) &= \sum_{\beta_1\ldots\beta_n} t^{\beta_1+\ldots+\beta_n} \lambda_1^{\beta_1}\ldots\lambda_n^{\beta_n} \\
&= \sum_{\beta_1\ldots\beta_n} (t\lambda_1)^{\beta_1} \ldots (t\lambda_n)^{\beta_n} \\
&= \sum_{\beta_1=0}^{\infty} (t\lambda_1)^{\beta_1} \ldots \sum_{\beta_n=0}^{\infty} (t\lambda_n)^{\beta_n} \\
&= \frac{1}{1-t\lambda_1} \cdot \ldots \cdot \frac{1}{1-t\lambda_n} \quad .
\end{aligned}$$

But the determinant of 1-tA is the product $(1-t\lambda_1)\ldots(1-t\lambda_n)$, hence formula (3.13).

In the general case, assume that A can be put in triangular form. It means that there exists a basis $e_1,\ldots,e_n$ of V and scalars $\lambda_1,\ldots,\lambda_n$ such that $Ae_i - \lambda e_i$ be a linear combination of the vectors e_j for $j < i$. The monomial $e_1^{\beta_1} \ldots e_n^{\beta_n}$ is transformed by S^kA into $(Ae_1)^{\beta_1} \ldots (Ae_n)^{\beta_n}$. Expanding this product, it is easy to see that the coefficient of $e_1^{\beta_1} \ldots e_n^{\beta_n}$ into $(Ae_1)^{\beta_1} \ldots (Ae_n)^{\beta_n}$ is equal to $\lambda_1^{\beta_1}\ldots\lambda_n^{\beta_n}$. Formula (3.14) is still valid, and the determinant of $1-tA$ is still the product $(1-t\lambda_1) \ldots (1-t\lambda_n)$. The proof is finished as before.

3.5. It is interesting to put together formulas (3.5) and (3.13). We get

$$1 = \{\sum_{m=0}^{\infty} t^m \mathrm{Tr}(S^mA)\}\cdot\{\sum_{k=0}^{n} (-1)^k t^k \mathrm{Tr}(\Lambda^kA)\} . \tag{3.15}$$

Comparing the coefficients of the powers of t, we get

$$\sum_{B+F=k} (-1)^F \mathrm{Tr}(S^BA)\cdot\mathrm{Tr}(\Lambda^FA) = 0 \quad \text{for } k \geq 1 . \tag{3.16}$$

This formula can be given a *supersymmetric* interpretation.

Denote by W the *double* V x V of V, with a parity operator given by $\varepsilon = \begin{pmatrix}1 & 0\\ 0 & -1\end{pmatrix}$ in matrix form, and a parity changing operator $\pi = \begin{pmatrix}0 & 1\\ 1 & 0\end{pmatrix}$. Hence π exchanges W^+ and W^-, and $\pi^2 = 1$. Both W^+ and W^- are copies of V. In physical slang, π associates to every boson state in W^+ its fermionic partner in W^-. From what we saw in section 2.9, we can identify Σ^kW with the direct sum of the spaces $\Sigma^{B,F}W = S^BW^+ \otimes \Lambda^FW^-$ (for $B+F = k$), and in turn to the sum of the spaces $S^BV \otimes \Lambda^FV$.

We extend A to W, in matrix form $\begin{pmatrix}A & 0\\ 0 & A\end{pmatrix}$, that is $A(x,y) = (Ax,Ay)$ for x,y in V. We then extend A to an operator $\tilde{A}$ acting on Σ^kW in such a way that

$$\tilde{A}(x_1 \ldots x_k) = Ax_1 \ldots Ax_k \tag{3.17}$$

for $x_1,\ldots,x_k$ in W. Then the subspaces $\Sigma^{B,F}W$ are stable under

$\tilde{A}$, and if we identify $\Sigma^{B,F}W$ to $S^B V \otimes \Lambda^F V$, the action of $\tilde{A}$ on $\Sigma^{B,F}W$ is given by the operator $S^B A \otimes \Lambda^F A$. Let now $\tilde{\varepsilon}$ be the operator on $\Sigma^k W$ which acts by multiplication by $(-1)^F$ on $\Sigma^{B,F}W$. It is related to the parity operator in W by

$$\tilde{\varepsilon}(x_1 \dots x_k) = \varepsilon x_1 \dots \varepsilon x_k \tag{3.18}$$

for $x_1,\dots,x_k$ in W.

After all these preparations, the identity (3.16) turns out as a *supertrace vanishing theorem*

$$\mathrm{Tr}(\tilde{\varepsilon}\tilde{A}) = 0 \qquad (\text{for } k \geqq 1) \tag{3.19}$$

(trace of operators acting on $\Sigma^k W$). A direct proof can be given using two new operators d and s acting on $\Sigma^k W$ by

$$d(x_1 \dots x_B \cdot y_1 \dots y_F) = \sum_{i=1}^{B} x_1 \dots x_{i-1} \pi x_i x_{i+1} \dots x_B \cdot y_1 \dots y_F \tag{3.20}$$

$$s(x_1 \dots x_B \cdot y_1 \dots y_F) = \sum_{j=1}^{F} (-1)^{j-1} x_1 \dots x_B \cdot y_1 \dots y_{j-1} \pi y_j y_{j+1} \dots y_F \tag{3.21}$$

where x_i's are *even* and the y_j's are *odd* . Since d changes the fermion number F by +1 and s changes it by -1, one gets

$$d\tilde{\varepsilon} = -\tilde{\varepsilon}d \quad , \quad s\tilde{\varepsilon} = -\tilde{\varepsilon}s \quad . \tag{3.22}$$

Moreover, from the previous definitions, one gets

$$\tilde{A}\tilde{\varepsilon} = \tilde{\varepsilon}\tilde{A} \quad , \quad \tilde{A}d = d\tilde{A} \quad , \quad \tilde{A}s = s\tilde{A} \quad . \tag{3.23}$$

It can also be proved that sd + ds multiplies every element of $\Sigma^k W$ by k (see for instance subsection
From these formulas, one derives

$$\begin{aligned}
\mathrm{Tr}(\tilde{\varepsilon}\tilde{A}sd) &= \mathrm{Tr}(\tilde{A}\tilde{\varepsilon}sd) && \text{because } \tilde{A}\tilde{\varepsilon} = \tilde{\varepsilon}\tilde{A} \\
&= -\mathrm{Tr}(\tilde{A}s\tilde{\varepsilon}d) && \text{because } \tilde{\varepsilon}s = -s\tilde{\varepsilon} \\
&= -\mathrm{Tr}(\tilde{\varepsilon}d\tilde{A}s) && \text{by cyclic invariance of the trace} \\
&= -\mathrm{Tr}(\tilde{\varepsilon}\tilde{A}ds) && \text{because } \tilde{A}d = d\tilde{A}
\end{aligned}$$

From $sd+ds = k$, one gets then $k\mathrm{Tr}(\tilde{\varepsilon}\tilde{A}) = \mathrm{Tr}(\tilde{\varepsilon}\tilde{A}(sd+ds)) = 0$ and $\mathrm{Tr}(\tilde{\varepsilon}\tilde{A}) = 0$ follows provided $k \geq 1$.

3.6. We derive as an application an important formula in group theory, known as *Cartan-Molien's formula*. Suppose G is a finite group, of order $|G|$, acting linearly on the vector space V. We let G act on the symmetric tensor space S^kV in such a way that $g.(x_1 \ldots x_k) = (gx_1) \ldots (gx_k)$ for the symmetric product of vectors $x_1,\ldots,x_k$ in V. Denote by I^k the subspace of S^kV consisting of the invariants of G. The dimension of these spaces is given by the following *generating series*

$$\sum_{k=0}^{\infty} t^k.\dim I^k = |G|^{-1} \sum_{g\in G} \det(1 - tg_V)^{-1} \quad . \tag{3.24}$$

For clarity, we denote by g_V the operator afforded by g on V. A similar formula with $|G|^{-1} \sum_{g\in G}$ replaced by $\int_G dg$ holds for a compact group G acting linearly on V.

The *proof* of formula (3.24) is as follows. Define an operator P_k in S^kV by $P_ku = |G|^{-1} \sum_{g\in G} g\,u$ for u in S^kV. Then P_k is a projection of S^kV onto I^k, hence its trace is equal to the dimension of I^k. But P_k is nothing else than $|G|^{-1} \sum_{g\in G} S^kg_V$, hence

$$\dim I^k = |G|^{-1} \sum_{g\in G} \mathrm{Tr}(S^kg_V) \quad . \tag{3.25}$$

One concludes using formula (3.13).

3.7. So far we have associated to an operator A acting on V the numbers $c_k(A) = \mathrm{Tr}(\Lambda^kA)$ and $h_k(A) = \mathrm{Tr}(S^kA)$. In terms of the eigenvalues $\lambda_1,\ldots,\lambda_n$ of A, $c_k(A)$, is the *elementary symmetric function of order* k and $h_k(A)$ is the *complete symmetric function of order* k, namely the sum of all monomials of degree k in $\lambda_1,\ldots,\lambda_n$ each with coefficient one. We found the generating series

$$\sum_{k\geq 0} c_k(A)t^k = \det(1 + tA) \quad , \tag{3.26}$$

$$\sum_{k\geq 0} h_k(A)t^k = \det(1 - tA)^{-1} \quad . \tag{3.27}$$

We introduce now the *sum-of-powers* $\tau_k(A) = \lambda_1^k+\ldots+\lambda_n^k$, or what is the same $\tau_k(A) = \mathrm{Tr}(A^k)$ (put again A in triangular form). The corresponding generating series is expressed in the following various forms:

$$\sum_{k\geq 1} \tau_k(A)t^k = \frac{\lambda_1 t}{1-\lambda_1 t} + \ldots + \frac{\lambda_n t}{1-\lambda_n t} \quad , \tag{3.28}$$

$$\sum_{k\geq 1} \tau_k(A)t^k = t\,\mathrm{Tr}(A\cdot(1 - tA)^{-1}) \quad , \tag{3.29}$$

$$\sum_{k\geq 1} \tau_k(A)t^k = -t\,\frac{d}{dt}\log\det(1 - tA) \quad ; \tag{3.30}$$

for the last equality, use the product formula

$$\det(1 - tA) = (1 - \lambda_1 t)\ldots(1 - \lambda_n t) \quad . \tag{3.31}$$

Formula (3.30) can easily be inverted to give

$$\det(1 - tA) = \exp - \sum_{k\geq 1} \tau_k(A)t^k/k \quad . \tag{3.32}$$

We can now restate formulas (3.26) and (3.27) in the following form

$$\sum_{k\geq 0} c_k(A)t^k = \exp \sum_{k\geq 1} (-1)^{k-1}\tau_k(A)t^k/k \quad , \tag{3.33}$$

$$\sum_{k\geq 0} h_k(A)t^k = \exp \sum_{k\geq 1} \tau_k(A)t^k/k \quad . \tag{3.34}$$

If we expand the exponentials and compare the coefficients of the various powers of t, we get the following variants of *Waring's formula*:

$$c_k(A) = \Sigma(-1)^{k+\alpha_1+\ldots+\alpha_k} \frac{\tau_1(A)^{\alpha_1}\tau_2(A)^{\alpha_2}\ldots\tau_k(A)^{\alpha_k}}{\alpha_1!1^{\alpha_1}\alpha_2!2^{\alpha_2}\ldots\alpha_k!k^{\alpha_k}} \quad , \tag{3.35}$$

$$h_k(A) = \Sigma \frac{\tau_1(A)^{\alpha_1}\tau_2(A)^{\alpha_2}\ \ldots\ \tau_k(A)^{\alpha_k}}{\alpha_1!1^{\alpha_1}\alpha_2!2^{\alpha_2}\ \ldots\ \alpha_k!k^{\alpha_k}} \quad . \tag{3.36}$$

In both cases, the summation is restricted to the systems $\alpha_1,\ldots,\alpha_k$ of positive integers (including 0) such that $1\cdot\alpha_1 + 2\cdot\alpha_2 + \ldots + k\cdot\alpha_k = k$. This latest restriction expresses the *homogeneity of the functions* c_k, h_k and τ_k: if we choose a basis of V and represent accordingly A by a matrix (a_{ij}), then $c_k(A)$, $h_k(A)$ and $\tau_k(A)$ are polynomials in the entries a_{ij}, homogeneous of degree k.

If U(t) and V(t) are power series such that $U(t)=\exp V(t)$, one gets $U'(t) = V'(t)U(t)$ for the derivatives. Using this remark in connection with formulas (3.33) and (3.34), we derive the following recursion formulas (due to Newton) (recall $c_0(A)= h_0(A) = 1$)

$$pc_p(A) = \sum_{j=1}^{p} (-1)^{j-1}\ \tau_j(A)\ c_{p-j}(A) \quad , \tag{3.37}$$

$$ph_p(A) = \sum_{j=1}^{p} \tau_j(A)\ h_{p-j}(A) \quad . \tag{3.38}$$

From these recursion formulas, one derives *determinantal formulas due to Plemelj*, namely

$$c_1 = \tau_1\ , \quad 2!c_2 = \det\begin{pmatrix}\tau_1 & \tau_2\\ 1 & \tau_1\end{pmatrix}, \quad 3!c_3 = \det\begin{pmatrix}\tau_1 & \tau_2 & \tau_3\\ 2 & \tau_1 & \tau_2\\ 0 & 1 & \tau_1\end{pmatrix}$$

and, in general, $p!c_p(A)$ is a p x p determinant whose nonzero entries are given by $u_{ij} = \tau_{j-i+1}(A)$ for $j \geq i$ and $u_{j+1,j}=p-j$ for $j=1,\ldots,p-1$. To prove this statement, develop the p-th determinant according to its first row and get back to formula (3.37). To get a similar formula for $h_p(A)$, the best is to use a *duality principle*, which follows obviously from formulas (3.33) and (3.34): *in every formula, one can exchange* $c_1(A)$ *with* $h_1(A),\ldots,c_k(A)$ *with* $h_k(A),\ldots$ *provided that at the same time one multiplies* $\tau_k(A)$ *by* $(-1)^{k-1}$.

So far, we considered the generating series as formal power series. If we care about *convergence*, the following has to be said:

(a) If n is the dimension of V, then $c_k(A) = 0$ for $k > n$. Hence the series $\sum_{k\geq 0} c_k(A)t^k$ *is a polynomial of degree* n *in* t (and so does $\det(1 + tA)$).

(b) If $\lambda_1,\ldots,\lambda_n$ are the eigenvalues of A, denote by $||A||_{sp}$ the maximum among the numbers $|\lambda_1|,\ldots,|\lambda_n|$ (the so-called *spectral radius* of A). Then the series $\sum_{k\geq 0} h_k(A)t^k$, $\sum_{k\geq 1} \tau_k(A)t^k$ and $\sum_{k\geq 1} \tau_k(A)t^k/k$ have the same radius of convergence equal to the inverse of $||A||_{sp}$.

3.8. In the preceding subsections, $c_k(A)$, $h_k(A)$ and $\tau_k(A)$ have been considered as symmetric functions of the eigenvalues $\lambda_1,\ldots,\lambda_n$ of A and our relations are in fact theorems about symmetric functions. An operator in the finite-dimensional case can be transformed into a triangular form, and we made extensive use of this possibility. Nothing of this sort exists for operators acting on Hilbert or Banch spaces. It is therefore not without interest to sketch alternate proofs of some of the previous results. In the infinite-dimensional case, formula (3.26) shall be a definition of the determinant, with a proper interpretation of $c_p(A)$. One convinces easily oneself that everything can be deduced from the Waring formulas (3.35) and (3.36).

As Cauchy remarked already, any permutation σ in the group S_k can be decomposed into *cycles*. If there are α_1 cycles of length $1,\ldots,\alpha_k$ cycles of length k, obviously $1\cdot\alpha_1+2\cdot\alpha_2+\ldots+k\cdot\alpha_k = k$ and the signature of σ is $(-1)^{k+\alpha_1+\ldots\alpha_k}$, or $(-1)^{k+s}$ where s is the total number of cycles. Moreover, the number of permutations σ corresponding to given values of $\alpha_1,\ldots,\alpha_k$ is $k!/\alpha_1!1^{\alpha_1}\ldots\alpha_k!k^{\alpha_k}$. Introduce a function on the group S_k whose value $I(\sigma)$ is equal to $\tau_1(A)^{\alpha_1}\ldots\tau_k(A)^{\alpha_k}$.

It is characterized by the following two properties.

(a) I *is a class function, that is* $I(\sigma) = I(\tau\sigma\tau^{-1})$ *for* σ,τ *in* S_k.

(b) *If* σ *is decomposed into cycles* $(1\ldots a)(a+1\ldots a+b)(a+b+1\ldots a+b+c)\ldots$, *then* $I(\sigma)$ *is equal to* $\tau_a(A)\tau_b(A)\tau_c(A)\ldots$

Moreover, our formulas take the following form:

$$k!c_k(A) = \sum_{\sigma\in S_k} (\mathrm{sgn}\sigma)\cdot I(\sigma) \quad , \tag{3.35bis}$$

$$k!h_k(A) = \sum_{\sigma\in S_k} I(\sigma) \quad . \tag{3.36bis}$$

Here is *a direct proof*. For clarity, denote by π_σ the action of the permutation σ on the space $V^{\otimes k}$ defined by formula (2.3), namely

$$\pi_\sigma(x_1\otimes \ldots \otimes x_k) = x_{\sigma^{-1}(1)}\otimes \ldots \otimes x_{\sigma^{-1}(k)} \quad . \tag{3.39}$$

The operator $P_+ = (k!)^{-1}\sum_{\sigma\in S_k}\pi_\sigma$ is a projection of $V^{\otimes k}$ onto S^kV and similarly $P_-=(k!)^{-1}\sum_{\sigma\in S_k}(\mathrm{sgn}\sigma)\cdot\pi_\sigma$ is a projection of $V^{\otimes k}$ onto Λ^kV; they both commute with $A^{\otimes k}$. The operator $P_+\cdot A^{\otimes k}$ coincides with S^kA on S^kV and with 0 on a subspace of $V^{\otimes k}$ supplementary to S^kV; hence the trace $h_k(A)$ of S^kA is equal to the trace of the operator $P_+\cdot A^{\otimes k}$ acting on $V^{\otimes k}$. Similarly $c_k(A)$ is equal to the trace of $P_-\cdot A^{\otimes k}$. Hence the formulas (3.35bis) and (3.36bis) are true with $I(\sigma)$ defined as $\mathrm{Tr}(\pi_\sigma\cdot A^{\otimes k})$. It remains to check that this function I enjoys properties (a) and (b) above.

As for (a), it follows from the cyclic invariance of the trace and the permutability of $A^{\otimes k}$ with π_τ , namely

$$I(\tau\sigma\tau^{-1}) = \mathrm{Tr}(\pi_\tau\pi_\sigma\pi_\tau^{-1}A^{\otimes k}) = \mathrm{Tr}(\pi_\sigma\pi_\tau^{-1}A^{\otimes k}\pi_\tau) =$$

$$= \mathrm{Tr}(\pi_\sigma A^{\otimes k}) = I(\sigma) \quad .$$

To prove (b), introduce a basis $e_1,\ldots,e_n$ of V, and denote by $A(i,j)$ the matrix of A with respect to this basis. The tensor products $e_{i_1}\otimes \ldots \otimes e_{i_k}$ make up a basis of $V^{\otimes k}$; the matrix of $A^{\otimes k}$ with respect to this basis is

$$A_k(i_1,\ldots,i_k;\ j_1,\ldots,j_k) = A(i_1;j_1)\ldots A(i_k;j_k) \tag{3.40}$$

and moreover π_σ transforms $e_{i_1}\otimes \ldots \otimes e_{i_k}$ into $e_{i_{\sigma^{-1}(1)}}\otimes\ldots e_{i_{\sigma^{-1}(k)}}$. It follows

$$I(\sigma) = \sum_{i_1 \ldots i_k} A(i_1, i_{\sigma(1)}) \ldots A(i_k, i_{\sigma(k)}) \quad . \tag{3.41}$$

If σ is decomposed into cycles $(1\ldots a)(a+1\ldots a+b)(a+b+1\ldots a+b+c)\ldots$, then the term under the summation sign can be written as the product of the factors

$$A(i_1,i_2)A(i_2,i_3)\ldots A(i_{a-1},i_a)A(i_a,i_1) = J_a$$

$$A(i_{a+1},i_{a+2})A(i_{a+2},i_{a+3})\ldots A(i_{a+b-1},i_{a+b})A(i_{a+b},i_{a+1})=J_b$$

. . . .

The sum in (3.41) breaks accordingly into a product. The sum of terms of the form J_a gives the trace of A^a, etc ... hence

$$\begin{aligned} I(\sigma) &= Tr(A^a)Tr(A^b)Tr(A^c)\ldots \\ &= \tau_a(A)\tau_b(A)\tau_c(A) \ldots \end{aligned}$$

This ends our proof.

One particular case of the previous results is worth mentioning. Namely, consider the cyclic permutation γ_k acting on $V^{\otimes k}$ by

$$\gamma_k(x_1 \otimes x_2 \otimes \ldots \otimes x_k) = x_k \otimes x_1 \otimes \ldots \otimes x_{k-1} \quad . \tag{3.42}$$

Then we get

$$Tr(A^k) = Tr(A^{\otimes k} \cdot \gamma_k) \quad . \tag{3.43}$$

More generally, let $A_1,\ldots,A_k$ be linear operators acting on V; define the operator $A_1 \otimes \ldots \otimes A_k$ acting on $V^{\otimes k}$ in the following way

$$(A_1 \otimes \ldots \otimes A_k)\cdot(x_1 \otimes \ldots \otimes x_k) = A_1x_1 \otimes \ldots \otimes A_kx_k \quad . \tag{3.44}$$

Then

$$Tr(A_1 \ldots A_k) = Tr((A_1 \otimes \ldots \otimes A_k)\cdot\gamma_k) \quad . \tag{3.45}$$

4. Some Integration Formulas

4.1. The *basic formula* reads as follows, in the simplest form

$$\int_{-\infty}^{+\infty} e^{-\pi x^2}\, dx = 1 \qquad . \tag{4.1}$$

The proof uses a well-known trick. Denoting by I the previous integral, we get

$$I^2 = \iint e^{-\pi(x^2+y^2)}\, dx\, dy$$

or, using polar coordinates,

$$I^2 = \int_0^\infty r dr \int_0^{2\pi} e^{-\pi r^2}\, d\theta = \int_0^\infty e^{-\pi r^2}\, 2\pi r dr = \int_0^\infty e^{-u} du = 1 \; .$$
$$(u = \pi r^2)$$

Since I is obviously positive, we get I = 1.

4.2. Let E be a *real euclidean space* of dimension n. We denote by $x \cdot y$ the dot product of two vectors. It is bilinear and symmetrical, and $x \cdot x > 0$ unless $x = 0$. Let the basis $e_1, \ldots, e_n$ of E be orthonormal; hence $e_i \cdot e_i = 1$ and $e_i \cdot e_j = 0$ for $i \neq j$. Consider the volume form ω such that $\omega(e_1, \ldots, e_n) = 1$. A classical calculation gives the following formula (where $x_1, \ldots, x_n$ are arbitrary vectors in E)

$$\omega(x_1, \ldots, x_n)^2 = \det(x_i \cdot x_j) \qquad . \tag{4.2}$$

In particular, one gets $\omega(e'_1, \ldots, e'_n)^2 = 1$ if $e'_1, \ldots, e'_n$ is another orthonormal basis of E. Up to a sign, ω is therefore independent of any orthonormal basis of E.

To the volume form ω is associated an integration element $d^n x$ in E. Explicitely, one gets

$$\int_E f(x) d^n x = \int \ldots \int f(x^1 e_1 + \ldots + x^n e_n) dx^1 \ldots dx^n \tag{4.3}$$

for every orthonormal basis $e_1, \ldots, e_n$ of E. More generally, from (4.2), one derives

$$\int_E f(x) d^n x = \int \ldots \int f(x^1 e_1 + \ldots + x^n e_n) g^{1/2}\, dx^1 \ldots dx^n \tag{4.4}$$

for an arbitrary basis (e_α) with $g_{\alpha\beta} = e_\alpha \cdot e_\beta$ and $g=\det(g_{\alpha\beta})$.

4.3. From the basic formula (4.1), one gets by multiplication

$$\int\ldots\int \exp\{-\pi \sum_{i=1}^{n} (x^i)^2\}\, dx^1 \ldots dx^n = 1 \quad . \tag{4.5}$$

To get an invariant formula, use (4.3) and get

$$\int_E e^{-\pi x\cdot x}\, d^n x = 1. \tag{4.6}$$

More generally, consider a symmetric positive-definite operator A in E. Hence $Ax\cdot x > 0$ unless $x = 0$, and $Ax\cdot y = x\cdot Ay$. It is legitimate to take as a new dot product of the vectors x and y the scalar $Ax\cdot y$. From subsection 4.2, one gets the existence of a volume form ω_A on E, associated to this dot product, and unique up to sign. From (4.2), one gets

$$\omega_A(x_1,\ldots,x_n)^2 = \det(Ax_i\cdot x_j) \tag{4.7}$$

for an arbitrary basis $x_1,\ldots,x_n$ of E. Specialize x_i to e_i and notice that the matrix (a_{ij}) of A with respect to the orthonormal basis $e_1,\ldots,e_n$ is given by $a_{ij} = Ae_i\cdot e_j$. We get

$$\omega_A(e_1,\ldots,e_n)^2 = \det(a_{ij}) = \det A. \tag{4.8}$$

Comparing with the definition of ω, we conclude

$$\omega_A = \pm (\det A)^{1/2}\, \omega \quad . \tag{4.9}$$

For the corresponding integration elements, we obtain $d^n_A x = (\det A)^{1/2} d^n x$. We can now replace $x\cdot x$ by $Ax\cdot x$ and $d^n x$ by $d^n_A x$ in formula (4.6). *Conclusion*

$$\int_E e^{-\pi Ax\cdot x}\, d^n x = (\det A)^{-1/2} \quad . \tag{4.10}$$

The most customary form is obtained by replacing A by $A/2\pi$ and reads as follows

$$\int_E \exp\{-\tfrac{1}{2} Ax\cdot x\}\, d^n x = (2\pi)^{n/2} (\det A)^{-1/2} \quad . \tag{4.11}$$

The normalization factor $(2\pi)^{n/2}$ may be troublesome when extending these formulas to the infinite-dimensional case.

4.4. We need a *complex form* of the previous formulas. We consider now a finite-dimensional Hilbert space H. That is, H is a complex vector space of dimension n, and there is given a scalar product $\langle x|y\rangle$ for vectors in H, with the following properties

$$\langle x|y_1 + y_2\rangle = \langle x|y_1\rangle + \langle x|y_2\rangle \quad , \tag{4.12}$$

$$\langle x|\lambda y\rangle = \lambda\langle x|y\rangle \quad , \tag{4.13}$$

$$\langle y|x\rangle = \langle x|y\rangle^* \quad , \quad , \tag{4.14}$$

$$\langle x|x\rangle > 0 \qquad \text{unless } x = 0 \quad , \tag{4.15}$$

where c^* is the complex-conjugate of a complex number c. We are following the physicist's convention (adopted also by Bourbaki!) that $\langle x|y\rangle$ is linear in y, and conjugate linear in x.

We denote by A^* the adjoint of any operator A in H, so

$$\langle Ax|y\rangle = \langle x|A^*y\rangle \quad . \tag{4.16}$$

Now assume that the operator A satisfies the inequality

$$\mathrm{Re}\langle z|Az\rangle > 0 \qquad \text{unless } z = 0 \quad , \tag{4.17}$$

that is $A = B + iC$ with $B^* = B$, $C^* = C$ and B is positive-definite $\langle z|Bz\rangle > 0$ for $z \neq 0$.

To define an integration element in H, let us remark that the dot product $x\cdot y = \mathrm{Re}\langle x|y\rangle$ enables one to consider the complex hilbertian space H of (complex) dimension n as a real euclidean space E of (real) dimension 2n. The dot product defines an integration element in E, (see section 4.2), which we denote by $d_H z$.

Let S be an invertible selfadjoint operator in H. It can be diagonalized, hence there exists an orthonormal basis $e_1,\ldots,e_n$ in H and nonzero real numbers $s_1,\ldots,s_n$ such that $Se_j = s_j e_j$ for $1 \leqq j \leqq n$. Put $e_{j+n} = ie_j$ and $s_{j+n} = s_j$

for $1 \leq j \leq n$. Hence $e_1,\dots,e_{2n}$ is an orthonormal basis of the real euclidean space E, and $Se_j = s_j e_j$ for $1 \leq j \leq 2n$. Since the determinant of S is equal to $s_1 \dots s_n$, we get the following change of variable formula

$$d_H(Sz) = (\det S)^2 d_H z \quad . \tag{4.18}$$

4.5. After these preparations, we state the *complex version of formula* (4.10), namely

$$\int_H e^{-\pi\langle z|Az\rangle} d_H z = (\det A)^{-1} \quad . \tag{4.19}$$

To prove it, we first remark that $A = B + iC$ where B is selfadjoint and positive-definite. Hence B can be diagonalized with strictly positive eigenvalues, and there exists a selfadjoint operator S in H such that $BS^2 = 1$. Set $D = SCS$; it is a selfadjoint operator. Using (4.18) we see that the integral $I = \int_H e^{-\pi\langle z|Az\rangle} d_H z$ is equal to

$$I = \int_H e^{-\pi\langle Sz|ASz\rangle} d_H(Sz)$$

$$= (\det S)^2 \int_H e^{-\pi\langle z|(1+iD)z\rangle} d_H z \quad .$$

Notice that $(\det S)^2 = (\det B)^{-1}$. Moreover, since D is selfadjoint, it is again diagonalizable with real eigenvalues $d_1 \dots d_n$. We obtain

$$I = (\det B)^{-1} \prod_{j=1}^{n} \int_{\mathbb{C}} e^{-\pi z^*(1+id_j)z} dz \quad . \tag{4.20}$$

Here dz is put for dxdy if $z = x + iy$. Now using polar coordinates in $\mathbb{C}$, one derives immediately

$$\int_{\mathbb{C}} e^{-\pi a|z|^2} dz = a^{-1} \tag{4.21}$$

for any complex number a such that $\operatorname{Re} a > 0$. Hence

$$I = (\det B)^{-1} \cdot \prod_{j=1}^{n} (1 + id_j)^{-1}$$

$$= (\det B)^{-1} \det (1 + iD)^{-1} \quad .$$

But

$$\det(1 + iD) = \det(1 + iSCS) = \det(1 + iS^2C)$$

$$= \det(1 + iB^{-1}C) = (\det B)^{-1}\det(B + iC) \quad .$$

This finishes the proof of (4.19).

4.6. To conclude this section, we shall rewrite (4.19) as follows

$$\int_{C^n} \exp -\{\sum_{j=1}^{n} \sum_{k=1}^{n} z_j^* a_{jk} z_k\} \, dz_1^* \wedge dz_1 \wedge \dots \wedge dz_n^* \wedge dz_n$$

$$= (2\pi i)^n (\det A)^{-1} \tag{4.22}$$

for every complex matrix $A = (a_{jk})$ whose selfadjoint part $B = \frac{1}{2}(A + A^*)$ is positive-definite. We remind the reader that for $z = x + iy$, with complex conjugate $z^* = x - iy$, one has

$$dz^* \wedge dz = 2i dx \wedge dy \quad . \tag{4.23}$$

PART TWO:

FREDHOLM DETERMINANTS

5. Fredholm Theory of Integral Equations

5.1. For orientation purposes, we record here *a few formulas in the finite-dimensional case*. We consider a vector space V with a finite basis $e_1,\ldots,e_n$ and a linear operator A acting on V, whose matrix with respect to the previous basis we denote by $(A(i,j))$. Let us denote by $x(i)$ (for $(1 \leqq i \leqq n)$ the coordinates of a vector x. Then the coordinates of the transformed vector Ax are given by

$$(Ax)(i) = \sum_{j=1}^{n} A(i,j)x(j) \qquad (1 \leqq i \leqq n) \quad . \tag{5.1}$$

Moreover, according to formula (3.10), the determinant of 1+A can be expanded as follows

$$\det(1+A) = \sum_{p \geqq 0} (p!)^{-1} \sum_{i_1 \ldots i_p} \Delta\begin{pmatrix} i_1 & \ldots & i_p \\ i_1 & \ldots & i_p \end{pmatrix} . \tag{5.2}$$

The series breaks up after the term for p = n and the *minors* of the matrix A are defined as follows

$$\Delta\begin{pmatrix} i_1 & \ldots & i_p \\ j_1 & \ldots & j_p \end{pmatrix} = \det_{\substack{1 \leq k \leq p \\ 1 \leq l \leq p}} (A(i_k, j_l)) \quad . \tag{5.3}$$

After some routine calculations with determinants, *Cramer's formula* for inverting a matrix takes the following form: the operator inverse to 1 + A exists iff the determinant Δ of 1+A does not vanish, and it is then of the form $1 - \Delta^{-1}B$ where the matrix B is given by the following power series expansion:

$$B(i,j) = \sum_{p \geqq 0} (p!)^{-1} \sum_{i_1 \ldots i_p} \Delta\begin{pmatrix} i & i_1 \ldots i_p \\ j & i_1 \ldots i_p \end{pmatrix} \tag{5.4}.$$

Again, the series breaks up after the term for $p = n-1$ since the minors of order $p+1 > n$ are all zero.

5.2. We now replace finite sums by integrals using the well-known analogy. Let us denote by Ω a compact subset of some euclidean space $\mathbb{R}^m$ and by dx the integration element in $\mathbb{R}^m$. An *integral operator with kernel* K is defined by the formula

$$Kf(x) = \int_\Omega K(x,y)\, f(y)\, dy\ . \tag{5.5}$$

Let us dnote by $C(\Omega)$ the Banach space of complex-valued continuous functions on Ω, with the norm $||f|| = \sup_{x\in\Omega} |f(x)|$. One defines $C(\Omega \times \Omega)$ similarly. Then, if K belongs to $C(\Omega \times \Omega)$, the function Kf defined by (5.5) belongs to $C(\Omega)$ if f does, and the linear operator $f \mapsto Kf$ in $C(\Omega)$ is bounded with norm $C \leq ||K||\cdot\mathrm{vol}\ (\Omega)$; here we denote by $\mathrm{vol}(\Omega)$ the volume $\int_\Omega dx$ of Ω.

Everything in the rest of this fifth section extends *verbatim* if one replaces Ω by a general Hausdorff compact space, integration being taken with respect to a regular Borel measure on Ω. This extension may be useful in some problems of classical statistical mechanics.

5.3. Let us denote by 1 the identity operator in $C(\Omega)$; it is not an integral operator of the sort previously defined, since we consider only *continuous kernels* $K(x,y)$ and not singular kernels like the Dirac $\delta(x-y)$.

Fredholm proposed (around 1900) to define the *determinant of the operator* $1+K$ (taking f into $f + Kf$) by the following formula, analogous to (5.2):

$$\det(1+K) = \sum_{p\geq 0} (p!)^{-1} \int_\Omega \dots \int_\Omega \Delta\begin{pmatrix} x_1 \dots x_p \\ x_1 \dots x_p \end{pmatrix} dx_1 \dots dx_p . \tag{5.6}$$

Again the minors are defined by analogy

$$\Delta\begin{pmatrix} x_1 \dots x_p \\ y_1 \dots y_p \end{pmatrix} = \det_{\substack{1\leq j\leq p \\ 1\leq k\leq p}} K(x_j, y_k) \quad ; \tag{5.7}$$

they are jointly continuous functions of their arguments. To prove the convergence of the series (5.6), one uses a *determinant inequality* discovered by Hadamard (and in a weaker form by Fredholm):

$$|\det A| \leq ||\vec{a}_1||_2 \cdots\cdots ||\vec{a}_p||_2 \quad . \tag{5.8}$$

Here A is any p x p matrix with complex entries, columns $\vec{a}_1,\ldots, \vec{a}_p$ in $\mathbb{C}^p$, and $||\vec{a}||_2$ is the hilbertian norm of a vector $\vec{a}$ in $\mathbb{C}^p$:

$$||\vec{a}|| = (|a_1|^2 + \ldots + |a_p|^2)^{1/2} \quad . \tag{5.9}$$

To prove (5.8), we may assume that $\det A \neq 0$, that is the columns $\vec{a}_1,\ldots,\vec{a}_p$ are linearly independent. By a well-known geometric construction, we may find new vectors $\vec{b}_1,\ldots,\vec{b}_p$ mutually orthogonal in $\mathbb{C}^p$, such that $\vec{b}_j - \vec{a}_j$ be a linear combination of the vectors $\vec{a}_1,\ldots,\vec{a}_{j-1}$. From the properties of volume forms explained at length in section 1, it follows that the matrices A with columns $\vec{a}_1,\ldots,\vec{a}_p$ and B with columns $\vec{b}_1,\ldots,\vec{b}_p$ have equal determinants. Moreover $\vec{a}_j - \vec{b}_j$ is also a linear combination of $\vec{b}_1,\ldots,\vec{b}_{j-1}$, hence it is orthogonal to $\vec{b}_j$; this implies the inequality $||\vec{b}_j|| \leq ||\vec{a}_j||$ by Pythagoras'theorem. Finally, the property that $\vec{b}_1,\ldots,\vec{b}_p$ are mutually orthogonal can be expressed by the fact that[1] B^*B is a diagonal matrix with diagonal entries $||\vec{b}_1||^2,\ldots,||\vec{b}_p||^2$. Hence

$$\begin{aligned} |\det A|^2 &= |\det B|^2 = \det(B^*B) = \\ &= ||\vec{b}_1||^2 \ldots ||\vec{b}_p||^2 \leq ||\vec{a}_1||^2 \ldots ||\vec{a}_p||^2 \end{aligned}$$

and Hadamard's inequality (5.8) follows.

Using Hadamard's inequality, we get the estimate

$$\left|\Delta\begin{pmatrix} x_1 \ldots x_p \\ y_1 \ldots y_p \end{pmatrix}\right|^2 \leq \prod_{j=1}^{p} \sum_{k=1}^{p} |K(x_j,y_k)|^2 \quad ,$$

hence

$$\left|\Delta\begin{pmatrix} x_1 \ldots x_p \\ y_1 \ldots y_p \end{pmatrix}\right| \leq p^{p/2}\, ||K||^p \quad . \tag{5.10}$$

---

[1]Here B^* is the matrix hermitian conjugate to B.

By definition, the determinant of $1+K$ is $\sum_{p\geq 0} c_p(K)$ with $c_o(K)=1$ and

$$c_p(K) = (p!)^{-1} \int_\Omega \dots \int_\Omega \Delta\begin{pmatrix} x_1 \dots x_p \\ x_1 \dots x_p \end{pmatrix} dx_1 \dots dx_p \qquad (5.11)$$
$$\text{for } p \geq 1.$$

Using (5.10), we obtain the inequality

$$|c_p(K)| \leq p^{p/2} \, ||K||^p \, \text{vol}(\Omega)^p/p! \qquad (\text{for } p \geq 1). \quad (5.12)$$

By Stirling's formula, there exists a constant C_o such that $0 < C_o < 1$ and $p! \geq C_o p^p e^{-p} \sqrt{2\pi p}$; therefore we get the estimate

$$|c_p(K)| \leq C_o^{-1} \{ ||K|| \, \text{vol}(\Omega) \, p^{-1/2} e \}^p \qquad (5.13)$$

for $p \geq 1$, hence $\lim_{p\to\infty} |c_p(K)|^{1/p} = 0$. The convergence of the series $\sum_{p\geq 0} c_p(K)$ follows.

5.4. We can view the Fredholm determinant as a *functional* $K \mapsto \det(1 + K)$ *on the Banach space* $C(\Omega \times \Omega)$. From the basic estimate (5.13), one infers that the series $\sum_{p\geq 0} c_p(K)$ converges uniformly on the set $||K|| \leq R$, for every constant $R > 0$. Hence $\det(1 + K)$ is a *continuous functional* of K.

It follows also from (5.13) that $\det(1 + zK) = \sum_{p\geq 0} c_p(K) z^p$ *is an entire function of the complex variable* z. More can be said about analyticity. For $K_1, \dots, K_p$ in $C(\Omega\times\Omega)$ let us define

$$c_p(K_1, \dots, K_p) = (p!)^{-1} \int_\Omega \dots \int_\Omega \{ \det_{\substack{1\leq j\leq p \\ 1\leq k\leq p}} K_j(x_j, x_k) \} dx_1 \dots dx_p. \qquad (5.14)$$

Obviously, $c_p(K_1, \dots, K_p)$ is a multilinear functional of $K_1, \dots, K_p$. Moreover, by construction,

$$\det(1 + K) = \sum_{p\geq 0} c_p(K, \dots, K) \qquad (5.15)$$

(with p arguments equal to K). If we set $C_p' = C_o^{-1} \text{vol}(\Omega)^p p^{-p/2} e^p$, one gets

$$\lim_{p\to\infty} (C_p')^{1/p} = 0 \quad . \qquad (5.16)$$

Moreover, from Hadamard's inequality (5.8), one gets the estimate

$$|c_p(K_1,\ldots,K_p)| \leq C_p' \; ||K_1|| \; \ldots \; ||K_p|| \qquad (5.17)$$

by a proof completely analogous to the proof of (5.13). The last three formulas express that $K \mapsto \det(1+K)$ *is an entire function on the Banach space* $C(\Omega)$ in the very strong sense used by Bourbaki [3. p.28].

As a corollary, suppose that $K_\lambda(x,y)$ is an *analytic family of kernels* in the following sense: D being a domain in the complex space $\mathbb{C}^r$, the function $(\lambda,x,y) \to K_\lambda(x,y)$ is continuous on $D \times \Omega \times \Omega$, and moreover $K_\lambda(x,y)$ is an holomorphic function of λ for x,y being fixed. Then *the determinant* $\det(1+K_\lambda)$ *is an homomorphic function of* λ *in* D.

5.5. We come now to the *multiplicative property* of Fredholm determinants. Using the analogy between finite sums and integrals, the usual matrix product suggests the following product for continuous kernels

$$(KL)(x,z) = \int_\Omega K(x,y)L(y,z)\,dy \quad . \qquad (5.18)$$

It is obviously linear in K and in L and possesses the expected associativity properties:

$$(KL)f = K(Lf) \qquad (5.19)$$

$$(KL)M = K(LM) \qquad (5.20)$$

where f is a continuous function on Ω and M another continuous kernel.

Let K and L be kernels, and consider the operators $U=1+K$ and $V=1+L$ acting on $C(\Omega)$. The product UV is of the form $1+M$ with a kernel $M = K + L + KL$. We claim:

$$\det(1+K)\det(1+L) = \det(1+K+L+KL) \quad . \qquad (5.21)$$

This could be proved by a brute force calculation. We prefer to resort to an *approximation procedure*.

A *decomposable kernel* is a function K in $C(\Omega \times \Omega)$ of the form $K(x,y) = \sum_{\alpha=1}^{r} f_\alpha(x)g_\alpha(y)$ for $f_1,\ldots,f_r$ and $g_1,\ldots,g_r$ in

$C(\Omega)$. Such functions are dense in the Banach space $C(\Omega\times\Omega)$ (a well-known lemma of Dieudonné [5, p.141],or an obvious corollary of Weierstrass' approximation theorem). Since the determinant is a continuous functional on $C(\Omega \times \Omega)$, it is enough to prove (5.21) for the case of decomposable kernels K and L.

Suppose now that K and L are decomposable kernels. Using *Schmidt's orthonormalization process*, we find a family of continous functions $f_1,\dots,f_r$ on Ω, orthonormal in the following sense:

$$\int_\Omega \overline{f_\alpha(x)} f_\beta(x)dx = \delta_{\alpha\beta} \qquad (\text{for } 1\leqq\alpha\leqq r,\ 1\leqq\beta\leqq r), \tag{5.22}$$

and complex matrices $A = (a_{\alpha\beta})$ and $B = (b_{\alpha\beta})$ of size $r \times r$ such that

$$K(x,y) = \sum_{\alpha=1}^{r} \sum_{\beta=1}^{r} f_\alpha(x)a_{\alpha\beta}\overline{f_\beta(y)} \quad , \tag{5.23}$$

$$L(x,y) = \sum_{\alpha=1}^{r} \sum_{\beta=1}^{r} f_\alpha(x)b_{\alpha\beta}\overline{f_\beta(y)} \quad . \tag{5.24}$$

The kernel $M = K + L + KL$ admits of a similar description

$$M(x,y) = \sum_{\alpha=1}^{r} \sum_{\beta=1}^{r} f_\alpha(x)\ c_{\alpha\beta}\overline{f_\beta(y)} \tag{5.25}$$

with the matrix $C = (c_{\alpha\beta})$ given by $C = A + B + AB$. By the multiplication rule for finite determinants, we obtain therefore

$$\det\ (1 + A)\ \det\ 1 + B) = \det\ (1 + C) \quad . \tag{5.26}$$

To establish the multiplicative property (5.21) for kernels, it suffices to prove the equality

$$\det\ (1 + K) = \det\ (1 + A) \tag{5.27}$$

for a kernel K as in formula (5.23).

The proof of (5.27) rests on a *generalization of the multiplicative property of determinants known as Binet-Cauchy formula*. Namely, let $U = (u_{ij})$ be a matrix of size $m \times n$ and $V = (v_{jk})$ a matrix of size $n \times p$, so that the matrix $W = UV$ is of size $m \times p$. Define the minors of r-th order of U by

$$U\binom{i_1 \dots i_r}{j_1 \dots j_r} = \det_{\substack{1 \le \alpha \le r \\ 1 \le \beta \le r}} u_{i_\alpha j_\beta} \tag{5.28}$$

and similarly for the minors of V and W. Then

$$W\binom{i_1 \dots i_r}{k_1 \dots k_r} = \sum_{j_1 < \dots < j_r} U\binom{i_1 \dots i_r}{j_1 \dots j_r} V\binom{j_1 \dots j_r}{k_1 \dots k_r} . \tag{5.29}$$

In invariant form, this can be expressed as follows. Set $E = \mathbb{C}^m$, $F = \mathbb{C}^n$, $G = \mathbb{C}^p$. Then the matrices U, V and W correspond to operators $u: F \to E$, $v: G \to F$, $w: G \to E$ such that $w = uv$. Moreover, there exists an operator $\Lambda^r u: \Lambda^r F \to \Lambda^r E$ characterized by $(\Lambda^r u)(x_1 \wedge \dots \wedge x_r) = u(x_1) \wedge \dots \wedge u(x_r)$ for vectors $x_1, \dots, x_r$ in F. One defines similarly $\Lambda^r v$ and $\Lambda^r w$. From $w = uv$ one derives immediately $\Lambda^r w = \Lambda^r u \cdot \Lambda^r v$. In natural basis for $\Lambda^r E$ and $\Lambda^r F$, the entries of the matrix of $\Lambda^r u$ consist of the minors $U\binom{i_1 \dots i_r}{j_1 \dots j_r}$. Hence formula (5.29) is the matrix version of $\Lambda^r w = \Lambda^r u \cdot \Lambda^r v$.

Introduce now continuous functions on $\Omega^p = \Omega \times \dots \times \Omega$ (p factors) as follows

$$f_{\alpha_1 \dots \alpha_p}(x_1, \dots, x_p) = \det f_{\alpha_i}(x_j) , \tag{5.30}$$

that is

$$f_{\alpha_1 \dots \alpha_p}(x_1, \dots, x_p) = \sum_{\sigma \in S_p} (\operatorname{sgn}\sigma) \cdot f_{\alpha_{\sigma(1)}}(x_1) \dots f_{\alpha_{\sigma(p)}}(x_p). \tag{5.31}$$

From the orthogonality property (5.22), one derives.

$$\int_\Omega \dots \int_\Omega \overline{f_{\alpha_1 \dots \alpha_p}(x_1 \dots x_p)}\, f_{\beta_1 \dots \beta_p}(x_1, \dots, x_p) dx_1 \dots dx_p =$$

$$= p!\, \delta_{\alpha_1 \beta_1} \dots \delta_{\alpha_p \beta_p} \tag{5.32}$$

for $\alpha_1 < \dots < \alpha_p$, $\beta_1 < \dots < \beta_p$. Moreover, from Cauchy's formula (5.29), one derives

$$\Delta\binom{x_1 \ldots x_p}{y_1 \ldots y_p} = \sum_{\substack{\alpha_1 < \ldots < \alpha_p \\ \beta_1 < \ldots < \beta_p}} A\binom{\alpha_1 \ldots \alpha_p}{\beta_1 \ldots \beta_p} f_{\alpha_1 \ldots \alpha_p}(x_1, \ldots, x_p) \cdot \overline{f_{\beta_1 \ldots \beta_p}(y_1, \ldots, y_p)} \qquad (5.33)$$

Using (5.32), one gets immediately

$$\int_\Omega \ldots \int_\Omega \Delta\binom{x_1 \ldots x_p}{x_1 \ldots x_p} dx_1 \ldots dx_p = p! \sum_{\alpha_1 < \ldots < \alpha_p} A\binom{\alpha_1 \ldots \alpha_p}{\alpha_1 \ldots \alpha_p}, \qquad (5.34)$$

and by summing over p, one obtains finally

$$\det(1 + K) = \sum_{p \geqq 0} \sum_{\alpha_1 < \ldots < \alpha_p} A\binom{\alpha_1 \ldots \alpha_p}{\alpha_1 \ldots \alpha_p} \quad . \qquad (5.35)$$

Our contention (5.27) follows by using formula (3.10) for matrices.

5.6. A crucial property of determinants in the finite-dimensional case is the following criterior: an operator A acting linearly on finite-dimensional vector space V is invertible iff its determinant is not zero. An analogous property holds for integral operators: *if* K *is a continuous kernel, the operator* 1+K *on* $C(\Omega)$ *possesses an inverse* (necessarily bounded by general results of functional analysis) *iff the Fredholm determinant* det (1+K) *is not zero.*

The proof consists of *three steps:*

(a) *Suppose that* T *is an inverse for* 1+K, hence T(1+K)=1, or T = 1 - TK. We claim that TK is an integral operator. Indeed, write $K_y(x)$ for $K(x,y)$. Then $y \mapsto K_y$ is a continuous map from Ω into the metric space $C(\Omega)$ (by uniform continuity of K). For y in Ω, set $L_y = T(K_y)$ and let $L(x,y) = L_y(x)$. Then $y \mapsto L_y$ is a continuous map from Ω into $C(\Omega)$ or, equivalently, the function L belongs to $C(\Omega \times \Omega)$. We claim that *the integral operator with kernel* L *is equal to* TK. This is easily verified if K is a decomposable kernel of the form $K(x,y) = \sum_{\alpha=1}^{r} f_\alpha(x)\, g_\alpha(y)$ (with $f_1, \ldots, f_r, g_1, \ldots, g_r$ in $C(\Omega)$). The general case is obtained by using a sequence $(K_n)_{n\geq 0}$ of decomposable kernels such that $\lim_{n\to\infty} \|K - K_n\| = 0$.

(b) *Suppose that* 1+K *is invertible*. By step (a), there exists a kernel L such that $(1+K)(1-L) = 1$. By the multiplication property (5.21), one gets

$$\det(1+K)\det(1-L) = 1, \quad ,$$

hence $\det(1+K) \neq 0$.

(c) It remains to prove that *when the Fredholm determinant of* 1+K *is not zero, the operator* 1+K *is invertible*. This will be done by providing an explicit formula for the inverse. For every integer $p \geq 0$, one defines a continuous kernel L_p by

$$L_p(x,y) = \int_\Omega \cdots \int_\Omega \Delta\begin{pmatrix} x & x_1 \cdots x_p \\ y & x_1 \cdots x_p \end{pmatrix} dx_1 \ldots dx_p \quad . \tag{5.36}$$

From the basic estimate (5.10), one derives

$$|L_p(x,y)| \leq (p+1)^{(p+1)/2} \|K\|^{p+1} \operatorname{vol}(\Omega)^p \quad . \tag{5.37}$$

Using Stirling's formula as in section 5.3, it follows that the series $\sum_{p\geq 0} (p!)^{-1} L_p(x,y)$ converges uniformly on $\Omega \times \Omega$, hence its sum $L(x,y)$ defines a continuous kernel.

From its definition, the Fredholm determinant $\Delta = \det(1+K)$ is given by the series $\sum_{p\geq 0} \gamma_p/p!$ where γ_p is defined as follows:

$$\gamma_p = \int_\Omega \cdots \int_\Omega \Delta\begin{pmatrix} x_1 \cdots x_p \\ x_1 \cdots x_p \end{pmatrix} dx_1 \ldots dx_p \quad . \tag{5.38}$$

Notice that $\Delta\begin{pmatrix} x & x_1 \cdots x_p \\ y & x_1 \cdots x_p \end{pmatrix}$ is the determinant of the matrix

$$\begin{pmatrix} K(x,y) & K(x,x_1) & \ldots\ldots & K(x,x_p) \\ K(x_1,y) & K(x_1,x_1) & \ldots\ldots & K(x_1,x_p) \\ \cdot & \cdot & \cdot & \cdot \\ K(x_p,y) & K(x_p,x_1) & \ldots\ldots & K(x_p,x_p) \end{pmatrix} .$$

Developing this determinant according to its first row, one obtains

$$\Delta\begin{pmatrix} x & x_1 \ldots x_p \\ y & x_1 \ldots x_p \end{pmatrix} = K(x,y)\Delta\begin{pmatrix} x_1 \ldots x_p \\ x_1 \ldots x_p \end{pmatrix} - \sum_{j=1}^{p} K(x,x_j)\Delta\begin{pmatrix} x_j & x_1 \ldots \hat{x}_j \ldots x_p \\ y & x_1 \ldots \hat{x}_j \ldots x_p \end{pmatrix} \quad . \tag{5.39}$$

We use the standard convention that a term with a carret $\hat{}$ has to be omitted. Integrating with respect to $x_1 \ldots x_p$ gives then

$$L_p(x,y) = \gamma_p K(x,y) - p \int_\Omega K(x,z) L_{p-1}(z,y)dz \quad ; \tag{5.40}$$

notice that the labelling of the integration variables being irrelevant the p terms in the summation of (5.39) give the same integral. Notice also the limiting case $L_0(x,y) = \gamma_0 K(x,y)$ for $p = 0$ (and $\gamma_0 = 1$!). Since $L(x,y)$ is given by the uniformly convergent series $\sum_{p \geq 0} L_p(x,y)/p!$, integration term by term is legitimate, and from (5.40) one gets

$$L(x,y) = \Delta K(x,y) - \int_\Omega K(x,z)L(z,y)\, dz \tag{5.41}$$

In terms of kernels, this formula can be stated as follows (where Δ is a constant):

$$L = \Delta K - KL \quad . \tag{5.42}$$

By a similar proof, expanding the determinant $\Delta\begin{pmatrix} x & x_1 \ldots x_p \\ y & x_1 \ldots x_p \end{pmatrix}$ according to its first column,we get

$$L = \Delta K - LK \quad . \tag{5.43}$$

It is time to assume $\Delta \neq 0$. The previous formulas (5.42) and (5.43) just mean that *the operator* $1 - \Delta^{-1}L$ *is an inverse for* $1 + K$. The reader will undoubtedly notice the analogy of this statement with formula (5.4).

Let us just add one *remark*. An inverse of $1+K$ is given by the geometris series $(1+K)^{-1} = 1-K+K^2 \ldots = 1 - \sum_{p \geq 0} (-1)^p K^{p+1}$.

By definition, K^{p+1} is given by a p-fold integral

$$K^{p+1}(x,y) = \int_\Omega \dots \int_\Omega K(x,x_1)K(x_1,x_2) \dots K(x_p,y)dx_1 .. dx_p \ , \tag{5.44}$$

hence the estimate $|K^{p+1}(x,y)| \leqq ||K||^{p+1} \operatorname{vol}(\Omega)^p$. *The convergence of the series* $M = \sum_{p\geqq 0} (-1)^p K^{p+1}$ is therefore guaranteed if $||K|| < \operatorname{vol}(\Omega)^{-1}$, *but may fail in general.* Multiply M by the convergent series $\Delta = \sum_{p\geqq 0} \gamma_p/p!$, and rearrange the terms according to Cauchy's rule for multiplying series. We get $\Delta M = \sum_{p\geqq 0} L'_p/p!$ with

$$L'_0 = \gamma_0 K \ , \quad L'_1 = \gamma_1 K - \gamma_0 K^2, \quad L'_2 = \gamma_2 K - 2\gamma_1 K^2 + 2\gamma_0 K^3, \dots$$

Formula (5.40) provides a recursive definition of the kernel L_p, and the equality $L'_p = L_p$ follows easily. Using Hadamard's inequality (5.10) as above then shows that the series $\sum_{p\geqq 0} L'_p/p!$ converges uniformly on $\Omega \times \Omega$. Hence, *after rearranging, the product* ΔM *is given by a convergent series, even if* M *does not.*

5.7. Using once more the analogy between sums and integrals, we are led to define *the trace of a kernel* K as the scalar

$$\operatorname{Tr}(K) = \int_\Omega K(x,x)\, dx \ . \tag{5.45}$$

The trace of a product of p kernels is given by the multiple integral

$$\operatorname{Tr}(K_1 \dots K_p) = \int_\Omega \dots \int_\Omega K_1(x_1,x_2)K_2(x_2,x_3) \dots K_{p-1}(x_{p-1},x_p)K_p(x_p,x_1)dx_1 \dots dx_p \ . \tag{5.46}$$

It is then obvious that *the trace* $\operatorname{Tr}(K_1 \dots K_p)$ *is invariant under cyclic permutations of* $K_1,\dots,K_p$.

We proceed now to the proof of an analogous to Waring's formula, namely

$$\det(1 - zK) = \exp\{-\sum_{n\geqq 1} \operatorname{Tr}(K^n)z^n/n\} \ . \tag{5.47}$$

Notice the estimate $|\mathrm{Tr}(K^n)| \leq ||K||^n \mathrm{vol}(\Omega)^n$, which follows from (5.46); the series $\sum_{n\geq 1} \mathrm{Tr}(K^n)z^n/n$ is then guaranteed to converge when $|z|<||K||^{-1}\mathrm{vol}(\Omega)^{-1}$ and the identity (5.47) holds under this assumption.

Denote by $I(\sigma)$ (for σ in the symmetric group S_p) the following integral

$$I(\sigma) = \int_\Omega \cdots \int_\Omega K(x_1,x_{\sigma(1)}) \cdots K(x_p,x_{\sigma(p)}) dx_1 \cdots dx_p \tag{5.48}$$

By expanding completely the determinant $\Delta\begin{pmatrix} x_1 \ldots x_p \\ x_1 \ldots x_p \end{pmatrix}$, one gets

$$\gamma_p = \sum_{\sigma\in S_p} (\mathrm{sgn}\sigma) \cdot I(\sigma) \quad . \tag{5.49}$$

Moreover, one does not change the integral (5.48) by relabelling the integration variables x_j as $y_{\tau(j)}$, for τ in S_p. It then follows easily that

$$I(\sigma) = I(\tau\sigma\tau^{-1}) \quad , \tag{5.50}$$

that is, $I(\sigma)$ is a *class function* of σ in S_p. Finally, if σ is decomposed into cycles $(1\ldots a)(a+1\ldots a+b)(a+b+1\ldots a+b+c)\ldots$ the integrand in (5.48) can be written as a product of the factors

$$K(x_1,x_2) \ldots K(x_{a-1},x_a)\, K(x_a,x_1) = J_a$$

$$K(x_{a+1},x_{a+2}) \ldots K(x_{a+b-1},x_{a+b})\, K(x_{a+b},x_{a+1}) = J_b$$

. . . .

The integral (5.48) splits accordingly, and by (5.46), we get

$$I(\sigma) = \mathrm{Tr}(K^a)\, \mathrm{Tr}(K^b)\, \mathrm{Tr}(K^c) \ldots \quad . \tag{5.51}$$

The rest of the proof of formula (5.47) is now completely similar to the proof in the finite-dimensional case (see subsection 3.8).

The consequences of formula (5.47) are derived as in subsection 3.7. Taking the logarithm, we get

$$\log \det(1 - zK) = -\sum_{n\geq 1} \mathrm{Tr}(K^n)z^n/n \tag{5.52}$$

for the principal branch, in the domain $|z| < ||K||^{-1} \operatorname{vol}(\Omega)^{-1}$. By derivation, this implies

$$\frac{d}{dz} \log \det (1-zK) = - \sum_{n \geq 1} \operatorname{Tr}(K^n) z^{n-1} = -\operatorname{Tr}(K(1-Kz)^{-1}) \tag{5.53}$$

in the same domain. Recall the series expansion

$$\det(1 + zK) = \sum_{p \geq 0} c_p(K)\, z^p \quad , \tag{5.54}$$

where $c_p(K)$ is defined by formula (5.11). We get an inductive definition of these coefficients

$$pc_p(K) = \sum_{j=1}^{p} (-1)^{j+1} \operatorname{Tr}(K^j) c_{p-j}(K) \qquad (\text{for } p \geq 1). \tag{5.55}$$

Finally, $c_p(K)$ can be given in *determinantal form* (with $\tau_j = \operatorname{Tr}(K^j)$):

$$p!c_p(K) = \det \begin{pmatrix} \tau_1 & \tau_2 & \tau_3 & \cdots & \tau_{p-2} & \tau_{p-1} & \tau_p \\ p-1 & \tau_1 & \tau_2 & \cdots & \tau_{p-3} & \tau_{p-2} & \tau_{p-1} \\ 0 & p-2 & \tau_1 & \cdots & \tau_{p-4} & \tau_{p-3} & \tau_{p-2} \\ \cdots & \cdots & \cdots & \cdots & \cdots & \cdots & \cdots \\ 0 & 0 & 0 & \cdots & 2 & \tau_1 & \tau_2 \\ 0 & 0 & 0 & \cdots & 0 & 1 & \tau_1 \end{pmatrix} . \tag{5.56}$$

5.8. *Fredholm's alternative* (or at least the main statement in in) reads as follows:

- either the operator 1+K is invertible,
- or there exists a nonzero function f in $\mathcal{C}(\Omega)$ such that $(1+K)f = 0$.

According to the theorem proved in subsection 5.6, the first case occurs if $\det(1+K) \neq 0$ and the second if $\det(1+K) = 0$. Otherwise stated, if the Fredholm determinant of 1+K is 0, then there exists a nonzero function f in $\mathcal{C}(\Omega)$ such that $(1+K)f = 0$.

Consider the entire function $D(z) = \det(1 - zK)$, with $D(0) = 1$. From the previous statement, one concludes that a nonzero complex number λ is a zero of this entire function, namely $D(\lambda) = 0$, iff there exists a nonzero function f in $C(\Omega)$ such that $(1-\lambda K)f = 0$, or equivalently $Kf = \lambda^{-1}f$. Hence *the zeroes of the entire function* $D(z)$ *are the inverses of the nonzero eigenvalues of the operator* K. In typical applications of Fredholm theory, the integral operator with continuous kernel K shall be the inverse of some differential operator P, and the equations $Kf = \lambda^{-1}f$ and $Pf = \lambda f$ are equivalent. In this case, *the zeroes of the entire function* $D(z)$ *are the eigenvalues of the operator* P.

The Fredholm alternative was established by Fredholm using computational methods. The method used in modern textbooks is due to F.Riesz and relies on a *compactness property* of the integral operators, namely: *the closure of the set of functions* Kf, *for* f *in* $C(\Omega)$ *of norm* ≤ 1, *is compact in the metric space* $C(\Omega)$. This is proved using a sequence of decomposable kernel K_n, converging uniformly to K on $\Omega \times \Omega$; then the integral operator with kernels K_n is of finite rank and converges in operator norm to the integral operator with kernel K.

5.9. Integral operators with continuous kernels are special cases of Hilbert-Schmidt operators (see subsection 7.16). The eigenvalues of these operators satisfy therefore the following properties:

(a) *the nonzero eigenvalues of the operator* K *can be arranged into a sequence* $(\lambda_j)_{j\geq 1}$ *with* $|\lambda_1| \leq |\lambda_2| \geq \ldots$ *and* $\lim_{j\to\infty} \lambda_j = 0$, *each eigenvalue being repeated according to a well-specified multiplicity.*

(b) *The sum* $\sum_{j\geq 1} |\lambda_j|^2$ *if finite, but in general* $\sum_{j\geq 1} |\lambda_j|$ *is not finite.*

(c) *For the traces of powers of* K, *one gets*

$$\mathrm{Tr}(K^n) = \sum_{j\geq 1} \lambda_j^n \qquad \textit{for every integer } n\geq 2 \ . \tag{5.57}$$

It may occur that there is only a finite number of nonzero eigenvalues for K, possibly none of them. In this case, formula (5.57) remains valid with an obvious interpretation, namely $Tr(K^n) = 0$ (for every integer $n \geq 2$) if there is no nonzero eigenvalue of K.

We can now derive a *product formula* for the Fredholm determinant, namely

$$\det(1 - zK) = e^{-zTr(K)} \prod_{j \geq 1} (1 - \lambda_j z) e^{\lambda_j z} \qquad (5.58)$$

for every complex number z. If there is no nonzero eigenvalue of K, the product in (5.58) has to be interpreted as 1, that is $\det(1 - zK) = e^{-zTr(K)}$ in this case. We mention also that the infinite product in (5.58) converges uniformly on every bounded domain of $\mathbb{C}$, since $\sum_{j \geq 1} |\lambda_j|^2$ is finite.

To prove formula (5.58) define two entire functions by the formulas

$$D_1(z) = e^{zTr(K)} \det(1-zK), \quad D_2(z) = \prod_{j \geq 1} (1-\lambda_j z) e^{\lambda_j z} \quad . \qquad (5.59)$$

We want to prove the equality $D_1(z) = D_2(z)$. By analytic continuation, it suffices to prove that these functions, with values $D_1(0) = D_2(0) = 1$ at the origin, have equal logarithms around the origin. By formula (5.52) one gets

$$-\log D_1(z) = \sum_{n \geq 2} Tr(K^n) \, z^n/n \qquad (5.60)$$

whenever $|z| < ||K||^{-1} vol\,(\Omega)^{-1}$. Moreover, using the usual Taylor series for the logarithm , one gets

$$-\log D_2(z) = \sum_{j \geq 1} \sum_{n \geq 2} (\lambda_j z)^n/n \qquad (5.61)$$

whenever $|z| < (\sum_{j \geq 1} |\lambda_j|^2)^{-1/2}$. This double series is then absolutely convergent, hence can be rearranged as

$$-\log D_2(z) = \sum_{n \geq 2} (\sum_{j \geq 1} \lambda_j^n) z^n/n \qquad (5.62)$$

Using now the equality of $Tr(K^n)$ and $\sum_{j \geq 1} \lambda_j^n$ (formula (5.57)), one deduces $-\log D_1(z) = -\log D_2(z)$, hence $D_1(z) = D_2(z)$ for $|z|$ small.

From the formula (5.58), one deduces that *the multiplicity of* λ *as a zero of the entire function* $D(z) = \det(1 - zK)$ *is equal to the multiplicity of* λ^{-1} *as an eigenvalue of* K.

5.10. *One central difficulty* in Fredholm's theory is that *the identity* (5.57) *is valid for* $n \geqq 2$, *but not in general for* $n = 1$. When the series $\sum_{j\geqq 1} \lambda_j$ converges to the sum $\mathrm{Tr}(K)$, formula (5.58) can be simplified, namely

$$\det(1 - zK) = \prod_{j\geqq 1} (1 - \lambda_j z) \quad . \tag{5.63}$$

That this can not be true in general can be seen using Fourier series. Namely, assume that our space Ω is the closed interval $[0,1]$ with the endpoints 0 and 1 identified and consider a kernel of the form $K(x,y) = k(x-y)$, where k is a continuous function on the real line with period one: $k(x+1)=k(x)$. Introduce the exponential functions e_n by $e_n(x) = e^{2\pi i n x}$. Any continuous function f on Ω with $f(0) = f(1)$ can be uniformly approximated by finite linear combinations of the e_n's. Moreover, one gets $Ke_n = c_n e_n$ where c_n is the usual Fourier coefficient $\int_0^1 k(x)e_n(-x)dx$ of k. The eigenvalues of the integral operator with kernel K are therefore the Fourier coefficients c_n. One gets

$$\mathrm{Tr}(K) = k(0) \quad , \qquad \mathrm{Tr}(K^2) = \int_0^1 k(x)k(-x)dx \ . \tag{5.64}$$

The identity $\mathrm{Tr}(K^2) = \sum_{n=-\infty}^{+\infty} c_n^2$ *is just another form of Parseval's identity. But there are well-known examples* (see Titchmarsh [14, p.416]) *of continuous periodic functions* $k(x)$ *whose Fourier series* $\sum_{n=-\infty}^{+\infty} c_n e_n(x)$ *fails to converge and represent* $k(x)$, *for* $x = 0$ *say.*

6. A Review of Operator Theory in Hilbert Spaces

6.1. We consider a Hilbert space H, and we assume that H is both infinite-dimensional and separable. Then there exists an orthonormal basis $(\psi_n)_{n\geq 0}$ in H. Every vector ψ in H is determined by its components $c_n = \langle\psi_n|\psi\rangle$, restricted only by the convergence of the series $\sum_{n\geq 0} |c_n|^2$ which represents $\langle\psi|\psi\rangle$, that is the square of the norm $||\psi||$. For scalar products, we follow the conventions introduced in subsection 4.4.

Let A be a bounded operator in H. We shall write $\langle\psi|A|\psi'\rangle$ for the scalar product of ψ with $A\psi'$. We say that A is *selfadjoint* in case $\langle\psi|A|\psi\rangle$ is real for every ψ, and that it is *positive* in case $\langle\psi|A|\psi\rangle \geq 0$ for every ψ in H.

Assume now A to be positive. We claim that *the sum* $\sum_{n\geq 0} \langle\psi_n|A|\psi_n\rangle$ (a positive number or $+\infty$) *is independent of the orthonormal basis* (ψ_n). Indeed, A possesses a square root B (also denoted by $A^{\frac{1}{2}}$) which is the unique positive operator B such that $B^2 = A$. Choose another orthonormal basis $(\theta_m)_{m\geq 0}$. Our contention follows from the following calculation:

$$\sum_n \langle\psi_n|A|\psi_n\rangle = \sum_n \langle\psi_n|B^*B|\psi_n\rangle = \sum_n \langle B\psi_n|B\psi_n\rangle =$$

$$= \sum_{n,m} |\langle\theta_m|B\psi_n\rangle|^2 = \sum_{n,m} |\langle\theta_m|B|\psi_n\rangle|^2 .$$

Indeed, since B is selfadjoint $\langle\theta_m|B|\psi_n\rangle$ is the complex-conjugate of $\langle\psi_n|B|\theta_m\rangle$ and a symmetrical calculation gives the result

$$\sum_{m\geq 0} \langle\theta_m|A|\theta_m\rangle = |\sum_{m,n} \langle\psi_n|B|\theta_m\rangle|^2 .$$

We define the *trace of the positive operator* A as the number $Tr(A) = \sum_{n\geq 0} \langle\psi_n|A|\psi_n\rangle$ in $[0,+\infty]$.

6.2. A fundamental theorem asserts that *a positive operator with a finite trace can be diagonalized.* More precisely, we can find an orthonormal set $(\psi_0,\psi_1,\ldots,\psi_n,\ldots)$ in H and a non-increasing sequence $(\lambda_0,\lambda_1,\lambda_2,\ldots,\lambda_n,\ldots)$ of strictly positive numbers such that $A\psi_n = \lambda_n\psi_n$ for every n, and $A\psi = 0$ for every vector ψ in H orthogonal to all ψ_n's. Both sequences can consist of finitely many terms $(\psi_0,\ldots,\psi_{N-1})$ and $(\lambda_0,\ldots,\lambda_{N-1})$ if A is of finite rank N, or be infinite. In the last case, completing the sequence $(\psi_n)_{n\geq 0}$ to an orthonormal basis of H, one concludes to the equality

$$\mathrm{Tr}(A) = \sum_{n=0}^{\infty} \lambda_n \quad . \tag{6.1}$$

A similar statement holds in the finite rank case.

The vectors ψ_n are not uniquely defined by the previous conditions. The scalars λ_n are unique because of the following peoperty: for any number $\lambda > 0$, the number of times λ occurs in the sequence $(\lambda_0,\lambda_1,\ldots)$ is equal to the dimension of the eigenspace corresponding to λ, that is the vectors ψ in H such that $A\psi = \lambda\psi$.

6.3. For the mathematically inclined reader, we sketch a proof of the diagonalization theorem based on *a compactness argument* (due essentially to Hilbert). Let A, B and θ_m as in subsection 6.1. Denote by H_1 the set of vectors ψ in H such that $||\psi|| \leq 1$. Associating to a vector ψ its components $c_m = \langle\theta_m|\psi\rangle$, we map bijectively H_1 onto the subset Σ of the sequence space $\mathbb{C}^{\mathbb{N}}$ consisting of the sequences $\vec{c} = (c_m)_{m\geq 0}$ such that $\sum_{m=0}^{\infty} |c_m|^2 \leq 1$. Endow $\mathbb{C}^{\mathbb{N}}$ with the product topology, which may be defined via the distance $\sum_{m\geq 0} 2^{-m}\inf(|c_m - c'_m|, 1)$ between two sequences $\vec{c} = (c_m)$ and $\vec{c}\,' = (c'_m)$. If $\vec{c} = (c_m)$ belongs to Σ, then $|c_m| \leq 1$ for every m; moreover $\Sigma = \bigcap_{m\geq 0} \Sigma_m$ where Σ_m is defined by the inequality $|c_0|^2 + \ldots + |c_m|^2 \leq 1$. It follows that Σ is bounded and closed in $\mathbb{C}^{\mathbb{N}}$, hence compact by Tychonov's theorem. Transporting the topology to H_1, we conclude that H_1 *is a compact Hausdorff space for what is the weak topology.*

Let θ be a fixed vector in H, with components d_m. Since a uniform limit of continuous functions in continuous, the inequality

$$|\langle\theta|\psi\rangle - (\bar{d}_0 c_0 + \ldots + \bar{d}_m c_m)| \leq \sum_{p=m+1}^{\infty} |d_p|^2 \quad .$$

for ψ in H_1, together with $\lim_{m\to\infty} \sum_{p=m+1}^{\infty} |d_p|^2 = 0$, shows that the function $\psi \mapsto \langle\theta|\psi\rangle$ is continuous on H_1.

For ψ in H_1, put $F(\psi) = \langle\psi|A|\psi\rangle$. A calculation similar to the one in subsection 6.1 gives $F(\psi) = \sum_{m=0}^{\infty} |\langle B\theta_m|\psi\rangle|^2$. Since $|\langle B\theta_m|\psi\rangle|^2$ is majorized by $||B\theta_m||^2$ and the series $\sum_{m=0}^{\infty} ||B\theta_m||^2$ converges to the finite limit $\sum_{m=0}^{\infty} \langle\theta_m|A|\theta_m\rangle = Tr(A)$, it follows, again by uniform convergence, that *the functional* F *is continuous on* H_1. Since the space H_1 is compact, F achieves its maximum at some point ψ_0 of H_1. If $\psi_0 = 0$, then F is identically 0 on H_1, hence $A = 0$. Otherwise, $||\psi_0||^{-1}\psi_0$ belongs to H_1, hence $F(\psi_0)$ majorizes $F(||\psi_0||^{-1}\psi_0) = F(\psi_0)/||\psi_0||^2$, hence $||\psi_0|| = 1$. Let ψ in H be orthogonal to ψ_0; comparing the values of F at ψ_0 and at the point $\psi(t) = (\psi_0 + t\psi)/\sqrt{1+t^2}$ for real t, one gets $\mathrm{Re}\langle\psi|A|\psi_0\rangle = 0$. Replacing ψ by $i\psi$ we prove that $\mathrm{Im}\langle\psi|A|\psi_0\rangle$ is also 0. *Conclusion:* $A\psi_0$ is orthogonal to any vector orthogonal to ψ_0, that is $A\psi_0 = \lambda_0\psi_0$ for some scalar λ_0. Notice that $F(\psi_0)=\langle\psi_0|A\psi_0\rangle = \lambda_0$, hence $\lambda_0 > 0$.

We repeat the same reasoning in the space $H(1)$ of vectors orthogonal to ψ_0. We get a vector ψ_1 in $H(1)$, of norm 1, such that ψ_1 attains on $H_1 \cap H(1)$ its maximum λ_1 at ψ_1; moreover $\lambda_1 \geq 0$ and $A\psi_1 = \lambda_1\psi_1$. Continuing in this way, we get an orthonormal sequence $(\psi_n)_{n\geq 0}$ of vectors and scalars $\lambda_n > 0$ such that $A\psi_n = \lambda_n\psi_n$, $\lambda_0 \geq \lambda_1 \geq \lambda_2 \geq \ldots \geq \lambda_n \geq \lambda_{n+1} \geq \ldots$ The process may break after a finite number of steps, or give an infinite sequence. Anyhow, if a vector in H_1 is orthogonal to $\psi_0, \psi_1, \ldots,$ one gets $F(\psi) = 0$, that is $B\psi = 0$, that is $A\psi = 0$ (*Hint:* use the definition of the λ_n as successive maxima !).

6.4. For any bounded operator A on H, the operator A^*A is (selfadjoint) positive, hence its square root $|A| = (A^*A)^{1/2}$ is defined. We set

$$||A||_1 = Tr(|A|) \tag{6.2}$$

and denote it the *trace-norm*. The basic property

$$||A + B||_1 \leq ||A||_1 + ||B||_1 \tag{6.3}$$

is not at all obvious. The main difficulty is to prove that if $||A||_1$ and $||B||_1$ are finite, then $|A + B|$ can be diagonalized; one can again rely on compactness arguments.

One denotes by $L^1(H)$ the set of bounded operators A in H for which $||A||_1$ is finite. According to (6.3), this is a subspace of the vector space $L(H)$ of all bounded operators. Moreover, any element A in $L^1(H)$ is a finite linear combination of positive operators A_j, with $\mathrm{Tr}(A_j)$ finite. It follows that the trace of A can be defined by

$$\mathrm{Tr}(A) = \sum_{n\geq 0} \langle\psi_n|A|\psi_n\rangle \quad ; \tag{6.4}$$

the series converges absolutely for every orthonormal basis $(\psi_n)_{n\geq 0}$ of H, and its sum does not depend on the particular basis chosen. Moreover, one proves the inequality

$$|\mathrm{Tr}(A)| \leq ||A||_1 \quad . \tag{6.5}$$

By formula (6.3), $A \mapsto ||A||_1$ is a norm on the space $L^1(H)$, *which is a Banach space*, that is satisfies Cauchy's convergence criterion.

Let A in $L^1(H)$. As we saw in subsection 6.2 and 6.3, the operator $|A|$ can be diagonalized. From the definition of $|A|$ by $|A|^2 = A^*A$, it follows that $A\psi$ and $|A|\psi$ have the same norm for every vector ψ in H. Suppose that $|A|$ is not of finite rank, and diagonalize it with eigenvectors ψ_n and eigenvalues λ_n as in subsection 6.2. Then

$$||A\psi_n|| = ||\ |A|\psi_n|| = ||\lambda_n\psi_n|| = \lambda_n$$

and, for $m \neq n$

$$\langle A\psi_n|A\psi_m\rangle = \langle\psi_n|A^*A\psi_m\rangle = \langle\psi_n|\ |A|^2\psi_m\rangle =$$

$$= \langle\psi_n|\lambda_m^2\psi_m\rangle = 0 \quad .$$

It follows that there exists an orthonormal sequence $(\theta_n)_{n\geq 0}$ such that $A\psi_n = \lambda_n\theta_n$ for every n, hence the following representation for the operator A

$$A\psi = \sum_{n=0}^{\infty} \lambda_n \theta_n \langle\psi_n|\psi\rangle \qquad (\text{for } \psi \text{ in } H) \quad . \qquad (6.6)$$

Notice that the series converges in norm since $\sum_n \lambda_n$ is finite. In Dirac's notation, this can be expressed as

$$A = \sum_{n=0}^{\infty} \lambda_n |\theta_n\rangle\langle\psi_n| \quad . \qquad (6.7)$$

When A is of finite rank N, just replace the summation symbol by $\sum_{n=0}^{N-1}$.

Conversely, if ψ_n and θ_n are vectors of norm 1 in H, and the scalars λ_n form an absolutely convergent series, then formula (6.7) defines an operator A in $L^1(H)$; we do not assume any orthogonality property of the vectors ψ_n and θ_n. In a series of operators like (6.7), the following estimate holds

$$||A - \sum_{n=0}^{N-1} \lambda_n |\theta_n\rangle\langle\psi_n| \,||_1 \leq \sum_{n=N}^{\infty} |\lambda_n|$$

and since $\sum_{n=N}^{\infty} |\lambda_n|$ tends to 0 with 1/N, it follows that *the operators of finite rank are dense in the Banach space* $L^1(H)$.

Define $\mu_n(A)$ as the eigenvalue of rank n of $|A|$. Hence

$$\mu_0(A) \geq \mu_1(A) \geq \ldots \geq \mu_n(A) \geq \mu_{n+1}(A) \geq \ldots \geq 0 \quad ,$$

and put $\mu_n(A) = 0$ for $n \geq N$ if A is of finite rank N. Moreover by definition

$$||A||_1 = \sum_{n=0}^{\infty} \mu_n(A) \quad . \qquad (6.8)$$

The inequality (6.3) can be strengthened to the sequence of inequalities

$$\mu_{n+m}(A+B) \leq \mu_n(A) + \mu_m(B) \qquad (6.9)$$

for $n > 0$, $m \geq 0$ and A,B in $L^1(H)$. For $n = 0$, $\mu_0(A)$ is equal to the *operator norm of* A, that is the smallest constant $||A||$ such that $||A\psi|| \leq ||A||\;||\psi||$ for all vectors ψ in H.

In general, $\mu_n(A)$ can be calculated using the *minimax principle*. Let V be a vector subspace of H, of finite dimension n; let us denote by $||A||_V$ the smallest constant such that $||A\psi|| \leq ||A||_V ||\psi||$ for each vector ψ in H orthogonal to V. Then the following inequality holds

$$||A||_V \geqq \mu_n(A) \quad , \tag{6.10}$$

with equality when V is spanned by the vectors $\psi_0, \ldots, \psi_{n-1}$, where $|A|\psi_k = \mu_k(A)\psi_k$ for any $k \geqq 0$.

6.5. The *trace class* of operators $L^1(H)$ is very important in theory, but there exists no easy criterion to decide whether a concrete operator is in $L^1(H)$. According to the connection between Fourier series and integral operators described in subsection 5.10, such a criterion would settle the question of characterizing the continuous functions with absolutely convergent Fourier series, a notably difficult question.

Contrasting with this situation, *Hilbert-Schmidt operators* are plentiful and easy to characterize. Define $L^2(H)$ as the class of operators A for which $\mathrm{Tr}(A^*A) = \mathrm{Tr}(|A|^2)$ is finite. This means that the positive operator $|A|$ can be diagonalized with eigenvectors $\psi_0, \psi_1, \ldots, \psi_n, \ldots$ and eigenvalues $\lambda_0, \lambda_1, \ldots, \lambda_n, \ldots$ such that $\sum_{n=0}^{\infty} |\lambda_n|^2$ be finite. As before, set $\mu_n(A) = \lambda_n$ and define the Hilbert-Schmidt norm by

$$||A||_2 = \mathrm{Tr}(A^*A)^{1/2} = (\sum_{n=0}^{\infty} \mu_n(A)^2)^{1/2} \quad . \tag{6.11}$$

The minimax principle holds again, as well as the representation (6.7) with two orthonormal sequences (ψ_n) and (θ_n).

Let $(e_n)_{n \geqq 0}$ be an arbitrary orthonormal basis of H. One gets, for any bounded operator A in $L(H)$

$$\mathrm{Tr}(A^*A) = \sum_n \langle e_n|A^*A\, e_n\rangle = \sum_n ||Ae_n||^2 =$$

$$= \sum_{m,n} |\langle e_m|A|e_n\rangle|^2 \quad .$$

If we associate to any operator A its matrix with entries $a_{mn} = \langle e_m|A|e_n\rangle$, we get an isomorphism of $L^2(H)$ with the set of matrices (a_{mn}) such that $\sum_{m,n} |a_{mn}|^2$ be finite. Otherwise stated $L^2(H)$ is a Hilbert space, and the operators $|e_m\rangle\langle e_n|$ form an orthonormal basis in $L^2(H)$.

Let A,B be operators in $L^2(H)$. From the *polarization formula*

$$4A^*B = (A+B)^*(A+B) - (A-B)^*(A-B) - i(A+iB)^*(A+iB) + i(A-iB)^*(A-iB) \quad ,$$

and the definition of $L^2(H)$, it follows that A^*B is in $L^1(H)$, and the trace of A^*B is defined. By repeating the calculation of $\mathrm{Tr}(A^*A)$, one gets

$$\mathrm{Tr}(A^*B) = \sum_{m,n} \overline{\langle e_m|A|e_n\rangle} \langle e_m|B|e_n\rangle \quad . \tag{6.12}$$

Otherwise stated, the scalar product in the Hilbert space $L^2(H)$ is given by

$$\langle A|B\rangle = \mathrm{Tr}(A^*B) \quad . \tag{6.13}$$

Cauchy-Schwarz inequality then holds:

$$|\mathrm{Tr}(A^*B)| \leq ||A||_2 ||B||_2 \quad . \tag{6.14}$$

It can be strengthened to

$$||AB||_1 \leq ||A||_2 ||B||_2 \tag{6.15}$$

(compare with formula (6.5) and notice that $L^2(H)$ is stable under $A \mapsto A^*$). It can be shown that conversely, any operator in $L^1(H)$ can be factored as the product of two operators in $L^2(H)$.

6.6. We conclude by two remarks:

(a) Suppose that Ω is any (measurable) subset of some euclidean space $\mathbb{R}^m$. Consider the Hilbert space $L^2(\Omega)$ of square-integrable functions on Ω with scalar product

$$\langle f|g\rangle = \int_\Omega \overline{f(x)}\, g(x)\, dx \quad . \tag{6.16}$$

The Hilbert-Schmidt operators in $L^2(\Omega)$ are then the operators given by a kernel K in $L^2(\Omega \times \Omega)$. More precisely, for f in $L^2(\Omega)$, and K in $L^2(\Omega \times \Omega)$, the integral

$$Kf(x) = \int_\Omega K(x,y)\, f(y)\, dy \tag{6.17}$$

converges for almost all x in Ω, the function Kf is in $L^2(\Omega)$, the operator $f \mapsto Kf$ is in $\mathcal{L}^2(L^2(\Omega))$ and

$$\mathrm{Tr}(K^*K) = \int_\Omega \int_\Omega |K(x,y)|^2\, dx\, dy \quad . \tag{6.18}$$

The proofs are easy consequences of Fubini's theorem about double integrals.

(b) For any bounded operator A acting on $\mathcal{H}$, we have

$$||A|| \leq ||A||_2 \leq ||A||_1 \quad , \tag{6.19}$$

hence hence in the opposite direction

$$\mathcal{L}(\mathcal{H}) \supset \mathcal{L}^2(\mathcal{H}) \supset \mathcal{L}^1(\mathcal{H}) \quad . \tag{6.20}$$

One can interpolate with the spaces $\mathcal{L}^p(\mathcal{H})$, introduced by Schatten around 1940, that is the set of operators A for which $|A|$ can be diagonalized with eigenvalues $\lambda_0, \lambda_1, \ldots, \lambda_n, \ldots$ such that $\sum_n |\lambda_n|^p$ be finite. One can mimic the basic properties of classical Lebesgue spaces $L^p(\Omega)$, for instance Minkowski and Hölder inequalities.

7. Fredholm Determinants in Hilbert Space

7.1. Suppose one wants to define the determinant of an operator B acting on some Hilbert space $\mathcal{H}$. Choose an orthonormal basis $(\psi_n)_{n\geq 0}$ of $\mathcal{H}$ and represent the operator by its matrix $b = (b_{mn})$ where $b_{mn} = \langle\psi_m|B|\psi_n\rangle$. To define the determinant Δ of this infinite matrix, a natural procedure is to truncate it to a finite determinant

$$\Delta_N = \det \begin{pmatrix} b_{oo} & \cdots\cdots & b_{oN} \\ b_{1o} & \cdots\cdots & b_{1N} \\ \cdot & \cdots\cdots & \cdot \\ b_{No} & \cdots\cdots & b_{NN} \end{pmatrix} \tag{7.1}$$

and to look for the limit $\Delta = \lim_{N\to\infty} \Delta_N$. If the matrix b is diagonal, we get $\Delta_N = b_{oo} \ldots b_{NN}$ and Δ is the infinite product $\prod_{n\geq 0} b_{nn}$. It is known that such an infinite product converges absolutely iff b_{nn} can be put in the form $b_{nn} = 1 + a_n$ where $\sum_{n\geq 0} |a_n|$ if finite. This remark led Poincaré and von Koch to assume that the operator B is of the form $1 + A$, where A is "small" in a suitable sense. Denoting as before by $\Delta\binom{i_1\ldots i_p}{j_1\ldots j_p}$ the minors of the matrix $a = (a_{mn})$ associated to the operator A, we get

$$\Delta_N = \sum_{p\geq 0} \quad \sum_{0\leq i_1<\ldots<i_p\leq N} \Delta\binom{i_1\ldots i_p}{i_1\ldots i_p} \tag{7.2}$$

by formula (5.2) (first noticed by von Koch). Hence, at least formally, we get

$$\Delta = \sum_{p\geq 0} \quad \sum_{i_1<\ldots<i_p} \Delta\binom{i_1\ldots i_p}{i_1\ldots i_p} \tag{7.3}$$

for the determinant of $1 + A$. It can be shown that the previous series converges absolutely provided the double sum $\sum_{m,n} |a_{mn}|$ is finite (Poincaré's criterion)[1]

---

[1] For a lively account of the prehistory of infinite-determinants, the reader may consult Dieudonné's book [4].

We shall follow this method, but to obtain a theory independent of the orthonormal basis chosen, we shall suppose that A is of trace class, a more general assumption than the mere convergence of $\sum_{m,n} |a_{mn}|$. There are a number of problems connected with such a definition:

- convergence of the expansion (7.3)
- the multiplicative property of determinants
- relations between the eigenvalues of A and the zeroes of the characteristic function $\det(1 - zA)$, where z is a complex variable.

7.2. Our method will be based on the construction of the *fermionic Fock space*. Let us extend to the Hilbert space set up the constructions of tensor spaces given in subsection 2. Let H_1 and H_2 be two Hilbert spaces. We propose to associate to H_1 and H_2 a new hilbert space H, to be denoted by $H_1 \otimes_2 H_2$, together with a map associating to a vector x_1 in H_1 and a vector x_2 in H_2 a vector $x_1 \otimes x_2$ in H and assume the following properties to hold:

(a) The vector $x_1 \otimes x_2$ depends linearly on x_1 for a fixed x_2, and symmetrically in x_2 for a fixed x_1.

(b) For the scalar products, one gets

$$\langle x_1 \otimes x_2 | y_1 \otimes y_2 \rangle = \langle x_1 | y_1 \rangle \langle x_2 | y_2 \rangle \quad . \tag{7.4}$$

(c) Any vector in H is a limit of finite linear combinations $\sum_{i=1}^{N} x_i \otimes y_i$. Equivalently, if z is any nonzero vector in H, there exist x_1 and x_2 such that $\langle x_1 \otimes x_2 | z \rangle \neq 0$.

As for the *existence of* H, choose an orthonormal basis $(\psi_n^{(i)})_{n \geq 0}$ in H_i (for i = 1 or 2), take any Hilbert space H with an orthonormal basis indexed as a double sequence $(e_{mn})_{m \geq 0, n \geq 0}$ and define the map $(x_1, x_2) \to x_1 \otimes x_2$ by the formula

$$x_1 \otimes x_2 = \sum_{m,n} \langle \psi_m^{(1)} | x_1 \rangle \langle \psi_n^{(2)} | x_2 \rangle e_{mn} \quad . \tag{7.5}$$

The properties (a), (b), (c) are easily checked. Moreover, according to this definition, one gets in particular $e_{mn} = \psi_m^{(1)} \otimes \psi_n^{(2)}$.

To prove the *uniqueness* of this construction, one first proves using (a), (b) and (c) that the tensor product $(\psi_m^{(1)} \otimes \psi_n^{(2)})$ of the orthonormal basis in H_1 and H_2 is an orthonormal basis in $H_1 \otimes_2 H_2$. Moreover by specializing formula (7.4), one gets the values $\langle\psi_m^{(1)}|x_1\rangle\langle\psi_n^{(2)}|x_2\rangle$ for the components of $x_1 \otimes x_2$ in the previous basis of $H_1 \otimes_2 H_2$, hence formula (7.5) is forced upon us.

By an obvious generalization, one defines the tensor product $H_1 \otimes_2 \ldots \otimes_2 H_k$ of any finite family of Hilbert spaces $H_1,\ldots,H_k$. For the record, notice the formula for the scalar product

$$\langle x_1 \otimes \ldots \otimes x_k | y_1 \otimes \ldots \otimes y_k\rangle = \prod_{i=1}^{k} \langle x_i|y_i\rangle \quad ; \tag{7.6}$$

moreover, if $(\psi_n^{(i)})_{n\geq 0}$ is any orthonormal basis in H_i (for $i=1,\ldots,k$) then the multiple sequence of tensor products $\psi_{n_1}^{(1)} \otimes \ldots \otimes \psi_{n_k}^{(k)}$ is an orthonormal basis in $H_1 \otimes_2 \ldots \otimes_2 H_k$. In the sequel, we restrict our attention to the tensor product $H^{\otimes k}$ of k identical spaces $H_1 = \ldots = H_k = H$. Whenever convenient, we assume that an orthonormal basis $(\psi_n)_{n\geq 0}$ has been chosen for H, hence the tensors $\psi_{n_1} \otimes \ldots \otimes \psi_{n_k}$ form an orthonormal basis in $H^{\otimes k}$.

7.3. The symmetry operators in $H^{\otimes k}$ are defined as follows. Given any permutation σ in S_k, there exists a unitary operator U_σ in $H^{\otimes k}$ permuting the basic tensors as follows

$$U_\sigma(\psi_{n_1} \otimes \ldots \otimes \psi_{n_k}) = \psi_{n'_1} \otimes \ldots \otimes \psi_{n'_k} \tag{7.7}$$

where $n_i = n'_{\sigma(i)}$ for $1 \leq i \leq k$. By linearity and continuity, one deduces the relation

$$U_\sigma(x_1 \otimes \ldots \otimes x_k) = x_{\sigma^{-1}(1)} \otimes \ldots \; x_{\sigma^{-1}(k)} \tag{7.8}$$

for $x_1,\ldots,x_k$ in H. Hence U_σ is invariantly defined, and the product rule $U_\sigma U_\tau = U_{\sigma\tau}$ holds. Otherwise stated, one obtains a unitary representation of the group S_k in the Hilbert space $H^{\otimes k}$.

As in the finite dimensional case, one introduces the subspace $\Lambda^k H$ of the antisymmetric tensors, elements t in $H^{\otimes k}$

such that $U_\sigma t = (\mathrm{sgn}\sigma).t$ for all σ in S_k. It is customary to modify the definition of the wedge product as follows

$$x_1\wedge\ldots\wedge x_k = (k!)^{-1/2} \sum_{\sigma\in S_k} (\mathrm{sgn}\sigma)\, U_\sigma(x_1\otimes\ldots\otimes x_k) \tag{7.9}$$

With this convention, the scalar product is given by

$$<x_1\wedge\ldots\wedge x_k|y_1\wedge\ldots\wedge y_k> = \det_{\substack{1\leq i\leq k\\ 1\leq j\leq k}} <x_i|y_j> \tag{7.10}$$

without normalization constant. It follows that *the wedge products* $\psi_{n_1}\wedge\ldots\wedge\psi_{n_k}$ *where* $0 \leq n_1<\ldots<n_k$ *form an orthonormal basis of* $\Lambda^k H$.

By the antisymmetry properties of the wedge product, it follows that $x_1\wedge\ldots\wedge x_k$ is equal to $x_1'\wedge\ldots\wedge x_k'$ where $x_1',\ldots,x_k'$ are deduced from $x_1,\ldots,x_k$ by the orthonormalisation process: $x_j' - x_j$ is a linear combination of $x_1,\ldots,x_{j-1}$ and $x_1',\ldots,x_k'$ are mutually orthogonal. By Pythagoras'theorem, one gets $||x_i'|| \leq ||x_i||$, and from (7.10) one gets

$$||x_1\wedge\ldots\wedge x_k||^2 = ||x_1'\wedge\ldots\wedge x_k'||^2 = ||x_1'||^2\ldots||x_k'||^2$$

hence

$$||x_1\wedge\ldots\wedge x_k|| \leq ||x_1||\ldots||x_k|| \quad . \tag{7.11}$$

Using Cauchy-Schwarz inequality, one derives the corollary

$$|<x_1\wedge\ldots\wedge x_k|y_1\wedge\ldots\wedge y_k>| \leq \prod_{i=1}^{k} ||x_i||\,||y_i|| \quad . \tag{7.12}$$

These formulas are just reincarnations of Hadamard's inequality for determinants.

In many applications, H is a Hilbert space $L^2(\Omega)$ of square-integrable functions. One identifies $H^{\otimes k}$ to the space $L^2(\Omega\times\ldots\times\Omega)$ (k factors) of square- integrable functions $f(t_1,\ldots,t_k)$ where the variables t_i run over Ω. The tensor product $f_1\otimes \ldots \otimes f_k$ of one variable functions in $L^2(\Omega)$ is then given by

$$(f_1\otimes \ldots \otimes f_k)(t_1,\ldots,t_k) = \prod_{i=1}^{k} f_i(t_i) \tag{7.13}$$

and the wedge product is a determinant

$$(f_1\wedge\ldots\wedge f_k)(t_1,\ldots,t_k) = (k!)^{-1/2} \det_{\substack{1\leq i\leq k\\ 1\leq j\leq k}} f_i(t_j) \quad . \tag{7.14}$$

7.4. Let us insert a few remarks.

(a) We remind the reader of the so-called *"reconstruction theorem"* for Hilbert spaces. Let I be any index set and let a_{ij} be complex numbers, for i and j in I. In order that there exist a Hilbert space H and vectors ψ_i in H (for i in I) such that $\langle\psi_i|\psi_j\rangle = a_{ij}$, it is necessary and sufficient that the following inequalities hold

$$\sum_{r=1}^{N}\sum_{s=1}^{N} \bar{z}_r a_{i_r i_s} z_s \geq 0 \tag{7.15}$$

for any set $\{i_1,\ldots,i_N\}$ of indices and any set $\{z_1,\ldots,z_N\}$ of complex numbers. We can always assume that the set of vectors ψ_i generates H, that is no vector $\psi \neq 0$ in H is orthogonal to all ψ_i's. If this is the case, the Hilbert space H is unique; namely if H' is any Hilbert space, generated by vectors ψ_i' such that $\langle\psi_i'|\psi_j'\rangle = a_{ij}$, there exists a unique unitary operator $U: H \to H'$ such that $U(\psi_i) = \psi_i'$ for all indices i.

The Hilbert spaces $H^{\otimes k}$ and $\Lambda^k H$ are respectively generated by vectors of the form $x_1\otimes\ldots\otimes x_k$ and $x_1\wedge\ldots\wedge x_k$. The scalar products are given by the formulas (7.6) and (7.10) respectively. One could therefore appeal to the reconstruction theorem to establish the existence of the spaces $H^{\otimes k}$ and $\Lambda^k H$. For this it would be necessary to establish the inequalities (7.15) directly.

(b) The concept of tensor product of Hilbert spaces is closely connected to the notion of Hilbert-Schmidt operators. Namely, let H be any Hilbert space. We associate to it a new Hilbert space H^* and a map $\psi \mapsto \psi^*$ from H onto H^* as follows: H^* consists of the bounded linear forms on H (i.e. linear mappings $h: H \to \mathbb{C}$ for which there exists a bound $C \geq 0$ with $|h(\psi)| \leq C||\psi||$). For ψ in H, we denote by ψ^* the linear form $\psi' \mapsto \langle\psi|\psi'\rangle$. It is a well-known result by F.Riesz that $\psi \mapsto \psi^*$ is a bijection of H onto H^*. The Hilbert space structure of H^*

is defined in such a way that $(c\psi)^* = \bar{c}\psi^*$ and $\langle\psi^*|\psi'^*\rangle = \langle\psi'|\psi\rangle$ for ψ, ψ' in H and a complex scalar c. In Dirac's notation, denoting ta vector ψ in H as a ket $|\psi\rangle$, the element ψ^* of H^* is the corresponding bra $\langle\psi|$.

There is a natural isomorphism of the Hilbert space $H \otimes_2 H^*$ onto the space $L^2(H)$ of Hilbert-Schmidt operators. It associates to a generator $\psi_1 \otimes \psi_2^*$ the operator $|\psi_1\rangle\langle\psi_2|$ (taking ψ into $\psi_1\langle\psi_2|\psi\rangle$ for ψ in H). This isomorphism is implicitely used in the Dirac notation of bras and kets.

7.5. We show how to extend a bounded operator A acting on H to the spaces $H^{\otimes k}$ and $\Lambda^k H$. First of all consider two Hilbert spaces H_1 and H_2, with orthonormal basis $(\psi_n^{(1)})$ and $(\psi_n^{(2)})$ respectively and bounded operators A_1 in $L(H_1)$ and A_2 in $L(H_2)$. Any vector in $H_1 \otimes_2 H_2$ can be uniquely expanded as t $t = \sum_n t_n \otimes \psi_n^{(2)}$ where $||t||^2 = \sum_n ||t_n||^2$ is finite; for instance, if $t = x_1 \otimes x_2$ then $t_n = \langle\psi_n^{(2)}|x_2\rangle \cdot x_1$. Extend A_1 to $H_1 \otimes_2 H_2$ by the rule

$$A_1^{(1)}t = \sum_n At_n \otimes \psi_n^{(2)} \qquad \text{for } t = \sum_n t_n \otimes \psi_n^{(2)} \quad . \tag{7.16}$$

This defines a bounded operator $A_1^{(1)}$ in $L(H_1 \otimes_2 H_2)$ according to the estimate

$$||A_1^{(1)}t||^2 = \sum_n ||At_n||^2 \le ||A||^2 \sum_n ||t_n||^2 = ||A||^2||t||^2 ,$$

hence $||A_1^{(1)}|| = ||A_1||$. Moreover, if $t = x_1 \otimes x_2$ we get $A_1^{(1)}t = A_1x_1 \otimes x_2$, thereby the definition of $A_1^{(1)}$ is independent of the basis $(\psi_n^{(2)})$. Similarly, there exists a bounded operator $A_2^{(2)}$ acting on $H_1 \otimes_2 H_2$ in such a way that $A_2^{(2)}(x_1 \otimes x_2) = x_1 \otimes A_2x_2$ with a bound $||A_2^{(2)}|| = ||A_2||$. Define the operator $A_1 \otimes A_2$ as the product $A_1^{(1)}A_2^{(2)}$, hence the rules

$$\begin{aligned} (A_1\otimes A_2)(x_1\otimes x_2) &= A_1x_1\otimes A_2x_2 \quad , \\ ||A_1 \otimes A_2|| &- ||A_1||\cdot||A_2|| \quad . \end{aligned} \tag{7.17}$$

The extension to the case of k spaces $H_1,\ldots,H_k$ is obvious.

Coming back to H and A, there exists therefore a bounded operator $A^{\otimes k}$ acting on $H^{\otimes k}$ in such a way that

$$A^{\otimes k}(x_1 \otimes \ldots \otimes x_k) = Ax_1 \otimes \ldots \otimes Ax_k \tag{7.18}$$

and $||A^{\otimes k}|| = ||A||^k$. From formula (7.8), it follows that $A^{\otimes k}$ and U_σ commute for σ in S_k, hence $A^{\otimes k}$ induces a bounded operator $\Lambda^k A$ in the closed subspace $\Lambda^k H$ of $H^{\otimes k}$. It is characterized by its action on the wedge products

$$\Lambda^k A\,(x_1 \wedge \ldots \wedge x_k) = Ax_1 \wedge \ldots \wedge Ax_k \quad . \tag{7.19}$$

7.6. THEOREM. *Suppose* A *is a trace-class operator on* H. *Then* $\Lambda^k A$ *is a trace-class operator on* $\Lambda^k H$; *its bound is given by* $||\Lambda^k A|| = \mu_0(A) \ldots \mu_{k-1}(A)$ *and moreover* $||\Lambda^k A||_1 \leq ||A||_1^k/k!$.

Proof: It follows from the construction of $\Lambda^k A$ that $\Lambda^k(A^*)$ is equal to $(\Lambda^k A)^*$ and $\Lambda^k(AB)$ to $\Lambda^k A \cdot \Lambda^k B$ for another operator B in $L(H)$. Since the positive operators are the operators of the form C^*C and $|C|$ is the unique positive operator such that $|C|^2 = C^*C$, we get $|\Lambda^k A| = \Lambda^k |A|$. Since A is trace-class, the operator $|A|$ can be diagonalized with eigenvectors ψ_n and eigenvalues $\nu_n \geq 0$. Then the wedge products $\psi_{n_1} \wedge \ldots \wedge \psi_{n_k}$ for $n_1 < n_2 < \ldots < n_k$ form an orthonormal basis of $\Lambda^k H$, and $\Lambda^k |A|$ multiplies it by $\nu_{n_1} \ldots \nu_{n_k}$. By definition, the sequence

$$\mu_0(A) \geq \mu_1(A) \geq \ldots \geq \mu_{k-1}(A) \geq \mu_k(A) \geq \ldots$$

is obtained by rearranging in descending order the nonzero eigenvalues ν_n. Therefore $\mu_0(A) \ldots \mu_{k-1}(A)$ is the largest among the products $\nu_{n_1} \ldots \nu_{n_k}$, hence is equal to $||\Lambda^k A||$. Finally $||\Lambda^k A||_1$ is the trace of $|\Lambda^k A|$, that is

$$||\Lambda^k A||_1 = \sum_{n_1 < \ldots < n_k} \nu_{n_1} \ldots \nu_{n_k} \tag{7.20}$$

In the same vein

$$||A||_1 = \sum_n \nu_n \tag{7.21}$$

hence the inequality $||\Lambda^k A||_1 \leq ||A||_1^k/k!$.

Q.E.D.

Let us add a few comments:

(a) Suppose the operator A is of finite rank N. Then the formula $||\Lambda^k A|| = \mu_0(A) \ldots \mu_{k-1}(A)$ holds for $0 \leq k \leq N$, moreover $\Lambda^k A = 0$ for $k > N$. This justifies the convention $\mu_k(A)=0$ for $k \geq N$.

(b) For a while, we do not assume that A is trace-class. It can be shown that the norm $||\Lambda^k A||$ can be defined as

$$||\Lambda^k A|| = \sup||Ax_1\wedge\ldots\wedge Ax_k|| \tag{7.22}$$

or

$$||\Lambda^k A|| = \sup\left|\det_{\substack{1\leq i\leq k\\1\leq j\leq k}} \langle x_i|A|y_j\rangle\right| \tag{7.23}$$

where $x_1,\ldots,x_k$ run over the vectors of norm 1 in H(for (7.22)) and similarly for $y_1,\ldots,y_k$ in (7.23). Define the numbers

$$\mu_k(A) = ||\Lambda^{k+1}A||\cdot||\Lambda^k A||^{-1} \quad . \tag{7.24}$$

By reduction to the finite-dimensional case (hint: use (7.22)), it can be shown that the sequence is non-increasing

$$\mu_0(A) \geq \mu_1(A) \geq \ldots$$

with $\mu_0(A) = ||A||$. It may very well be that $\mu_k(A) = ||A||$ for all k's.

(c) By a proof similar to that of theorem 7.6, one gets that $A^{\otimes k}$ is trace-class if A is. Moreover

$$||A^{\otimes k}|| = ||A||^k \quad , \quad ||A^{\otimes k}||_1 \leq ||A||_1^k \quad , \tag{7.25}$$

and the inequality is optimal. The antisymmetric part $\Lambda^k H$ of $H^{\otimes k}$ is the *fermionic Fock space*. One can define in a similar way the symmetric part $S^k H$ of $H^{\otimes k}$, consisting of the element t such that $U_\sigma t = t$ for σ in S_k. It is the *bosonic Fock space*.

The operator $A^{\otimes k}$ induces an operator $S^k A$ in $S^k H$, and we get the estimates

$$||S^k A|| \leq ||A||^k \quad , \quad ||S^k A||_1 \leq ||A||_1^k \tag{7.26}$$

which are again optimal. The constant $1/k!$ in the estimate $||\Lambda^k A||_1 \leq (1/k!)||A||_1^k$ has therefore no counterpart in the spaces $H^{\otimes k}$ and $S^k H$. It is one of the deepest manifestations of *Pauli's exclusion principle*.

7.7. We are ready to define the determinant $\det(1+A)$ in case A is trace-class. Indeed, define $c_k(A)$ as the trace of $\Lambda^k A$. By theorem 7.6 one gets the estimate

$$|c_k(A)| = |\mathrm{Tr}\Lambda^k A| \leq ||\Lambda^k A||_1 \leq ||A||_1^k/k! \quad .$$

The series $\sum_{k\geq 0} c_k(A)$ is therefore absolutely convergent and we put

$$\det(1+A) = \sum_{k\geq 0} c_k(A) \quad . \tag{7.27}$$

Replacing A by zA, where z is a complex variable, one defines the *characteristic determinant*

$$\det(1-zA) = \sum_{k\geq 0} (-1)^k c_k(A) z^k \quad , \tag{7.28}$$

which is an entire function of z by the basic estimate

$$|c_k(A)| \leq ||A||_1^k/k! \quad . \tag{7.29}$$

Put in another form, introduce the *total Fock space* ΛH, orthogonal sum of the Hilbert spaces $\Lambda^0 H$, $\Lambda^1 H$, ... It is generated by the wedge products $x_1\wedge\ldots\wedge x_k$ of varying order k, with scalar product given by

$$\langle x_1\wedge\ldots\wedge x_k | y_1\wedge\ldots\wedge y_k\rangle = \det\langle x_i | y_j\rangle \tag{7.30}$$

$$\langle x_1\wedge\ldots\wedge x_k | y_1\wedge\ldots\wedge y_l\rangle = 0 \qquad \text{if } k \neq l \; . \tag{7.31}$$

The various operators $\Lambda^k A$ extend to an operator ΛA acting on ΛH and mapping $x_1\wedge\ldots\wedge x_k$ into $Ax_1\wedge\ldots\wedge Ax_k$. From theorem 7.6, it follows that ΛA is a trace-class operator such that $||\Lambda A||_1 = \sum_{k\geq 0} ||\Lambda^k A||_1$ hence

$$||\Lambda A||_1 \leq \exp||A||_1 \quad . \tag{7.32}$$

From (7.27) we get the *compact definition*

$$\det(1+A) = \mathrm{Tr}(\Lambda A) \quad . \tag{7.33}$$

Introduce an orthonormal basis (ψ_n) for H and the matrix a corresponding to A with elements $a_{mn} = \langle\psi_m|A|\psi_n\rangle$. Then the wedge products $\psi_{i_1}\wedge\ldots\wedge\psi_{i_k}$ for $i_1<\ldots<i_k$ (k variable) form an orthonormal basis of ΛH. Moreover the minors $\Delta\binom{i_1\ldots i_k}{j_1\ldots j_k}$ are simply scalar products, namely

$$\Delta\binom{i_1\ldots i_k}{j_1\ldots j_k} = \langle\psi_{i_1}\wedge\ldots\wedge\psi_{i_k}|\Lambda A|\psi_{j_1}\wedge\ldots\wedge\psi_{j_k}\rangle \quad . \tag{7.34}$$

The trace of an operator can be calculated using any orthonormal basis, hence the *absolutely convergent expansions*

$$c_k(A) = \mathrm{Tr}(\Lambda^k A) = \sum_{i_1<\ldots<i_k} \Delta\binom{i_1\ldots i_k}{i_1\ldots i_k} \tag{7.35}$$

and

$$\det(1-zA) = \sum_{k\geq 0} (-1)^k z^k \sum_{i_1<\ldots<i_k} \Delta\binom{i_1\ldots i_k}{i_1\ldots i_k} \quad . \tag{7.36}$$

Von Koch' definition (7.3) is fully justified !

7.8. The previous definitions can be illustrated with the help of some considerations of *quantum statistical mechanics*. So suppose H corresponds to a quantum-mechanical particle, with hamiltonian operator $H^{(1)}$. Then the space ΛH corresponds to an assembly of particles obeying Fermi-Dirac statistics. The total hamiltonian H_F acts on ΛH in such a way that

$$H_F(x_1\wedge\ldots\wedge x_k) = \sum_{i=1}^{k} x_1\wedge\ldots\wedge H\, x_i\wedge\ldots\wedge x_k \tag{7.37}$$

(non-interacting system, the total energy is the sum of the energies of the k constituents). We do not discuss the question of domains, the self-adjoint operator $H^{(1)}$ being in general unbounded. At least formally, one gets

$$e^{-\beta H_F}(x_1\wedge\ldots\wedge x_k) = e^{-\beta H^{(1)}}x_1\wedge\ldots\wedge e^{-\beta H^{(1)}}x_k \qquad (7.38)$$

for $\beta > 0$. So the last definition of the total hamiltonian H_F is as the infinitesimal generator of the semi-group of operators $e^{-\beta H_F} = \Lambda(e^{-\beta H^{(1)}})$. In the cases of physical interest, the spectrum of $H^{(1)}$ has a lower bound, hence the operator $e^{-\beta H^{(1)}}$ is bounded.

The particle number operator N is given by $Nt = kt$ for t in $\Lambda^k H$. In statistical mechanics, one introduces the inverse temperature $\beta = 1/kT$ where k is Boltzmann's constant, and the chemical potential μ. According to Gibbs and Boltzmann, the thermodynamical quantities can be calculated using the so-called (fermionic) *partition function* $Z_F(\beta,\mu)=\mathrm{Tr}(e^{-\beta(H_F+\mu N)})$. From our definitions, one gets

$$Z_F(\beta,\mu) = \det(1 + e^{-\beta(H^{(1)}+\mu)}) \quad . \qquad (7.39)$$

These definitions make sense provided the operator $e^{-\beta H^{(1)}}$ be of trace-class for $\beta > 0$. In this case, the operator $H^{(1)}$ can be diagonalized with eigenvalues $E_0 \leq E_1 \leq E_2 \leq \ldots$ and eigenvectors $\psi_0, \psi_1, \psi_2, \ldots$. For k fixed, the state with lowest energy is $\psi_0\wedge\ldots\wedge\psi_{k-1}$ corresponding to the eigenvalue $E_0 + \ldots + E_{k-1}$ of H_F. Put $A = e^{-\beta(H^{(1)}+\mu)}$, hence $\Lambda A = e^{-\beta(H_F+\mu N)}$. Then $\psi_0\wedge\ldots\wedge\psi_{k-1}$ corresponds to the largest eigenvalue $\exp -\beta(E_0+\ldots+E_{k-1}+k\mu)$ of $\Lambda^k A$, hence we get a "physical" interpretation of the formula

$$||\Lambda^k A|| = \mu_0(A)\ \ldots\ \mu_{k-1}(A)$$

where $\mu_i(A) = \exp -\beta(E_i+\mu)$. Since the operator ΛA is now in diagonal form, it is easy to calculate its trace, that is $\det(1+A)$, and we get the well-known formula

$$Z_F(\beta,\mu) = \prod_{i\geq 0}(1 + e^{-\beta(E_i+\mu)}) \quad . \qquad (7.40)$$

7.9. We can repeat almost *verbatim* the considerations in section 3. For instance, the operator

$$P_-^k = (k!)^{-1} \sum_{\sigma \in S_k} (\operatorname{sgn}\sigma) \cdot U_\sigma \tag{7.41}$$

is the orthogonal projection of $H^{\otimes k}$ onto the closed subspace $\Lambda^k H$. For σ in S_k, put

$$I(\sigma) = \operatorname{Tr}(U_\sigma \cdot A^{\otimes k}) \quad . \tag{7.42}$$

The trace of $\Lambda^k A$ is equal to $\operatorname{Tr}(P_-^k \cdot A^{\otimes k})$ since $A^{\otimes k}$ restricts to $\Lambda^k A$ on $\Lambda^k H$; hence we get

$$k!\, c_k(A) = \sum_{\sigma \in S_k} (\operatorname{sgn}\sigma)\, I(\sigma) \quad . \tag{7.43}$$

We have now to check the following properties:

(a) for σ, τ *in* S_k, *we get* $I(\tau\sigma\tau^{-1}) = I(\sigma)$;

(b) *if* σ *is decomposed into cycles* $(1 \ldots a)(a+1 \ldots a+b)$ $(a+b+1 \ldots a+b+c) \ldots$, *then* $I(\sigma)$ *is equal to* $\operatorname{Tr}(A^a).\operatorname{Tr}(A^b).\operatorname{Tr}(A^c) \ldots$

The proof of (a) rests on the fact that U_τ commutes to $A^{\otimes k}$ and on the *unitary invariance of the trace*

$$\operatorname{Tr}(UBU^{-1}) = \operatorname{Tr}(B) \quad , \tag{7.44}$$

where U is any unitary operator (notice that the trace can be calculated using *any* orthonormal basis). For the proof of (b), introduce an orthonormal basis (ψ_n) in H; since the tensors $\psi_{n_1} \otimes \ldots \otimes \psi_{n_k}$ form an orthonormal basis of $H^{\otimes k}$, we get the *absolutely convergent expansion*.

$$I(\sigma) = \sum_{n_1 \ldots n_k} \prod_{i=1}^{k} \langle \psi_{n_i} | A | \psi_{n_{\sigma(i)}} \rangle \quad . \tag{7.45}$$

If σ is decomposed into cycles as in (b), this summation breaks into a product of sums like

$$J_a = \sum_{n_1 \ldots n_a} \langle \psi_{n_1} | A | \psi_{n_2} \rangle \langle \psi_{n_2} | A | \psi_{n_3} \rangle \ldots \langle \psi_{n_a} | A | \psi_{n_1} \rangle .$$

From Parseval theorem, one gets in general the matrix rule

$$\sum_s \langle\psi_r|B|\psi_s\rangle\langle\psi_s|C|\psi_t\rangle = \langle\psi_r|BC|\psi_t\rangle \tag{7.46}$$

hence $J_a = \sum_{n_1} \langle\psi_{n_1}|A^a|\psi_{n_1}\rangle = \mathrm{Tr}(A^a)$. Hence (b) follows. A similar proof would give

$$\mathrm{Tr}((A_1 \otimes \ldots \otimes A_k)\gamma_k) = \mathrm{Tr}(A_1 \ldots A_k) \tag{7.47}$$

for trace-class operators $A_1,\ldots,A_k$ in H and the cyclic permutation γ_k.

In a similar vein, introduce the trace $h_k(A)$ of S^kA acting on the bosonic Fock space S^kH. Since the orthogonal projection of $H^{\otimes k}$ onto S^kH is given by

$$P_+^k = (k!)^{-1} \sum_{\sigma\in S_k} U_\sigma \quad , \tag{7.48}$$

one gets

$$k!h_k(A) = \sum_{\sigma\in S_k} I(\sigma) \quad . \tag{7.49}$$

Putting $\tau_k(A) = \mathrm{Tr}(A^k)$, the formulas (3.33) to (3.38) as well as Plemelj's determinantal formulas (5.56) can be taken *verbatim* in our new context. Let us also mention the analogue of formula (3.27)

$$\sum_{k\geq 0} h_k(A)\, z^k = \det(1-zA)^{-1} \tag{7.50}$$

as well as the logarithmic derivative

$$\frac{d}{dz} \log \det(1 + zA) = \mathrm{Tr}(A(1 + zA)^{-1}) \ . \tag{7.51}$$

Notice that formula (7.50) holds for $|z|$ small enough and that both sides in (7.51) are meromorphic functions of z.

Here is a "physical" interpretation of formula (7.50). Suppose again that H is the one-particle state space with hamiltonian operator $H^{(1)}$. Introduce the total bosonic Fock space SH as the orthogonal sum of the spaces S^oH, S^1H, $S^2H,\ldots$ The total hamiltonian H acts on SH in such a way that

$$H_B(x_1 \ldots x_k) = \sum_{i=1}^{k} x_1 \ldots H^{(1)}x_i \ldots x_k \tag{7.52}$$

(symmetric product of vectors !). The bosonic partition function $Z_B(\beta,\mu) = \mathrm{Tr}(e^{-\beta(H_B+\mu N)})$ is then given in invariant form as

$$Z_B(\beta,\mu) = \det(1 - e^{-\beta(H^{(1)}+\mu)})^{-1} \tag{7.53}$$

provided μ is large enough. This corresponds to the customary relation of Planck-Einstein

$$Z_B(\beta,\mu) = \prod_{i \geq 0} \frac{1}{1 - e^{-\beta(E_i+\mu)}} \tag{7.54}$$

(valid whenever $E_0+\mu > 0$) in terms of the energy levels $E_0 \leq E_1 \leq \ldots$

7.10. Recall that the space of all trace-class operators is a Banach space $L^1(H)$ with norm $||A||_1 = \mathrm{Tr}(|A|)$, and that the operators of finite rank are dense in $L^1(H)$.

By definition, $\det(1+A)$ is given by the series $\sum_{k \geq 0} c_k(A)$ with the estimate $|c_k(A)| \leq ||A||_1^k/k!$. It follows that this series converges uniformly on any set of operators with $||A||_1 \leq R$, where R is a fixed bound. Hence the functional $A \mapsto \det(1+A)$ is *continuous* on the Banach space $L^1(H)$. It is even *holomorphic*. Indeed as in subsection 5.4, introduce a multilinear form on $L^1(H)$ by

$$c_k(A_1,\ldots,A_k) = \mathrm{Tr}(P_-^k\,(A_1 \otimes \ldots \otimes A_k)) \tag{7.55}$$

hence

$$\det(1+A) = \sum_{k \geq 0} c_k(A,\ \ldots,\ A) \quad . \tag{7.56}$$

Use now the *polarization formula*

$$2^k k!\, c_k(A_1,\ldots,A_k) = \sum_{\varepsilon_1,\ldots,\varepsilon_k} c_k(\varepsilon_1 A_1 + \ldots + \varepsilon_k A_k)\varepsilon_1 \ldots \varepsilon_k \tag{7.57}$$

where $\varepsilon_1,\ldots,\varepsilon_k$ take independently the values 1 and -1. By the estimate $|c_k(A)| \leq ||A||_1^k/k!$, one obtains easily

$$|c_k(A_1,\ldots,A_k)| \leqq \gamma_k \; ||A_1||\ldots||A_k|| \tag{7.58}$$

with a constant

$$\gamma_k = 2^{-k}k^k/(k!)^2 \quad . \tag{7.59}$$

From Stirling's formula,one gets $\lim_{k\to\infty} \gamma_k^{1/k} = 0$ and one concludes as in subsection 5.4.

As a corollary, suppose $(A_\lambda)_{\lambda\in D}$ is a family of operators in $L^1(H)$, bounded in norm $||A_\lambda||_1 \leqq R$ for a fixed constant R, and holomorphic in λ (D is a domain in a complex space C^r) in the sense that the matrix elements $\langle\psi|A|\psi'\rangle$ are holomorphic in λ for fixed vectors ψ and ψ'. Then the determinant $\det(1+A_\lambda)$ is a holomorphic function of λ in D.

7.11. The *multiplicative property* is now easy to prove. Let A and B be trace-class operators acting on H. Then $(1+A)(1+B)$ is equal to $1+C$ with $C = A + B + AB$. The product of a trace-class operator and a bounded operator is again trace-class, hence C is trace-class. We state

$$\det(1+A)\ \det(1+B) = \det(1 + A + B + AB) \quad . \tag{7.60}$$

Any trace-class operator can be approximated by finite rank operators in the Banach space $L^1(H)$ and the determinant is a continuous functional on $L^1(H)$. Hence it suffices to consider the case where A and B are of finite rank. We may then choose an orthonormal basis (ψ_n) of H such that A and B map H into the subspace K generated by $\psi_1,\ldots,\psi_N$ for a suitable N. Let the operators A_0 and B_0 in the finite-dimensional space K be obtained by restricting A and B respectively. Since the multiplicative rule holds for determinants of operators in K, it suffices to check that $1+A$ and $1+A_0$ have the same determinant (and similarly for $1+B$ and $1+B_0$). The matrix elements $a_{mn} = \langle\psi_m|A|\psi_n\rangle$ of A agree with those of A_0 for $1 \leqq m \leqq N$, $1 \leqq n \leqq N$ and are zero otherwise. The minors

$\Delta\begin{pmatrix} i_1\ldots i_k \\ i_1\ldots i_k \end{pmatrix}$ of A agree therefore with those of A_0 when

$i_1,\ldots,i_k$ lie between 1 and N, and are 0 otherwise. By using formula (7.36) for both A and A_0, we get $\det(1+A) = \det(1+A_0)$ as asserted.

7.12. We derive a few consequences of the multiplicativity of determinants. Let A be again a trace-class operator in H. According to formula (6.6), the operator A can be approximated in operator norm by finite rank operators, hence is compact. By F.Riesz'theory, *Fredholm's alternative* is valid:

(a) *either the operator* 1+A *has a bounded inverse,*
(b) *or there exists a nonzero vector* ψ *in* H *such that* $(1+A)\psi=0$.

In case (a), let T be a bounded inverse to 1+A. Then T=1-AT and AT is trace-class again. By the multiplicativity of determinants, we get

$$\det(1 + A) \quad \det(1 - AT) \quad = \quad \det(1) \quad = 1$$

hence $\det(1+A) \neq 0$.

In case (b) choose an orthonormal basis $\psi_1,\psi_2,\ldots$ of H with $\psi = \psi_1$ and represent A by its matrix (a_{mn}). Then the determinant of 1+A is the limit of the truncated tdeterminants

$$D_N = \det \begin{pmatrix} 1+a_{11} & a_{12} & \cdots & a_{1N} \\ a_{21} & 1+a_{22} & \cdots & a_{2N} \\ \cdot & \cdot & \cdot & \cdot \\ a_{N1} & a_{N2} & \cdots & 1+a_{NN} \end{pmatrix} \quad ;$$

the hypothesis $(1+A)\psi_1 = 0$ means that the first columns of the previous determinant consists of 0's, hence $D_N = 0$ and in the limit $\det(1+A) = 0$.

Hence, the cases (a) and (b) of the alternative correspond to $\det(1+A)\neq 0$ and $\det(1+A)=0$ respectively.

7.13. Put in another way, the *inverse eigenvalues* $1/\lambda_n$ corresponding to the nonzero eigenvalues λ_n of A are the roots of the equation $\det(1 - zA) = 0$. We want now to express the determinant itself in terms of the eigenvalues; the question is very easy when A is selfadjoint and lies much deeper in the general case.

Assume first that A is trace class and *selfadjoint*. There exists an orthonormal basis (ψ_n) diagonalizing A, that is $A\psi_n = \lambda_n\psi_n$. The matrix elementsof A are given by

$$<\psi_m|A|\psi_n> = \begin{cases} \lambda_n & \text{if } m = n \\ 0 & \text{otherwise} \end{cases} . \tag{7.61}$$

The minors are those of a diagonal matrix, hence

$$\Delta\binom{i_1 \dots i_k}{i_1 \dots i_k} = \lambda_{i_1} \dots \lambda_{i_k} \tag{7.62}$$

for $i_1 < \dots < i_k$. Since A is trace-class, the series of diagonal elements $\sum_n <\psi_n|A|\psi_n> = \sum_n \lambda_n$ converges absolutely, hence the infinite product $\prod_n (1-\lambda_n z)$ converges absolutely and can be expanded as follows

$$\prod_n (1 - \lambda_n z) = \sum_{k \geq 0} (-z)^k \sum_{i_1 < \dots < i_k} \lambda_{i_1} \dots \lambda_{i_k} . \tag{7.63}$$

Comparing with (7.36), one concludes

$$\det(1 - zA) = \prod_{n \geq 1} (1 - \lambda_n z) . \tag{7.64}$$

We come to the general case and denote by D(z) the entire function det(1-zA) of the complex variable z. As it is customary, let us introduce the *growth indicator*

$$M(R) = \sup_{|z|=R} |D(z)| . \tag{7.65}$$

To estimate M(R), *the strategy consists of comparing the eigenvalues of* A *and* $|A|$. Indeed, one gets

$$|c_k(A)| = |\mathrm{Tr}(\Lambda^k A)| \leq ||\Lambda^k A||_1 = \mathrm{Tr}(\Lambda^k |A|) = c_k(|A|).$$

Since $D(z) = \sum_{k \geq 0} (-1)^k c_k(A)\, z^k$, using a term-by-term estimate, one gets

$$M(R) \leq \sum_{k \geq 0} |c_k(A)| R^k \leq \sum_{k \geq 0} c_k(|A|) R^k = \det(1+R|A|).$$

Introduce now the eigenvalues $\mu_0, \mu_1, \ldots$ of $|A|$; they are positive and $\sum_{n \geq 0} \mu_n$ is finite. Hence

$$M(R) \leq \prod_{n \geq 0} (1 + \mu_n R) \qquad . \tag{7.66}$$

Taking the logarithm, one gets

$$\frac{1}{R} \log M(R) \leq \sum_{n \geq 0} \frac{1}{R} \log(1 + \mu_n R) \qquad . \tag{7.67}$$

Each term in the left-hand side converges to 0 with 1/R, and is majorized by μ_n. Since $\sum_n \mu_n$ is finite, one gets by dominated convergence

$$\lim_{R \to \infty} \frac{1}{R} \log M(R) = 0 \qquad .$$

(Notice that M(R) tends to $+\infty$ with R, unless D(z) is a constant).

We can now appeal to *Hadamard's factorization theorem.* (see Titchmarsh [14], th.8.24). We know already that a number z_0 satisfies $D(z_0) = 0$ iff there exists an eigenvalue $\lambda \neq 0$ of A with $z_0 = 1/\lambda$. Hence there are two possible cases:

(a) *The characteristic determinant* $\det(1 - zA)$ *is a polynomial in* z, *and can be expanded as* $(1 - \lambda_1 z)\ldots(1 - \lambda_N z)$ *where* $\lambda_1, \ldots, \lambda_N$ *are the nonzero eigenvalues of* A (with possible repetition). The limiting case D(z) = 1, that is $\mathrm{Tr}(A^k) = 0$ for each $k \geq 1$, occurs iff there is no eigenvalue of A, except possibly 0. In the self-adjoint case, this would mean A = 0, but not necessarily here.

(b) *There is an infinite sequence* $(\lambda_n)_{n \geq 1}$ *tending to* 0 *such that* $\sum_n |\lambda_n|$ *converges and*

$$\det(1 - zA) = \prod_{n \geq 1} (1 - \lambda_n z) \qquad . \tag{7.69}$$

7.14. We know that the λ_n's are the eigenvalues of A. There remains to settle the *question of multiplicities*. Let $\lambda \neq 0$ be any eigenvalue of A and let m be the number of times λ occurs among the λ_n's, that is the multiplicity of $1/\lambda$ as a zero of the entire function D(z) = det(1-zA). We need a *refinement of Fredholm's alternative*. For every integer $p \geq 0$, let L_p be the set of vectors ψ in H such that $(1-\lambda^{-1}A)^p\psi = 0$. Then L_p is *finite-dimensional*, and there exists an integer N such that $L_p = L_N$ for $p \geq N$ (notice that $L_0 \subset L_1 \subset \ldots \subset L_p \subset L_{p+1} \subset \ldots$). Moreover, let M_N be the set of vectors of the form $(1-\lambda^{-1}A)^N\psi$. Then M_N *is a closed vector subspace of* H *, the space* H *is the direct sum of* L_N *and* M_N *and* $1-\lambda^{-1}A$ *induces an operator in* M_N *with a bounded inverse*. There is a slight difficulty, namely L_N and M_N are not necessarily orthogonal. So let us introduce the orthogonal complement L_N' of L_N, so that the operator A can be expressed in block form

$$A = \begin{pmatrix} E & F \\ 0 & G \end{pmatrix} ,$$

where E is an operator acting on the finite-dimensional space L_N and G is a trace-class operator acting on the Hilbert space L_N'. Since $1-\lambda^{-1}A$ has a bounded inverse, so does $1-\lambda^{-1}G$. Calculating the determinant of $1 - zA$ as a limit of finite determinants, and using an orthonormal basis $(\psi_n)_{n\geq 1}$ of H such that $(\psi_1,\ldots,\psi_d)$ forms a basis of L_N, one gets

$$\det(1 - zA) = \det(1 - zE)\det(1 - zG) .$$

From our previous description, one gets $(1-\lambda^{-1}E)^N = 0$, hence λ is the only eigenvalue of E and therefore $\det(1 - zE) = (1-\lambda z)^d$. Define $D_0(z)$ as $\det(1-zG)$, so that $D(z) = (1-\lambda z)^d D_0(z)$. Then $D_0(z)$ is an entire function of z, and $D_0(\lambda^{-1}) = \det(1-\lambda^{-1}G)$ is not zero, since the operator $1-\lambda^{-1}G$ has a bounded inverse. Hence $1/\lambda$ is a zero of $D(z)$ of multiplicity d.

Summing up: *an eigenvalue* λ *of* A *occurs in the list* $\lambda_1, \lambda_2, \ldots$ *a number of times equal to the* (finite) *dimension of the subspace* H_λ *of* H *consisting of the vectors* ψ *for which there exists an integer* $p>0$ *with* $(A-\lambda)^p\psi = 0$.

The previous subtleties are connected with the so-called *Jordan normal form* of a matrix, when it can not be put in diagonal form. In most cases the space H_λ defined above consists of the eigenvectors, that is the vectors ψ such that $(A-\lambda)\psi = 0$.

We know that for small $|z|$, the logarithmic derivative $D'(z)/D(z)$ can be expanded as $\sum_{k\geq 1}(-1)^k \mathrm{Tr}(A^k)z^{k-1}$. Since $D(z)$

is equal to the product $\prod_{n\geq 1}(1-\lambda_n z)$ where $\sum_n|\lambda_n|$ is finite, $D'(z)/D(z)$, can be expanded as an absolutely convergent double series $\sum_{n\geq 1}\sum_{k\geq 1}(-1)^k\lambda_n^k z^{k-1}$. Comparing the two expansions of $D'(z)/D(z)$, one gets

$$\mathrm{Tr}(A^k) = \sum_{n\geq 1}\lambda_n^k \tag{7.70}$$

and in particular

$$\mathrm{Tr}(A) = \sum_{n\geq 1}\lambda_n \quad . \tag{7.71}$$

These relations are obvious in the selfadjoint case and were established by Dikii in 1957 for the general case.

7.15. We consider now the case of a *Hilbert-Schmidt operator* A in $L^2(H)$. In general we can define the traces of $A^2, A^3, \ldots$ but not of A itself, and so we need a modification of the determinant; this way introduced by Carleman around 1930, and the method simplified by Seiler in 1972.

We introduce a map Φ from $L^2(H)$ into $L^1(H)$ by $\Phi(A) = 1-(1+A)e^{-A}$. Indeed, expanding the exponential into the familiar power series, we get

$$\Phi(A) = \sum_{k\geq 2}(-1)^k(k-1)\,A^k/k!$$

Notice that for A in $L^2(H)$, its square is in $L^1(H)$ and $||A^2||_1 = ||A||_2^2$. For $k \geq 2$, one gets therefore

$$||A^k||_1 \leq ||A^2||_1||A^{k-2}|| \leq ||A||_2^2||A||^{k-2} \leq ||A||_2^k$$

since $||A|| \leq ||A||_2$. Hence the general term in the series for $\Phi(A)$ is bounded in L^1-norm by $||A||_2^k/(k-2)!$ This is enough to show that this series is absolutely convergent in the Banach space $L^1(H)$, and that Φ is a continuous map from $L^2(H)$ into $L^1(H)$. We set

$$\det_2(1+A) = \det(1-\Phi(A)) = \det(e^{-A}(1+A)) \tag{7.72}$$

where the determinant of $1-\Phi(A)$ is the one considered before. We state the main properties of the modified determinant:

(a) *The functional* $A \mapsto \det_2(1+A)$ *is continuous on the Hilbert space* $L^2(H)$, as a composition of continuous maps

(b) *The operator* $1+A$ *is invertible iff* $\det_2(1+A) \neq 0$; indeed e^{-A} has a bounded inverse e^A hence $1+A$ is invertible iff $1 - \Phi(A)$ is, that is iff $\det(1 - \Phi(A)) \neq 0$.

(c) *For a trace-class operator* A, *one gets*

$$\det_2(1+A) = e^{-\mathrm{Tr}(A)} \det(1+A) \quad . \tag{7.73}$$

Indeed $L^1(H)$ is a subspace of $L^2(H)$ and the embedding is continuous since $||A||_2 \leq ||A||_1$. Therefore in formula (7.73) both sides depend continuously on A in $L^1(H)$ and it suffices to check it for a finite rank operator A. But then we can split H as $F \oplus G$ where F is finite-dimensional, A induces an operator A_0 in F, and vanishes on the orthogonal complement G of F. We are left with the verification of $\det(e^{-A_0}(1+A_0)) = e^{-\mathrm{Tr}(A_0)} \det(1+A_0)$, and using the multiplicativity of determinants, everything is reduced to the proof of $\det(e^{-A_0}) = e^{-\mathrm{Tr}(A_0)}$. Putting A_0 in triangular form, we find

$$\det(e^{-A_0}) = e^{-\lambda_1} \ldots e^{-\lambda_N}$$

and $\mathrm{Tr}(A_0) = \lambda_1 + \ldots + \lambda_N$ where $\lambda_1, \ldots, \lambda_N$ are the eigenvalues of A_0. Done !

(d) *If* A *and* B *are Hilbert-Schmidt operators, then*

$$\det_2((1+A)(1+B)) = \det_2(1+A)\det_2(1+B)\, e^{-\mathrm{Tr}(AB)} \quad . \tag{7.74}$$

Notice that AB is trace-class, hence $\mathrm{Tr}(AB)$ is defined. Arguing by continuity, we need only to prove (7.74) for finite rank operators, and this case follows from (7.73) by an easy calculation.

7.16. A Hilbert-Schmidt operator is compact, hence its eigenvalues are described qualitatively by F.Riesz'theory. So let Σ be the set of nonzero eigenvalues of A, and for λ in Σ its multiplicity $m(\lambda)$ be defined as the (finite) dimension of the space H_λ of vectors annihilated by some power of the operator $A-\lambda$. Alternatively, we can rearrange the eigenvalues in a sequence $\lambda_1, \lambda_2, \ldots$ such that $|\lambda_1| \geq |\lambda_2| \geq \ldots$ and each eigenvalue $\lambda \neq 0$ occurs $m(\lambda)$ times in the sequence (λ_n). If A is of finite rank, there are only a finite number of nonzero eigenvalues; otherwise, $|\lambda_n|$ tends to 0 with 1/n.

Fix an integer $k \geq 2$, so that the operator A^k is trace-class. Let $\Sigma(k)$ be the set of nonzero eigenvalues of A^k, and $m_k(\lambda)$ be the multiplicity of λ. By an algebraic reasoning, it can be shown that a complex number $\lambda \neq 0$ belongs to $\Sigma(k)$ iff it is the k-th power of an element of Σ. Moreover, the set of vectors annihilated by some power of $A^k - \lambda$ is the direct sum $\bigoplus_\mu H_\mu$ where μ runs over the k-th roots of λ belonging to Σ. Hence $m_k(\lambda) = \sum_{\mu^k=\lambda} m(\mu)$, and this implies

$$\sum_{\lambda \in \Sigma(k)} \lambda m_k(\lambda) = \sum_{\mu \in \Sigma} \mu^k \, m(\mu) \quad . \tag{7.75}$$

By Dikii's theorem (7.71), the left-hand side represents the trace of A^k. Hence we get

$$\mathrm{Tr}(A^k) = \sum_{\mu \in \Sigma} \mu^k \, m(\mu) = \sum_{n \geq 1} \lambda_n^k \quad (\text{for } k \geq 2) \quad . \tag{7.76}$$

Of course, when A itself is trace-class, the same formula remains true for k = 1. Notice that the series in (7.75) are absolutely convergent and in particular $\sum_{n\geq 1} |\lambda_n|^2$ *is finite.*

Let us introduce the characteristic determinant

$$D_2(z) = \det_2(1 - zA) \tag{7.77}$$

where z is a complex variable. Notice the series expansion

$$e^{zA}(1 - zA) = 1 - \sum_{k\geq 2} (k-1)(k!)^{-1} A^k z^k \tag{7.78}$$

which converges absolutely in the Banach space $L^1(H)$. Since the map $B \mapsto \det(1+B)$ from $L^1(H)$ to $\mathbb{C}$ is holomorphic, it follows by composition that $D_2(z)$ is an entire function of z. By property (b) of subsection 7.15, *the roots of the equation* $D_2(z) = 0$ *are the inverses of the eigenvalues* $\lambda \neq 0$ *of* A. More precisely, we claim

$$D_2(z) = \prod_{n\geq 1} (1 - \lambda_n z) \, e^{\lambda_n z} \quad , \tag{7.79}$$

where the infinite product converges absolutely since $\sum_n |\lambda_n|^2$ is finite. To prove this formula, let us remark that both sides are entire functions of z, hence it suffices to prove it for small $|z|$. Moreover, both sides take the value 1 for z = 0, hence it suffices to consider the logarithmic derivatives, that is to prove

$$D_2'(z)/D_2(z) = -\sum_{n\geq 1} \frac{\lambda_n^2 z}{1-\lambda_n z} . \tag{7.80}$$

The right-hand side can be expanded into a double series, and taking (7.76) into account we get

$$\begin{aligned}-\sum_{n\geq 1} \frac{\lambda_n^2 z}{1-\lambda_n z} &= -\sum_{n\geq 1}\sum_{k\geq 2} \lambda_n^k z^{k-1} = -\sum_{k\geq 2} z^{k-1}\mathrm{Tr}(A^k)\\ &= -z\,\mathrm{Tr}\Big(\sum_{k\geq 2} z^{k-2}A^{k-2}A^2\Big) = -z\mathrm{Tr}(A^2(1-zA)^{-1}).\end{aligned}$$

The formal manipulations are easily justified for $|z|$ small. On the other hand, using the modified multiplicative property (7.74) and putting $R(z) = A(1-zA)^{-1}$, one gets

$$\begin{aligned}D_2'(z)/D_2(z) &= \lim_{\varepsilon\to 0} \{\det_2(1-zA-\varepsilon A)\det_2(1-zA)^{-1} - 1\}\\ &= \lim_{\varepsilon\to 0} \{\det(e^{\varepsilon R(z)}(1-\varepsilon R(z)))\, e^{-\varepsilon z\mathrm{Tr}(A\cdot R(z))} - 1\}\end{aligned}$$

Since the expansion of $e^{\varepsilon R(z)}(1-\varepsilon R(z))$ into powers of ε has no term of degree 1, one gets $\det_2(e^{\varepsilon R(z)}(1-\varepsilon R(z))) = 1 + O(\varepsilon^2)$ hence $D_2'(z)/D_2(z) = -z\mathrm{Tr}(A\cdot R(z))$ and we are done.

7.17. We noticed already that the determinant $\det_2(1+zA)$ is an entire function of z when A is in $L^2(H)$. Introduce the power series expansion

$$\det_2(1+zA) = \sum_{k\geq 0} b_k(A)\, z^k . \tag{7.81}$$

The coefficients $b_k(A)$ can be calculated as follows

$$b_k(A) = (2\pi)^{-1} R^{-k} \int_0^{2\pi} \det_2(1+Re^{i\theta}A)e^{-ik\theta}\, d\theta . \tag{7.82}$$

It is now easy to estimate $b_k(A)$. Indeed using the product expansion (7.79) and the elementary inequality $|(1+\xi)e^{-\xi}| \leq e^{|\xi|^2/2}$ (for a complex number ξ), one gets

$$\det_2(1+zA) \leq \exp \frac{1}{2} \sum_n |\lambda_n|^2 \, |z|^2 \quad . \tag{7.83}$$

Using (7.82) and choosing R as the square root of $k/\sum_n |\lambda_n|^2$, one gets

$$|b_k(A)| \leqq (e/k)^{k/2} \left(\sum_n |\lambda_n|^2\right)^{k/2} \quad . \tag{7.84}$$

A *crucial estimate, due to H.Weyl,* asserts

$$\sum_n |\lambda_n|^2 \leqq ||A||_2^2 \quad . \tag{7.85}$$

It is in turn derived from a similar statement for trace-class operators, which can be proved using the comparison of eigenvalues of $\Lambda^k B$ and $\Lambda^k |B|$ as in subsection 7.6. Hence we can derive from (7.84) the final estimate

$$|b_k(A)| \leqq (e/k)^{k/2} \, ||A||_2^k \quad . \tag{7.86}$$

We offer now a method to calculate the coefficients $b_k(A)$. First of all, use polarization as in subsection 7.10 and define $b_k(A_1,\ldots,A_k)$ for $A_1,\ldots,A_k$ in $L^2(H)$ by

$$2^k k! b_k(A_1,\ldots,A_k) = \sum_{\varepsilon_1\ldots\varepsilon_k} b_k(\varepsilon_1 A_1+\ldots+\varepsilon_k A_k)\varepsilon_1\ldots\varepsilon_k \tag{7.87}$$

($\varepsilon_1,\ldots,\varepsilon_k$ take independently the values 1 and -1). From (7.86) we get the estimate

$$|b_k(A_1,\ldots,A_k)| \leq \gamma'_k \, ||A_1||_2 \ldots ||A_k||_2 \quad , \tag{7.88}$$

with a constant

$$\gamma'_k = (ek/4)^{k/2}/k! \quad . \tag{7.89}$$

Since the functional $A \mapsto \det_2(1+A)$ is continuous on the Hilbert space $L^2(H)$, it follows from (7.82) and (7.87) that the functional $b_k(A_1,\ldots,A_k)$ is jointly continuous on $L^2(H) \times \ldots \times L^2(H)$. By continuity, it suffices to consider the case where $A_1,\ldots,A_k$ are of finite rank. When A is of finite rank, we know that $\det_2(1+zA)$ is equal to $e^{-z\mathrm{Tr}(A)}\det(1+zA)$, hence we get

$$b_k(A) = \sum_{j=0}^{k} (-1)^j \mathrm{Tr}(A)^j \, c_{k-j}\,(A,\ldots,A)/j! \tag{7.90}$$

using the multilinear form

$$c_r(A_1,\ldots,A_r) = (r!)^{-1} \sum_{\sigma\in S_r} (\mathrm{sgn}\sigma)\ \mathrm{Tr}((A_1\otimes\ldots\otimes A_r)U_\sigma) \tag{7.91}$$

as in subsection 7.10. It follows immediately that $b_k(A_1,\ldots,A_k)$ is a continuous multilinear form on $L^2(H) \times \ldots \times L^2(H)$. Moreover, by some group-theoretical calculations, one derives

$$k!\,b_k(|\psi_1\rangle\langle\theta_1|,\ldots,|\psi_k\rangle\langle\theta_k|) = \det_{\substack{1\le i\le k\\ 1\le j\le k}} a_{ij} \quad , \tag{7.92}$$

with

$$a_{ij} = \begin{cases} \langle\theta_i|\psi_j\rangle & \text{if } i \neq j \\ 0 & \text{if } i = j \quad . \end{cases} \tag{7.93}$$

Problem: derive the estimate (7.88) from the definition of $b_k(A_1,\ldots,A_k)$ by the previous formulas.

To conclude this subsection, let us remark that the formula for the logarithmic derivative of $D_2(z)$ given at the end of subsection 7.16 can be rewritten as follows

$$\det_2(1 + zA) = \exp \sum_{k\ge 2} (-1)^{k-1}\, z^k \mathrm{Tr}(A^k)/k \quad . \tag{7.94}$$

Notice the similarity with the formula for trace-class operators

$$\det(1 + zA) = \exp \sum_{k\ge 1} (-1)^{k-1}\, z^k \mathrm{Tr}(A^k)/k \quad . \tag{7.95}$$

Since Tr *makes in general no sense for* $k = 1$, *when* A *is Hilbert-Schmidt, just omit it !* We leave it to the reader to modify accordingly Waring's formula.

7.18. Let us go back to integral operators. So let Ω and the continuous kernel $K(x,y)$ be as in section 5. Since a continuous kernel $K(x,y)$ is a square-integrable function on $\Omega \times \Omega$, the operator $f \mapsto Kf$ acting on the space $C(\Omega)$ of continuous functions is the restriction of a Hilbert-Schmidt operator A_k acting in $L^2(\Omega)$. The main results in subsection 5.9 can be expressed as follows:

$$\det(1 + zK) = e^{z\tau} \det_2(1 + zA_K) \tag{7.96}$$

where the left-hand side if Fredholm's determinant and $\tau = \int_\Omega K(x,x)\,dx$.

The question about the trace can be reformulated as follows: *suppose the continuous kernel* K *on* $\Omega \times \Omega$ *is such that* A_k *is a trace-class operator in* $L^2(\Omega)$. *Is it true that Fredholm's determinant* $\det(1+zK)$ *agrees with the determinant* $\det(1 + zA_K)$ *of Hilbert space operators ?* Or, according to formula (7.96), do we have in this case

$$\mathrm{Tr}(A_k) = \int_\Omega K(x,x)\,dx. \tag{7.97}$$

The answer is yes, by Mercer's theorem, if A_k is a positive operator. I do not know the answer in general.

From (7.96), one can derive a power series expansion for $\det_2(1 + zA_k)$. Indeed recall the definition of Fredholm's determinant

$$\det(1+zK) = \sum_{k\geq 0} (z^k/k!) \int_\Omega \ldots \int_\Omega \Delta\begin{pmatrix} x_1 \ldots x_k \\ x_1 \ldots x_k \end{pmatrix} dx_1 \ldots dx_k \tag{7.98}$$

where $\Delta\begin{pmatrix} x_1 \ldots x_k \\ x_1 \ldots x_k \end{pmatrix}$ is the determinant of the matrix with elements $K(x_i,x_j)$ for $1 \leq i \leq k$, $1 \leq j \leq k$. Define similarly $\Delta'\begin{pmatrix} x_1 \ldots x_k \\ x_1 \ldots x_k \end{pmatrix}$ as the modified determinant where we replace the diagonal elements $K(x_i,x_i)$ by 0. Then we get

$$\det_2(1+zA_k) = \sum_{k\geq 0} (z^k/k!) \int_\Omega \ldots \int_\Omega \Delta'\begin{pmatrix} x_1 \ldots x_k \\ x_1 \ldots x_k \end{pmatrix} dx_1 \ldots dx_k \quad . \tag{7.99}$$

This formula remains valid if we suppose only that K belongs to $L^2(\Omega \times \Omega)$, hence A_K is Hilbert-Schmidt. This modification of Fredholm's definition was first proposed by Hilbert and Harnack in 1906.

8. Fredholm Determinants: the Case of Banach Spaces

8.1. Let E be a Banach space, with norm $||x||$ for vectors. If A is a bounded operator in E, its norm $||A||$ is the smallest constant $C \geqq 0$ such that $||Ax|| \leqq C\,||x||$ for all x in E. We owe to Grothendieck the following definition of a *nuclear* (or trace-class) operator in E: any operator which can be represented as a series $A = \sum_n h_n$ where each h_n is of rank 1, and $\sum_n ||h_n||$ is finite (such a series converges in operator norm). *The nuclear norm* $||A||_1$ is the infimum of the set of numbers $\sum_n ||h_n||$ for all such decompositions $A = \sum_n h_n$. Using the fact that a normed space is complete (hence a Banach space) iff any absolutely convergent series has a sum, it follows easily that the set $L^1(E)$ of nuclear operators is a Banach space for the nuclear norm.

Introduce the *dual space* E' of E and denote by $\langle x'|x\rangle$ the natural pairing between E' and E (for x in E and x' in E') More explicitely, a nuclear operator can be represented as follows

$$Ax = \sum_n \lambda_n \langle x_n'|x\rangle\, x_n \tag{8.1}$$

where x_n are vectors in E and x_n' in E', λ_n are complex numbers and $||x_n|| = ||x_n'|| = 1$, $\sum_n |\lambda_n|$ finite. In Dirac's notation, where the elements of E' are bras and those of E are kets, this is expressed as

$$A = \sum_n \lambda_n\, |x_n\rangle\langle x_n'| \quad . \tag{8.2}$$

According to subsection 6.4, when E *is a Hilbert space*, the nuclear operators are exactly the trace-class operators. Moreover, in this case, we can compute the trace of A using any orthonormal basis (ψ_m), hence

$$\mathrm{Tr}(A) = \sum_m \langle\psi_m|A|\psi_m\rangle = \sum_{m,n} \lambda_n \langle\psi_m|x_n\rangle\langle x'_n|\psi_m\rangle$$

and using Parseval's equality, we conclude

$$\mathrm{Tr}(A) = \sum_n \lambda_n \langle x'_n|x_n\rangle \quad . \tag{8.3}$$

8.2. There arises the question whether a similar definition of trace works for nuclear operators in a Banach space. The series (8.3) converges absolutely since $\sum_n |\lambda_n|$ is finite and $|\langle x'_n|x_n\rangle| \leqq ||x'_n||\ ||x_n|| \leqq 1$. The question is whether the value given by (8.3) is independent of the chosen decomposition (8.2) for A. First of all, the trace is well-defined for *finite rank operators*. Indeed, let B be such an operator, represented as a sum $\sum_{n=1}^{N} |x_n\rangle\langle x'_n|$. Let us choose a set of linearly independent vectors $e_1,\dots,e_M$ such that every x_n is a linear combination of the e_m's. Now any vector of the form Bx is a unique linear combination of $e_1,\dots,e_M$, hence there exist uniquely defined elements $e'_1,\dots,e'_M$ of E' such that $B = \sum_{m=1}^{M} |e_m\rangle\langle e'_m|$. Write $x_n = \sum_{m=1}^{M} u_{nm} e_m$ with complex numbers u_{nm}. Then

$$B = \sum_n |x_n\rangle\langle x'_n| = \sum_{n,m} u_{nm}\ |e_m\rangle\langle x'_n|$$

and, by the uniqueness of the expansion $B = \sum_m |e_m\rangle\langle e'_m|$, one gets $e'_m = \sum_n u_{nm} x'_n$. Therefore

$$\sum_n \langle x'_n|x_n\rangle = \sum_{n,m} u_{nm} \langle x'_n|e_m\rangle = \sum_m \langle e'_m|e_m\rangle \ .$$

If we start from two decompositions

$$B = \sum_n |x_n\rangle\langle x'_n| = \sum_r |y_r\rangle\langle y'_r| \quad ,$$

we can use the same e_m's for both, hence $\sum_n \langle x'_n|x_n\rangle$ and $\sum_r \langle y'_r|y_r\rangle$ are equal. *Conclusion: The finite rank operators form a dense linear subspace* $L_f(E)$ *of* $L^1(E)$ *and a trace is defined on* $L_f(E)$ *which is linear and takes* $|x\rangle\langle x'|$ *into* $\langle x'|x\rangle$.

We can extend the trace to the whole of $L^1(E)$ when E satisfies *Banach approximation property:* given any compact set K in E, there exists a sequence of finite rank operators p_k such that $\sup_k ||p_k||$ be finite and $\lim_{k\to\infty} ||p_k(x) - x|| = 0$ uniformly for x in K. The standard Banach spaces (continuous functions, Hilbert spaces, L^p spaces, Sobolev spaces) satisfy the property. At the time Grothendieck considered these problems (around 1950), it was still conjectured that every Banach space does. Counter examples were discovered much later the most remarkable being the space $L(H)$ of all bounded operators in a Hilbert space H, with the operator norm $||A||$. Suppose that E satisfies the approximation property and let $A = \sum_n \lambda_n |x_n><x'_n|$ be a nuclear operator. Since $\sum_n |\lambda_n|$ is finite, we can write $\lambda_n = \mu_n \cdot \nu_n$ with $\sum_n |\mu_n|$ finite and $\lim_{n\to\infty} \nu_n = 0$. Choose a compact set K in E containing the vectors $\nu_n x_n$, and operators p_k adapted to K. Then $p_k A$ are finite rank operators, hence their traces are defined; the formula

$$\mathrm{Tr}(p_k A) = \sum_n \mu_n <x'_n|p_k(\nu_n x_n)> \qquad (8.4)$$

is easily checked. By dominated convergence for series, one deduces

$$\lim_{k\to\infty} \mathrm{Tr}(p_k A) = \sum_n \mu_n <x'_n|\nu_n x_n> = \sum_n \lambda_n <x'_n|x_n> \quad . \qquad (8.5)$$

As in the finite rank case, the same set K and the same p_k's can be used simultaneously for two decompositions

$$A = \sum_n \lambda_n |x_n><x'_n| = \sum_r \pi_r |y_r><y'_r| \quad , \qquad (8.6)$$

hence

$$\sum_n \lambda_n <x'_n|x_n> = \sum_r \pi_r <y'_r|y_r> \quad . \qquad (8.7)$$

Conclusion: when E *satisfies the approximation property, there exists a linear form* Tr *on* $L^1(E)$ *such that* $|\mathrm{Tr}(A)| \leq ||A||_1$ *and taking* $|x><x'|$ *into* $<x'|x>$.

8.3. We come now to the general case. It turns out that nuclear operators in Banach spaces are better compared to Hilbert-Schmidt operators than to trace-class operators in Hilbert spaces. We proceed to define *the trace of the square of a nuclear operator.*

Let B and C be nuclear operators in E, with decompositions

$$B = \sum_n \lambda_n |x_n\rangle\langle x_n'| \ , \qquad C = \sum_m \mu_m |y_m\rangle\langle y_m'| \quad . \tag{8.8}$$

Then one gets absolutely convergent series

$$\sum_{n,m} \lambda_n \mu_m \langle x_n'|y_m\rangle\langle y_m'|x_n\rangle = \sum_n \lambda_n \langle x_n'|Cx_n\rangle = \sum_m \mu_m \langle y_m'|By_m\rangle \ .$$

The second expression does not depend on the decomposition chosen for C and the third one does the same for B. Hence these expressions depend solely on B and C. It follows that *there exists a bilinear form* Tr(B;C) *on* $L^1(E) \times L^1(E)$ *with the following properties:*

(a) *the inequality* $|\mathrm{Tr}(B;C)| \leq ||B||_1 \, ||C||_1$;

(b) *when* $B = |x\rangle\langle x'|$ *and* $C = |y\rangle\langle y'|$ *are decomposable, then* $\mathrm{Tr}(B;C) = \langle x'|y\rangle\langle y'|x\rangle$;

(c) *symmetry* $\mathrm{Tr}(B;C) = \mathrm{Tr}(C;B)$.

We prove now that Tr(B;C) *depends solely on the product* A = BC. Let us introduce an auxiliary Hilbert space H with an orthonormal basis (ψ_r). We factor every λ_n as $\beta_n \beta_n'$ with $|\beta_n| = |\beta_n'|$, hence $\sum_n |\beta_n|^2 = \sum_n |\beta_n'|^2$ is equal to $\sum_n |\lambda_n|$, hence finite. Define operators $\beta: H \to E$ and $\beta': E \to H$ by

$$\beta = \sum_n \beta_n |x_n\rangle\langle\psi_n| \tag{8.9}$$

$$\beta' = \sum_n \beta_n' |\psi_n\rangle\langle x_n'| \tag{8.10}$$

(by convention, our indices run over the integers 1,2,3...). More explicitely, for ψ in H, one has

$$\beta(\psi) = \sum_n \beta_n \langle\psi_n|\psi\rangle \, x_n \tag{8.11}$$

and the series converges absolutely in E since $||x_n|| = 1$ and

$\sum_n |\beta_n|^2$, $\sum_n |\langle\psi_n|\psi\rangle|^2$ are finite (*Hint:* use Cauchy-Schwarz inequality). Similarly

$$\beta'(x) = \sum_n \beta'_n \langle x'_n|x\rangle\ \psi_n \tag{8.12}$$

and $\sum_n |\beta'_n\langle x'_n|x\rangle|^2$ is bounded by $||x||^2 \sum_n|\beta'_n|^2$. According to these calculations, we get the norm estimates

$$||\beta||^2 \leq \sum_n|\beta_n|^2 \quad , \ ||\beta'||^2 \leq \sum_n |\beta'_n|^2 \quad . \tag{8.13}$$

From the construction of β and β', we get $B = \beta\beta'$ since $\beta(\psi_n) = \beta_n x_n$ by (8.11). On the other hand, $\beta'\beta$ is a bounded operator in the Hilbert space H, with matrix elements

$$\langle\psi_m|\beta'\beta|\psi_n\rangle = \beta_n\beta'_m \langle x'_m|x_n\rangle \quad . \tag{8.14}$$

Since $|\langle x'_m|x_n\rangle|$ is bounded by 1, and $\sum_n |\beta_n|^2$, $\sum_m |\beta'_m|^2$ are finite, it follows that $\sum_{m,n} |\langle\psi_m|\beta'\beta|\psi_n\rangle|^2$ is finite. Taking into account the definition of the nuclear norm, we conclude: *given the nuclear operator* B *and any* $\varepsilon > 0$, *there exists a decomposition* $B = \beta\beta'$ *with bounded operators* $\beta: H \to E$ *and* $\beta': E \to H$, *while* $\beta'\beta$ *is a Hilbert-Schmidt operator in* H *with* $||\beta'\beta||_2 \leq ||B||_1 + \varepsilon$.

Introduce a similar decomposition $C = \gamma\gamma'$ where $\gamma'\gamma$ is a Hilbert-Schmidt operator in H with $||\gamma'\gamma||_2 \leq ||C||_1 + \varepsilon$. By calculations similar to the previous ones, one shows that $\beta'\gamma$ and $\gamma'\beta$ are Hilbert-Schmidt operators in H. Putting $\delta = \beta'\gamma\gamma'$ and noticing that $\delta\beta = (\beta'\gamma)(\gamma'\beta)$, we conclude: *the operator* $A = BC$ *in* E *can be factored as* VU *with bounded operators* $U: E \to H$ *and* $V: H \to E$, *in such a way that* UV *be a trace-class operator in* H *with trace given by* $\mathrm{Tr}(UV)=\mathrm{Tr}(B;C)$.

We can calculate the trace of UV in terms of its eigenvalues. By easy calculations, one shows that for any $\lambda \neq 0$, and any integer $N > 0$, U maps the set of solutions ψ in H of $(UV - \lambda)^N\psi = 0$ isomorphically onto the set of solutions x in E of $(VU - \lambda)^N x = 0$. A similar statement holds with U, V interchanged and E,H interchanged. Hence the operators UV in H and VU in E have the same eigenvalues $\lambda \neq 0$, with a common multiplicity $m(\lambda)$. Borrowing from the spectral theory of trace-class operators in Hilbert spaces (see subsection 7.14), we can conclude:

Let the operator A *in* E *be factored as* BC , *where* B *and* C *are nuclear. Let* Σ *be the set of nonzero eigenvalues of* A , *and* $m(\lambda)$ *the multiplicity of* λ *in* Σ. *Then the series* $\sum_{\lambda\in\Sigma} m(\lambda)\lambda$ *is absolutely convergent and its sum is equal to* $\mathrm{Tr}(B;C)$. In particular $\mathrm{Tr}(B;C)$ depends only on BC, as asserted.

8.4. From now on, it is very easy to transfer the properties of Hilbert-Schmidt operators into properties of nuclear operators. Taking B = C in the previous result, we get:

(a) *Let* B *be a nuclear operator, not of finite rank. Then its eigenvalues (multiplicities included) can be arranged as a sequence* $(\lambda_n)_{n\geq 1}$, *tending to zero, with* $|\lambda_1| \geq |\lambda_2| \geq \ldots$ *Moreover* $\sum_n |\lambda_n|^2$ *is finite and bounded by* $||B||_1^2$.

Then taking $C = B^{k-1}$, with $k \geq 2$:

(b) *For every integer* $k \geq 2$, *one gets*

$$\mathrm{Tr}(B;B^{k-1}) = \sum_{n\geq 1} \lambda_n^k \quad . \tag{8.15}$$

The (modified) determinant of 1 + B can be defined as follows

$$\det_2(1+B) = \prod_{n\geq 1} (1+\lambda_n)\, e^{-\lambda_n} \quad ; \tag{8.16}$$

the convergence of the infinite product is guaranteed since $\sum_n |\lambda_n|^2$ is finite. The *characteristic determinant of* B is the function $D_2(z) = \det_2(1 - zB)$, that is

$$D_2(z) = \prod_{n\geq 1} (1-\lambda_n z)\, e^{\lambda_n z} \quad . \tag{8.17}$$

Hence $D_2(z)$ is an entire function of the complex variable z. We can expand it as a power series

$$D_2(z) = \sum_{k\geq 0} (-1)^k\, b_k(B)\, z^k \quad . \tag{8.18}$$

Recall the existence of a factorization $E \xrightarrow{\beta'} H \xrightarrow{\beta} E$ of B such that $\beta'\beta$ be a Hilbert-Schmidt operator in H. By a previous remark, the operators $B = \beta\beta'$ and $T = \beta'\beta$ have the same eigenvavalues, with equal multiplicities. Hence one gets

$$\mathrm{Tr}(B;B^{k-1}) = \mathrm{Tr}(T^k) \tag{8.19}$$

$$\det_2(1 - zB) = \det_2(1 - zT) \quad . \tag{8.20}$$

Notice also that, given $\varepsilon > 0$, we can choose β and β' such that $||T||_2 \leq ||B||_1 + \varepsilon$. We can derive the properties of $D_2(z)$ from those of T without any new calculation. For instance, we get

$$D_2(z) = \exp - \sum_{k\geq 2} \mathrm{Tr}(B;B^{k-1})\, z^k/k \tag{8.21}$$

for $|z| < 1/||B||_1$ as well as the estimate

$$|b_k(B)| \leq (e/k)^{k/2}\, ||B||_1^k \quad . \tag{8.22}$$

From (8.21) flow the usual corollaries, such as Waring formulas and Plemelj determinants. From (8.21) and the bilinearity of Tr(B;C) it follows that there exists continuous multilinear forms $b_k(B_1,\ldots,B_k)$ on $L^1(E)\times \ldots \times L^1(E)$ such that $b_k(B) = b_k(B, \ldots, B)$. We can take $b_k(B_1,\ldots,B_k)$ as symmetrical, hence given by the polarization formula (7.87). This provides the following bound, implying that $B \mapsto \det_2(1+B)$ is *holomorphic* on the Banach space $L^1(E)$:

$$|b_k(B_1,\ldots,B_k)| \leq (ek/4)^{k/2}(k!)^{-1}||B_1||_1\ldots||B_k||_1 \quad . \tag{8.23}$$

Let us mention the following analogue of formula (7.92)

$$k!\,b_k(B_1,\ldots,B_k) = \det\begin{pmatrix} 0 & \langle x_1'|x_2\rangle & \ldots & \langle x_1'|x_k\rangle \\ \langle x_2'|x_1\rangle & 0 & & \langle x_2'|x_k\rangle \\ \cdot & \cdot & \cdot & \cdot \\ \langle x_k'|x_1\rangle & \langle x_k'|x_2\rangle & \ldots & 0 \end{pmatrix} \tag{8.24}$$

for $B_1 = |x_1\rangle\langle x_1'|, \ldots, B_k = |x_k\rangle\langle x_k'|$. We could use this formula to estimate $b_k(B_1,\ldots,B_k)$; using Hadamard's estimate on determinants, we recover (8.23) with a slightly larger constant, namely $k^{k/2}/k!$.

To conclude this subsection, let us mention the *modified multiplicative rule*

$$\det_2((1+B)(1+C)) = \det_2(1+B)\det_2(1+C)e^{-\mathrm{Tr}(B;C)} \qquad (8.25)$$

which is analogous to (7.74). The simplest way to derive it, is by noticing that both sides are continuous functions of B and C in the Banach space $L^1(E)$ and that finite rank operators are dense in $L^1(E)$. This reduces the proof to the finite dimensional case where $\det_2(1+B) = e^{-\mathrm{Tr}(B)}\det(1+B)$ and $\mathrm{Tr}(B;C) = \mathrm{Tr}(BC)$; the rest of the calculation is easy.

8.5. The theory behaves in a much smoother way when the Banach space E enjoys the approximation property. In this case, the trace of a nuclear operator is defined and Tr(B;C) is the trace of the product BC. We then define the determinant by

$$\det(1+B) = e^{\mathrm{Tr}(B)} \det_2(1+B) \quad . \qquad (8.26)$$

For the characteristic determinant we get

$$D(z) = \det(1 - zB) = e^{z\tau} \prod_{n\geq 1} (1 - \lambda_n z)\, e^{\lambda_n z} \qquad (8.27)$$

with $\tau = \mathrm{Tr}(B)$. In general, the series $\sum_n \lambda_n$ is not convergent, and even if it converges, may fail to sum to the trace τ of B.

The formulas of the previous subsections admit of the following variants:

$$D(z) = \exp - \sum_{k\geq 1} \mathrm{Tr}(B^k)\, z^k/k \qquad (8.28)$$

(for $|z| < 1/||B||_1$) and

$$D(z) = \sum_{k\geq 0} (-1)^k\, c_k(B)\, z^k \quad . \qquad (8.29)$$

Here again $c_k(B)$ is obtained by putting $B_1 = \ldots = B_k = B$ in a continuous symmetrical multilinear form $c_k(B_1,\ldots,B_k)$. It is characterized by the formula

$$k!\,c_k(B_1,\ldots,B_k) = \det_{\substack{1\leq i\leq k\\ 1\leq j\leq k}} \langle x_i^{\prime} | x_j \rangle \qquad (8.30$$

for decomposable operators $B_1 = |x_1\rangle\langle x_1'|, \ldots, B_k = |x_k\rangle\langle x_k'|$. Using Hadamard's estimate for determinants, one gets the estimate

$$|c_k(B_1,\ldots,B_k)| \leq k^{k/2}(k!)^{-1}||B_1||_1 \ldots ||B_k||_1 \quad . \qquad (8.31)$$

Grothendieck originally developed his theory of Fredholm determinants in Banach space by using formulas (8.29) and (8.30) as a starting point.

As we expect, the operator $1+B$ is invertible iff $\det(1+B) \neq 0$ and from (8.25) one derives immediately the multiplicative rule

$$\det((1+B)(1+C)) = \det(1+B)\det(1+C) \quad . \qquad (8.32)$$

8.6. To make the story complete, let us specialize our theory to the Banach space $E = C(\Omega)$ of continuous functions. For every integer $r \geq 1$, choose a finite open covering $(U_\alpha)_{\alpha\in I(r)}$ of Ω by sets of diameter $< 1/r$. For each α, choose a point x_α in U_α and a continuous function φ_α, taking positive values and vanishing outside U_α, in such a way that $\sum_{\alpha\in I(r)} \varphi_\alpha = 1$ ("partition of unity"). Define the finite rank linear operator p_r in $C(\Omega)$ by

$$(p_r f)(x) = \sum_{\alpha\in I(r)} f(x_\alpha)\varphi_\alpha(x) \quad . \qquad (8.33)$$

Any continuous function f being uniformly continuous on the compact space Ω, the sequence of functions $p_r f$ converges uniformly to f on Ω. Moreover, by Ascoli theorem, the convergence is uniform in f when f runs over a compact subset of $C(\Omega)$. The space $C(\Omega)$ enjoys therefore the approximation property. It can be shown that an integral operator A_K with continuous kernel K *acts on $C(\Omega)$ as a nuclear operator,* although *it does not act on $L^2(\Omega)$ as a trace-class operator,* generally speaking. With the previous notations, the finite rank operator $p_r A_K$ transforms a function f in $C(\Omega)$ into $\sum_\alpha \varphi_\alpha \mu_\alpha(f)$ where $\mu_\alpha(f) = \int_\Omega K(x_\alpha,y)f(y)\,dy$. Therefore $\mathrm{Tr}(p_k A_K)$ is equal to $\sum_\alpha \mu_\alpha(\varphi_\alpha) =$

$\sum_\alpha \int_\Omega K(x_\alpha, x)\varphi_\alpha(x)\,dx$. This converges to $\int_\Omega K(x,x)\,dx$ hence *the trace of* A_K *as a nuclear operator in* $C(\Omega)$ *is equal to the "naive" trace* $\int_\Omega K(x,x)\,dx$.

From this fact and the results given in subsection 7.18, it follows that *the Fredholm determinant* $\det(1+K)$ *is equal to the Grothendieck determinant* $\det(1+A_K)$ *associated to the nuclear operator* A_K.

PART THREE:

OVERVIEW OF RECENT DEVELOPMENTS

9. Grassmann Calculus and Berezin Determinants

9.1. Let V be a complex vector space of finite dimension n, and choose a basis $e_1,\ldots,e_n$ of V. We introduced in subsection 3.4 the symmetric algebra SV and remarked that its elements can be put in bijective correspondence with the polynomials in n variables $x_1,\ldots,x_n$. In this correspondence, the vector e_i corresponds to the variable x_i. The multiplication obeys the *commutative law* $x_i x_j = + x_j x_i$. From the variables we built the monomials, products of variables, which can be put in the normal form $x_1^{\alpha_1} \ldots x_n^{\alpha_n}$ because of commutativity. From the monomials, one builds the polynomials by using linear combinations with complex coefficients.

Consider now the exterior algebra ΛV built on V. In a similar way, it can be considered as the algebra of Grassmann variables $\xi_1,\ldots,\xi_n$, the vector e_i corresponding to ξ_i. These variables obey the *anticommutative law* $\xi_i\xi_j = - \xi_j\xi_i$. This implies $\xi_i\xi_i = -\xi_i\xi_i$, hence $\xi_i^2 = 0$. The monomials in $\xi_1,\ldots,\xi_n$ can therefore be normalized as $\xi_{i_1}\ldots\xi_{i_k}$ with indices $i_1,\ldots,i_k$ in strictly increasing order. A Grassmann polynomial in the variable $\xi_1,\ldots,\xi_n$ can be expressed in a unique way as a linear combination of the 2^n monomials with complex coefficients.

A monomial $\xi_{i_1}\ldots\xi_{i_k}$ is called *even* or *odd* if k is respectively even or odd. An even (odd) polynomial is a linear combination containing only even (odd) monomials.

There is only one monomial of degree n, namely $\xi_1 \ldots \xi_n$. Let $A = (a_{ij})$ be any matrix of size n x n, and introduce the Grassmann polynomials $\eta_i = \sum_{j=1}^{n} a_{ij}\xi_j$ of degree 1. Then the anticommutativity rule $\eta_i\eta_j = -\eta_j\eta_i$ holds and the product $\eta_1\ldots\eta_n$ is homogeneous of degree n, hence a scalar multiple of $\xi_1 \ldots\xi_n$. Since the Grassmann polynomials are just another way of denoting the elements of ΛV, their product corresponds to the wedge product, hence formula (2.16) can be rewritten as

$$\eta_1\ldots\eta_n = (\det A)\ \xi_1 \ldots \xi_n \quad . \tag{9.1}$$

We can develop the product $\eta_1\ldots\eta_n$ as the sum of the n^n products $a_{1j_1}\ldots a_{nj_n}\ \xi_{j_1}\ldots\xi_{j_n}$. The monomial $\xi_{j_1}\ldots\xi_{j_n}$ is 0 unless the indices $j_1\ldots j_n$ form a permutation σ of 1 ... n, and in this case is equal to $(\operatorname{sgn}\sigma)\xi_1\ldots\xi_n$. Hence we get the familiar complete expansion of the determinant

$$\det A = \sum_{\sigma\in S_n} (\operatorname{sgn}\sigma)\ a_{1\,\sigma(1)}\cdots a_{n\sigma(n)} \quad . \tag{9.2}$$

9.2. Berezin's very original idea was to define a differential and integral calculus of Grassmann polynomials. Consider first the derivatives. Choose an index i between 1 and n. Then any given Grassmann polynomial $P(\xi_1,\ldots,\xi_n)$ can be written uniquely as $A + \xi_i B$, where A and B are Grassmann polynomials in the variables different from ξ_i. We define the partial derivative $\delta_i P = \delta P/\delta\xi_i$ as the coefficient B of ξ_i in P. Notice that, due to anticommutativity we have to distinguish $\xi_i B$ from $B\xi_i$. For this reason, $\delta_i P$ is called the *forward derivative* with respect to ξ_i.

We record here a few basic formulas, which except for the minus signs, are similar to familiar formulas

$$\delta_i\delta_j = -\delta_j\delta_i \quad , \quad \xi_i\xi_j = -\xi_j\xi_i \tag{9.3}$$

$$\delta_i\xi_j + \xi_j\delta_i = \delta_{ij} \tag{9.4}$$

$$\delta_i(PQ) = \delta_i P\cdot Q \pm P\cdot\delta_i Q \quad . \tag{9.5}$$

In formula (9.4), ξ_i is interpreted as the operator mapping a Grassman polynomial P into $\xi_i P$ ("forward multiplication" by ξ_i); hence, we get more explicitely

$$\delta_i(\xi_j P) + \xi_j\cdot(\delta_i P) = \delta_{ij} P \quad . \tag{9.6}$$

In formula (9.5), the sign is + or - if P is even or odd respectively. Notice that $\delta_i P$ is odd (even) when P is even(odd), hence δ_i changes the parity. According to the sign rule for the parity of products

even x even = even ; even x odd = odd ;

odd x even = odd ; odd x odd = even ,

(see formula (3.3)), the operators δ_i (as well as ξ_i) are to be considered as *odd*. Formula (9.4) can be rewritten as $\delta_i\xi_j = -\xi_j\delta_i$ for $i \neq j$ in analogy with (9.3). Moreover in formula (9.5), a minus sign occurs only at the place where δ_i and P are interchanged and only when P is odd. All this agrees fully with Koszul's sign rule (see subsection 2.9)).

If I is any subset of the set $\{1,2,\ldots,n\}$ with elements $i_1,\ldots,i_k$ arranged in increasing order, we set

$$\xi_I = \xi_{i_1}\ldots\xi_{i_k} \quad , \qquad \delta_I = \delta_{i_1}\ldots\delta_{i_k} \quad . \tag{9.7}$$

According to the rules (9.3) and (9.4), we can shift in any product of factors ξ_i and δ_j the factors δ_j to the right and the factors ξ_i to the left. Hence a differential operator acting on the Grassman polynomials can be written in the normal form

$$D = \sum_{I,J} a_{I,J}\,\xi_I\delta_J \quad . \tag{9.8}$$

Moreover the monomials ξ_K form a basis of the Grassman algebra $\Lambda(\xi_1,\ldots,\xi_n)$. The action of the operator $\xi_I\delta_J$ is given as follows

$$(\xi_I\delta_J)(\xi_K) = \begin{cases} \pm\,\xi_L & \text{if } J \subset K \text{ and } I \cap (K\setminus J) = \emptyset \\ 0 & \text{otherwise} \end{cases} \tag{9.9}$$

(where $L = I \cup (K \setminus J)$). It is then easy to prove that any linear operator acting on $\Lambda(\xi_1, \ldots, \xi_n)$ can be uniquely written in the form (9.8). In particular, *any operator is a differential operator*.

Let us add two remarks. When $n = 1$, write ξ_1 as ξ. Hence the Grassmann algebra $\Lambda(\xi)$ has a basis $1, \xi$ in which the operators $\xi = \xi_1$ and $\delta = \delta_1$ are expressed by the matrices

$$\xi = \begin{pmatrix} 0 & 0 \\ 1 & 0 \end{pmatrix} \quad , \qquad \delta = \begin{pmatrix} 0 & 1 \\ 0 & 0 \end{pmatrix} \quad . \tag{9.10}$$

The relations (9.3) and (9.4) take the form

$$\xi\xi = \delta\delta = 0, \qquad \xi\delta + \delta\xi = 1 \quad , \tag{9.11}$$

which are easy to check on (9.10). Notice that the Pauli matrices are $\sigma_+ = \delta, \sigma_- = \xi$, $\sigma_3 = \delta\xi - \xi\delta$.

Moreover, the relations (9.3) and (9.4) define a Clifford algebra with generators $\xi_1, \ldots, \xi_n$, $\delta_1, \ldots, \delta_n$ and our result about differential operators corresponds to the well-known fact that such an algebra is an algebra $M_{2^n}(\mathbb{C})$ of complex matrices of size $2^n \times 2^n$.

9.3. We come now to *Berezin integral*. Write a Grassmann polynomial in $\Lambda(\xi_1, \ldots, \xi_n)$ as

$$P = c_0 + \sum_{i=1}^{n} c_i \xi_i + \ldots + c_{12\ldots n} \xi_1 \cdots \xi_n \quad .$$

The coefficient $c_{12\ldots n}$ is called the *Berezin integral* of P, to be denoted by $\int P \cdot \delta^n \xi$, or $\int P \cdot \delta\xi_1 \cdots \delta\xi_n$. This integral can be calculated as a repeated integral

$$\int \delta\xi_1 \int \delta\xi_2 \ldots \int \delta\xi_n \; P(\xi_1, \ldots, \xi_n) \quad .$$

In this kind of calculation, we assume that the $\delta\xi_i$ anticommute with each other, and *commute* with the ξ_j (there is no unanimous agreement on this last point). The basic rules are as follows:

$$\int P(\xi_1, \ldots, \xi_n) \delta\xi_1 \ldots \delta\xi_n = \int Q(\xi_1, \ldots, \xi_q) \delta\xi_1 \cdots \delta\xi_q \cdot \int R(\xi_{q+1}, \ldots, \xi_n) \delta\xi_{q+1} \cdots \delta\xi_n \tag{9.12}$$

if $P(\xi_1,\ldots,\xi_n)$ splits as $Q(\xi_1,\ldots,\xi_q)R(\xi_{q+1},\ldots,\xi_n)$

$$\int \delta\xi_i = 0 \quad , \qquad \int \xi_i \delta\xi_i = 1 \quad . \tag{9.13}$$

In the derivative $\delta_i P$, we have a Grassmann polynomial of $\xi_1,\ldots,\xi_{i-1},\xi_{i+1},\ldots,\xi_n$ hence there can be no term proportional to $\xi_1,\ldots,\xi_n$. Hence we get

$$\int \delta_i P \cdot \delta^n \xi = 0 \quad , \tag{9.14}$$

and from (9.5) one derives the rule of *integration by parts*

$$\int \delta_i P \cdot Q \delta^n \xi = \mp \int P \cdot \delta_i Q \delta^n \xi \tag{9.15}$$

where the sign + (-) holds for P odd(even).

9.4. Let us mention the formula for a *linear change of variables in a Berezin integral*. Recall first that given a real function $f(x)$ on $\mathbb{R}^n$ and an invertible linear map A from $\mathbb{R}^n$ to $\mathbb{R}^n$, one gets

$$\int_{\mathbb{R}^n} f(x) d^n x = \det A \int_{\mathbb{R}^n} f(Ay) d^n y \quad . \tag{9.16}$$

Hence the rule: the integral is unchanged under the simultaneous substitutions $x \mapsto Ay$, $d^n x \mapsto (\det A)\cdot d^n y$.

Consider now a Grassmann polynomial P in $\Lambda(\xi_1,\ldots,\xi_n)$ and express the ξ_i as linear combinations of new Grassmann variables

$$\xi_i = \sum_{j=1}^{n} a_{ij}\eta_j \quad , \text{ or in shorter form } \xi = A\eta , \tag{9.17}$$

where the matrix $A = (a_{ij})$ is invertible. Then we get

$$\int P(\xi)\delta^n \xi = (\det A)^{-1} \int P(A\eta)\delta^n \eta \tag{9.18}$$

or in symbolic form

$$\delta^n \xi = (\det A)^{-1} \delta^n \eta \qquad \text{for } \xi = A\eta \quad . \tag{9.19}$$

To prove (9.18), let us remark that for $P(\xi)$ homogeneous of degree $k < n$ in ξ, the Grassmann polynomial $P(A\eta)$ is homogeneous of degree $k < n$ in η, hence both integrals in (9.18) are 0.

It remains the case $P(\xi) = \xi_1 \cdots \xi_n$, hence $P(A\eta) = (\det A)\eta_1 \cdots \eta_n$ and our formula follows from

$$\int \xi_1 \ldots \xi_n \, \delta\xi_1 \ldots \delta\xi_n = \int \eta_1 \ldots \eta_n \delta\eta_1 \ldots \delta\eta_n = 1 .$$

9.5. We come to the *exponential*. If P and Q are even Grassmann polynomials, we get the commutativity rule $PQ = QP$. We define the exponential of P by the familiar power series

$$\exp P = \sum_{m=0}^{\infty} P^m/m! \tag{9.20}$$

provided it converges. The convergence can be proved as follows. Write P as $c + Q$ where c is the constant term of P. Then Q having no constant term begins with terms of degree ≥ 2, hence $Q^m = 0$ for $m > \frac{n}{2}$. The series for exp Q breaks down, namely

$$\exp Q = \sum_{m=0}^{N} Q^m/m! \quad , \tag{9.21}$$

where N is the integral part of $\frac{n}{2}$. Moreover, by the binomial theorem, one obtains

$$P^m/m! = \sum_{r=0}^{N} (c^{m-r}/(m-r)!)\cdot(Q^r/r!) \quad . \tag{9.22}$$

By the convergence of the ordinary exponential series for exp c, we get, after rearranging, the convergence of the series for exp P, and the formula

$$\exp P = \exp c \cdot \exp Q \quad . \tag{9.23}$$

The functional equation

$$\exp(P + P') = \exp P \cdot \exp P' \tag{9.24}$$

can now be proved by expanding the exponentials in power series and using the binomial theorem to calculate $(P+P')^m/m!$. The algebra works because P and P' commute, and the calculus goes on because of the convergence of the series. One could also use formula (9.23).

Notice that the square of any monomial in $\xi_1,\ldots,\xi_n$ is 0. Hence, any even element Q of $\Lambda(\xi_1,\ldots,\xi_n)$ without constant term, can be written as $Q = \lambda_1+\ldots+\lambda_r$ where

$\lambda_1^2 = \ldots = \lambda_r^2 = 0$ and of course $\lambda_i\lambda_j = \lambda_j\lambda_i$. If $\lambda^2 = 0$ then $\exp\lambda = 1 + \lambda$, and from the functional equation (9.24), one concludes

$$\exp Q = \prod_{i=1}^{r} (1 + \lambda_i) \quad . \tag{9.25}$$

9.6. We derive now the Grassmann analogues of the gaussian integrals (see section 4). Consider an element Q of $\Lambda(\xi_1,\ldots,\xi_n)$ homogeneous of degree 2, hence $Q = \sum_{i,j} q_{ij}\xi_i\xi_j$, with a skew-symmetric matrix (q_{ij}). We can write also $\frac{1}{2}Q = \sum_{i<j} q_{ij}\xi_i\xi_j$ and from (9.25),one derives

$$\exp \tfrac{1}{2}Q = \prod_{i<j} (1 + q_{ij}\xi_i\xi_j) \quad . \tag{9.26}$$

We have to calculate the coefficient c of $\xi_1 \ldots \xi_n$ in exp Q. It is obviously 0 if n is odd. Suppose n even, n = 2m say. Then c is obtained as follows: one considers all possible partitions of the set $\{1,2,\ldots,2m\}$ into m pairs $\{i_1,j_1\},\ldots,\{i_m,j_m\}$, denotes by ε the sign of the permutation sending $12 \ldots 2m$ into $i_1j_1i_2j_2 \ldots i_mj_m$, and multiplies it by $q_{i_1j_1}\ldots q_{i_mj_m}$. Then make the sum of all such contributions, two partitions into m pairs differing by the order of the pairs being considered as identical. This is the so-called *Pfaffian* of the matrix (q_{ij})

Examples: a) for m = 1 , $c = q_{12}$

b) for m = 2 , $c = q_{12}q_{34} - q_{13}q_{24} + q_{14}q_{23}$.

This Pfaffian, to be denoted Pf(Q) or $\mathrm{Pf}(q_{ij})$ is a Berezin integral

$$\mathrm{Pf}(Q) = \int \exp \tfrac{1}{2}Q(\xi)\, \delta^n\xi \qquad (n \text{ even}) \quad . \tag{9.27}$$

The determinant of a square matrix $A = (a_{ij})$ of size n x n can also be interpreted as a Berezin integral. Namely introduce Grassman variables $\xi_1,\ldots,\xi_n$, $\eta_1,\ldots,\eta_n$ and the bilinear form $A(\xi,\eta) = \sum_{i,j} a_{ij}\xi_i\eta_j$. We claim

$$\iint \exp A(\xi,\eta)\, \delta^n\xi\delta^n\eta = (-1)^{n(n-1)/2} \cdot \det A \quad . \tag{9.28}$$

Indeed develop $A(\xi,\eta)$ as $\xi_1 u_1(\eta) + \ldots + \xi_n u_n(\eta)$ with $u_i(\eta) = \sum_{j=1}^{n} a_{ij}\eta_j$. Each of the terms $\xi_i u_i(\eta)$ has a square equal to 0, hence we have to calculate the coefficient of $\xi_1 \ldots \xi_n \eta_1 \ldots \eta_n$ in $\prod_{i=1}^{n}(1 + \xi_i u_i(\eta))$. The relevant term in the product is $\xi_1 u_1(\eta) \ldots \xi_n u_n(\eta)$, which can be rearranged as

$$(-1)^{n(n-1)/2}\ \xi_1 \ldots \xi_n\ u_1(\eta) \ldots u_n(\eta)$$

We remarked already that $u_1(\eta) \ldots u_n(\eta)$ is equal to $(\det A)\cdot\eta_1 \ldots \eta_n$, hence formula (9.28) is proved. We could rewrite this formula as

$$\iint \exp\{\sum_{i,j} \xi_i a_{ij} \eta_j\}\ \delta\xi_1 \delta\eta_1 \ldots \delta\xi_n \delta\eta_n = \det A \qquad (9.29)$$

in analogy with formula (4.22). To make the analogy more complete, some authors have proposed the normalization $\int \xi\delta\xi = (-2\pi i)^{1/2}$ instead of $\int \xi\delta\xi = 1$.

To get a formula analogous to (4.10), namely

$$\int \exp \tfrac{1}{2}Q(\xi)\delta^n\xi = (\det Q)^{1/2} \quad , \qquad (9.30)$$

we need only to prove the classical result that the determinant of the skew-symmetric matrix Q is the square of its Pfaffian (for n even). This can be done by a trick similar to the one used in subsection 4.1. Namely, from (9.12) and (9.27) we get

$$(-1)^{n/2}\ \mathrm{Pf}(Q)^2 = \iint \exp \tfrac{1}{2}(Q(\xi) - Q(\eta))\ \delta^n\xi\delta^n\eta \quad . \qquad (9.31)$$

Introduce the bilinear form $Q(\xi,\eta) = \sum_{i,j} q_{ij}\xi_i\eta_j$. From (9.28) one gets

$$(-1)^{n(n-1)/2} \det Q = \iint \exp Q(\xi,\eta)\delta^n\xi\delta^n\eta \quad . \qquad (9.32)$$

But the integral is transformed into the previous one, if one makes the substitution $(\xi,\eta) \to (\xi-\eta, \tfrac{1}{2}(\xi+\eta))$, of determinant 1 (Notice that n is even, hence $(-1)^{n^2/2} = 1$). The important fact about formula (9.30) is that it chooses one of the square roots of det Q.

9.7. We shall now mix ordinary variables with Grassmann variables. So consider vectors $x = (x_1,\dots,x_p)$ in $\mathbb{R}^p$ and Grassmann variables $\xi = (\xi_1,\dots,\xi_q)$. By a *superfunction* we mean a Grassmann polynomial in the variables ξ whose coefficients depend on x, namely

$$F(x;\xi) = \sum_I F_I(x)\cdot\xi_I \tag{9.33}$$

(summation over all subsets I of $\{1,\dots,q\}$). On such a superfunction, we operate with ordinary derivatives $\partial_i = \partial/\partial x_i$ in the parameter x, and Grassmann derivatives $\delta_j = \delta/\delta\xi_j$. The rules (9.3) and (9.4) are supplemented by the classical ones

$$\partial_i\partial_{i'} = \partial_{i'}\partial_i \quad , \qquad x_i x_{i'} = x_{i'}x_i \tag{9.34}$$

$$\partial_i x_{i'} - x_{i'}\partial_i = \delta_{ii'} \quad . \tag{9.35}$$

Moreover any operator in the *bosonic family* $x_1,\dots,x_p$, $\partial_1,\dots,\partial_p$ commutes with any operator in the *fermionic family* $\xi_1,\dots,\xi_q$, $\delta_1,\dots,\delta_q$.

Berezin's integral can be defined in the general case

$$\iint F(x;\xi)\, d^px\, d^q\xi = \int F_{12\dots q}(x)\, d^px \ . \tag{9.36}$$

Integration by part with respect to ordinary variables as well as to Grassmann variables is now permitted.

In what sense if $F(x;\xi)$ a function ? The question has been much debated. Here is a simple answer. Consider an auxiliary Grassman algebra $\Lambda = \Lambda(\eta_1,\dots,\eta_r)$ with *real* coefficients. Consider even elements $a_1,\dots,a_p$ of Λ and odd elements $\alpha_1,\dots,\alpha_q$ of Λ . For any subset I of $\{1,2,\dots,q\}$ with elements $j_1 < j_2 < \dots < j_s$, the substitution of α_j to ξ_j in $\xi_I = \xi_{j_1}\dots\xi_{j_s}$ gives obviously $\alpha_I = \alpha_{j_1}\dots\alpha_{j_s}$. We propose to define $F(a_1,\dots,a_p;\alpha_1,\dots,\alpha_q)$ as $\sum_I F_I(a_1,\dots,a_p)\alpha_I$ provided we give a rule to substitute even elements $a_1,\dots,a_p$ of Λ into an ordinary function $G(x_1,\dots,x_p)$. Remark that a_j can be written as $a_j = a_j^o + b_j$ where the constant term a_j^o is a real number. Since $b_1,\dots,b_p$ begin with Grassmann monomials of degree ≥ 2, any monomial in $b_1,\dots,b_p$ (which commute two by two !) of degree $d > \frac{r}{2}$ will vanish. Hence we can use a truncated Taylor series to define $G(a_1,\dots,a_p)$ as an element of Λ, namely

$$G(a_1^o+b_1,\dots,a_p^o+b_p) = \sum_{\lambda_1+\dots+\lambda_p\leq\frac{r}{2}} \partial_1^{\lambda_1}\dots\partial_p^{\lambda_p} G(a_1^o,\dots,a_p^o) \cdot b_1^{\lambda_1}\dots b_p^{\lambda_p}/\lambda_1!\ \dots\ \lambda_p! \qquad (9.37)$$

9.8. Let us return to our discussion in subsection 3.5. Introduce a basis $e_1,\dots,e_n$ of V. The algebra ΣW, direct sum of the spaces $\Sigma^k W$, is also the sum of the spaces $\Sigma^{B,F}W=S^BV \otimes \Lambda^F V$. It follows that ΣW is an algebra of mixed polynomials in commuting variables $x_1,\dots,x_n$ and anticommuting variables $\xi_1,\dots,\xi_n$. They can be considered as superfunctions $F(x;\xi) = \sum_I F_I(x)\xi_I$ where each component $F_I(x)$ is a polynomial in $x = (x_1,\dots,x_n)$. On these mixed polynomials, we can operate with the operators $x_i, \xi_j, \partial/\partial x_i, \delta/\delta\xi_j$. We consider the two differential operators

$$d = \sum_{j=1}^{n} \xi_j \frac{\partial}{\partial x_j} \quad , \qquad s = \sum_{j=1}^{n} x_j \frac{\delta}{\delta\xi_j} \quad . \qquad (9.38)$$

We get the Leibnitz rules

$$d(FG) = dF\cdot G \pm F\cdot dG$$

$$s(FG) = sF\cdot G \pm F\cdot sG$$

(with a plus (minus) sign for F even (odd)). Since $\partial/\partial x_j$ is an even operator, and $\delta/\delta\xi_j$ and odd operator, both d *and* s *are odd operators*. Let us calculate $sd + ds$:

$$\begin{aligned} sd+ds &= \sum_{i,j}(x_i\delta_i\xi_j\partial_j + \xi_j\partial_j x_i\delta_i) \\ &= \sum_{i,j} x_i(\delta_{ij} - \xi_j\delta_i)\partial_j + \xi_j(x_i\partial_j+\delta_{ij})\delta_i \\ &= \sum_i x_i\partial_i + \sum_i \xi_i\delta_i \end{aligned}$$

(notice that $x_i\xi_j\delta_i\partial_j$ is equal to $\xi_j x_i\partial_j\delta_i$ since x_i commutes to ξ_j and ∂_j commutes to δ_i). An element H in $\Sigma^{B,F}W$ is a homogeneous polynomial of degree B in x, hence by Euler classical result, one gets $\sum_i x_i \partial_i H = B \cdot H$. Similarly, H is a homogeneous polynomial of degree F in ξ and the formula $\sum_i \xi_i \delta_i H = F \cdot H$ is easily proved. Hence sd + ds multiplies H by the total degree $k = B + F$.

It remains to prove that our operators satisfy the rules (3.20) and (3.21). Any element a in W^+ is a linear combination with complex coefficients of $x_1, \dots, x_n$, hence $da = \pi a$. From Leibnitz rule one derives

$$d(a_1 \dots a_B) = \sum_{i=1}^{B} a_1 \dots a_{i-1} \, da_i a_{i+1} \dots a_B$$

and moreover elements of W^- act as scalars with respect to the derivation d. Formula (3.20) follows at once, and the proof of (3.21) is similar.

9.9. The basic idea in supersymmetry is to consider transformations mixing ordinary variables $x_1, \dots, x_p$ (hereafter called bosonic variables) with Grassmann variables $\xi_1, \dots, \xi_q$ (the "fermionic" variables). A linear transformation will take the matrix form

$$\begin{pmatrix} x' \\ \xi' \end{pmatrix} = T \begin{pmatrix} x \\ \xi \end{pmatrix}$$

where T is written in block form

$$T = \begin{pmatrix} M & N \\ P & Q \end{pmatrix}$$

For instance, one gets $x_i' = \sum_{j=1}^{p} m_{ij} x_i + \sum_{k=1}^{q} n_{ik} \xi_k$.

If we insist that x_i' should be bosonic (that is even), we cannot achieve such a transformation with ordinary numbers unless the n_{ik}'s are 0, that is N = 0, P = 0. If T is a matrix with complex coefficients, it will have the form $T = \begin{pmatrix} M & 0 \\ 0 & Q \end{pmatrix}$ and we shall get no mixing

$$x' = Mx \quad , \quad \xi' = Q\xi \quad .$$

The trick is to introduce an auxiliary Grassmann algebra $\Lambda = \Lambda(\eta_1, \dots, \eta_r)$ and to assume the following parity rule:

the elements of M and Q are even
the elements of N and P are odd.

We shall define the (Berezin) determinant of such a matrix as an even element of Λ .

Let us preface the definition by a few remarks:

(a) Let us denote by a^0 the constant term of an element a of Λ; we can say that a^0 is obtained by putting $\eta_1=\ldots=\eta_r=0$ in a. One checks that $(a+b)^0 = a^0 + b^0$, and $(ab)^0 = a^0b^0$ for a,b in Λ. So if for instance $P(t_1,\ldots,t_s)$ is a polynomial and $a_1,\ldots,a_s$ are even elements of Λ , the constant term of $P(a_1,\ldots,a_s)$ will be equal to $P(a_1^0,\ldots,a_s^0)$.

(b) Suppose an element a of Λ has an inverse b, that is $ab = 1$; hence $a^0b^0 = 1$ holds. Therefore, a^0 is not zero. Conversely, if this is so, write $a = a^0(1 - \lambda)$ where λ has a zero constant term. Then one gets $\lambda^{r+1} = 0$ and $(a^0)^{-1}(1+\lambda+\ldots+\lambda^r)$ is an inverse of a.

Let us assume that $T = \begin{pmatrix} M & N \\ P & Q \end{pmatrix}$ is as before and that the constant terms of detM and detQ are not zero. Then the matrix M has an inverse, and the matrices Q and $Q - PM^{-1}N$ have the same constant term (since P and N have odd elements, and the constant term of an odd element is 0). Hence the determinant of $Q - PM^{-1}N$ has the same constant term as detQ, which is not 0, and our determinant has an inverse in Λ. We are justified to define the *superdeterminant* (or Berezin determinant) of T as

$$\text{sdet } T = \det M \cdot \det(Q - PM^{-1}N)^{-1} \quad . \tag{9.39}$$

This formula should be compared to formula (1.24).

9.10. The justification of the previous definition comes from the possibility to extend to superfunctions a number of classical formulas.

We begin with Gaussian integrals. Consider a *mixed quadratic form* in the variables $(x,\xi) = (x_1,\ldots,x_p,\xi_1\ldots\xi_q)$ namely

$$Q(x,\xi) = \sum_{i,i'} a_{ii'}x_ix_{i'} + 2\sum_{i,j} b_{ij}x_i\xi_j + \sum_{j,j'} c_{jj'}\xi_j\xi_{j'} \quad . \tag{9.40}$$

Here we assume $a_{ii'} = a_{i'i}$ and $c_{jj'} = -c_{j'j}$. We assume furthermore that $a_{ii'}$ and $c_{jj'}$ are even and that b_{ij} is odd. Hence $Q(x,\xi)$ will take even values when we replace $x_1,\ldots,x_p$ by even elements in Λ, and $\xi_1,\ldots,\xi_q$ by odd elements in Λ. Introduce the matrix of coefficients of $Q(x,\xi)$, namely $Q = \begin{pmatrix} A & B \\ {}^tB & C \end{pmatrix}$ where $A = (a_{ii'})$ etc. Assume that the constant term of $\sum_{i,i'} a_{ii'}x_i x_{i'}$ has a strictly positive real part when $x_1,\ldots,x_p$ are real. Furthermore, assume the matrix C has an inverse (that is $\det(C)^o \neq 0$). Notice that this implies that the integer q is even. Under these hypothesis, one gets

$$\iint \exp - \frac{1}{2} Q(x,\xi) d^p x \delta^q \xi = (2\pi)^{p/2} (-1)^{q/2} (\text{sdet } Q)^{-1/2} \tag{9.41}$$

a common generalization of formulas (4.11) and (9.30). The integration variables $x_1,\ldots,x_p$ are *real* .

In a similar way associate to the matrix $T = \begin{pmatrix} M & N \\ P & Q \end{pmatrix}$ a bilinear form

$$S(x^*,\xi^*;x,\xi) = \sum_{i,i'} x_i^* m_{ii'} x_{i'} + \sum_{i,j} x_i^* n_{ij} \xi_j + \sum_{j,i} \xi_j^* p_{ji} x_i + \sum_{j,j'} q_{jj'} \xi_j^* \xi_{j'} \quad . \tag{9.42}$$

Assume that the selfadjoint part $\frac{1}{2}(M + M^*)$ of M be positive definite and that Q be invertible as a matrix. Then we get the following generalization of formulas (4.22) and (9.29)

$$\int \exp - S(x^*,\xi^*;x,\xi) dx_1^* dx_1 \ldots dx_p^* dx_p \delta\xi_1^* \delta\xi_1 \ldots \delta\xi_q^* \delta\xi_q = (2\pi i)^p (\text{s det } T)^{-1} \quad . \tag{9.43}$$

In this integral, $x_1,\ldots,x_p$ are considered as complex variables, with x_j^* complex conjugate to x_j, and $dx_j^* dx_j$ interpreted as $2i du_j dv_j$ if $x_j = u_j + iv_j$ and u_j, v_j are real (see formula (4.23)).

We can also generalize the formula for nonlinear changes of coordinates. We consider again variables $x_1,\ldots,x_p$, $\xi_1,\ldots,\xi_q$ where the x_i's are even (or bosonic) and the ξ_j's odd (or fermionic). A superchange of variables is of the form

$$x_i = U_i(y;\eta) \quad , \qquad \xi_j = V_j(y;\eta) \qquad .$$

Here $y = (y_1,\ldots,y_p)$ are even variables and $\eta = (\eta_1,\ldots,\eta_q)$ are odd variables. Moreover, the superfunction $U_i(y;\eta)$ is even containing only terms with a product of an even number of odd variables η_j, and $V_j(y;\eta)$ is odd with a similar definition.

Denote by $\partial U/\partial y$ the matrix $M = (m_{ii'})$ with entries $m_{ii'} = \partial U_i(y;\eta)/\partial y_{i'}$ and use similar notations in the case of the other partial derivatives. The matrix

$$T = \begin{pmatrix} \partial U/\partial y & \delta U/\delta\eta \\ \partial V/\partial y & \delta V/\delta\eta \end{pmatrix}$$

has the required properties: $\partial U/\partial y$ and $\delta V/\delta\eta$ are even and $\delta U/\delta\eta$, $\partial V/\partial y$ are odd. Assume furthermore that the matrices $\partial U/\partial y$ and $\delta V/\delta\eta$ are invertible. Then the superjacobian determinant is defined as the superfunction $J(y;\eta) = \text{s det } T$. Given any superfunction $G(x;\xi)$ we have the following integration formula

$$\iint G(x;\xi)\, d^p x\, \delta^q \xi = \iint G(U(y;\eta);V(y;\eta))J(y;\eta)\; d^p y\, \delta^q \eta. \qquad (9.44)$$

Symbolically, we have

$$d^p x \xi^q = J(y;\eta) d^p y \delta^q \eta \qquad \text{for} \quad x = U(y;\eta),\ \xi = V(y;\eta). \qquad (9.45)$$

9.11. As with any definition of determinant, there arises the question of the validity of the multiplicative rule. The definition of sdet T can be recast as follows (see our calculations in subsection 1.8). Write T as a product

$$T = \begin{pmatrix} I_p & 0 \\ X & I_q \end{pmatrix} \begin{pmatrix} A & 0 \\ 0 & B \end{pmatrix} \begin{pmatrix} I_p & Y \\ 0 & I_q \end{pmatrix} \quad . \qquad (9.46)$$

Then we get

$$\text{s det } (T) = (\det A)\cdot(\det B)^{-1} \quad . \qquad (9.47)$$

In particular, the matrices of the form $\begin{pmatrix} I_p & 0 \\ X & I_q \end{pmatrix}$ or $\begin{pmatrix} I_p & Y \\ 0 & I_q \end{pmatrix}$ have a superdeterminant equal to 1, and the formula

$$s\det(TT') = s\det(T)\cdot s\det(T') \qquad (9.48)$$

holds if T is of the form $\begin{pmatrix} I_p & 0 \\ X & I_q \end{pmatrix}$ or $\begin{pmatrix} A & 0 \\ 0 & B \end{pmatrix}$, or else if T' is of the form $\begin{pmatrix} I_p & Y' \\ 0 & I_q \end{pmatrix}$ or $\begin{pmatrix} A' & 0 \\ 0 & B' \end{pmatrix}$. To prove the multiplicative rule (9.48) in general, we need only to settle the case where

$$T = \begin{pmatrix} I_p & Y \\ 0 & I_q \end{pmatrix} \qquad T' = \begin{pmatrix} I_p & 0 \\ X & I_q \end{pmatrix} \quad .$$

It then reduces to the proof of the identity

$$\det(I_p + YX) = \det(I_q - X(I_p + YX)^{-1}Y) \quad . \qquad (9.49)$$

Using the identity

$$I_p - (I_p + YX)^{-1}YX = (I_p + YX)^{-1} \quad , \qquad (9.50)$$

it suffices to prove the relation

$$\det(I_p + VU) = \det(I_q + UV)^{-1} \quad , \qquad (9.51)$$

if U and V are matrices of respective sizes q x p and p x q with odd elements from Λ. This should be compared to the standard formula

$$\det(I_p + VU) = \det(I_q + UV) \quad , \qquad (9.52)$$

where U and V are matrices with elements from a commutative ring, for instance *even* elements from Λ. We leave it to the reader to prove formula (9.51) directly.

9.12. Let us sketch an "invariant" definition of the superdeterminant, due to I.Manin. Consider first polynomials $P(x;\xi)$ in one even variable x and one odd variable ξ. Such a polynomial can be written as $P(x;\xi) = A(x) + B(x)\xi$. Denote by δ the odd polynomial $x\xi$. Then the product $\delta P(x;\xi)$ is equal to $xA(x)\xi$ because $\xi^2 = 0$. Hence if $\delta P = 0$, we get $xA(x) = 0$, hence $A(x) = 0$, hence $P(x;\xi) = B(x)\xi$. In turn, $B(x)$ can be written as $B(0) + xC(x)$, where $C(x)$ is a polynomial, and finally

$$P(x;\xi) = B(0)\xi + \delta C(x) \quad . \tag{9.53}$$

This can be generalized to any number of even variables $x_1,\ldots,x_n$ and odd variables $\xi_1,\ldots,\xi_n$. Consider the odd polynomial $\delta = x_1\xi_1 + \ldots + x_n\xi_n$. Setting $\delta_j = x_j\xi_j$, we get $\delta = \delta_1 + \ldots + \delta_n$ and $\delta_i\delta_j = -\delta_j\delta_i$ since $x_ix_j = x_jx_i$ and $\xi_i\xi_j = -\xi_j\xi_i$. Therefore $\delta^2 = 0$. Then by induction on n, one proves that a polynomial $P = P(x;\xi)$ satisfies the equation $\delta P = 0$ iff it can be written as

$$P(x;\xi) = c\,\xi_1\ldots\xi_n + \delta Q(x,\xi) \tag{9.54}$$

for some polynomial $Q(x;\xi)$. The constant c is uniquely defined by P since $c\xi_1\ldots\xi_n = P(0;\xi)$.

Consider again variables $x_1,\ldots,x_p$, $\xi_1,\ldots,\xi_q$ and a linear transformation with matrix $T = \begin{pmatrix} M & N \\ P & Q \end{pmatrix}$ as before. Introduce new variables $y_1,\ldots,y_q$, $\eta_1,\ldots,\eta_p$ with y_j even and η_k odd, and the odd polynomial

$$\delta = x_1\eta_1 + \ldots + x_p\eta_p + y_1\xi_1 + \ldots + y_q\xi_q \quad . \tag{9.55}$$

Let us denote by $\begin{pmatrix} M' & N' \\ P' & Q' \end{pmatrix}$ the inverse of the transpose of T and define

$$T^* = \begin{pmatrix} Q' & P' \\ N' & M' \end{pmatrix} \quad . \tag{9.56}$$

If we act simultaneously on (x,ξ) by the linear transformation with matrix T, and on (y,η) by the matrix T^*, we see that δ remains invariant.

Put $\omega = \xi_1\ldots\xi_q\ \eta_1\ldots\eta_p$. Then $\delta\omega = 0$, and any solution $P = P(x,\xi,y,\eta)$ of $\delta P = 0$ is of the form $c\cdot\omega + \delta Q$ for some constant c and some polynomial $Q = Q(x,\xi,y,\eta)$. The constant c is equal to $P(0,\xi,0,\eta)$. Now transform simultaneously x,ξ by T and y,η by T^*. Let φ be the transform of ω. From $\delta\omega = 0$ and the invariance of δ, we get $\delta\varphi = 0$, hence

$$\varphi = c\omega + \delta Q(x,\xi,y,\eta) \quad . \tag{9.57}$$

The constant c is obtained by putting x and y equal to 0, that is neglecting the terms containing these variables. But then, ξ transforms into $Q\xi$ and η into $M'\eta$, hence by the Grassmann definition of the determinant (formula (9.1)), one gets $c = \det Q \cdot \det M'$. We want to show that *the constant* c, which we denote for a moment by D(T), *is the inverse of the superdeterminant of* T.

By definition, T transforms ω in $D(T)\cdot\omega + \delta Q$ for some Q. Acting now by the transformation associated to a matrix T', we transform ω in $D(T')\omega + \delta Q'$ for some Q' and Q into Q", hence we transform $D(T)\omega + \delta Q$ into $D(T)D(T')\omega + \delta Q'''$ for a suitable Q'". The multiplicative rule follows

$$D(TT') = D(T) \cdot D(T') \quad . \tag{9.58}$$

It remains to consider the elementary cases:

(a) If $T = \begin{pmatrix} I_p & 0 \\ X & I_q \end{pmatrix}$ then $T^* = \begin{pmatrix} I_q & 0 \\ -{}^tX & I_p \end{pmatrix}$, hence $Q = I_q$, $M' = I_p$ and $D(T) = 1$.

(b) If $T = \begin{pmatrix} I_p & Y \\ 0 & I_q \end{pmatrix}$ then $T^* = \begin{pmatrix} I_q & -{}^tY \\ 0 & I_p \end{pmatrix}$, hence $Q = I_q$, $M' = I_p$ and $D(T) = 1$.

(c) If $T = \begin{pmatrix} A & 0 \\ 0 & B \end{pmatrix}$ then $T^* = \begin{pmatrix} {}^tB^{-1} & 0 \\ 0 & {}^tA^{-1} \end{pmatrix}$, hence $Q = B$, $M' = {}^tA^{-1}$, hence $D(T) = (\det B)\cdot(\det A)^{-1}$.

The formula $D(T) = (s \det T)^{-1}$ follows now from the remarks at the beginning of subsection 9.11.

Since s det T is equal to $D(T)^{-1}$, the multiplicative rule follows at once from formula (9.58).

9.13. We conclude by a few remarks.

(a) For $T = \begin{pmatrix} M & N \\ P & Q \end{pmatrix}$ as before, put $T^\pi = \begin{pmatrix} Q & P \\ N & M \end{pmatrix}$. Then $s \det (T^\pi)$ is the inverse of $s \det (T)$. This is the general symmetry rule for changing the parity. Indeed T acting on variables x_i, ξ_j, replace the even variables x_i by odd variables

η_i and the odd variables ξ_j by even variables y_j. Then T is transformed into T^{π}.

(b) Same notations as before. Consider the vector space of polynomials in x,ξ homogeneous of degree B in x and F in ξ; it is the same as the space denoted $\Sigma^{B,F}V$ in subsection 2.9, where V is the vector space generated by $x_1,\ldots,x_p$, $\xi_1,\ldots,\xi_q$ with obvious subspaces V^+ and V^-. Introduce an auxiliary variable t. We get the following general form of the Master theorem

$$s\det(1 + tT) = \sum_{B,F} (-1)^F t^{B+F}\ \mathrm{Tr}(T|\Sigma^{B,F}V) \quad , \tag{9.59}$$

where the last term is the trace of the operator induced on $\Sigma^{B,F}V$ by the transformation of the variables x,ξ by T in the polynomials $P(x;\xi)$ in $\Sigma^{B,F}V$. The Master theorem in the form (3.16) reads as follows

$$s\det\left(1 + t\begin{pmatrix} A & 0 \\ 0 & A\end{pmatrix}\right) = 1. \tag{9.60}$$

(c) For a matrix $T = \begin{pmatrix} M & N \\ P & Q\end{pmatrix}$, we define its supertrace as $\mathrm{str}(T) = \mathrm{Tr}(M) - \mathrm{Tr}(Q)$. The following formulas

$$d\log s\det T = s\,\mathrm{tr}(T^{-1}dT) \quad , \tag{9.61}$$

$$s\det(1 + zT) = \exp\{\sum_{n\geq 1} (-1)^{n-1}\ s\,\mathrm{tr}(T^n)z^n/n\}$$

are the analogues of formulas for determinants considered before.

References

[1]. F.A.Berezin, *The mathematics of second quantization*, Academic Press, New York, 1966.

[2]. N.Bourbaki, *Espaces Vectoriels Topologiques*, Masson, Paris, 1981.

[3]. N.Bourbaki, *Varietés différentielles et analytiques*, Masson (CCLS), Paris, 1983.

[4]. J.Dieudonné, *History of functional analysis*, North Holland, Amsterdam, 1981.

[5]. J.Dieudonné, *Choix d'Oeuvres mathématiques*, Hermann, Paris, 1981.

[6]. I.Fredholm, Sur une classe d'équations fonctionnelles, Acta Math. 27 (1903), p.365-390.

[7]. F.R.Gantmacher, *The theory of matrices*, 2 vol., Chelsea, New York, 1959.

[8]. A.Grothendieck, Produits tensoriels topologiques et espaces nucléaires, Mem.Amer.Math.Soc., no. 16, Providence, 1953.

[9]. A.Grothendieck, La théorie de Fredholm, Bull.Soc.Math. France 84(1956), p.319-384.

[10]. D.Hilbert, *Grundzüge einer allgemeinen Theorie der linearen Integralgleichungen*, Chelsea, New York, 1953.

[11]. H.Hochstadt, *Integral equations*, Wiley/Interscience, New York, 1973.

[12]. A.F.Ruston, *Fredholm theory in Banach spaces*, Cambridge, Univ.Press, Cambridge, 1986.

[13]. B.Simon, *Trace ideals and their applications*, London Math.Soc.Lecture Notes, vol. 35, Cambridge, 1979.

[14]. E.C.Titchmarsh, *The theory of functions*, (2nd edition), Oxford Univ. Press, 1975.

Ecole Polytechnique, Centre de Mathématiques

F-91128, Palaiseau, France

Perspectives in Physics

Huzihiro Araki, A. Libchaber, and Giorgio Parisi, editors

Frances Bauer, Octavio Betancourt, Paul Garabedian, Masahiro Wakatani, *The Beta Equilibrium, Stability, and Transport Codes: Applications to the Design of Stellarators*

Pierre Pelcé, editor, *Dynamics of Curved Fronts*

Petre Dita and Vladimir Georgescu, editors, *Conformal Invariance and String Theory*